Wearable Biosensing in Medicine and Healthcare

Kohji Mitsubayashi
Editor

Wearable Biosensing in Medicine and Healthcare

Springer

Editor
Kohji Mitsubayashi
Institute of Biomaterials
and Bioengineering
Tokyo Medical and Dental University
Tokyo, Japan

ISBN 978-981-99-8121-2 ISBN 978-981-99-8122-9 (eBook)
https://doi.org/10.1007/978-981-99-8122-9

This Springer imprint is published by the registered company Springer Nature Singapore Pte Ltd.
The registered company address is: 152 Beach Road, #21-01/04 Gateway East, Singapore 189721, Singapore

Paper in this product is recyclable.

Preface

Along with the development of smartphones, the market for wearable devices has expanded rapidly since around 2010. In particular, the largest area of use for wearable devices is human sensing in the medical, healthcare, and welfare fields. To date, various sensor elements (temperature, acceleration, gyro, pressure, etc.) have been implemented in wearable devices, and in combination with IoT and AI technologies, a huge amount of human physical information (i.e. heartbeat, pulse wave, and activity level) can be measured in real time. The sensing technology has affected beyond diagnosis and health management to the point where it has led to changes in human behavior patterns. Of course, in the medical field, biological and chemical information contained in body fluids such as blood and urine is extremely important for diagnosis and testing (it accounts for over 80% of important items in medical checkups). Currently, it is not easy to measure the bio/chemical information non-invasively. However, there is no doubt that bio/chemical sensing will become a major medical and social need for wearable measurements in the future, and such wearable biomedical devices will be needed. In addition, the development of wearable biosensing requires not only sensor technology but also various related technologies such as materials, device production, mobile energy supply (wireless power supply, kinetic and bio/chemical power generation), and methods for identifying body parts suitable for wearable measurement etc.

This book contains the following parts and chapters on wearable biomedical sensors and their assistive technologies for promoting behavior change in medical and health care. The first part reviews several wearable biomedical sensors based on biocompatible materials and nano and MEMS technologies in the medical and dental fields (soft contact lens biosensors, smart textile wears, transdermal gas-sensing, cuff-less blood pressure monitoring, mouthguard sensors, etc.). The second part introduces the latest approaches to wearable biosensing using unique devices (microneedle, suture and thread, and headphone) for various skin targets such as sweat, interstitial fluid, and transcutaneous gases. The third part presents technologies supporting wearable sensors, including soft and flexible materials, manufacturing methods, skin volatile-marker imaging, and energy harvesting devices (bio-fuel cells, wireless power transfer, energy collection by chewing force, etc.).

This book is contributed to graduate students, academic researchers, and professors who work in the field of medical and environmental research, and for industry professionals involved in the development of wearable and medical devices and biosensing systems for human bio/chemical measurement, medical monitoring, and healthcare services with the Internet technologies. We would like to sincerely express our appreciation to the distinguished authors of the chapters whose expertise has certainly contributed significantly to the book. We hope that this book can shed light on various technological aspects-related wearable biosensing and related applications for the healthcare context and stimulate further research in this field.

Tokyo, Japan Kohji Mitsubayashi

Contents

Wearable Biomedical Sensors

Cavitas Biosensors (Body Cavity Sensors)

Takahiro Arakawa, Kenta Iitani, Koji Toma, and Kohji Mitsubayashi

1 Introduction

Non-invasive measurement of human body has been extensively investigated and advanced in the fields of medical and healthcare. Wearable sensors have emerged as promising tools for monitoring relevant parameters related to health care, sports, and medical applications, resulting in their widespread development and commercialization globally [1–5]. Most of the existing commercialized wearable devices primarily focus on measuring and assessing physical parameters. The investigation of non-invasive measurement of chemical substances from the human body has been comparatively limited, yet it plays a crucial role in obtaining comprehensive biological information in the fields of medicine and healthcare. The development of non-invasive biosensing techniques could significantly impact the management of patients with lifestyle diseases. For instance, self-monitoring of blood glucose (SMBG) traditionally involves invasive finger-prick testing using a blood glucose meter for diabetic patients [6]. However, this approach often leads to compliance issues due to its unpleasantness, pain, risk of infection, and potential anxiety or fear

T. Arakawa · K. Mitsubayashi (✉)
Department of Electric and Electronic Engineering, Tokyo University of Technology, 1404-1 Katakura, Hachioji City, Tokyo 192-0982, Japan
e-mail: m.bdi@tmd.ac.jp

T. Arakawa
e-mail: arakawath@stf.teu.ac.jp

K. Iitani
Department of Biomedical Devices and Instrumentation, Institute of Biomaterials and Bioengineering, Tokyo Medical and Dental University (TMDU), 2-3-10 Kanda-Surugadai, Chiyoda-Ku, Tokyo 101-0062, Japan

K. Toma
Department of Electronic Engineering, Shibaura Institute of Technology, 3-7-5 Toyosu, Koto-Ku, Tokyo 135-8547, Japan

K. Mitsubayashi (ed.), *Wearable Biosensing in Medicine and Healthcare*,
https://doi.org/10.1007/978-981-99-8122-9_1

it induces. Furthermore, finger-prick testing provides only a single measurement, limiting the ability to capture short-term fluctuations and immediate effects of interventions. Continuous glucose monitoring addresses these limitations by providing real-time monitoring, enabling better glycemic control, and alerting patients to hyper- or hypo-glycemia through alarms. Nevertheless, many commercial devices remain invasive and may only serve as adjuncts to finger-prick testing. The Freestyle Libre, developed by Abbott in the U.S., represents a notable advancement in continuous measurement of glucose concentration in the intercellular fluid, alleviating the burden on individuals with diabetes [6, 7].

The bio/chemical samples excreted from the human body encompass not only urine and feces, but also sweat, exhaled breath, saliva, nasal secretion, body odor, and tears. While urine and feces have been extensively utilized in clinical settings; they do not yield comprehensive biochemical information for routine health care. Human secretions carry valuable insights into the health and well-being of an individual, and developing non-invasive techniques for the extraction of this information represents a significant goal in the field of diagnosis and health care monitoring. In this regard, wearable sensors hold immense potential for evaluating chemical and biochemical markers relevant to health care or fitness [8–12].

Since the 1990s, researchers have also focused on human body cavities. The human body has several cavities, including cavitas oris (oral cavity), cavitas pharyngis (pharyngeal cavity), saccus lacrimalis (tear sac), cavitas nasi (nasal cavity), cavitas abdominalis (abdominal cavity), cavitas infraglotticum (infraglottic cavity), cavitas larynges (laryngeal cavity), cavitas oris propria (oral cavity proper), cavitas peritonealis (peritoneal cavity), cavitas thoracis (thoracic cavity), and cavitas tympanica (tympanic cavity). The term cavitas originates from the Latin word "cavity". Researchers have developed many sensor devices to apply to human cavities (cavitas oris, cavitas pharyngis, conjunctival sac) for non-invasive monitoring of biomedical information in the permanent body fluid in the human cavities. The research group of Prof. Mitsubayashi called these sensors "Cavitas sensors," a new category of detachable medical sensors between "Implantable" and "Wearable" (Fig. 1). While the implantable sensor is a medical device applied via surgery (non-detachable by human subjects), the wearable sensor is a detachable device applied by subjects themselves but not adequate for collecting significant medical information. In recent years, many cavitas sensors have been developed and commercialized worldwide [13–17]. The authors believe that new awareness of daily medicine (healthcare, presymptomatic test and preventive medicine) with cavitas and wearable sensors is necessary to improve the quality of life regarding the aging of society and rapid changes in living environments. In this chapter, we present the concept and future prospect of "cavitas sensors" in the form of contact lens-type biosensors in the orbital cavity, as well as mouthguard (MG)-type biosensors in the oral cavity.

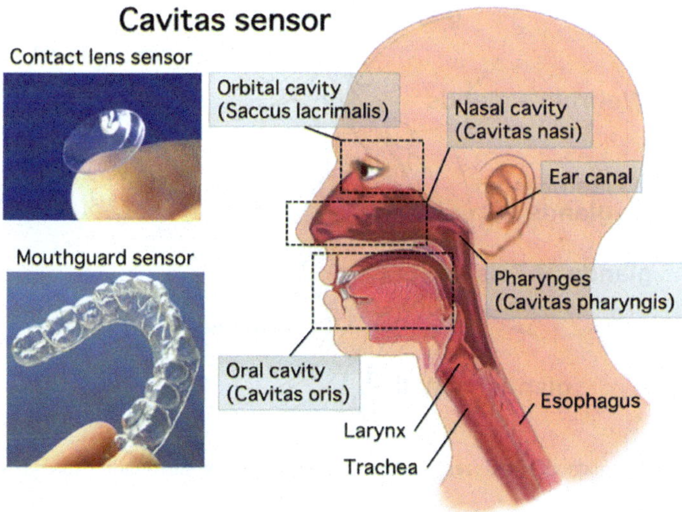

Fig. 1 Conceptual image of detachable "cavitas sensors" for non-invasive monitoring in human body cavities

2 Contact Lens Sensors in the Orbital Cavity

2.1 Tear Fluid

The pre-corneal and conjunctival tear film constitutes an interface between the ocular tissues and the air (refer to Fig. 2). The lacrimal secretory system is comprised of two main components: the larger orbital and smaller palpebral portions of the lacrimal gland, which together account for approximately 95% of the aqueous component of tears, and the accessory lacrimal glands of Krause and Wolfring, situated in the conjunctival stroma [18]. Tears serve several important functions, including lubrication of the eyelids, formation of a smooth and even layer over the irregular corneal surface, and provision of antibacterial systems for the ocular surface and nutrients for the corneal epithelium [19].

The lacrimal glands receive innervation from parasympathetic fibers carried in the lacrimal nerve, a branch of the ophthalmic nerve. Excessive lacrimation can be triggered by abnormal stimuli to the cornea or conjunctiva, as well as by sneezing, coughing, and psychic stimuli. Tears have a specific gravity of 1.01 g/mL, and their pH is generally around 7.4, although values ranging from 5.2 to 8.3 have been observed, as alkaline tears are shed in response to corneal injuries [20]. The electrolyte composition of tears is primarily composed of Na^+ and Cl^-, although significant levels of K^+ may also be present (Na^+: 120–165 mmol/L; K^+: 20–42 mmol/L; Cl^-: 118–135 mmol/L) [21]. The three main tear proteins are lysozyme, lactoferrin, and tear-specific pre-albumin. Comparison with serum levels reveals similar concentrations

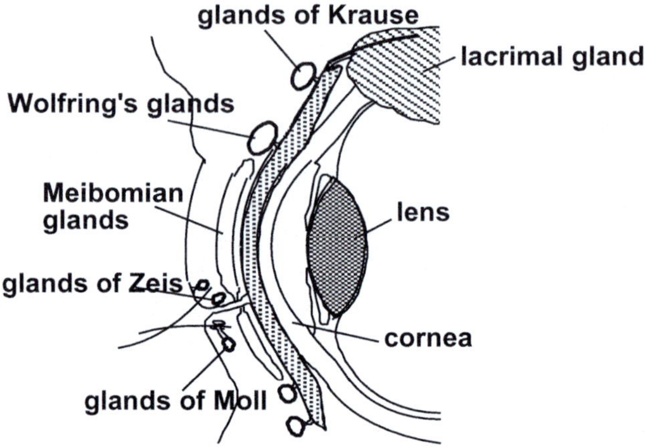

Fig. 2 Schematic image of glands responsible for tear production surrounding the eye when eyelids are closed

for Na^+, Cl^-, HCO_3^- (20–42 mmol/L), Mg^{2+} (0.5–0.9 mmol/L), and urea. However, other electrolytes are present at markedly different levels in tears compared to serum: K^+ and lactate (2–5 mmol/L) are higher, while Ca^{2+} (0.4–1.1 mmol/L), glucose (0.1–0.60 mmol/L), and protein (5–9 mg/mL) are lower in tears [18].

Tear fluid, which serves as a vital specimen for chemical analysis, is conventionally collected using either filter paper or glass capillary pipettes. The collected samples are typically pooled and subsequently analyzed using diverse techniques. Nevertheless, it is widely acknowledged that obtaining precise estimations of the chemical composition or physical properties of the tear fluid is challenging due to various factors, including the small sample size, evaporation during collection, wide inter-individual variability, diurnal variations, and notably, the method of collection. The eye site encompassing the tear and conjunctival sac plays a crucial role in monitoring physiological chemicals and blood gases in relation to health conditions and disease states. To overcome the challenges associated with conventional tear collection methods, an eye-cavity sensor placed at the conjunctival sac could potentially be applied directly on the surface of the cornea, which is the most sensitive organ, for monitoring tear analytes without encountering the problems.

2.2 *Eyelid Transcutaneous Gas Sensor*

A transcutaneous oxygen sensor has been employed for monitoring arterial oxygen pressure in premature infants to prevent retinopathy of prematurity in neonatal intensive care units [22]. However, commercially available sensors with rigid cylindrical cells affixed to the infant's skin using adhesive plaster often cause skin rashes and discomfort. Hence, there is a need for a new oxygen sensor that offers good flexibility and wearability, akin to a clinical wet-pack, for comfortable transcutaneous monitoring.

The conjunctiva, which exhibits high gas permeability and supplies oxygen to the cornea, has been considered as a potential site for oxygen monitoring using a conjunctival oxygen sensor that eliminates the need for heating. Isenberg et al. conducted a pilot study in 2002, evaluating a conjunctival oxygen monitor on 10 newborn subjects, and found a significant correlation coefficient between conjunctival oxygen tension and pulse oximetry ($p < 0.001$) [23]. However, the size of the sensor used in their study was not suitable for general application due to its large dimensions.

To address this limitation, a thinner and flexible oxygen sensor, developed as a Soft-MEMS (microelectromechanical system) device, was designed for transcutaneous oxygen tension monitoring from the conjunctiva [24–26]. The wearable oxygen sensor with a membrane structure was constructed by encapsulating KCl electrolyte solution with non-permeable and gas-permeable membranes, and patterning Pt and Ag/AgCl electrodes using photolithography and sputtering methods. The dimensions of the wearable oxygen sensor were 3 mm in width and 84 μm in thickness. The electrochemical measurement was conducted at a fixed potential of −550 mV versus Ag/AgCl, yielding a calibration range for dissolved oxygen from 0.01 to 8.0 mg/L. The sensor was also evaluated in the gas phase by purging with 10% oxygen gas, and the response time to reach 90% of the steady current after purging was approximately 45 s. The sensor outputs and responses remained stable during repeated measurements, with a coefficient of variance of 3.66%. As a physiological application, the wearable sensor was placed on the conjunctiva of a Japanese white rabbit without thermoregulation. In an experiment, the rabbit inhaled standard air (20.9%) and high concentration oxygen (60 and 90%) (Fig. 3). The sensor output increased and decreased synchronously with inhalation of high concentration oxygen and standard air, respectively, suggesting that the sensor could serve as a novel transcutaneous oxygen sensor.

2.3 *Soft Contact Lens (SCL) Biosensors for Tear Chemicals*

Continuous glucose monitoring typically does not directly measure blood glucose, but instead relies on measuring glucose concentration in other biological fluids. In the context of continuous monitoring, relationships between general physiological conditions and the constituents of biological fluids such as tears, mucus, sweat, and saliva have been reported. For example, a correlation between glucose concentration in tears and blood glucose has been reported [20], with tear glucose levels showing a delay of approximately 5 min compared to blood glucose levels. In addition, flexible electrochemical conductimetric sensors and oxygen sensors for bioinstrumentation on the ocular surface have been reported in previous works [27]. In 1995, a flexible glucose sensor was developed by immobilizing GOD within a gold-coated,

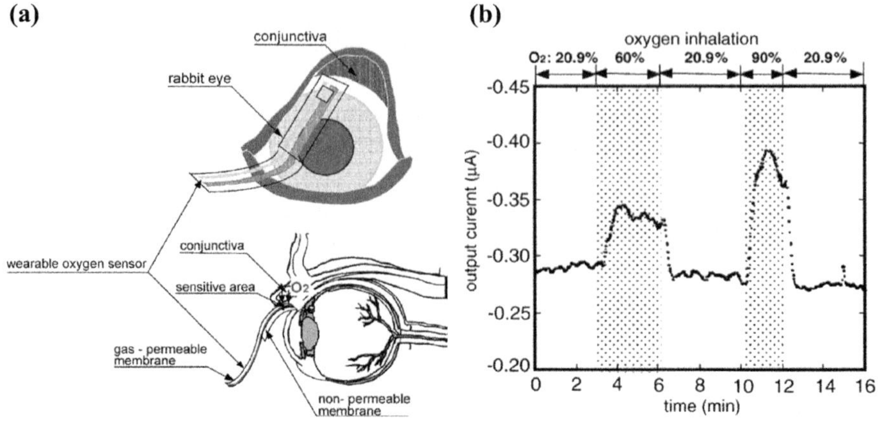

Fig. 3 **a** Illustration depicting the attachment of the wearable oxygen sensor to the eyelid conjunctiva and a cross-section of the attached sensor. **b** Typical response of the flexible sensor for transcutaneous oxygen monitoring (inhaled O_2 concentrations: 20.9%, 60%, 90%) at the conjunctiva of a rabbit [24]. (Permission of Sensors and Actuators B: Chemical)

hydrophilic polytetrafluoroethylene membrane without the use of harmful substances in a simple fabrication process [28]. The sensor was calibrated against glucose solutions ranging from 6.7 to 662 mg/L, including tear glucose levels. Non-uniform deposition of the gold layer using a mesh mask was found to be more effective in immobilizing the enzyme within the PVA-SbQ (Polyvinyl alcohol bearing styrylpyridinium group) matrix, resulting in a device with high current output and excellent calibration characteristics. Ethanol sterilization had a negligible effect on the output current and is therefore a suitable method for use in tear glucose sensing applications.

As part of the physiological study, a tear glucose sensor was constructed for on-site monitoring of tear glucose at the ocular surface [29, 30]. The glucose sensor is constructed by immobilizing GOD onto a flexible hydrogen peroxide electrode, consisting of a Pt working electrode and Ag/AgCl counter/reference electrode, fabricated using ion beam sputtering technique on these functional polymers. The sensor exhibits unprecedented flexibility and stretchability, enabling it to function normally even when stretched to 120% of its original length. Subsequently, GOD is immobilized onto the surface of the sensor using the 2-methacryloyloxyethyl phosphoryl choline (MPC) polymer (Fig. 5). The salient feature of the biosensor lies in the utilization of flexible and biocompatible polymers for all structural components. Notably, the sensing region employs a phospholipid polymer known as MPC polymer, which mimics the molecular configuration of a cell membrane. This unique configuration is achieved through the application of polymer chemistry techniques [31, 32]. By employing such a polymer as the contacting component at the measurement site, a biocompatible sensor with superior properties can be realized. The novel biosensor employs hydrophobic polydimethyl siloxane (PDMS) and hydrophilic MPC copolymerized with dodecyl methacrylate (DMA) to capitalize on the physical and chemical functionalities of these materials.

To pursue a physiological approach, a soft contact lens (CL) biosensor for in situ monitoring of tear glucose was fabricated using the biocompatible PMEH polymer in combination with PDMS [33]. The CL biosensor was designed specifically for in situ monitoring of tear glucose levels. PDMS was used as the material for the body of the sensor. The soft PDMS contact lens (with a base curve radius of 8.6) was preliminarily tested using a wired line in a preclinical experiment. The ultra-soft CL sensor exhibited excellent flexibility and soft characteristics, comparable to commercially available lenses. This device maintained its electrical performance without experiencing electrical breakdown even after mechanical bending. The soft CL biosensor demonstrated a robust linear relationship between the output current and glucose concentration within the range of 0.03–5.0 mmol/L, with a high correlation coefficient of 0.999 [29, 30]. In vivo testing of the CL biosensor in rabbits for tear glucose monitoring showed successful real-time measurements of tear glucose concentration, which was approximately one-tenth of the blood glucose concentration. Basal tear glucose was estimated to be 0.11 mmol/L. Furthermore, changes in tear glucose induced by alterations in blood sugar levels were assessed using an oral glucose tolerance test (OGTT), revealing a time delay of 10 min between tear glucose and blood glucose levels (Fig. 4). These results highlight the significant relationship between tear glucose and blood glucose levels with temporal dynamics, providing detailed insights into tear glucose dynamics using the CL biosensor. These findings indicate the utility of the SCL biosensor for advanced biomonitoring of the eye. Given the flexibility and versatility of PDMS, the CL biosensor can be optimized for application to various surfaces and cavitas of the human body.

3 Biosensors in an Oral Cavity

3.1 Human Salivary Fluids: An Overview of Composition and Potential Applications

Analytes present in saliva mainly come from capillaries that supply blood to the salivary glands. These capillaries pass through the interstitial fluid (ISF) before reaching the glands. According to these results, it is believed that certain components of the blood are present in saliva. Although there are numerous small glands distributed over the buccal mucosa, saliva is primarily secreted by the parotid, submaxillary, and sublingual glands. Parotid saliva is non-viscous, while sublingual and submaxillary saliva are viscous due to their mucoprotein content. In general, saliva comprises a diverse array of compositions, including amino acids, ions, proteins, sugars, nucleotides, microorganisms, and more. Saliva is a complex biofluid that contains numerous constituents that permeate from blood through transcellular or paracellular pathways. Saliva has garnered significant attention from researchers for the development of portable in vitro salivary diagnostic tools [34, 35]. Saliva serves as a promising diagnostic fluid that offers an alternative to direct blood analysis, as it

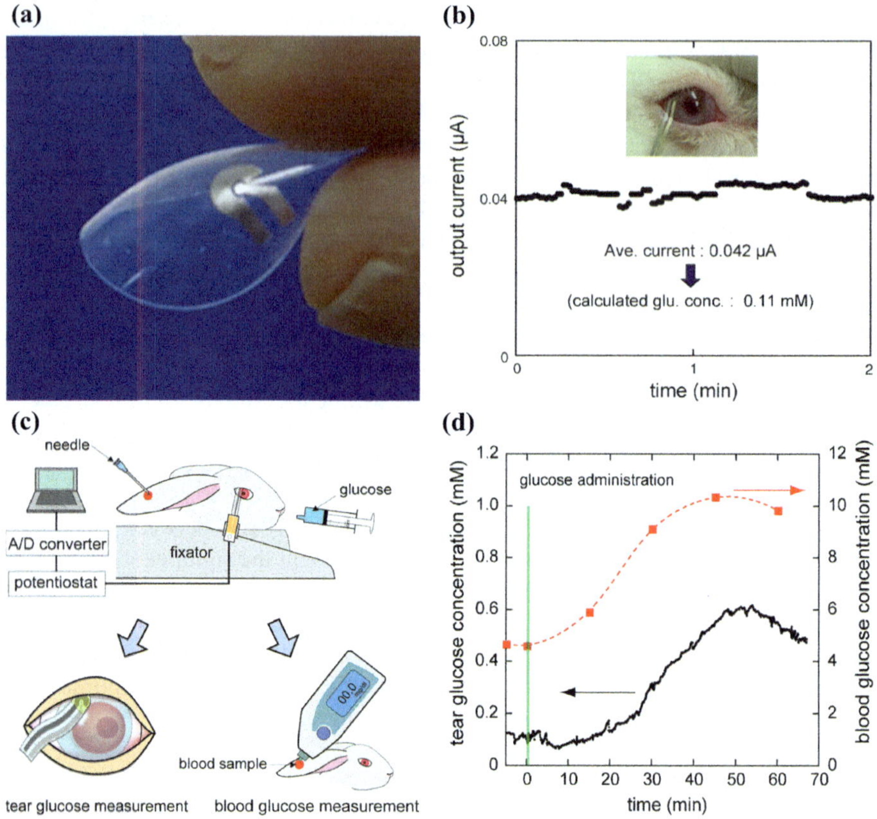

Fig. 4 **a** Photographs of SCL biosensor showing the flexibility not only lens but also coated electrodes. **b** Tear glucose monitoring using the CL biosensor on eye site. A stable output current was observed and the mean output current value was 0.042 μA. **c** OGTT Comparison experiments between tear glucose by CL biosensor and blood one by a commercially available test-kit. Upper right: Stable tear glucose monitoring using CL biosensor on eye site (estimated to 0.11 mmol/L). **d** Temporal changes both in tear and blood glucose[30]. (Permission of Talanta)

allows for the permeation of blood constituents without the need for skin piercing during blood sampling.

3.2 pH Sensing

Continuous monitoring of salivary pH holds immense potential for maintaining optimal oral health by preventing tooth enamel degradation, detecting gastroesophageal reflux disease, and monitoring drug activity, as saliva pH can impact drug efficacy in certain cases [36, 37]. Zuliani et al. demonstrated the development

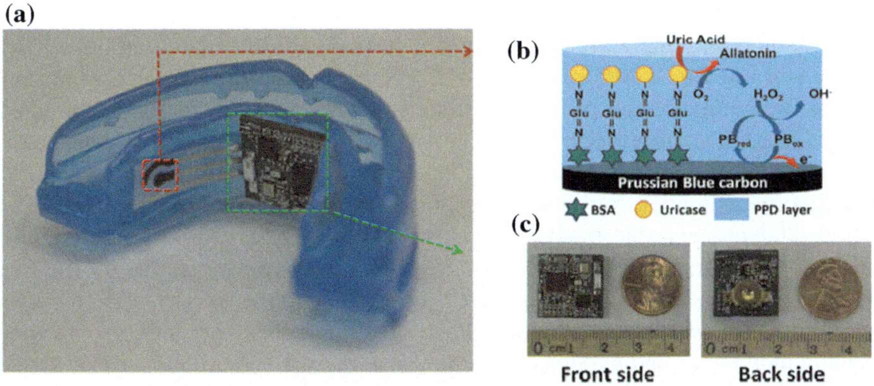

Fig. 5 **a** Photograph of the MG biosensor integrated with a wireless amperometric circuit board. **b** Reagent layer of the Prussian-Blue carbon working electrode containing uricase for the salivary uric acid biosensor. **c** Photograph of the wireless amperometric circuit board: front (left) and back (right) [43]. (Permission of Biosensors and Bioelectronics)

of a potentiometric strip utilizing a planar screen printed substrate for direct pH measurement in saliva samples [38]. The potentiometric strip comprises of a solid contact pH-selective electrode and a solid-contact ion gel reference electrode, both prepared on a dual screen-printed substrate. The pH of real saliva samples was accurately monitored using the optimized potentiometric strip. Ratanaporncharoen et al. developed the preparation and application of Ir/IrO_x pH sensors, which are used to measure the surface pH of dental caries [39]. After visually examining and categorizing 18 extracted dentinal caries in different locations as either active or arrested caries, the surface pH values of both sound and carious areas were directly measured using an Ir/IrO_x pH sensor, which had a diameter of 300 μm and functioned like a dental explorer. Thus, miniaturized pH sensors have also been developed, and studies are being conducted to mount these sensors in the oral cavity. Monitoring the pH in teeth and saliva is expected to contribute to the early diagnosis of caries and oral diseases.

3.3 MG Biosensor for Monitoring of Saliva Uric Acid

Salivary uric acid (UA) serves as a bio-indicator for various diseases, including hyperuricemia, gout, Lesch–Nyhan syndrome, and renal syndrome [40]. Notably, elevated levels of UA are associated with an increased risk of type 2 diabetes, as well as its severity and complications. UA can also function as an indicator of physical stress-induced reactive oxygen species, acting as a scavenger of free radicals. Unlike blood UA measurements, which require invasive blood collection, salivary UA measurements can be conducted non-invasively and in real-time. Shibasaki et al. and Soukup et al. have established a significant correlation between UA levels in blood and saliva, demonstrating that this metabolite can be monitored in saliva without the need for blood sampling [41, 42]. Kim et al. have demonstrated a wearable salivary UA sensor utilizing a MG platform [43]. The MG platform integrates an enzyme (uricase)-modified screen-printed electrode system, along with anatomically miniaturized instrumentation electronics comprising a potentiostat, microcontroller, and a Bluetooth Low Energy (BLE) chipset (Fig. 5). The adoption of a BLE chipset enables wireless connectivity to various BLE-enabled devices, such as smartwatches, smartphones, tablets, portable media players, laptops, and others. Two watch batteries (1.55 V, 33 mAh each) connected in series serve as the power source for the MG. The biosensor system integrated into the MG exhibits high sensitivity, selectivity, and stability for UA detection in human saliva, covering concentration ranges for both healthy individuals and hyperuricemia patients. The wireless MG biosensor system has been reported to provide real-time monitoring of saliva UA levels.

3.4 MG Biosensor for Monitoring of Saliva Glucose

We present the development of "Cavitas sensors" for non-invasive monitoring of salivary glucose in the human oral cavity [13, 44]. Salivary glucose concentrations, which closely correlate with circadian fluctuations in blood glucose levels, typically range from approximately 20 to 200 μmol/L in normal and diabetic individuals [45], offering promising prospects for non-invasive glucose monitoring. Several studies have demonstrated reasonable correlation between salivary and blood glucose levels in different individuals [46, 47]. Particularly, a stronger correlation has been observed within the same individual, enabling estimation of blood glucose concentrations based on salivary glucose measurements.

Prof. Mitsubayashi's group reported the successful measurement of salivary glucose using a developed MG glucose sensor without the need for pretreatment of human saliva [44]. Salivary glucose, as a non-invasive monitoring target, has demonstrated correlation with blood glucose levels, holding promise for clinical applications. However, the accuracy of glucose measurement in saliva can be affected by interfering components, such as ascorbic acid (AA) and UA. To address this challenge, we have developed a strategy to coat the electrodes with cellulose acetate (CA) in order to investigate its interference-rejection properties. Moreover, we have conducted a comprehensive assessment of the effects of AA and UA on the measurement of hydrogen peroxide, a reaction product of GOD, to thoroughly evaluate the performance of the sensor.

Polyethylene terephthalate glycol (PETG) from Erkodent Erich Kopp GmbH was selected as the material for the supporting structure of the MG glucose sensor. PETG, along with the Pt and Ag electrodes, was chosen due to its biocompatibility and safety for human use. The working electrode and counter/reference electrode of the glucose sensor were fabricated by coating Pt and Ag electrodes with a thickness of 200 nm and 300 nm, respectively, using a parallel-plate sputtering machine. For the immobilization of GOD, a high-performance biocompatible copolymer, poly

(MPC-co-EHMA-co-MBP) (PMEHB), known for its excellent biocompatibility and mechanical strength, was used. GOD was mixed with a solution of PMEHB (5 wt% in ethanol), and the enzyme solution was applied onto the Pt electrode coated with a contaminant-suppression membrane. Subsequently, GOD was immobilized comprehensively through UV irradiation-induced crosslinking. The fabrication process of the MG glucose sensor involved the production of a dental cast, vacuum forming of the PETG MG support, electrode sensor fabrication on the MG support, installation of a wireless transmitter, and packaging (Fig. 6).

We aimed to suppress the interference of contamination components, such as UA and AA, in saliva during glucose measurement using a MG biosensor. To achieve this, we coated the sensitive electrode area with a polymeric material, cellulose acetate (CA), as a contamination suppression membrane. The effect of the CA membrane was compared, and the concentration of CA was optimized. The glucose sensor exhibited high selectivity for glucose, utilizing the substrate specificity of the GOD enzyme. Subsequently, the developed MG biosensor was installed in the oral cavity to measure saliva glucose levels in vivo. During in vivo measurements, the current value stabilized in pure water before being placed in the subject's mouth to observe the output signal. The output current rapidly increased to around 30 nA upon installation in the oral cavity and then gradually decreased over time, eventually stabilizing at around 20 min. Importantly, no significant disturbance in the output current due to the MG device was observed. By calculating the difference between the steady value in deionized water and the stable value after wearing as the sensor output, we estimated the glucose concentration in saliva in the oral cavity to be 21.1 μmol/L using the calibration curve for artificial saliva. This estimation was consistent with the glucose concentration obtained from the saliva sample collected before wearing, as determined by a glucose measurement kit and spectrophotometer, which showed a concentration of 17.6 μmol/L. The results obtained using the MG sensor in the oral cavity were in good agreement with the experimental results from the glucose measurement kit.

4 Summary

The human secretions, such as tears, saliva, sweat, and body gases, contain valuable information about an individual's health and well-being. Extracting this information through non-invasive diagnostic techniques is a crucial goal in medical research. Among various body cavities, the use of "cavitas sensors," such as CL-type and MG-type devices, represents a new category of detachable devices for daily medical monitoring that are neither implantable nor wearable. In the future, a wide range of cavitas sensors are expected to be developed and commercialized to address the healthcare needs of an aging global population. To ensure the safe and effective fabrication of cavitas sensors, it is essential to use non-toxic and harmless chemicals, materials, and techniques, and to carefully select materials suitable for human body sites, considering factors such as flexibility for skin and mucosa, as well as rigidity for

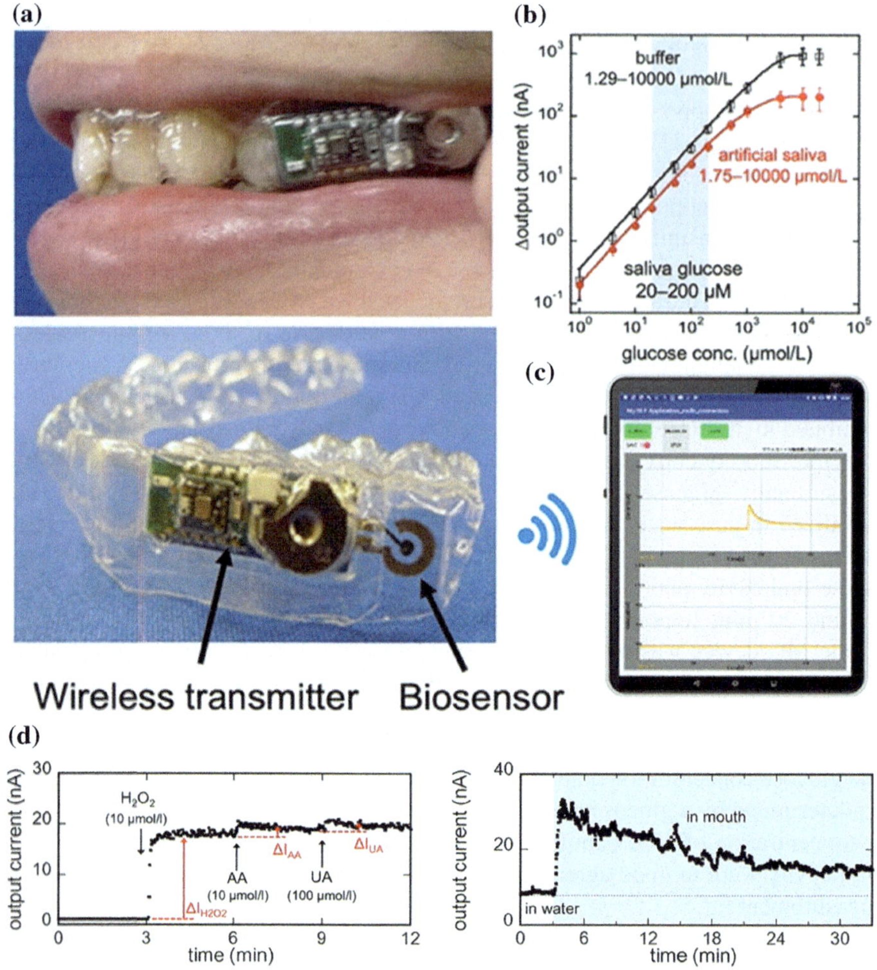

Fig. 6 **a** Photograph of the fabricated MG sensor integrated with a wireless module and battery. Wearing of the MG sensor. **b** Quantitative characteristics of glucose sensor in phosphate buffer and artificial saliva. **c** Portable wireless measurement system using Android devices. **d** Comparison of the output of each component and noise ratio against CA concentration (hydrogen peroxide, AA, and UA). Changes in CA (5%, w/w) coating electrode changes over time when each component was loaded (lower left). (e) Output current response when MG biosensor was installed and mounted in the human oral cavity [44]. (Permission of Analytical Chemistry)

tooth and bone applications. The authors firmly believe that incorporating cavitas and wearable sensors into daily medical practice, including healthcare, pre-symptomatic, and preventive medicine, is imperative to enhance the quality of life in the context of aging societies and changing living environments.

References

1. Bandodkar, A.J., Wang, J.: Non-invasive wearable electrochemical sensors: a review. Trends Biotechnol. **32**, 363–371 (2014). https://doi.org/10.1016/j.tibtech.2014.04.005
2. Kim, J., Campbell, A.S., Ávila, B.E.D., Wang, J.: Wearable biosensors for healthcare monitoring. Nat. Biotechnol. **37**, 389–406 (2019). https://doi.org/10.1038/s41587-019-0045-y
3. Salim, A., Lim, S.: Recent advances in noninvasive flexible and wearable wireless biosensors. Biosens. Bioelectron. **141**, 111422 (2019). https://doi.org/10.1016/j.bios.2019.111422
4. Min, J., Sempionatto, J.R., Teymourian, H., Wang, J., Gao, W.: Wearable electrochemical biosensors in North America. Biosens. Bioelectron. **172**, 112750 (2021). https://doi.org/10.1016/j.bios.2020.112750
5. Zhu, P., Peng, H., Rwei, A.Y.: Flexible, wearable biosensors for digital health. Med Nov Technol Devices **14**, 100118 (2022). https://doi.org/10.1016/j.medntd.2022.100118
6. Yoo, E.H., Lee, S.Y.: Glucose biosensors: an overview of use in clinical practice. Sensors. **10**, 4558–4576 (2010). https://doi.org/10.3390/s100504558
7. Bailey, T., Bode, B.W., Christiansen, M.P., Klaff, L.J., Alva, S.: The performance and usability of a factory-calibrated flash glucose monitoring system. Diabetes Technol. Ther. **17**, 787–794 (2015). https://doi.org/10.1089/dia.2014.0378
8. Trung, T.Q., Lee, N.E.: Flexible and stretchable physical sensor integrated platforms for wearable human-activity monitoringand personal healthcare. Adv. Mater. **28**, 4338–4372 (2016). https://doi.org/10.1002/adma.201504244
9. Lou, Z., Wang, L., Jiang, K., Wei, Z., Shen, G.: Reviews of wearable healthcare systems: materials, devices and system integration. Mater. Sci. Eng. R. Rep. **140**, 100523 (2020). https://doi.org/10.1016/j.mser.2019.100523
10. Sharma, A., Badea, M., Tiwari, S., Marty, J.L.: Wearable biosensors: an alternative and practical approach in healthcare and disease monitoring. Molecules **26**, 748 (2021). https://doi.org/10.3390/molecules26030748
11. Tu, J., Gao, W.: Ethical considerations of wearable technologies in human research. Adv Healthc Mater. **2100127**, 2100127 (2021). https://doi.org/10.1002/adhm.202100127
12. Verma, D., Singh, K.R., Yadav, A.K., Nayak, V., Singh, J., Solanki, P.R., Singh, R.P.: Internet of things (IoT) in nano-integrated wearable biosensor devices for healthcare applications. Biosens Bioelectron X **11**, 100153 (2022). https://doi.org/10.1016/j.biosx.2022.100153
13. Arakawa, T., Kuroki, Y., Nitta, H., Chouhan, P., Toma, K., Sawada, S., Takeuchi, S., Sekita, T., Akiyoshi, K., Minakuchi, S., Mitsubayashi, K.: Mouthguard biosensor with telemetry system for monitoring of saliva glucose: a novel cavitas sensor. Biosens. Bioelectron. **84**, 106–111 (2016). https://doi.org/10.1016/j.bios.2015.12.014
14. Arakawa, T., Dao, D.V., Mitsubayashi, K.: Biosensors and chemical sensors for healthcare monitoring: a review. IEEJ Trans. Electr. Electron. Eng. **17**, 626–636 (2022). https://doi.org/10.1002/tee.23580
15. Mitsubayashi, K., Arakawa, T.: Cavitas sensors: contact lens type sensors and mouthguard sensors. Electroanalysis **28**, 1170–1187 (2016). https://doi.org/10.1002/elan.201600083
16. Ciui, B., Tertis, M., Feurdean, C.N., Ilea, A., Sandulescu, R., Wang, J., Cristea, C.: Cavitas electrochemical sensor toward detection of N-epsilon (carboxymethyl)lysine in oral cavity. Sens. Actuators B Chem. **281**, 399–407 (2019). https://doi.org/10.1016/j.snb.2018.10.096
17. Sangsawang, R., Buranachai, C., Thavarungkul, P., Kanatharana, P., Jeerapan, I.: Cavitas electrochemical sensors for the direct determination of salivary thiocyanate levels. Microchim. Acta **188**, 1–12 (2021). https://doi.org/10.1007/s00604-021-05067-7
18. Berman, E.R.: Tears. In: Biochemistry of the Eye, pp. 63–88. Springer US, Boston, MA (1991)
19. Masoudi, S.: Biochemistry of human tear film: a review. Exp. Eye Res. **220**, 109101 (2022). https://doi.org/10.1016/j.exer.2022.109101
20. Sen, D.K., Sarin, G.S.: Tear glucose levels in normal people and in diabetic patients. Br. J. Ophthalmol. **64**, 693–695 (1980). https://doi.org/10.1136/bjo.64.9.693
21. Van Haeringen, N.J.: Clinical biochemistry of tears. Surv. Ophthalmol. **26**, 84–96 (1981). https://doi.org/10.1016/0039-6257(81)90145-4

22. Shoemaker, W.C., Lawner, P.M.: Method for continuous conjunctival oxygen monitoring during carotid artery surgery. Crit. Care Med. **11**, 946–947 (1983). https://doi.org/10.1097/00003246-198312000-00010
23. Isenberg, S.J., Neumann, D., Fink, S., Rich, R.: Continuous oxygen monitoring of the conjunctiva in neonates. J. Perinatol. **22**, 46–49 (2002). https://doi.org/10.1038/sj.jp.7210602
24. Iguchi, S., Mitsubayashi, K., Uehara, T., Ogawa, M.: A wearable oxygen sensor for transcutaneous blood gas monitoring at the conjunctiva. Sens. Actuators B Chem. **108**, 733–737 (2005). https://doi.org/10.1016/j.snb.2004.12.099
25. Iguchi, S., Chu, M.K., Takahashi, D., Arakawa, T., Kudo, H., Mitsubayashi, K.: Soft-MEMS glucose sensor with functional polymers. J. Photopolym. Sci. Technol. **23**, 167–170 (2010). https://doi.org/10.2494/photopolymer.23.167
26. Kudo, H., Iguchi, S., Yamada, T., Kawase, T., Saito, H., Otsuka, K., Mitsubayashi, K.: A flexible transcutaneous oxygen sensor using polymer membranes. Biomed. Microdevices **9**, 1–6 (2007). https://doi.org/10.1007/s10544-006-9000-z
27. Mitsubayashi, K., Wakabayashi, Y., Tanimoto, S., Murotomi, D., Endo, T.: Optical-transparent and flexible glucose sensor with ITO electrode. Biosens. Bioelectron. **19**, 67–71 (2003). https://doi.org/10.1016/S0956-5663(03)00130-1
28. Mitsubayashi, K., Dicks, J.M., Yokoyama, K., Takeuchi, T., Tamiya, E., Karube, I.: A flexible biosensor for glucose. Electroanalysis **7**, 83–87 (1995). https://doi.org/10.1002/elan.1140070110
29. Chu, M.X., Shirai, T., Takahashi, D., Arakawa, T., Kudo, H., Sano, K., Sawada, S.I., Yano, K., Iwasaki, Y., Akiyoshi, K., Mochizuki, M., Mitsubayashi, K.: Biomedical soft contact-lens sensor for in situ ocular biomonitoring of tear contents. Biomed. Microdevices **13**, 603–611 (2011). https://doi.org/10.1007/s10544-011-9530-x
30. Chu, M.X., Miyajima, K., Takahashi, D., Arakawa, T., Sano, K., Sawada, S.I., Kudo, H., Iwasaki, Y., Akiyoshi, K., Mochizuki, M., Mitsubayashi, K.: Soft contact lens biosensor for in situ monitoring of tear glucose as non-invasive blood sugar assessment. Talanta **83**, 960–965 (2011). https://doi.org/10.1016/j.talanta.2010.10.055
31. Ishihara, K.: Bioinspired phospholipid polymer biomaterials for making high performance artificial organs. Sci. Technol. Adv. Mater. **1**, 131–138 (2000). https://doi.org/10.1016/S1468-6996(00)00012-7
32. Iwasaki, Y., Ishihara, K.: Cell membrane-inspired phospholipid polymers for developing medical devices with excellent biointerfaces. Sci. Technol. Adv. Mater. **13** (2012). https://doi.org/10.1088/1468-6996/13/6/064101
33. Kudo, H., Sawada, T., Kazawa, E., Yoshida, H., Iwasaki, Y., Mitsubayashi, K.: A flexible and wearable glucose sensor based on functional polymers with soft-MEMS techniques. Biosens. Bioelectron. **22**, 558–562 (2006). https://doi.org/10.1016/j.bios.2006.05.006
34. Kaufman, E., Lamster, I.B.: The diagnostic applications of saliva—a review. Crit. Rev. Oral Biol. Med. **13**, 197–212 (2002). https://doi.org/10.1177/154411130201300209
35. Zhang, C.Z., Cheng, X.Q., Li, J.Y., Zhang, P., Yi, P., Xu, X., Zhou, X.D.: Saliva in the diagnosis of diseases. Int. J. Oral Sci. **8**, 133–137 (2016). https://doi.org/10.1038/ijos.2016.38
36. Dodds, M.W.J., Johnson, D.A., Yeh, C.-K.: Health benefits of saliva: a review. J. Dent. **33**, 223–233 (2005). https://doi.org/10.1016/j.jdent.2004.10.009
37. Shieh, W.Y., Wang, C.M., Cheng, H.Y.K., Imbang, T.I.: Noninvasive measurement of tongue pressure and its correlation with swallowing and respiration. Sensors **21**, 1–22 (2021). https://doi.org/10.3390/s21082603
38. Zuliani, C., Matzeu, G., Diamond, D.: A potentiometric disposable sensor strip for measuring pH in saliva. Electrochim. Acta **132**, 292–296 (2014). https://doi.org/10.1016/j.electacta.2014.03.140
39. Ratanaporncharoen, C., Tabata, M., Kitasako, Y., Ikeda, M., Goda, T., Matsumoto, A., Tagami, J., Miyahara, Y.: PH mapping on tooth surfaces for quantitative caries diagnosis using micro Ir/IrOx pH sensor. Anal. Chem. **90**, 4925–4931 (2018). https://doi.org/10.1021/acs.analchem.8b00867

40. Guan, Y., Chu, Q., Ye, J.: Determination of uric acid in human saliva by capillary electrophoresis with electrochemical detection: potential application in fast diagnosis of gout. Anal. Bioanal. Chem. **380**, 913–917 (2004). https://doi.org/10.1007/s00216-004-2848-y
41. Shibasaki, K., Kimura, M., Ikarashi, R., Yamaguchi, A., Watanabe, T.: Uric acid concentration in saliva and its changes with the patients receiving treatment for hyperuricemia. Metabolomics **8**, 484–491 (2012). https://doi.org/10.1007/s11306-011-0334-z
42. Soukup, M., Biesiada, I., Henderson, A., Idowu, B., Rodeback, D., Ridpath, L., Bridges, E.G., Nazar, A.M., Bridges, K.G.: Salivary uric acid as a noninvasive biomarker of metabolic syndrome. Diabetol. Metab. Syndr. **4**, 1–5 (2012). https://doi.org/10.1186/1758-5996-4-14
43. Kim, J., Imani, S., de Araujo, W.R., Warchall, J., Valdés-Ramírez, G., Paixão, T.R.L.C., Mercier, P.P., Wang, J.: Wearable salivary uric acid mouthguard biosensor with integrated wireless electronics. Biosens. Bioelectron. **74**, 1061–1068 (2015). https://doi.org/10.1016/j.bios.2015.07.039
44. Arakawa, T., Tomoto, K., Nitta, H., Toma, K., Takeuchi, S., Sekita, T., Minakuchi, S., Mitsubayashi, K.: A wearable cellulose acetate-coated mouthguard biosensor for in vivo salivary glucose measurement. Anal. Chem. **92**, 12201–12207 (2020). https://doi.org/10.1021/acs.analchem.0c01201
45. Yamaguchi, M., Mitsumori, M., Kano, Y.: Noninvasively measuring blood glucose using saliva. IEEE Eng. Med. Biol. Mag. **17**, 59–63 (1998). https://doi.org/10.1109/51.677170
46. Soni, A., Jha, S.K.: A paper strip based non-invasive glucose biosensor for salivary analysis. Biosens. Bioelectron. **67**, 763–768 (2015). https://doi.org/10.1016/j.bios.2014.09.042
47. Soni, A., Jha, S.K.: Smartphone based non-invasive salivary glucose biosensor. Anal. Chim. Acta **996**, 54–63 (2017). https://doi.org/10.1016/j.aca.2017.10.003

Wearable Sensors in the Medical Field

Yusuke Inoue, Tomoyuki Yokota, and Yoshiaki Takewa

Wearable sensors are playing an increasingly important role in our health and disease management. Developments in this technology allow us to monitor our bodies more closely, enabling us to take early detection and preventive measures. In recent years, with the proliferation of wearable devices, more and more people are wearing body surface sensors. These devices allow us to obtain real-time health information such as heart rate, number of steps taken, and calories burned. The latest research shows that this technology can be used to collect even more information relevant to our health and lives. For example, body surface sensors can measure respiration rates, skin temperature, sleep quality, muscle activity, and more. Such information can be very useful to health care professionals and may be used to monitor a patient's condition. They can also provide the general public with a more detailed understanding of their own health status and provide useful information for self-improvement. In addition, wearable sensors are suitable for monitoring in hospitals and at home. In hospitals, wearable sensors can reduce the burden on nurses and doctors and provide real-time information on patient conditions, potentially leading to lower healthcare costs. At home, they are expected to assist the elderly and chronically ill in leading better lives. However, body surface sensors have several problems. There are limitations in the accuracy and precision of the sensors. To address these issues, the design and manufacture of sensors requires a high degree of sophistication and requires precise calibration and inspection.

Y. Inoue (✉) · Y. Takewa
Advanced Medical Engineering Research Center, Asahikawa Medical University, Midorigaoka-Higashi 2-1-1-1, Asahikawa, Hokkaido 078-8510, Japan
e-mail: inoue@asahikawa-med.ac.jp

Y. Inoue · T. Yokota
Graduate School of Engineering, The University of Tokyo, 7-3-1 Hongo, Bunkyo, Tokyo 113-8656, Japan
e-mail: yokota@ntech.t.u-tokyo.ac.jp

K. Mitsubayashi (ed.), *Wearable Biosensing in Medicine and Healthcare*,
https://doi.org/10.1007/978-981-99-8122-9_2

Recent technological innovations not only enable the use of sensors in medical settings, but also new medical services such as home health care and telehealth. These services enable early detection of illness and disease and real-time monitoring of patient health status. For example, in home healthcare for the elderly, regular monitoring of vital signs such as body temperature, heart rate, and blood pressure enables urgent medical treatment. Protection of personal information is also an important issue in home healthcare. The data generated by body surface sensors can be very personally identifiable, such as personal health information and behavior patterns. Therefore, appropriate systems and legal regulations are needed to handle such data securely while ensuring the protection of personal information. In addition, the volume of data generated by these new technologies and services is enormous, making analysis and management of such data extremely difficult using conventional methods. For this reason, new technologies such as data analysis and artificial intelligence are also attracting attention. These technologies may lead to the provision of new medical services, such as early detection of diseases and development of treatments, by analyzing large amounts of data.

As described above, body surface sensors have sufficient potential to play an important role in medicine. It is expected that the development of more advanced sensor technology and data analysis technology will lead to higher quality medical services in the future. In addition, miniaturization and cost reduction of sensors may make them accessible not only to the medical community but also to the general public on a daily basis. Furthermore, more advanced medical sensor technologies are developing, including not only body surface sensors but also internal sensors and biometric sensors.

The history of modern medical measurement can be traced back to the 1800s, when thermometers, stethoscopes, and sphygmomanometers were invented to quantitatively measure body information for the diagnosis and treatment of diseases. 1900s saw a major development in electrical engineering and medical sensors. The successful electrical measurement of heart and muscle activity, particularly electrocardiography, became widely used as an indispensable test for the diagnosis and treatment of heart disease. In the 1950s, semiconductors were used to create more advanced medical sensors. Ultrasonography was invented to obtain images of internal organs and tissues in a noninvasive manner. Computed tomography (CT scan) was invented to obtain three-dimensional internal images. Magnetic resonance imaging (MRI) was invented to obtain high-resolution images of internal organs and tissues. Thus, in medicine, treatment and measurement are interrelated, and the process of treatment has become more effective as the necessary measurements are made in the process of treatment and the treatment is modified based on the measurement results, a process that is repeated. Therefore, in the medical field, it is hoped that advances in measurement technology will lead to improvements in treatment methods.

In Japan and many other countries, the aging of the population is making home health monitoring increasingly important relative to treatment at medical institutions. In Japan, the percentage of the elderly aged 65 and over has already exceeded 30% by 2020. Predictions also estimate that it will exceed 40% by 2050. In such an aging society, it is important to monitor the health of healthy people with wearable sensors, rather than dealing with diseases after they are detected. The measurement items required and the acceptable wearing comfort are different between a sensor worn by a patient in a hospital for treatment and a sensor worn by a healthy person all the time. We will discuss these issues in this chapter, citing actual examples.

The technology has been applied to the driving circuits of artificial hearts and medical devices, and all kinds of devices have been miniaturized year by year. These devices are made of inorganic semiconductor materials such as silicon, which excel in high density packing of sensors and actuators, and portable and wearable devices have begun to be developed and many devices are benefiting from microfabrication technologies. On the other hand, high-density circuits are hard and inflexible, so while there is no major impact when used outside the body, they are not suitable for use in the softness of the body. In the next generation of medical sensing, non-invasive and highly accurate measurements are required.

The harmony required between the sensor and the wearer includes many factors such as heat reduction and biocompatibility, but this chapter discusses the physical affinity between the device and the living body. We will introduce research trends and future prospects related to the medical field, focusing on soft organic thin-film electronics technology.

1 Hardware: Flexible Sensors with Organic Materials

Organic materials are basically softer than inorganic materials and maintain excellent bending properties. Organic polymeric composite materials such as graphene and carbon nanotubes, which are composed of carbon, have attracted particular attention because of their abundant resources and easy fabrication process. Already, electroluminescent (OLED) displays [1] and thin-film transistor electronics [2] , which can be bent at the product level, have appeared [3–9]. New applications for electronics using organic materials are expected to take advantage of their softness and lightness. The softness of electronics used in living organisms, which are mainly composed of organic materials, can be a great advantage. We have successfully fabricated high-performance organic transistors using an insulating film and a low-molecular-weight organic semiconductor as the channel layer on a plastic substrate [10]. Excellent bending characteristics have been achieved by providing an intermediate layer structure that mitigates distortion during bending. The bending radius is less than 10 μm, and the transistor does not break even when folded, dropped, or crumpled, and it has been subjected to repeated bending tests and has been tested for over 100,000 times [11]. Thinness and bendability are major advantages for medical devices, and if the device is softer than the living body when applied in vivo, it is less likely to cause separation of the device from the living body, thereby greatly reducing the risk of infection and inflammation. Figure 1 shows a highly durable electronic circuit that is light, thin, flexible, and adheres closely to curved and pulsating surfaces [12]. The world's lightest (3 g/m^2) and thinnest (2 μm) high-performance organic transistor integrated circuit was fabricated on a 1.2 μm-thick polyethylene terephthalate substrate, successfully creating an ultra-flexible and highly durable circuit. The device is one-fifth the thinness of plastic wrap and one-fifth the lightness of standard office paper. Integrated circuit sensors of organic transistors are configured on the device, with one cell of each element measuring 4 mm^2 and 144 cells mounted on an active area of 48×48 mm^2. This flexible sensor system can be affixed to free-form surfaces such as human skin and organs, and no degradation in electrical properties was observed after the devices were immersed in saline solution for more than two weeks.

In addition to transistors, many other flexible devices have been developed, and besides discrete devices such as resistors and capacitors, pressure sensors and temperature sensors can also be configured. Figure 2 shows an example of a high-brightness (100 cd/m^2) red light-emitting polymer organic LED (OLED) that can be applied to sensors for biometric measurement. Like the transistor, this OLED is very thin (2 μm), extremely lightweight (4 g/m^3), flexible, and does not break even when bent while energized, demonstrating high durability [13].

Flexibility and Stretchability

In addition to being flexible, it is also important for sensor devices applied to living bodies to have some degree of elasticity. In order to attach sensors to organs with volume changes such as the heart, lungs, and bladder, or to joints such as knees and elbows that extend due to bending, elasticity as well as flexibility is an essential component. Figure 3 shows an example of organic electronics with elasticity [13]. This is an organic LED that is more than twice as stretchable and was fabricated by applying the aforementioned thin-film OLED. When the ultra-thin, highly flexible OLED is attached to stretched rubber and the tension applied to the rubber is released, the OLED forms a wavy structure on the rubber surface and becomes stretchable (Fig. 4) [14]. This is a principle that is impossible with thicker devices, and the thin film and high mechanical durability of the OLED are what made it possible

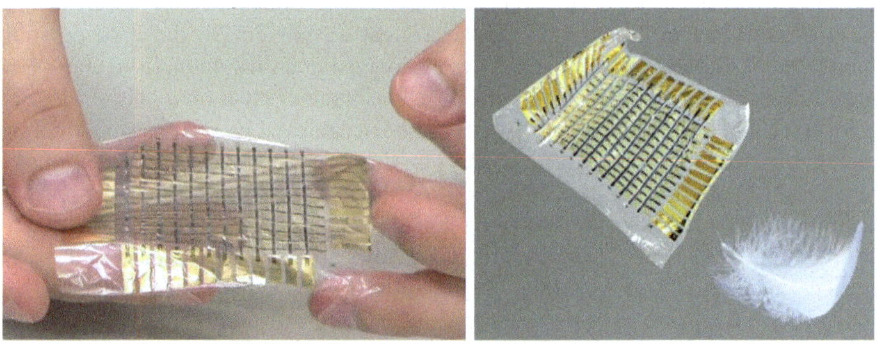

Fig. 1 Organic electronic circuits that are light, thin, flexible, and capable of sensing with high sensitivity by adhering closely to curved surfaces and the expanding and contracting human body [10]

Fig. 2 Thin-film flexible organic LEDs with high durability and unbreakable even when flexed [13]

to create such a stretchable device. Although stretchable devices have existed in previous research, they were mainly based on a mechanism in which only the wiring expands or contracts, and there were no devices in which the functional surface of the device (in the case of LEDs, the light-emitting surface) itself changes area. This technology can be applied to a variety of applications. Since this OLED has the flexibility to withstand bending at the elbow and hand joints, LEDs can be attached to clothing to show various designs, advertisements, and safety signs on the garment. Similarly, since it has the same elongation and flexibility as human skin, it can be applied to artificial skin and smart adhesive plasters. In the future, it is expected to create lights and displays that are so inexpensive that they can be purchased by the roll, allowing them to display any message they wish in the event of a traffic accident or other emergency. In the future, the characteristics and stability of this sensor device will be improved so that it can be incorporated into more complex systems and mass-produced.

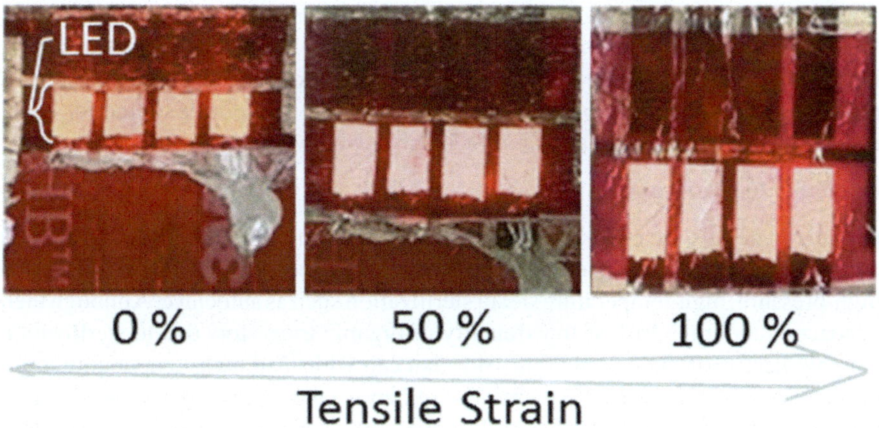

Fig. 3 OLED with extendable and retractable emitting surface [13]

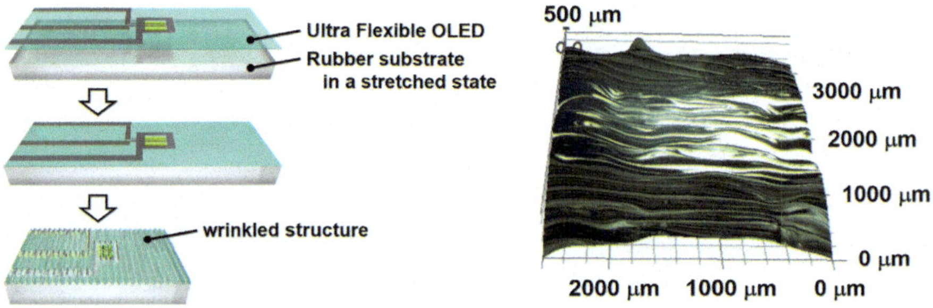

Fig. 4 Fabrication process of stretch OLEDs, which are fabricated by combining a thin OLED with a stretchable substrate, and a magnified image [14]

Figure 5 shows the organic transistor matrix and organic solar cells [4]. The world's thinnest and lightest organic solar cell was successfully fabricated using an ultra-thin polymer film 1.4 μm thick as a substrate. The energy conversion efficiency of this solar cell was 4.2%, which is comparable to those fabricated on glass substrates using the same materials, and its electrical and mechanical properties did not deteriorate even when the device was stretched and shrunk by 300%. Solar cells using organic semiconductors can be easily fabricated on polymer films and are expected to simultaneously realize large-area, low-cost, and lightweight solar cells. However, fabricating organic solar cells with energy conversion efficiency as high as that on glass substrates on a flexible thin polymer film was difficult, and a solution was sought. In this research, we developed a process technology to uniformly form organic semiconductor thin films on ultra-thin polymer films with a thickness of 1.4 μm, and succeeded in fabricating the world's thinnest and lightest organic solar cells on polymer films. Because the cell is fabricated on an ultra-thin film, it has excellent flexibility, and its electrical and mechanical properties do not deteriorate even when stretched by 300%, making it possible to wrap it around a hair as thin as 35 μm in diameter. Each gram of this organic solar cell is equivalent to 10 W, the highest power generation per weight of any solar cell. This achievement is expected to lead to the expansion of new applications of solar cells, such as in portable information and communication devices, and as a power source for health care and medical devices that do not feel heavy even when worn.

Durability in Sterilization

Organic materials are considered less sensitive to heat than inorganic materials. In contrast, the sterilization method most commonly used for medical sterilization is high-pressure, high-temperature steam sterilization such as autoclave. Although there is concern about the loss of functionality of organic transistors due to sterilization, a highly heat-resistant organic transistor that can withstand high-temperature, high-pressure sterilization at 150 °C, which is well above the basic autoclave temperature of 121 °C, has now been realized [3]. This transistor was realized by using high

Fig. 5 Stretchable organic transistor (left) and stretchable organic solar cell (right) [4]

mobility dinaphthothienothiophene (DNTT) as the organic semiconductor and a 4-nm-thick aluminum oxide film and self-assembled monolayer (SAM) as the gate insulator.

1.1 Sensors Used on the Body Surface

This section presents research on wearable biosensors useful in medicine and health care, using various examples. The area of application where wearable sensors are best suited is the body surface. There are various advantages to placing sensors on the body surface. The advantages include the ease of mounting and dismounting the sensor, the fact that the sensor is almost noninvasive, and the fact that the measurement range is large since the body surface area of an adult is more than 15,000 cm^2. Another advantage of the skin as a measurement site is that the biological information obtained from the skin is extremely multitemporal. Three types of signals (electrical, physical, and chemical) can be obtained from the body surface. By obtaining electrical signals, electroencephalogram (EEG), electrocardiogram (ECG), and electromyogram (EMG) can be obtained. EEG is used to evaluate various medical conditions such as seizures, epilepsy, dizziness, trauma, brain tumors, and sleep disorders. Electrocardiograms can be used for cardiac and psychiatric disorders, while electromyograms can be used for information applicable to motor control and neurological disorders. Various information can also be obtained from physical signals such as strain, pressure, temperature, light, and sound. Strain can be used to sense body movement, heart rate, and respiratory rate; pressure can measure blood pressure, intraocular pressure, body movement, and postural position; temperature can measure inflammation, infection, and blood flow as well as body temperature; light can measure pulse wave, blood vessel location, blood pressure, and blood oxygen saturation; and sound can measure dysphagia, heart valve disorders, and vocal cord abnormalities. Sound can be used to detect swallowing disorders, heart valve disorders, vocal cord abnormalities, and so on. By sensing chemical signals, blood glucose can be detected to control diabetes, lactic acid can be sensed to evaluate muscle fatigue, and sodium ions and chloride ions can be used to control dehydration and other symptoms. As such, a variety of information can be sensed from the body surface, and the use of wearable sensors is expected to be very promising.

When sensing on the body surface, the affinity between the sensor and the body surface is very important. The closer the sensor is placed to the body surface and the smaller the gap between the sensor and the body surface, the more sensitive the sensing can be. John A. Rogers et al. at Northwestern University reported that the affinity of a sensor to a living body changes depending on the thickness of the base material, even if the same material is used [15]. They evaluated the gap between the substrate and the living body when PVA, which is widely used as a base material for biosensors, was applied to the body surface with varying thicknesses from 5 to 500 μm [16]. To further reduce the gap between the nanomesh electrode and the living body, Someya-Yokota et al. of the University of Tokyo developed nanomesh electrode [17] by layering PVA into a fiber shape using an electrospinning method and depositing gold on the PVA fibers (Fig. 6). After being attached to the body surface, the PVA is dissolved by spraying it with water, leaving only the gold thin film body, which fits perfectly into the living body. The nanomesh electrode showed the best results compared to silicone and parylene, with no inflammation. In addition, to evaluate the robustness of the nanomesh electrode, a 10,000 bend and stretch test was conducted with the electrode attached to a finger joint, and it showed high mechanical properties without breaking. Currently, this electrode is being applied as an electrode for electrocardiographs that can be used for long periods of time, and we have succeeded in obtaining electrocardiograms over a long period of time.

Our group and Kohei Ishii's group at Kagawa National College of Technology discovered that the pawl twitches slightly in synchronization with the heartbeat and

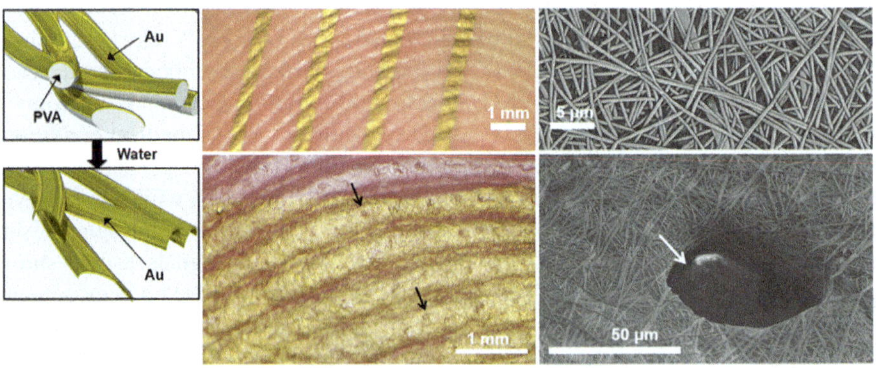

Fig. 6 Fabrication process of nanomesh electrodes and their application to the body surface: arrows indicate sweat glands

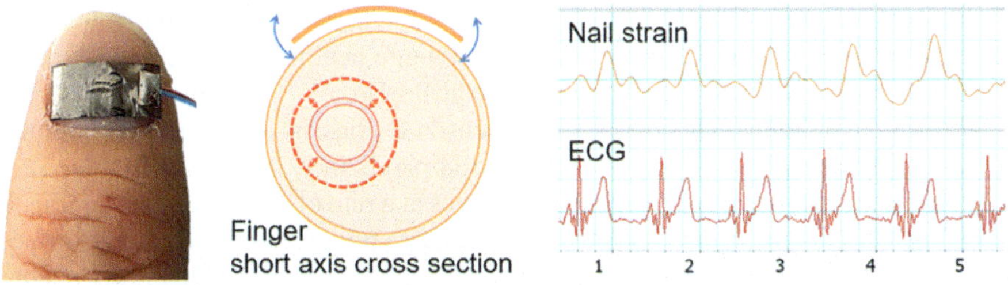

Fig. 7 A piezoelectric nail sensor that measures minute deformations of the nail synchronized with the heartbeat

succeeded in detecting the heartbeat and respiration by attaching a strain sensor to the nail (Fig. 7) [18]. The amount of microstrain of the nail synchronized with the heartbeat is a displacement of several hundred nanometers, which, although extremely small, is within the range of measurement possible with current piezoelectric devices. Although the mechanism of nail microstrain is not fully understood, we hypothesize that the blood flow into the finger causes a slight increase in finger volume, which in turn deforms the nail. This nail displacement has been shown to be measurable regardless of gender or age, and is gaining attention as a possible site for new wearable sensors. In addition, the absence of sweat glands and nerves makes fingernails suitable for sensing because they do not cause steam or inflammation even if sensors are worn, and they can be worn for a long period of time because of their low wearing comfort. On the other hand, its disadvantage is that it requires the wearer to be at rest in order to measure the heart rate in detail, as it is greatly affected by external disturbances such as body movement. The LED light method has been commonly used to detect the heart rate from the fingers, but LEDs consume large amounts of power, and long-term use requires either a wired power supply or battery operation with a large device. On the other hand, the piezoelectric element can be used as a sensor for detecting nail strain, and since the piezoelectric element generates a gas voltage due to strain, it does not consume power for detection, which is a great advantage. While smartwatches and other devices can detect the heart rate of healthy people, it is believed that the false fingernail type heart rate sensor can play a significant role in situations where the sensor may be unintentionally removed, such as by the elderly or infants.

1.2 Imaging and Optical Sensors

Flexible image sensors are attracting attention as a new imaging device because of their lightness, softness, and bendability. Since light can measure information outside the body, optical imaging, as typified by X-rays, is widely used for diagnosis of diseases in hospitals. Unlike conventional sensors, flexible image sensors are soft and can be directly attached to curved surfaces such as skin to continuously measure biological information with high precision. As such, they are expected to find a wide range of applications in wearable devices and home healthcare. Here, we introduce the application of such sensors in the biomedical field.

We are developing a sheet-type image sensor. It consists of an infrared photodetector and transistors with a resolution of 508 pixels/inch in a 15 μm thick, 12 × 12 mm sheet, and is capable of acquiring images at a rate of 41 flames per sec (Fig. 8) [19]. With this single sensor, it is possible to acquire fingerprints on the body surface, blood vessels inside the body, and pulse waveforms. The sensor's flexibility enables continuous sensing without invasing the living body, and yet it can acquire vital information at high speed, so that a very large amount of information can also be obtained in a short period of time. In addition to medical applications, applications in the fields of biometric authentication and security can be expected soon.

Someya and his group have applied flexible OLEDs to research using the body surface itself as a display. The skin display shown in Fig. 9 consists of 24 × 16 full-color micro LEDs arranged to display ECG waveforms and vital information [14, 20]. The three primary colors of red, green, and blue are realized with OLEDs, enabling full-color display [14, 21]. Seven-segment OLEDs can also be used on the body surface to display numerals and alphabets. These skin displays are expected to save more patients by using the body surface not only as a sensing target but also as a display location where information obtained by sensors can be quickly presented to anyone in an easy-to-understand manner.

In addition to LEDs, photodiodes can also be fabricated flexibly using organic materials. Yokota et al. combined an organic LED and an organic PD to develop a flexible pulse oximeter that fits tightly to the finger (Fig. 10) [14, 22]. The fabricated device is a reflective pulse oximeter. Green and red light irradiated onto the finger by the OLED is reflected and diffused through the in vivo tissue and returns to the OPD. Since red and green light have different absorption characteristics depending on the oxygen saturation of red blood cells, the amount of light entering the OPD makes it possible to measure oxygen saturation. Most common pulse oximeters are worn on the finger, but the issue is that pressure is applied to the finger because of the increased measurement sensitivity achieved by increasing the degree of adhesion between the living body and the sensor. Patients with circulatory insufficiency who need to wear a pulse oximeter may have low blood pressure, and the pressure of

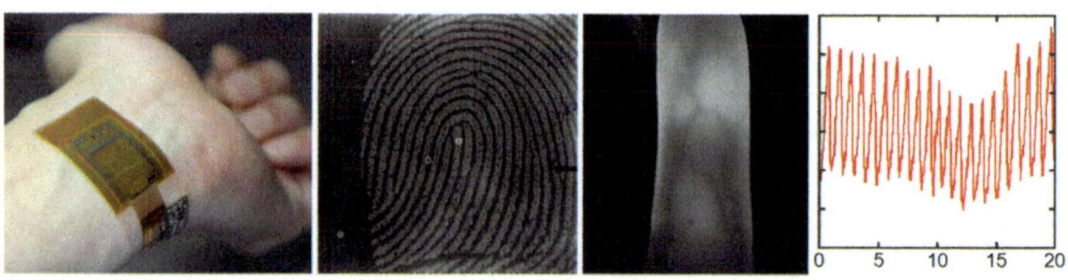

Fig. 8 Bendable sheet image sensor capable of capturing fingerprints and veins and measuring pulse waves

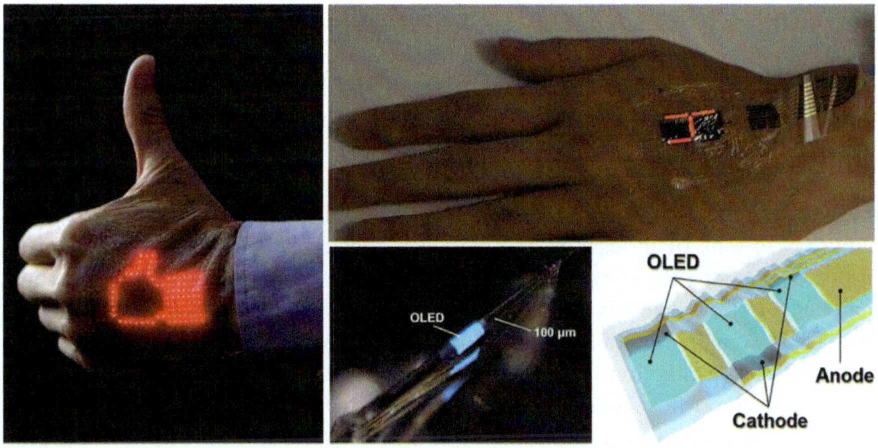

Fig. 9 Development of skin display by application of full-color OLED

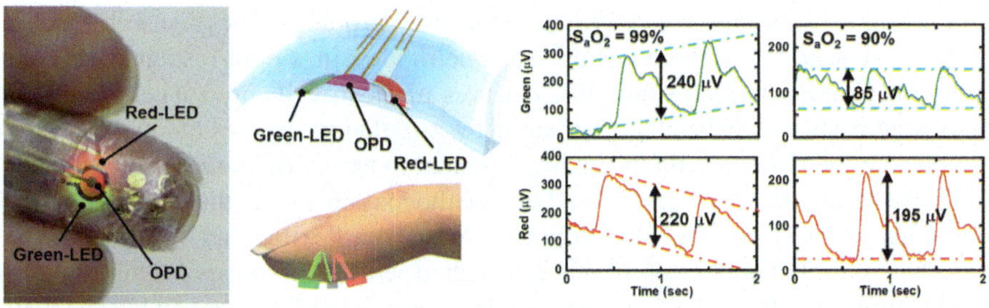

Fig. 10 Development of flexible Pulse Oximeter with Organic Photodiode and Organic LED

the pulse oximeter may block blood flow. With blood flow blocked, it is of course impossible to correctly measure the patient's vital information, but the film-type pulse oximeter developed by Yokota et al. is very good in that it does not apply pressure to the finger, so correct vital information can always be obtained.

1.3 In Vivo Sensor

Sensing using organic electronic sensors can be used not only on the body surface but also inside the body. The sensor used inside the body is temporarily invasive when it is implanted in the body, but by applying the new flexible sensor to the surface of organs, it is possible to sense with high sensitivity without invading the organ. The disadvantages of invasive implantation are greatly outweighed by the advantages of the data obtained from sensing. In conventional cardiac surgery, a needle electrode is inserted into the heart after opening the chest, which is often an invasive procedure for both the body surface and organs. In contrast, the method using a thin, flexible sensor allows the sensor to be inserted into a catheter with a diameter of about 5 mm, and the catheter can be inserted through a small opening on the body surface to apply the sensor to the organ surface. These methods are not completely noninvasive, but they are much less invasive than existing methods and nearly noninvasive to the organ being measured. While sensing inside the body has the disadvantage of being invasive, it also has the great advantage of improving measurement sensitivity by hundreds to tens of thousands of times over that of the body surface, and if the disadvantages greatly outweigh the advantages, the use of sensors inside the body may be an option. The advantage of placing sensors inside the body is not only the increase in signal strength but also the increase in response speed. In the process of measuring brain function using fMRI, brain cells consume oxygen when they are active, and blood flow increases to compensate for the lack of oxygen, and brain function is measured from the changes in blood flow. The time lag in this process is thought to be about 10 s. If flexible electrodes that can be directly attached to the brain are realized, the electrical activity of brain cells can be sensed directly [23], and the response speed will be improved 100 times.

Flexible sensors and OLEDs are also being used in optogenetics research. Optogenetics is a research technique that uses channelrhodopsin, a light-responsive protein, to control neural activity by light. When nerves are electrically fired, it is very difficult to distinguish between the potential caused by stimulation and the potential caused by neural activity. However, if a nerve is stimulated to fire by light, the resulting potential is the result of neural activity, and this method is superior in that it produces very low-noise measurement results. Yokota et al. used a flexible and very thin OLED to induce muscle contraction in rats transfected with a photoresponsive protein by light stimulation (Fig. 11) [21]. Compared to existing hard silicon LEDs, the OLED is thin and highly flexible, allowing MRI imaging with the OLED inside the body, making this device possible to measure brain function by optical stimulation and fMRI simultaneously.

It is possible to develop a temperature sensor by using a polymer whose volume changes with temperature. By mixing conductive microparticles and a temperature-functional resin into a paste, a sensor whose resistance changes with temperature can be developed [24]. Using this sensor principle, it can be formed into various

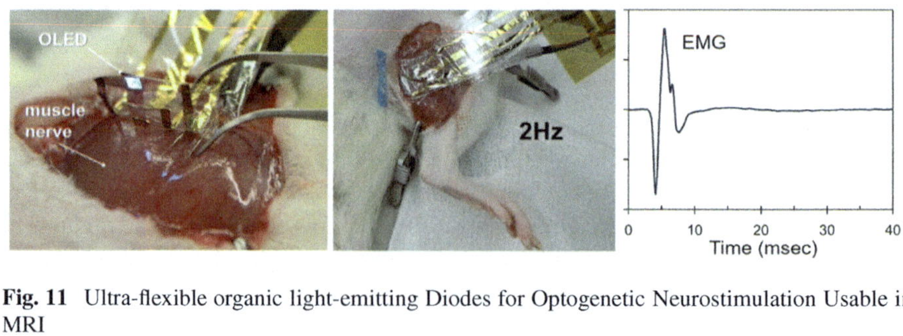

Fig. 11 Ultra-flexible organic light-emitting Diodes for Optogenetic Neurostimulation Usable in MRI

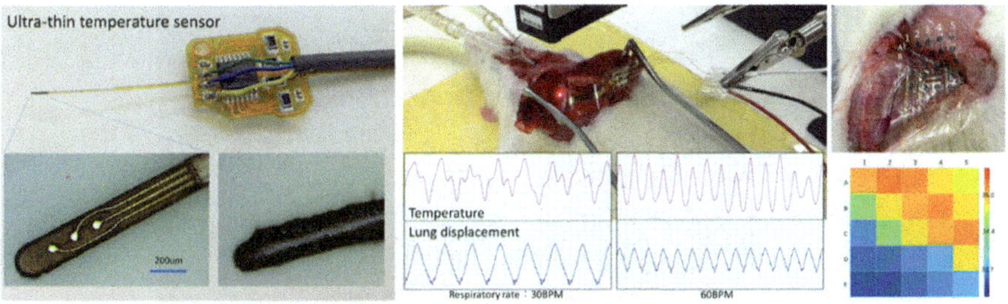

Fig. 12 Micro temperature sensor with ultra-fast response time

shapes, such as a needle or a multi-point film (Fig. 12). In addition to the advantage of being small and thin, the greatest advantage of this sensor is its high time response. Contact-type temperature sensors require contact between the object and the sensor until they reach the same temperature, the same principle as clinical thermometers, which require time for accurate measurement. On the other hand, this ultra-compact temperature sensor has a needle shape of about 0.2 mm^2, which has a very small heat capacity, resulting in a much faster response time. Figure 12 shows the temperature variation of the lung surface during respiration. The temperature difference was about 0.2 °C. The sensor captured the phenomenon in which the lung temperature decreases when inhaling an ambient temperature of 24 °C and rises when exhaling due to blood flow. This sensor can measure lung temperature without open chest surgery, and is useful for identifying abnormal tissue such as tumors. In addition, by constructing a temperature sensor with multiple points on a film substrate, it is possible to map minute temperature changes. Figure 12 right shows the results of recording temperature changes when the femoral artery of a rat was ligated to create an ischemic site, and these results show that this temperature sensor can be used to identify blood flow disorders and tumor sites such as skin cancer.

Multipoint flexible electrodes have the potential to change conventional cardiac surgery. In current surgical procedures to measure myocardial action potentials and regulate the rhythm of the heartbeat by defibrillation and pacemakers, needle electrodes are commonly sewn into the myocardium or fixed using alligator clips. While these methods provide stable fixation, they are problematic because they cause damage to the myocardium. Our flexible electrodes adhere to the surface of the heart and can be fixed in any position without invading the heart (Fig. 13) [25–27]. The electrode was made by depositing gold as a conductor on a 1.5 μm PEN film, followed by a 1.5 μm layer of parylene as an encapsulating film. By sandwiching the conductor between polymers of the same thickness, the electrode has high strength without microscopic fracture of the conductor, even when the electrode is bent and stretched. The rat heart has a diameter of approximately 10 mm and a heart rate of 300 bpm, which is larger than that of a human heart in terms of curvature and vibration, but the electrode did not fracture even after long-time measurement.

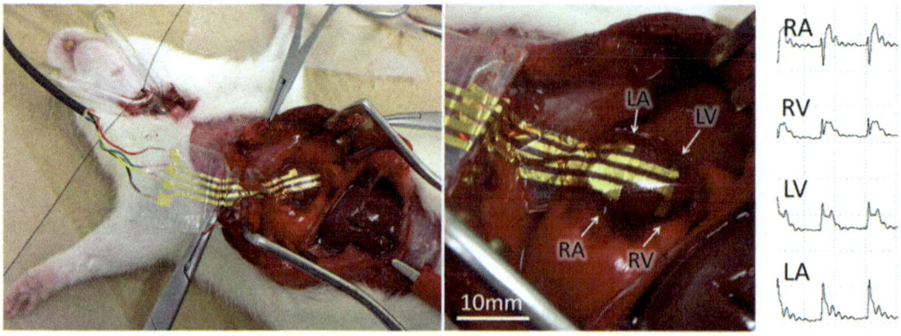

Fig. 13 Cardiac surface potential measurement using flexible multipoint electrodes

2 Conclusions

This chapter introduces the needs and development status of wearable sensors, which are playing an increasingly important role in our health and disease management. Until now, attempts to apply these sensors to the human body have mainly been based on small electronic devices that are mainly made of silicon. In recent years, the development of electronics using organic semiconductors has been promoted in place of silicon. Although organic semiconductors are not as sensitive as silicon semiconductors, the flexibility of the element itself can be maintained, making it much more applicable to the human body, and soft sensors can achieve physical harmony between the wearer and the sensor. It is also very important that the sensor itself be stretchable, not only because of the flexibility achieved by its thinness, but also when the sensor is applied to joints, rib cage, and other parts of the body where shape and volume change. These studies will elucidate the mechanisms of diseases that have not been elucidated so far in biomedical measurements in the medical field. These wearable sensors will also contribute to home and telemedicine, effectively utilizing medical costs and improving the quality of life of patients. The basic principles, measurable parameters, and application examples of body surface sensors were explained. In the future, further advances are expected through the evolution of sensor technology and integration with AI. Challenges exist in the widespread use of body surface sensors, such as improving accuracy, data interpretation issues, and privacy protection. It is important to address these challenges. In addition, when using sensors, collaboration with medical professionals and appropriate data management are required.

Of the five senses, vision is currently the most greatly developed sensor. The resolution of the best image receivers and display monitors is higher than that of the human eye. In the near future, it will be possible to construct a pressure-sensitive sensor with a density that exceeds the 10^7 pain senses in the skin, which are said to be present throughout the body. The problem that will arise at that time is how to process the data output from a large number of sensor elements. In order to handle a large amount of data, it is necessary to either send all the information to the brain (processor) and then process it with software using AI, or to selectively transmit the information by bundling it near the receptors before sending it to the brain, as human nerves do. Just as the human neural network has a sub-pathway called reflection, it may become important to have a mechanism that changes the sensor pathway depending on the level of importance.

The current engineering technology has not caught up with the sensors of living organisms, and is still in the stage of developing science and technology by imitating living organisms as Biomimetics. However, the author is convinced that the engineering technology will surpass the sensing capabilities of all living organisms in the near future, and that the day is approaching when the sensor technology will enhance living organisms. In the future, there is a great possibility that wearable sensors will be able to detect physical phenomena that are inherently imperceptible to humans. Contact lenses can visualize radiation, and essential for medical diagnosis but harmful to the human body, and clothing and gloves that can detect harmful viruses and molds may be developed.

Finally, the evolution of body surface sensors will have a significant impact on our healthcare and medical care. Through this publication, our goal is to disseminate knowledge about body surface sensors and promote their proper use and application. Your interest and commitment are important to maximize the potential of body surface sensors and realize a better future for health and medicine.

References

1. Gelinck, G.H., Huitema, H.E., van Veenendaal, E., Cantatore, E., Schrijnemakers, L., van der Putten, J.B., Geuns, T.C., Beenhakkers, M., Giesbers, J.B., Huisman, B.-H., Meijer, E.J., Benito, E.M., Touwslager, F.J., Marsman, A.W., van Rens, B.J., de Leeuw, D.M.: Flexible active-matrix displays and shift registers based on solution-processed organic transistors. Nature Materials. 3, 106–110 (2004)
2. Klauk, H., Zschieschang, U., Pflaum, J., Halik, M.: Ultralow-power organic complementary circuits. Nature **445**, 745–748 (2007)
3. Kuribara, K., Wang, H., Uchiyama, N., Fukuda, K., Yokota, T., Zschieschang, U., Jaye, C., Fischer, D., Klauk, H., Yamamoto, T., Takimiya, K., Ikeda, M., Kuwabara, H., Sekitani, T., Loo, Y.-L., Someya, T.: Organic transistors with high thermal stability for medical applications. Nat. Commun. **3**, Article number: 723 (2012)
4. Someya, T., Someya, T.: Stretchable Electronics. Wiley-VCH, Weinheim (2013). Print ISBN:9783527329786, ISBN:9783527646982. https://doi.org/10.1002/9783527646982
5. Lanzani, G.: Organic electronics meets biology. Nat. Mater. **13**, 775–776 (2014)
6. Liao, C., Zhang, M., Yao, M.Y., Hua, T., Li, L., Yan, F.: Flexible organic electronics in biology: materials and devices. Adv. Mater. **27**, 7493–7527 (2014)
7. Sekitani, T., Yokota, T., Someya, T.: Large-area, printed organic circuits for ambient electronics. Geol. Ecol. Landsc. 365–380 (2015).
8. Sekitani, T., Someya, T.: Ambient electronics. Jpn. J. Appl. Phys. **51**, 100001 (2012)
9. Sekitani, T., Zaitsu, K., Noguchi, Y., Ishibe, K., Takamiya, M., Sakurai, T., Someya, T.: Printed nonvolatile memory for a sheet-type communication system. IEEE Trans. Electron Devices **56**, 1027–1035 (2009)

10. Kaltenbrunner, M., Sekitani, T., Reeder, J., Yokota, T., Kuribara, K., Tokuhara, T., Drack, M., Schwödiauer, R., Graz, I., Bauer-Gogonea, S., Bauer, S., Someya, T.: An ultra-lightweight design for imperceptible plastic electronics. Nature **499**, 458–463 (2013)
11. Iba, S., Sekitani, T., Kato, Y., Someya, T., Kawaguchi, H., Takamiya, M., Sakurai, T., Takagi, S.: Control of threshold voltage of organic field-effect transistors with double-gate structures. Appl. Phys. Lett. **87** (2005)
12. Kato, Y., Iba, S., Teramoto, R., Sekitani, T., Someya, T., Kawaguchi, H., Sakurai, T.: High mobility of pentacene field-effect transistors with polyimide gate dielectric layers. Appl. Phys. Lett. **84**, 3789–3791 (2004)
13. Inoue, Y.: Current and future of the electrical circuit technology—SPECIAL ISSUE on the artificial heart which turns into an electronic device. J. Inst. Electr. Eng. Jpn. **135**(9), 626–628 (2015)
14. Yokota, T., Zalar, P., Kaltenbrunner, M., Jinno, H., Matsuhisa, N., Kitanosako, H., Tachibana, Y., Yukita, W., Koizumi, M., Someya, T.: Ultraflexible organic photonic skin. Sci. Adv. **2** (2016)
15. https://rogersgroup.northwestern.edu
16. Jeong, J.-W., Yeo, W.-H., Akhtar, A., Norton, J.J., Kwack, Y.-J., Li, S., Jung, S.-Y., Su, Y., Lee, W., Xia, J., Cheng, H., Huang, Y., Choi, W.-S., Bretl, T., Rogers, J.A.: Materials and optimized designs for human-machine interfaces via epidermal electronics. Adv. Mater. **25**, 6839–6846 (2013)
17. Miyamoto, A., Lee, S., Cooray, N.F., Lee, S., Mori, M., Matsuhisa, N., Jin, H., Yoda, L., Yokota, T., Itoh, A., Sekino, M., Kawasaki, H., Ebihara, T., Amagai, M., Someya, T.: Inflammation-free, gas-permeable, lightweight, stretchable on-skin electronics with nanomeshes. Nat. Nanotechnol. **12**, 907–913 (2017)
18. Ishii, K., Saito, I., Fujii, J., Oga, Y., Nakai, S., Iima, S., Hiraoka, N.: Measurement of micro-strain in nail caused by Pulse Wave. Adv. Biomed. Eng. **9**, 31–34 (2020)
19. Yokota, T., Nakamura, T., Kato, H., Mochizuki, M., Tada, M., Uchida, M., Lee, S., Koizumi, M., Yukita, W., Takimoto, A., Someya, T.: A conformable imager for biometric authentication and vital sign measurement. Nat. Electron. **3**, 113–121 (2020)
20. Someya, T.: Continuous health-monitoring with ultraflexible on-skin sensors. https://aaas.confex.com/aaas/2018/meetingapp.cgi/Paper/20827.
21. Kim, D., Yokota, T., Suzuki, T., Lee, S., Woo, T., Yukita, W., Koizumi, M., Tachibana, Y., Yawo, H., Onodera, H., Sekino, M., Someya, T.: Ultraflexible organic light-emitting diodes for optogenetic nerve stimulation. Proc. Natl. Acad. Sci. **117**, 21138–21146 (2020)
22. Park, S., Fukuda, K., Wang, M., Lee, C., Yokota, T., Jin, H., Jinno, H., Kimura, H., Zalar, P., Matsuhisa, N., Umezu, S., Bazan, G.C., Someya, T.: Ultraflexible near-infrared organic photodetectors for conformal photoplethysmogram sensors. Adv. Mater. **30**, 1802359 (2018)
23. Woo, T., Kim, D., Someya, T., Sekino, M.: Spatial resolution and maximum compensation factor of two-dimensional selective excitation pulses for MRI of objects containing conductive implants. AIP Adv. **7** (2017)
24. Yokota, T., Inoue, Y., Terakawa, Y., Reeder, J., Kaltenbrunner, M., Ware, T., Yang, K., Mabuchi, K., Murakawa, T., Sekino, M., Voit, W., Sekitani, T., Someya, T.: Ultraflexible, large-area, physiological temperature sensors for multipoint measurements. Proc. Natl. Acad. Sci. **112**, 14533–14538 (2015)
25. Sekitani, T., Yokota, T., Kuribara, K., Kaltenbrunner, M., Fukushima, T., Inoue, Y., Sekino, M., Isoyama, T., Abe, Y., Onodera, H., Someya, T.: Ultraflexible organic amplifier with biocompatible gel electrodes. Nat. Commun. **7** (2016)
26. Lee, S., Inoue, Y., Kim, D., Reuveny, A., Kuribara, K., Yokota, T., Reeder, J., Sekino, M., Sekitani, T., Abe, Y., Someya, T.: A strain-absorbing design for tissue–machine interfaces using a tunable adhesive gel. Nat. Commun. **5** (2014)
27. Reeder, J., Kaltenbrunner, M., Ware, T., Arreaga-Salas, D., Avendano-Bolivar, A., Yokota, T., Inoue, Y., Sekino, M., Voit, W., Sekitani, T., Someya, T.: Mechanically adaptive organic transistors for implantable electronics. Adv. Mater. **26**, 4967–4973 (2014)

Wearable Electrochemical Biosensors for Glucose Monitoring

Marjan Majdinasab

1 Introduction

Diabetes, also known as diabetes mellitus, is the most common metabolic disease related to carbohydrate metabolism. It is a group of endocrine disorders resulting from insulin deficiency and characterized by hyperglycemia or sustained high blood sugar (glucose) levels as a common effect of uncontrolled diabetes. Blood glucose concentrations in diabetic patients are higher than the normal range of 80–120 mg/dL (4.4–6.6 mM) [1]. Diabetes mellitus is one of the main causes of mortality and disability worldwide and a main health problem for most developed societies. According to the World Health Organization (WHO) report the number of diabetic individuals worldwide was approximately 171 million in 2000, and this is estimated to increase to 366 million by 2030 [2]. High levels of blood glucose lead to numerous complications such as higher risks of kidney failure, heart disease, stroke, blindness, and in some cases amputation [3]. Considering that there is no permanent treatment for diabetes, the only way to prevent these unwanted complications is disease management with routine blood glucose monitoring. Although laboratory tests are available for glucose monitoring in patients with diabetes but it requires frequent visits of patients to the laboratory, which is time-consuming and costly. To address this issue, self-monitoring of blood glucose has been established as a valuable tool for the management of diabetes. In order to be replaced by clinical tests, self-monitoring tools must meet some characteristics such as high accuracy and sensitivity, quick response, easy to use and low cost. This demand has been met by excessive research on glucose sensors using different transducer systems, i.e., electrochemical, acoustic, thermal, optical, and magnetic. Electrochemical biosensors specifically have been investigated extremely since the first generation of glucose sensors due to numerous

M. Majdinasab (✉)
Department of Food Science and Technology, School of Agriculture, Shiraz University, 71441-65186 Shiraz, Iran
e-mail: majdinasab@shirazu.ac.ir

K. Mitsubayashi (ed.), *Wearable Biosensing in Medicine and Healthcare*,
https://doi.org/10.1007/978-981-99-8122-9_3

advantages such as high accuracy and reproducibility, better sensitivity, simplicity, and low cost [2, 3]. Enzymes are the most common recognition element not only in electrochemical sensors but also in other types of glucose sensors due to their high selectivity towards glucose substrate. Enzymatic amperometric glucose biosensors, inspired by the first glucose biosensor developed by Clark and Lyons in 1962, are the most popular devices commercially available, and have been extensively studied in the last few decades. Currently, relying on the numerous researches conducted in the field of enzymatic electrochemical glucose biosensors, finger glucometers are commercially available as the common diagnostic standard tools for self-monitoring of blood glucose. Despite many advantages of finger glucometers, they are painful and invasive and also they don't have the ability to continuously monitor glucose [4]. Therefore, the efforts to improve glucose biosensors are still ongoing. Different approaches have been investigated in the performance of glucose enzyme electrodes. Moreover, one of the most interesting topics in recent years is the effort to develop wearable non-invasive or minimally invasive glucose monitoring devices, which have attracted the attention of many researchers in the field of biosensors. Wearable devices are able to receive physiological information worn on the human body by measuring dynamically changing contents of biomarkers in biofluids including interstitial fluid (ISF), sweat, saliva, and tears, while analyzing and storing the collected data by smart transmission and information processing systems [5, 6]. Among different kinds of glucose monitoring devices developed so far, wearable non-invasive and minimally invasive glucose sensors are of interest for diabetes management due to their ability to continuously monitor glucose levels. These devices are generally based on different techniques including optical, colorimetric, ultrasound, and electrochemical approaches [4]. Each of these techniques has its own pros and cons. However, wearable electrochemical sensors are exclusively attractive due to their noticeable advantages such as high selectivity and sensitivity, inherent miniaturization, low cost, high speed, low power requirements, and highly scalable fabrication by the use of screen-printing technology. Figure 1 shows the wide field of glucose detection approaches and distinguishes three different groups including invasive, minimally invasive, and non-invasive techniques. In invasive methods, there is direct detection of glucose in the blood like the technology used in glucose self-monitoring devices (finger glucometer). Minimally invasive methods are those that require extracting some body fluids (i.e., ISF, tears, saliva and sweat), with a single insertion into the skin and minimal pain, to measure the glucose concentration using an enzymatic reaction. In non-invasive techniques, biological fluids containing glucose at concentrations that are proportional to its concentration in the blood are collected without puncturing the skin [7].

One of the most effective wearable glucose monitoring systems that has recently been successfully implemented in several commercial platforms is wearable minimally invasive systems developed for continuous glucose monitoring. One of the most appealing wearable minimally invasive systems developed in recent years is microneedle-based systems to monitor glucose levels in the ISF using enzymatic reactions [8]. Although due to their small size, they have been considered to be minimally invasive, but in some cases, they can be associated with challenges such as tissue

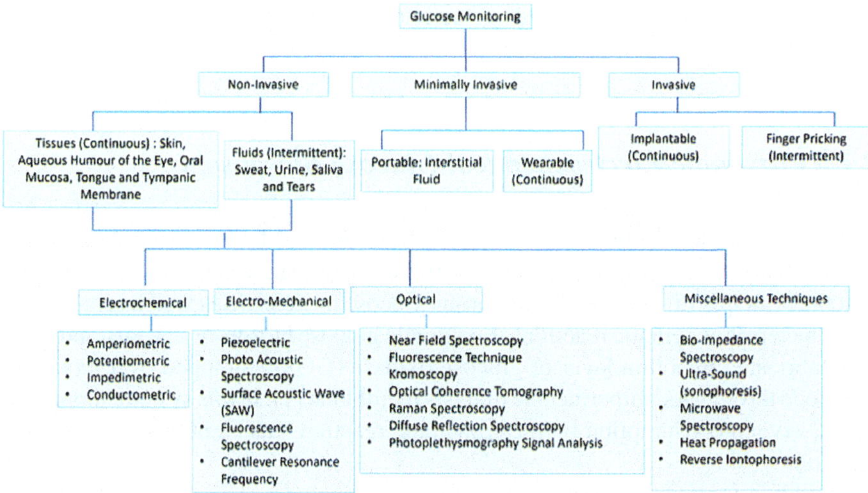

Fig. 1 A review of various techniques for glucose measurement. Reprinted from Ref. [7] with permission

damage and microbial infection. Therefore, the development of fully non-invasive devices for continuous monitoring of glucose is the main goal of new generation of wearable biosensors. These new generation of biosensors should be non-invasive and able to measure glucose levels on a daily basis. Moreover, the results of glucose measurement can be transmitted wirelessly to the patient's physician.

The main goal of this chapter is to evaluate the history and current status of electrochemical glucose biosensors focusing on wearable electrochemical biosensors that are a growing trend of recent publications. Two types of wearable electrochemical glucose biosensors including microneedle-based sensors as minimally invasive devices, and flexible electrodes as non-invasive devices are evaluated. Principles of operation and various materials applied in these two groups of wearable electrochemical glucose biosensors with their advantages and limitations are discussed. Moreover, the advantages of self-powered wearable devices are explained.

2 Electrochemical Glucose Biosensors

The principle of electrochemistry-based transducers is the electron flow generated by a chemical reaction. Regarding electrochemical detection of glucose, this process mainly relies on oxidation/reduction reactions. During these reactions, glucose is converted to gluconic acid (gluconate). Electrochemical glucose biosensors are classified into two groups including enzymatic and non-enzymatic. In enzymatic biosensors, glucose oxidation can occur in the presence of enzyme while, in non-enzymatic group glucose detection is based on activities of metal-centered electrocatalysts [3, 9, 10].

2.1 Enzymatic Electrochemical Glucose Biosensors

Since the first glucose biosensor proposed by Clark and Lyons, enzymatic glucose sensing has been employed ever since with more popularity and high potential such that the principle of detection in numerous glucose biosensors as well as commercial glucometers is enzymatic reactions. Enzymatic glucose biosensor of Clark and Lyons was fabricated by a thin layer of glucose oxidase (GOx) entrapped over an oxygen electrode through a semipermeable dialysis membrane [1]. In this system, monitoring of the oxygen consumption by the enzymatic reaction was an indicator for glucose detection.

In addition to GOx, two other kinds of enzymes including hexokinase and glucose-1-dehydrogenase (GDH) are used for glucose measurement. While the hexokinase method is used as the reference method in many clinical laboratories for accurate estimation of glucose in serum or plasma, GDH and GOx are two common enzymes employed in glucose biosensors and self-monitoring of blood glucose [11]. GDH and GOx differ in some features including selectivity towards glucose, cofactors, Michaelis constants (K_m), redox potential, cosubstrates, and turnover rate [12]. GOx is more common and considered as gold standard enzyme for glucose sensing due to its unique characteristics such as higher selectivity for glucose, relatively low cost, and good stability towards environmental conditions including high temperature, pH and ionic strength [2, 13]. Glucose detection using this enzyme relies on the following three reactions:

(1) Glucose + $GOx\text{–}FAD^+ \rightarrow$ Gluconic acid + $GOx\text{–}FADH_2$
(2) $GOx\text{–}FADH_2 + O_2 \rightarrow GOx\text{–}FAD + H_2O_2$
(3) $H_2O_2 \rightarrow 2H^+ + O_2 + 2e$

Since GOx requires a redox cofactor for the catalysis of glucose oxidation, flavin adenine dinucleotide (FAD) plays this specific role as the initial electron acceptor. Therefore, in the first step, FAD is strongly bound to GOx. Then, GOx-FAD catalyzes the oxidation of β-D-glucose to gluconic acid. The produced $FADH_2$ is regenerated by reacting with molecular oxygen, resulting in the formation of hydrogen peroxides. On the other hand, H_2O_2 is oxidized at a catalytic, classically platinum anode. The electrode quickly detects the number of electron transfers, and the electron current corresponds to the number of glucose molecules in the blood [2, 12].

Three common strategies are employed for electrochemical detection of glucose including (1) measurement of glucose utilization; (2) measurement of the concentration of produced hydrogen peroxide; or (3) application of a permeable or immobilized mediator to transport electrons from the GOx to the electrode.

Despite a lot of advantages of GOx, this enzyme uses oxygen as an external electron acceptor in the oxidation reaction, thus device efficiency is sensitive to and variable depending on the atmospheric oxygen level [14]. To overcome this limitation and increase sensing reliability, GDH does not require oxygen and is being employed in different types of glucose sensors. Recently, GDH-based biosensors relying on amperometric detection are increasing. GDHs are subdivided based on their redox cofactors; The GDHs utilizing the cofactor nicotine adenine dinucleotide (NAD) or nicotine adenine dinucleotide phosphate (NADP), GDHs utilizing the pyrroloquinoline quinone (PQQ) cofactor and FAD-dependent GDHs [15]. PQQ-GDH catalyzes not only the oxidation of glucose, but also of other sugars. Among these three members of GDHs family, FAD-GDH represents high thermal stability and substrate selectivity and does not require further cofactors or active catalysts. However, FAD-GDH shows ineffective electron transfer with the commercial $Ru(NH_3)_6$ mediator that is employed with GOx, and thus $Ru(dmo\text{-}bpy)_2Cl_2$ has been evaluated to be an efficient redox mediator for electron transfer with FAD-GDH [16].

2.2 *Different Generations of Enzymatic Electrochemical Glucose Biosensors*

Three kinds of enzymatic glucose biosensors have been proposed and are classified as first, second, and third-generation. Briefly, the first and second generation types are based on amperometric detection by monitoring electrochemical oxidation of hydrogen peroxide or reduction of oxygen for first generation-based sensors, or electrochemical oxidation of reduced mediators in the second generation types [17]

First-generation glucose sensors are enzymatic devices based on oxidizing glucose using GOx as a specific bioreceptor [1]. They have relied on the utilization of natural oxygen co-substrate and the production and recognition of H_2O_2 (Fig. 2) [18]. Although detection of hydrogen peroxide is simple and miniaturization of such systems is easy, but measurement of H_2O_2 requires a relatively high reduction potential in which other endogenous reducing species (e.g., uric acid, ascorbic acids and some drugs) in body fluids are also electroactive and can interfere with glucose detection [19, 20]. Another challenge of this system is the limitation of oxygen solubility in body fluids, which can lead to errors during measurement and low reproducibility of the test [1]. To address these challenges, oxygen was replaced with a non-physiological electron acceptor in second generation of glucose biosensors. In this system, the chemical mediator was able to shuttle electrons from the redox center of the GOx to the electrode surface in order to decrease detection potential (Fig. 2) [21]. Different kinds of artificial mediators including ferricyanide, quinines, tetrathiafulvalene (TTF), ferrocene, tetracyanoquinodimethane (TCNQ), methylene blue, thionine and methyl viologen were investigated to increase sensor efficiency [22–24]. Among these, ferrocene showed all the desirable characteristics of a mediator and was applied in the first commercial mediator-based glucometers [25, 26]. Due to high toxicity of the mentioned mediators, the third-generation glucose biosensors without the use of any chemical mediator and based on direct transfer between the enzyme and

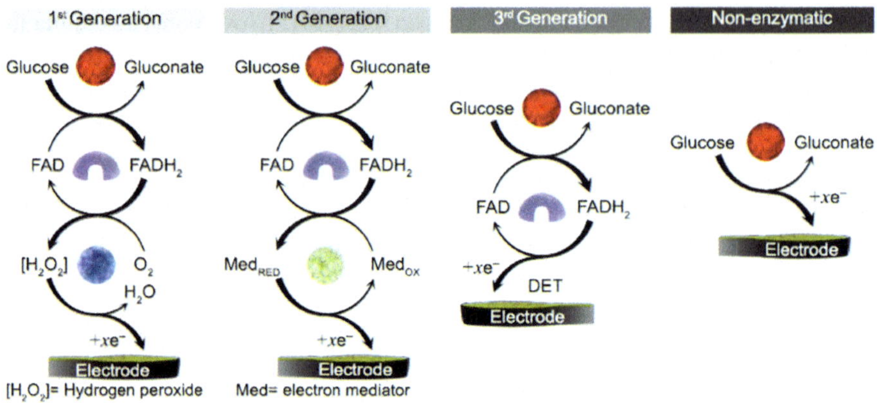

Fig. 2 A schematic representation of different generations of electrochemical glucose biosensors. Reprinted from Ref. [40] with permission

electrode were developed (Fig. 2) [4, 20]. In these systems, the electrode can perform direct electron transfers using organic conducting materials based on charge-transfer complexes. Hence, the third-generation glucose sensors have resulted in implantable, needle-type systems for continuous in vivo monitoring of glucose. Conducting organic salts including tetrathiafulvalene-tetracyanoquinodimethane (TTF-TCNQ) are known as intermediates in the electrochemistry of GDH-PQQ enzymes as well as flavoproteins (GOx) [12]. Direct electron transfer in third-generation glucose sensors enhances electron transfer and provides a simple system for glucose monitoring [2].

2.3 *Non-enzymatic Electrochemical Glucose Biosensors*

Despite many advantages of enzymatic glucose biosensors such as high specific catalytic action excellent selectivity and sensitivity, high efficiency and mild measurement conditions, they suffer from several limitations that arise from proteinaceous nature of enzymes [10, 27]. The performance of enzymatic group is highly dependent on environmental conditions including temperature, pH, humidity and toxic chemicals limiting their application [3, 10]. To address these challenges, non-enzymatic glucose sensors have been proposed based on the electrocatalytic oxidation of glucose, which can be detected optically or electrochemically (Fig. 2). Compared to optical sensors, non-enzymatic electrochemical sensors offer higher sensitivity, fast response, simpler operation, ease of miniaturization and scalability, lower cost, low power requirement, and portability [28]. Noble metals and their alloys (e.g., Pt, Au, Pd, and Rh) [29, 30], metal oxide (e.g., NiO and CuO) [31–33], and transition metals (e.g., Cu, Ni, Zn, and Mn) [34, 35], are mostly used materials in electrochemical glucose sensors due to their electrocatalytic activity. Moreover, they are mostly integrated with other nanostructures such as graphene [36], carbon nanotubes [37],

MoS_2 [38], and conducting polymers [39] to increase the sensitivity. Noble metals and their composites have been widely applied as the electrode materials for non-enzymatic glucose sensors due to their high electrocatalytic activity, better stability compared to enzymes and high sensitivity to the electro-oxidation of glucose [20].

Glucose oxidation through synthetic electrocatalysts is explained by two models including activated chemisorption model and incipient hydrous oxide adatom mediator (IHOAM) model. In activated chemisorption model oxidation process is initiated by the glucose adsorption onto catalyst (electrode) surface followed by breaking bonds and new intermediate formation. After glucose adsorption on the metal surface, it is oxidized to glucono-δ-lactone, which is further converted to gluconic acid. IHOAM model relies on reactive hydroxide species on the electrode surface generated during the electrocatalysis and their effect on many organic molecules' redox reactions [10]. Both models can be considered for catalytic reaction on noble metal electrodes, but they are not qualified to explain glucose oxidation on transition metals or metal oxides [3].

The main drawback of non-enzymatic glucose sensors is the absorption of glucose oxidation intermediates or solution active species which leads to blockage of electrode activity for direct glucose electro-oxidation. Moreover, non-enzymatic amperometric glucose sensors suffer from lower selectivity in comparison with enzymatic amperometric glucose biosensors due to the weakness of the electrocatalytic materials to specifically catalyze glucose oxidation [20].

2.4 Continuous Monitoring of Glucose Using Wearable Sensors

Wearable sensing method is an emerging technology relying on a combination of chemistry, engineering, computer sciences, and wireless technology. This technology has made a considerable progress in the aspect of digital quantification of our daily health continuously. In this regard, continuous glucose monitoring (CGM) in different body fluids using wearable devices has gained considerable attention. CGM sensors are wearable non/minimally invasive tools that measure glucose concentration almost constantly (1–5 min sampling period) for multiple consecutive days/weeks reducing the need of the self-monitoring glucometers (Fig. 3a) and significantly increasing the information on blood glucose fluctuations [41]. In recent years, different mechanisms for non-invasive or at least minimally invasive CGM devices [42–45] have been investigated to meet all the basic needs of a sensor such as sensitivity, specificity, linearity within biological relevant range (Fig. 3b, c) [41]. Among different proposed techniques (i.e. optical, electrochemical, and piezoelectric), the one that is today used by most of the commercialized CGM devices is based on the concept of 1st or 2nd generation of electrochemical glucose biosensors relying on GOx reaction [1]. Commercial CGM device based on 3rd generation glucose sensors has been yet to be done.

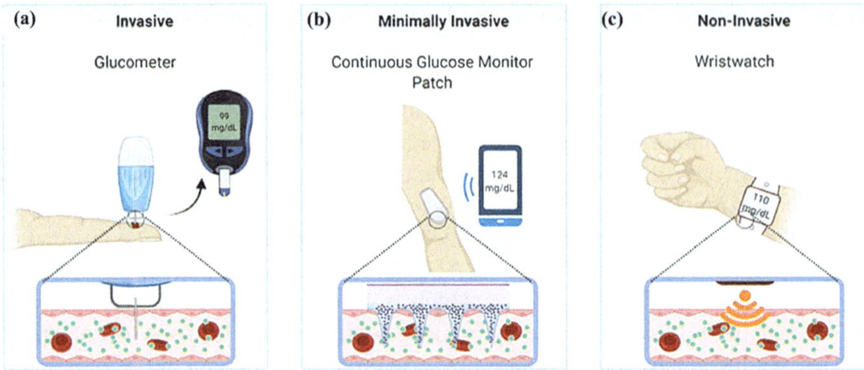

Fig. 3 Different types of invasive, minimally invasive and non-invasive glucose monitoring devices. Reprinted from Ref. [49] with permission

The first prototype of wearable CGM sensor, Medtronic Minimed Gold, was introduced and approved by FDA in 1999 [46] and devices have evolved rapidly since then. The system was a wired-type glucose electrochemical needle-type sensor inserted in the subcutaneous tissue in the abdomen or on the arm, in a minimally invasive manner, and measured an electrical current signal produced by enzymatic reaction. The electrical signal is proportional to the glucose content in the interstitial fluid, which is then converted into a glucose concentration by a calibration method normally performed twice a day. Current Medtronic CGM devices are all equipped with wireless transmission. Dexcom Inc. launched their first CGM system in 2006, which also utilized 1st generation basis, and was equipped with wireless transmission. Abbott introduced the first 2nd generation type CGM device in 2008, by using their iconic technology redox-polymer hydrogel, that consisted osmium complex as electron mediator incorporated with GOx [17]. In recent years, many non-invasive (e.g., flexible electrodes) or minimally invasive (e.g., microneedles) electrochemical monitoring devices that perform glucose measurements in other biological fluids instead of blood have attracted the attention of research topics. It is worth noticing that although glucose is detectable in different body fluids such as sweat, saliva, urine, tears and ISF and thus glucose levels in these matrixes have the potential to correspond to blood glucose content [47] but, the glucose concentrations in these biofluids are much lower than those in blood, leading to large deviations in the measurement results from the actual blood glucose content. However, among the wearable devices, minimally invasive devices are preferred due to the use of ISF whose glucose concentration is closer to the blood plasma and has a reliable correlation with blood glucose levels.

Non-invasive CGM techniques have strict requirements for the environment and must overcome challenges such as those concerning noise interference, detection sensitivity, and stability. Most non-invasive sensing techniques are only employed for "proof of concept" and have not yet been used in clinicalapplications [48]. The most extensively used CGM devices in clinical applications are currently based on minimally invasive electrochemical techniques. The minimally invasive electrochemical CGM sensors show several advantages of high sensitivity and stability, short response time, simple operation, and easy miniaturization. The features of these two categories of wearable CGM sensors are explained in more detail below.

2.4.1 Wearable Minimally Invasive CGM Systems

Primary CGM devices are blinded systems which record retrospective glucose data for professional use only. In these systems, a sensor probe ranging from 5 to 13 mm in length is implanted under the skin of a diabetic individual to continuously collect glucose data. A statistical report of the glucose concentrations is produced at the end of the wearing. A retrospective CGM sensor can represent the variations in a diabetic's glucose contents on several consecutive days and make a reference for physicians to investigate the diabetes treatment plan. The limitations are that the data processing is complicated and diabetics are not able to check their own glucose data in real-time. A real-time CGM sensor updates glucose information every 1 to 5 min [50]. Therefore, the glucose content and fluctuation graphs are created at any time, and the history of the results can also be reviewed. Moreover, the device is easy to wear and does not interfere with the normal daily life of diabetic person. Whenever the glucose concentration exceeds the threshold level set for hyperglycemia or hypoglycemia, the device will generate an alarm. Some products of Medtronic's Guardian™ Connect and Guardian™ Sensor 3 even are able to predict high or low glucose excursions up to 60 min in advance. The latest development in commercial CGM sensors is that the device is integrated with an insulin pump to create an artificial pancreas, so that insulin injection can be done automatically if necessary [48].

There are several critical technologies for the construction of minimally invasive electrochemical-based CGM systems. In this regard, several criteria should be considered to provide the best performance and minimal invasiveness and pain: (1) the size of implanted sensor probe should be small enough to minimize the incision zone and reduce implantation pain and the infection risk; (2) the enzyme activity should be preserved as much as possible to inhibit large enzyme losses during long-term operation, thereby increasing the sensitivity and stability of the sensor; (3) the sensor should be more biocompatible to reduce the effect of allogeneic reaction. In this regard, application of microneedles and biocompatible membranes and also immobilization of enzymes in sensor development can meet these requirements to a great extent.

Microneedle-Based Electrochemical CGM Systems

Microneedle-based CGM sensors have emerged in order to access ISF in an efficient, non-invasive and pain-free way and to reduce infection risk associated with implantable probe. These systems are safer for patients by reducing the possibility of infections [51]. Due to the micron size (< 1 mm in length), they can penetrate the epidermis to reach the dermis without contacting the dermal neurons and blood vessels [52, 53]. Furthermore, microneedles possess a larger electrode surface area and show high sensitivity in both in vitro and in vivo tests. The type of microneedle and the materials applied for its fabrication play a main role in the design of microneedle-based biosensors to ensure full compatibility with both sensing elements integration and implementation for transdermal investigation [54]. Three main classes of materials including polymers, metals and silicon are generally used to fabricate microneedles. Polymers are usually used more than other materials due to their low cost and ease of construction. Different kinds of polymers such as epoxy-based negative photoresists [55, 56], acrylate-based polymers [57, 58] and polycarbonate [59, 60] have been investigated. In the case of electrochemical sensing, to increase electrical conductivity, polymeric microneedle is coated with metals, metal oxides, alloys (e.g., Au, Pt/IrOx, Ti/Au or Cr/Pt) or even conductive forms of carbon (e.g., carbon paste, carbon fibers or carbon nanotubes). Concerning metallic microneedles, the most popular materials include Pt, Au and W because of their superior conductivity and good biocompatibility. Silicon with the feature of excellent is extensively used to fabricate microneedles. Microneedles fabricated with hollow silicon can passively extract ISF by capillary action. Depending on the method of forming the materials, four different fabrication techniques can be used to develop microneedles. These techniques include lithography, micromachining, mold-filling, and computer assisted polymerization [54].

In microneedle-based electrochemical sensors for glucose monitoring, generally enzyme, in particular GOx, is used as a bioreceptor to modify microneedle electrodes for selective monitoring of glucose. For example, in several microneedle-based glucose sensors, poly(3,4-ethylenedioxythiophene) (PEDOT) was employed for the immobilization of GOx on the surface of platinum-coated stainless steel microneedle electrode. PEDOT also played the role of electrical mediator. On the other hand, Pt was applied due to high electrical conductivity and chemical stability. Therefore, with the combination of PEDOT, Pt and GOx, a highly sensitive and selective glucose sensor was obtained [61]. In another effort, Sharma et al. reported a minimally invasive glucose biosensor based on a microneedle array electrode fabricated from an epoxy-based negative photoresist (SU8 50) and developed for continuous measurement in the dermal compartment with minimal pain (Fig. 4a) [55]. The device was constructed by casting the structures in SU8 50, crosslinking and then metallizing with Pt or Ag to produce the working and reference electrodes, respectively. The metallized microneedle array electrodes were subsequently functionalized by entrapping GOx in electropolymerized polyphenol (PP) film. In vitro studies showed that the developed devise is able to measure glucose concentrations less than 0.5 mM with a response time of 15 s. After the microneedle insertion into the forearm of several volunteers for 24 h, no inflammation or skin irritation was reported by them.

Carbon-based nanomaterials are very interesting for the coverage of microneedles in order to increase electrical conductivity. In this regard, a microneedle-based

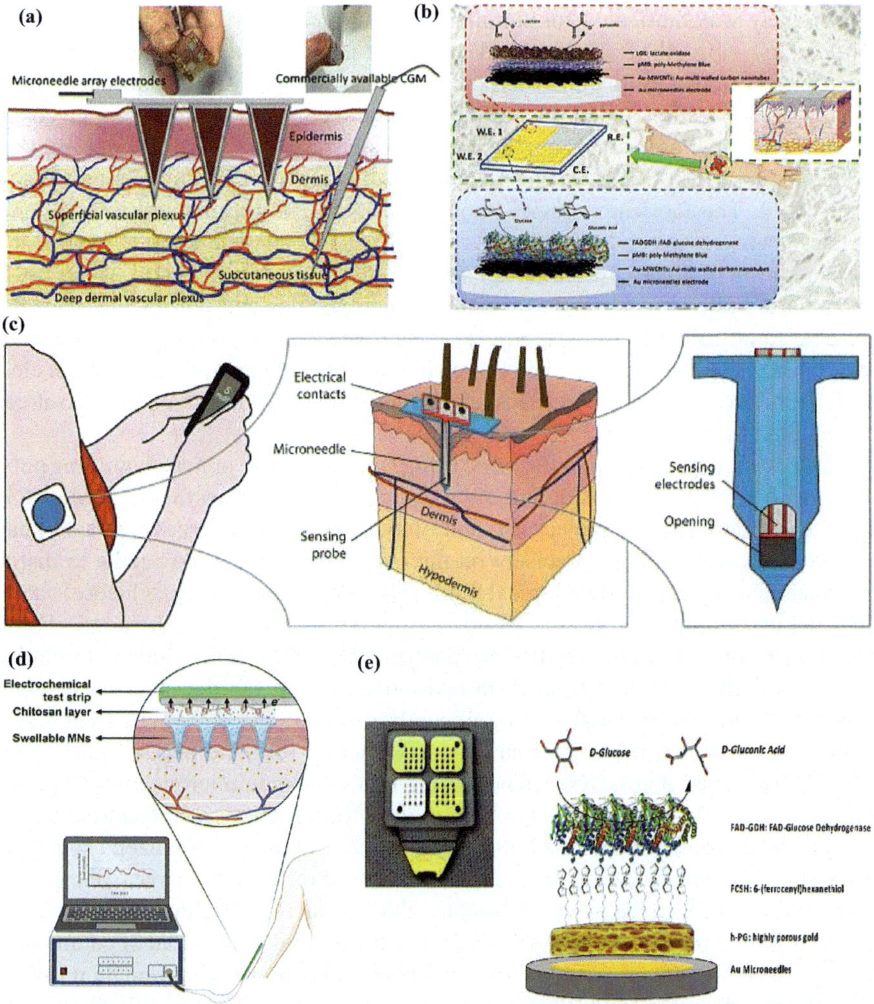

Fig. 4 **a** Comparison of microneedle array electrodes and commercial CGM and the two areas of the skin cavity applied to reach the ISF. Reprinted from Ref. [55] with permission; **b** A microneedle-based biosensor array for the continuous and simultaneous detection of lactate and glucose in the artificial ISF. W.E.: working electrode; R.E.: reference electrode; C.E.: counter electrode. Ref [62]; **c** A CGM device based on hollow silicon microneedle. Reprinted from Ref. [65] with permission; **d** A CGM device combining swellable microneedles and electrochemical test strips for the evaluation of glucose or alcohol content in the skin's ISF. Reprinted from Ref. [68]; **e** A highly porous gold/6-(ferrocenyl)hexanethiol (FcSH)/FAD-GDH/Au microneedles-based biosensor for glucose monitoring in ISF. Reprinted from Ref. [59]

sensor array was fabricated for simultaneous monitoring of lactate and glucose in artificial ISF (Fig. 4b) [62]. The gold surface of the microneedles was modified by electrodeposition of multiwalled carbon nanotubes (MWCNTs) and subsequently by electropolymerization of methylene blue as the redox mediator to form second generation of glucose biosensor. Then, the modified microneedle was functionalized with lactate oxidase (LOX) in working electrode 1, and with the FAD-GDH in working electrode 2. The fabricated device showed a high sensitivity towards glucose and lactate.

In addition to glucose monitoring, microneedles can also act as ISF collector. In this regard, hollow microneedles, swelling microneedles and porous microneedles are able to collect ISF. In these kinds of microneedles, the sensing electrodes are integrated inside the microneedle. Therefore, microneedle not only can protect electrode from breakage or enzyme loss during insertion into the skin, but also it can collect glucose of ISF [48].

Hollow microneedles, manufactured using silicon, metals, or non-dissolving polymers, with empty space inside are a type of conventional transfusion needle [63]. They are widely used to collect ISF and blood. For sensing purpose, a rapid analysis system can be placed precisely on the outlet of hollow microneedle to distinguish analytes in collected biological fluids. Therefore, analyte (e.g., glucose) can be detected by hollow microneedle-based sensing device. However, the possibility of blocking the opening of the needle during the piercing of the skin is a major limitation for hollow MNs, which will result in resistance to flow. On the other hand, fabrication of hollow microneedles is significantly difficult, and those with high aspect ratios don't have the usual internal support structure for solid needles [64]. Ribet et al. (2018) fabricated a CGM device using hollow silicon microneedles (Fig. 4c). The device was based on the combination of an ultra-miniaturized electrochemical sensing probe inside of a single hollow microneedle, separately realized using standard silicon microfabrication procedures. The device relied only on passive capillary lumen filling without the need for complex fluid extraction mechanisms. Moreover, the transdermal portion of the system was fifty times smaller than that of commercial devices [65]. Since long-term stability and continuous monitoring remain unsolved challenges associated with electrochemical sensors, Parrilla et al. (2022) tried to overcome these challenges by developing a highly stable redox mediator bilayer for electrochemical sensing of glucose in ISF [66]. The biosensor was composed of screen-printed carbon electrodes (SPCE) modified with GOx and redox mediator bilayer of Prussian blue (PB)/nickel hexacyanoferrate (NiHCF) system to enhance the stability of the redox mediator layer. In addition, a diffusion-limited Nafion layer is applied to the electrode to increase the linearity of the biosensor to millimolar levels required for diabetes monitoring. After assessment of hollow MNs array for easy skin piercing, the biosensor was attached to the back of the hollow MNs array leaving a microfluidic cell. Finally, the hollow MNs sensor array was coupled to a syringe using microfluidic tubing. This design provided rapid ISF extraction and glucose analysis, as only the glucose diffusion into the SPCE results in long-term operation.

The swellable microneedles have obtained considerable attention due to swelling under the skin after absorption of ISF and also maintaining the structure without dissolving [67]. Swellable MNs are commonly constructed by physical crosslinking methods such as exposure to ultraviolet (UV) light with photoinitiators [63]. The major limitation of swellable microneedles is their lower mechanical strength. By selection of suitable polymers and fabrication method, a balance can be established between swelling and mechanical strength [67]. Zheng et al. (2022) developed a skin patch integrating swellable microneedles, made of swellable methacrylated hyaluronic acid, and electrochemical test strips for glucose and alcohol measurement in ISF [68]. This device was fabricated by adhering microneedle patch on the electrochemical strips using the chitosan as the connecting layer. The electrochemical strip was a screen-printed three-electrode strip which working electrode was first modified with a Prussian blue as mediator layer and subsequently functionalized with Gox. The microneedle penetrated the skin for extraction of ISF that flowed to the backing layer of microneedles and was analyzed by the test strip (Fig. 4d). In another example, Zheng et al. (2022) embedded wire electrodes in the tips of silk fibroin-microneedle and fabricated a swellable microneedle patch for in situ real-time detection of glucose in the ISF of skin or plant tissues [69]. Working and counter wire electrodes were made of platinum (Pt), while wire reference electrode was made of Ag/AgCl. Chitosan was used as the binder to immobilize GOx onto the surface of Pt wires. The wettable yet insoluble fibroin hydrogel provided good mechanical strength in dry conditions for easy tissue penetration. After insertion, fibroin MNs swelled and extracted biomarker-containing ISF into their polymeric matrix to reach the embedded electrodes. Rapid detection was achieved due to the immediate contact between the electrodes and the ISF. This microneedle strategy is suitable for long-term monitoring of glucose as well as other biomarkers.

Porous microneedles possess interconnected micro-sized pores throughout the entire structure of microneedle. They are usually fabricated from biocompatible metals, ceramics, or polymers with small randomly distributed and interconnected pores. Due to the continuous pores, the rapid absorption and collection of ISF can be reached through capillary action. Therefore, ISF extraction and the subsequent monitoring of glucose in fluids can be obtained using porous microneedles integrated with lab-on-chip biosensing devices. Increased fragility due to the large volume of holes is one of the disadvantages of porous microneedles [63]. Bollella et al. (2019) developed a highly porous gold microneedles-based second generation biosensor for minimally invasive monitoring of glucose in artificial ISF [59]. The highly porous microneedles-based electrode was fabricated using a simple electrochemical self-templating strategy that included two steps, gold electrodeposition and hydrogen bubbling at the electrode (Fig. 4e). The highly porous gold surface of the microneedles was modified by immobilization of 6-(ferrocenyl)hexanethiol as a redox mediator and then by immobilization of FAD-GDH enzyme using a drop-casting method. The device showed high sensitivity and stability. Porous metals have been investigated as highly active catalysts as well as sensitive electrodes due to their large surface area. Kai et al. (2021) developed porous microneedle electrodes (with a pore diameter of approximately 1 μm) by electroless plating of nickel and gold on the porous polymer microneedles of poly(glycidyl methacrylate), and applied them to an enzymatic glucose sensor [70]. The GOx was immobilized on the gold-plated microneedle electrodes. The amperometry of glucose content in solution was determined using the fabricated electrode as the working electrode, together with the Ag/AgCl reference electrode and the gold counter electrode, both of which were made of microneedles. The porous structures of microneedle electrode resulted in increased sensitivity.

2.4.2 Wearable Non-invasive CGM Systems

As mentioned earlier, development of wearable non-invasive needle-free CGM devices provides a convenient tool to monitor the glucose concentration of diabetes patients without discomfortability and risk of infection. In recent years, a lot of wearable devices in different forms of watches, glasses and wrist/armbands have been developed for real-time and continuous monitoring of glucose level. However, current commercial devices are generally just miniaturized forms of conventional electronics that include rigid, bulky, and flat components. Such stiffness and bulkiness can not only disturb and annoy the devices, but also limit the matching contact of the devices with the human body [71]. Moreover, due to the need for advanced technologies and hardware combinations the production cost increases, which will be one of the major limitations of this kind of biosensors. To overcome these challenges and develop a fully wearable sensor system several important factors including the application of soft and flexible materials, wireless technologies, and high integrity should be considered. With recent advances in soft, stretchable, and flexible conducting materials, development of wearable biosensors has become easier. On the other hand, advances in wireless communication technologies and electronic engineering simplify data processing and transmission with a miniaturized electronic chip. The main advantage of wireless power transfer is that it just needs a simple antenna circuit, which provides a vast choice of materials and structures. Using tribo and piezoelectric mechanisms, energy harvesters enable charging of electricity through body movement, thus providing an easy way to charge regardless of environmental conditions [72, 73]. The combination of all functional ingredients with different functions, characteristics and shapes in small and compact devices should include rigorous methodological approaches.

Non-invasive and on-body analysis of glucose can be performed in different biofluids such as tear, saliva, sweat, and ISF.

Tears include various biomarkers such as electrolytes, metabolites (e.g., glucose, lactate, and ascorbate), and proteins (e.g., lysozyme, lactoferrin, lipocalin, and albumin). Tear biomarkers directly diffuse from blood, resulting in a close correlation between the biomarker level in tears and blood [74]. Moreover, accessibility of tear is very easy. However, sample collection of tears is challenging due to the small sample volume, rapid evaporation during sample preparation, eye irritation, and deficiency of tear fluids of individuals who suffer from dry eye syndrome. Wearable tear sensors consist of eyeglasses and contact lenses [74].

Saliva contains proteins, hormones and small metabolites such as glucose. A main advantage of saliva is the possibility of collecting a large volume of sample in a non-invasive, low cost and simple way. However, the concentration of biomarkers in saliva is generally lower than blood and other non-invasive biofluids. Moreover, due to presence of food particles, bacteria and other contaminants, some sample preparation steps such as filtering or centrifuging are required before analysis. One of saliva wearable sensors include electrodes well-equipped in mouthguard [74].

Sweat as an epidermally available biofluid contains electrolytes, metabolites (e.g., glucose), proteins, small molecules and disease biomarkers. Analysis of sweat has several limitations such as variation of sweat secretion (upon weather condition, physical activity, stress, and chemical stimulus), sweat rate, sample evaporation, and skin surface contamination. Tattoo biosensors including conductive inks attached on the wearable flexible materials are an example of electrochemical biosensors for the sweat analyzing [74].

ISF, an extracellular fluid surrounding tissue cells, contains proteins, electrolytes and metabolites. Concentration of analytes in ISF is similar to blood plasma leading to reliable correlation between ISF and blood plasma. However, ISF components should be excreted at the surface of skin by an additional approach such as reverse iontophoresis (RI) and sonophoresis for non-invasive sampling [75]. RI is the most widely used technique in electrochemical glucose sensing relying on ISF sample. This technique led to the development of the GlucoWatch® as a commercial wearable non-invasive electrochemical glucose monitoring device [76]. RI is a needle-free method for extracting biomolecules through intact skin to gain the goal of blood glucose detection. RI consists of two mechanisms: (1) electromigration, the direct interaction between charged ions and applied electric field, and (2) electroosmosis, a connective solvent flow from anode to cathode direction. For the charged compounds, electromigration is the major transport mechanism [77]. The total quantity of charge transferred depends on the strength and duration of the electric field. For the uncharged compounds (e.g., glucose), electroosmosis driven by voltage is the main extraction mechanism. The shape, size and position of electrodes are considered as important parameters for RI. Different shapes of electrodes such as thin disks (same as GlucoWatch) to snakes, ridges or otherwise have been used. It is not only necessary that the electrode can be closely connected to the human skin, but also keeping a suitable micro-distance from the skin surface. Therefore, flexible materials are generally used for the electrode fabrication [77].

Flexible Electrodes: Materials and Fabrication Methods

Selection of flexible materials with appropriate characteristics such as stretching, bending, high compression and tensile strength, and twisting is of particular importance in design of a wearable system. The most common flexible materials employed for wearable glucose biosensors include polyethylene terephthalate (PET) and polydimethylsiloxane (PDMS). The most important features of these materials are their transparency, wide commercial availability, and the ability to print with a wide range of electrically conductive materials. In addition, fibrous materials including cellulose, polyurethane, polyamide, and cotton fibers, particularly moisture management as a kind of cotton material with 100% moisture absorption, are suggested as inexpensive, available and flexible materials in wearable electrochemical CGM devices. A flexible and wearable glucose biosensor was fabricated using hydrophobic PDMS and hydrophilic 2-methacryloyloxyethyl phosphorylcholine (MPC) copolymerized with dodecyl methacrylate (DMA) by photolithography and ion beam sputtering techniques [78]. Electrodes (Pt working electrode and Ag/AgCl reference/counter electrode) were formed on the PDMS using microelectromechanical (Soft-MEMS) techniques. Thus, a flexible hydrogen peroxide electrode was obtained. The gas-permeable poly (MPC-co-DMA) (PMD) membrane was placed on the other surface of the electrodes. The GOx was also immobilized using PMD solution casting onto hydrogen peroxide electrode. The calibration range for glucose solution includes the tear glucose concentration of normal subject (0.14 mmol/l). Using the same materials and procedure, this group developed soft contact lens for non-invasive and in situ monitoring of tear glucose [79]. The contact lens biosensor showed a good relationship between the output current and glucose concentration in a range of 0.03–5.0 mM, with a correlation coefficient of 0.999. The biosensor was applied to a rabbit for tear glucose monitoring. The results of tear glucose monitoring showed that tear glucose level increased with a delay of 10 min from blood sugar level.

In a recent study, a breathable flexible glucose biosensor with excellent flexibility and moisture/air permeability was developed [80]. The biosensor's electrodes were fabricated by inkjet printing method onto a flexible porous nanofiber polyimide substrate obtained through electrospinning (Fig. 5a). Graphene and platinum nanoparticles (PtNPs) were used to build three-dimensional nanostructures on the surface of the working electrode to increase specific surface area and enhance the sensing performance. Prussian blue was used to reduce the working potential, shielding the effect of interferents. The porous structure of the nanofiber substrate made the biosensor breathable, usefully inhibiting the body-fluid accumulation that could remarkably affect measurement accuracy in long-term wearable glucose monitoring. Moreover, the porous structure allowed the electrodes to be embedded within it. The array substrate was attached to a volunteer's leg skin for 12 h and no sweat accumulation was observed.

Using one-step laser-scribed PtNPs nanostructured 3D porous laser-scribed graphene (LSG), a flexible electrochemical sensing platform was developed for glucose and pH detection [81]. The prepared nanocomposites simultaneously had a hierarchical porous graphene architecture, while PtNPs significantly increased their sensitivity and electrocatalytic activity. For the fabrication of electrodes, Pt-hydroxyethyl cellulose (HEC) hydrogel was prepared and coated on the polyimide (PI) film. Laser scribing was used to fabricate super hydrophilic and co-doped LSG PtNPs. The Pt-HEC/LSG electrode was immersed in ultrapure water to remove the unetched hydrogel. The PtNP-LSG electrode was patterned by a CO_2 direct scribing laser machine. Finally, GOx was immobilized on the Pt-HEC/LSG working electrode and biosensor was used for glucose monitoring in sweat.

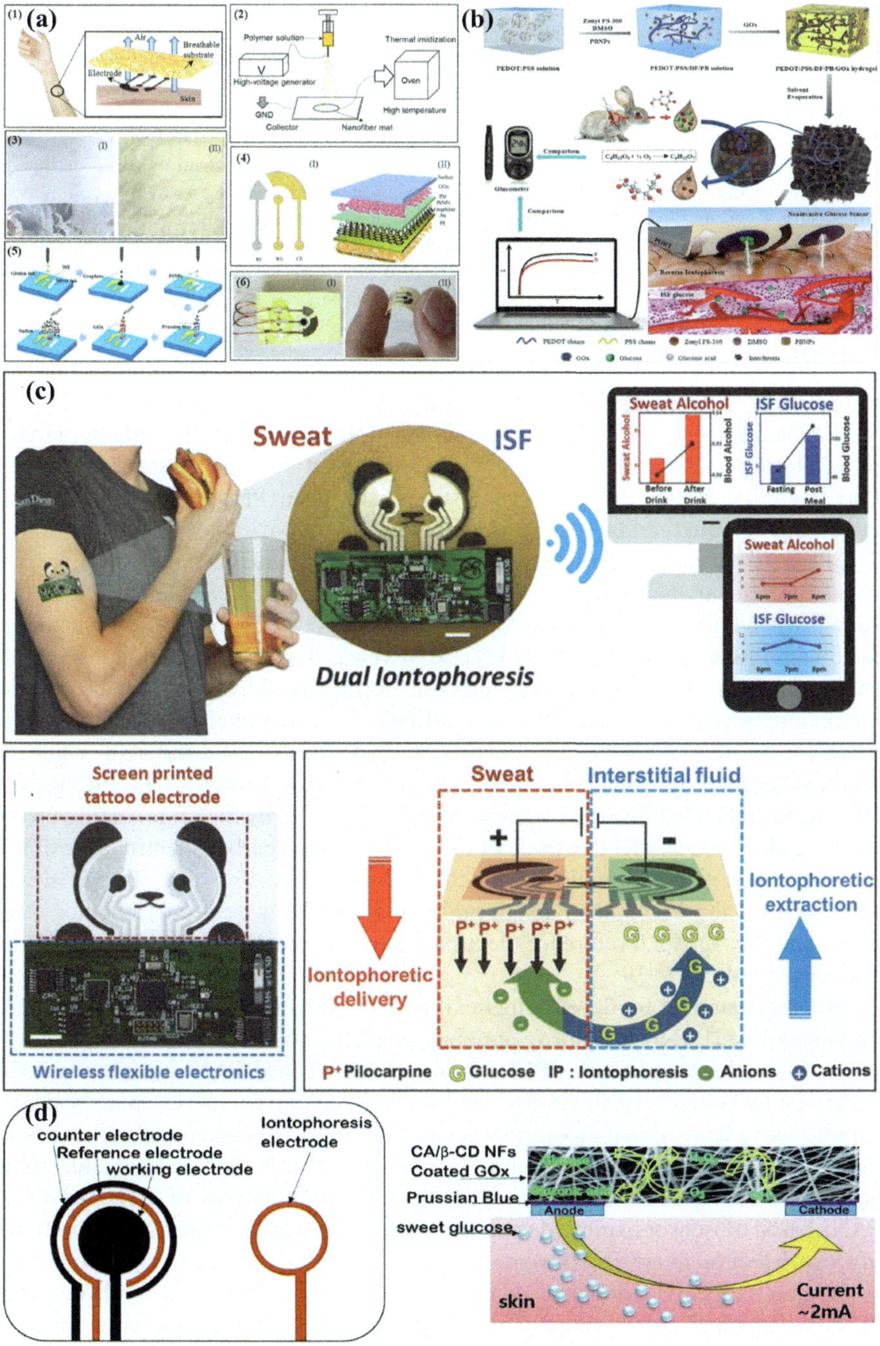
(a)
Air
Breathable substrate
Electrode
Skin
Polymer solution
Thermal imidization
High-voltage generator
Oven
GND
High temperature
Collector
Nanofiber mat
(b)
Comparison
Glucometer
(c)
Sweat
ISF
Dual Iontophoresis
Sweat Alcohol
ISF Glucose
Sweat Alcohol
Blood Alcohol
Before Drink
After Drink
ISF Glucose
Blood Glucose
Fasting
Post Meal
Screen printed tattoo electrode
Wireless flexible electronics
Sweat
Interstitial fluid
Iontophoretic extraction
Iontophoretic delivery
P+ Pilocarpine
G Glucose
IP : Iontophoresis
Anions
Cations
(d)
counter electrode
Reference electrode
working electrode
Iontophoresis electrode
CA/β-CD NFs
Coated GOx
Prussian Blue
Anode
Cathode
sweet glucose
skin
Current ~2mA

◀**Fig. 5** **a** A schematic representation of different steps of fabricating flexible glucose biosensor using porous nanofiber polyimide substrate, (1) the glucose biosensor, (2) substrate development, (3) images of the substrate (I) before and (II) after thermal imidization, (4) the structure of (I) three-electrode and (II) working electrode, (5) the electrode construction steps, (6) physical images that exhibit the (I) glucose biosensor and (II) flexibility of the biosensor. Reprinted from Ref. [80] with permission; **b** A schematic representation of the fabrication of PEDOT:PSS/DF/PB/GOx biosensor for the glucose monitoring in serum and the non-invasive glucose determination in ISF on human skin relying on the reverse iontophoresis, and a comparison with the glucose assay by a commercial glucometer. Reprinted from Ref. [83] with permission; **c** Simultaneous non-invasive sampling and monitoring of ISF and sweat using a wearable iontophoretic biosensor on a printed tattoo platform for glucose and alcohol measurement. Reprinted from Ref. [88] with permission; **d** A schematic illustration of iontophoresis printable electrodes and iontophoresis mechanism on the epidermal cellulose/ β-cyclodextrin /GOx nanofibers (NFs) glucose biosensor. Reprinted from Ref. [91] with permission

Conducting polymers with high flexibility and stretchability close to those of traditional plastics, and also high speed of electron transfer are interesting substrate [82]. Poly (3,4-ethylenedioxythiophene) polystyrene sulfonate (PEDOT:PSS) is one of the most widely studied conducting polymer for flexible and wearable glucose sensor [83, 84]. PEDOT partially aggregates and forms a "hard" conductive network in the soft PSS matrix due to the weak electrostatic interaction between PSS and PEDOT [71]. An electrochemical biosensor based on PEDOT:PSS conductive hydrogel incorporated with Prussian blue nanoparticles (PBNPs) was developed for the non-invasive and continuous monitoring of ISF glucose on human skin based on the reverse iontophoresis (RI) extraction [83]. The PEDOT:PSS hydrogel was used as a skin patch for the improvement of contact between the sensor and the skin, while its high conductivity resulted in the achievement of high sensitivity. The PBNPs and GOx were embedded into the hydrogel solution followed by the drop-coating on the working electrode and RI electrode of the screen-printed carbon electrodes (SCPE), which resulted in the formation of glucose sensing films (Fig. 5b). The SPCE was fabricated on the PET substrate. In addition to working electrode, counter electrode, and the Ag/AgCl reference electrode, the SCPE included a RI electrode with a diameter of 1 cm.

In another attempt for developing flexible electrodes, a nanocomposite film based on polyaniline (PANI) and nitrogen-doped graphene quantum dots (N-GQDs) was prepared and deposited on the SPCE fabricated on PET [85]. PANI was chosen as the matrix for GOx due to its flexibility and excellent binding for GOx, where N-GQDs supplied N-rich functional groups that act as electron transfer agents. This overcomes the electrical conductivity issues in neutral electrolytes and increases the electrical activity towards glucose detection. After the GOx immobilization, GOx/N-GQDs/PANI-based biosensor exhibited excellent performance and enhanced sensitivity for glucose detection in artificial sweat.

Wearable electrochemical biosensors have also been integrated with low cost paper-based materials and directly applied on human skin for non-invasive analyzes [86]. The use of a flexible and hydrophilic cellulose paper substrate has a significant effect on the absorption and dispersion of active substances and increases the permeation of sweat as well as its diffusion from the human-device interface [87]. In the case of paper, tattoo-based paper sensors show high flexibility, printability, stability, and adhesion ability to human skin. They can be employed as wearable epidermal sensing devices. Kim et al. (2018) developed a flexible tattoo paper dual-electrode wearable biosensor for simultaneous and independent sampling and analysis of two epidermal biofluids (ISF and sweat) for the monitoring of glucose and alcohol [88]. The dual sampling and analysis were performed through the parallel operation of reverse iontophoretic ISF extraction across the skin and iontophoretic delivery of a sweat-inducing drug (pilocarpine) into the skin at separate positions (Fig. 5c). The wearable system was fabricated on epidermal temporary tattoo platform through screen-printing technique for disposable single use. This tattoo paper biosensor was able to measure ISF glucose at the cathode side using a glucose GOx and sweat alcohol at the anode side using an alcohol oxidase (AOx). In addition to the previously mentioned advantages of paper substrate, other advantages include the self-absorption of sweat through the created microfluidic channels in paper-based devices using the hydrophilic and hydrophobic material modification, wax printing, inkjet printing, and photolithography. Moreover, the softness and folding ability of papers are also suitable for reducing the size of the device and making devices with multiple functions [87]. In this context, a highly integrated sensing paper (HIS paper) was developed for fast sweat collection and glucose determination [87]. The HIS paper was assembled from a combination of hydrophobic protective wax, conductive electrodes, and MXene/methylene blue ($Ti_3C_2T_x$/MB) hybrid active materials by using a simple printing process. Carbon paste and silver paste were employed as counter electrode and reference electrode, respectively. Carbon paste was used to construct screen-printing carbon working electrode (SPCE). The printed paper was folded into a multi-layered structure, where a sensible three-dimensional perspiration diffusion pathway was created by connecting the hydrophilic zones of each layer, providing efficient paths for perspiration collection and diffusion of sweat along the vertical direction of the folded HIS paper. Moreover, the independent three-dimensional positioning of the three electrodes facilitated the decoration and immobilization of enzymes (GOx and lactate oxidase) as well as the accessibility of electrolytes. The unique design of the three-dimensional structure enhanced the absorption of sweat from the surface of the skin, and prevented the accumulation of liquid, thereby reducing the discomfort caused by sweat in the human–machine interface.

Hydrogels can have a homologous contact with the human body because their mechanical features are similar to the interfaces between human skin and underlying tissue [89]. Hydrogel-fibrous materials with high surface-to-volume ratio, flexibility, and high-water absorption properties are good candidates in glucose sensing wearable devices. They can be fabricated by electrospinning technique with a variety of porosity and diameters [4]. In this regard, Kim et al. (2019) developed poly(vinyl alcohol)/β-cyclodextrin hydrogels cross-linked by citric acid for glucose measurement. The GOx was physically entrapped within the β-cyclodextrin cavity for accurate detection of ISF glucose levels. The developed hydrogel system showed great potential as a patch sensor for non-invasive glucose monitoring [90].

Another advantage of hydrogels is that electrodes could be easily printed on the surface of fibrous hydrogels for wearable electrochemical devices. Kim et al. (2019) developed a cellulose/β-cyclodextrin electrospun immobilized GOx patch with reverse iontophoresis for non-invasive monitoring of ISF glucose levels. Carbon-Ag/AgCl-carbon: working-reference-counter electrode coated Prussian Blue were screen-printed on the surface of the fabricated path (Fig. 5d). The nanoporous structure of the patch was able to keep the extracted solution on the body [91]. In an interesting study with a combination of hydrogel and paper, a hydrogel paper patch (HPP) with highly porous conductive PEDOT:PSS hydrogel self-assembled on paper fiber was developed [92]. To fabricate electrode, HPP with microfluidic channels was prepared on a filter paper using a wax printer. Carbon paste and Ag/AgCl paste were printed individually on the specific areas of filter paper using screen-printing method to fabricate the counter and reference electrodes respectively. To prepare the working electrode of the glucose sensor, PEDOT:PSS solution was casted on a rectangular area. After the filter paper was completely filled with the PEDOT:PSS solution, the filter paper was soaked in dodecylbenzene sulfonic acid (DBSA) to form the PEDOT:PSS conductive hydrogel. Then, PtNPs were electrodeposited onto paper-based hydrogel. Finally, GOx was dropped on the working electrode and cross-linked with glutaraldehyde. Due to the easy processability of paper and the conductivity of PEDOT:PSS hydrogels, the proposed HPP conductivity sensing system is promising as a disposable and low cost sensor kit for health care and glucose monitoring.

Several other examples of flexible electrodes employed in wearable CGM devices are listed in Table 1.

Besides selecting the flexible substrate, electrode fabrication technique is of great importance. The electrically conductive or non-conductive substrates can be modified with different methods to create sensor electrodes. Electrodes including sensing (working), counter and reference electrodes are designed and fabricated on a flexible substrate. The materials chosen to make the electrode should have good adhesion to the substrate to create a flexible sensor. Furthermore, different kinds of nanomaterials can be used in the sensing zone of sensor to increase sensitivity and performance. Various nano/microfabrication techniques including printing (e.g., screen-printing, inkjet printing and 3D printing), laser induction, photolithography, electrodeposition, seed mediated and dealloying methods have been studied and used for the fabrication of wearable devices [4]. Among these methods, screen-printing techniques are the most common methods in the fabrication of wearable electrochemical glucose biosensors [93]. This method is extensively applied for printing conductive or non-conductive inks on flexible substrates.

3 Wearable Self-powered Electrochemical CGM Systems

Although conventional electrochemical batteries are the important source of the electricity required for wearable devices, they suffer from several limitations such as large size, discomfortability to wear the sensor, need to recharge and regular

Table 1 Several examples of flexible electrodes employed in wearable CGM devices

Flexible substrate	Working electrode	Fabrication method	Biofluid	Sensitivity	Linear range	Refs.
Polyethylene terephthalate (PET)	GOx-Chitosan/ CNTs-PB-Gold	UV mediated chemical plating	Sweat	22.05 μA mM^{-1} cm^{-2}	0.02– 1.11 mM	[104]
Paper and PET	GOx-Chitosan-Nafion-PB/ Graphite ink- Ag/AgCl ink	Screen-printing	Sweat	35.7 μA mM^{-1} cm^{-2}	0–1.9 mM	[105]
Polyimide	Nafion-GOx/ Chitosan-AuPtNPs-rGO	Photolithography and wet etching	Sweat	45 μA mM^{-1} cm^{-2}	0–2.4 mM	[106]
Polyimide	GOx-PB-Au NPs-Gold	Electrodeposition and photolithography	Sweat	26.31 μA mM^{-1} cm^{-2}	1–600 μM	[107]
EcoFlex elastomer	Chitosan-GOx/ CNTs-PB-AuNWs	Seed mediated method	Sweat	23.72 μA mM^{-1} cm^{-2}	0–800 μM	[108]
Polyurethane modified by poly(ethylene glycol)	GOx/Chitosan- PtNPs-Gold	Photolithography and magnetron sputtering	Serum	3.418 μA mM^{-1}	0.5–2.5 mM	[109]
PDMS	GOx/Chitosan/PB-Ni–Co MOF nanosheets-Gold	Chemical deposition	Sweat	205.1 μA mM^{-1} cm^{-2}	0.02–0.79 mM	[110]
Agarose hydrogel	Nafion- GOx/Chitosan-PB-PEDOT NC-SPCE	Electrochemical deposition	Sweat	0.2 mA	6.25 μM–0.8 mM	[111]
AuNWs/SEBS fiber	GOx/Chitosan/PB	Dry spinning and electroless plating	Sweat	11.7 μA mM^{-1} cm^{-2}	0–500 μM	[112]
Parylene	Gox-AgNWs-Graphene	Photolithography	Tear	–	1 μM–10 mM	[113]

CNTs: carbon nanotubes; PB: Prussian blue; rGO: reduced graphene oxide; AuNWs: gold nanowires; MOF: Metal–organic framework; PB-PEDOT NC: Prussian blue-doped poly(3,4-ethylenedioxythiophene nanocomposite; SPCE: Screen-printed carbon electrode; SEBS: Styrene-ethylene/butylene-styrene replacement. Therefore, multiple wearable energy harvesters including electrostatic, thermoelectric, electromagnetic, piezoelectric and triboelectric power have been developed. Self-powered energy harvesters may convert numerous forms of environmental energy into electrical energy and can partially replace batteries. Compared to the conventional batteries, self-powered energy harvesters are inexpensive, compact, lightweight and environmentally friendly. The energy that is harvested can come from body movements, body temperature, human physiological processes, surrounding environment, solar energy and artificial light [94, 95]. In addition to mechanical energy, in self-powered wearable biosensors, chemical energy from the components secreted from body fluids such as glucose or lactate can be used as energy source and converted to electrical energy. Since fiber-based textiles are the most popular material for human clothing, they can be used as interesting materials for wearable biosensors. Micro-cable structured textiles with light weight and high flexibility have been used to harvest energy from mechanical excitation, wind blowing and chemical energy (from body secreted materials). Textile- and paper-based chemical energy converters are also known as biofuel cells. A wearable and flexible textile-based glucose biofuel cell by polyester-based employing moisture management fabric (MMF), which is extensively used in sportswear, was proposed by Wang et al. (2020). In this system, MMF as electrolyte and biofuel transport media were located between two carbon-based components that served as cathode and anode. PB and GOx were immobilized onto cathode and anode, respectively (Fig. 6a). The reduction of PB-modified cathode was driven by the oxidation of glucose catalyzed by GOx-modified anode, and this enables a single-compartment structure where MMF acts as biofuel transport media. Because MMF was made of polyester, it naturally induced a continuous and high speed flow which facilitated molecule transport for efficient chemical reactions without an additional pump. With this system, glucose was detected in the range of 0.1–1 mM [96]. In another study, a self-powered glucose biosensor was developed by embedding self-powered bioelectronics on a mask that measured the biological glucose signal with continuous energy (Fig. 6b) [97]. The fabrication of mask-based printed electrodes was based on screen-printing technique. The bioanode and cathode of the printed electrode on the medical mask were modified with highly capacitive materials including conducting polymers of PEDOT and PANI and nanomaterials of carbon nanotubes (CNTs). The electrode modification significantly increased capacitive properties on the electrode surfaces to store charges. The PANI-modified bioanode was functionalized with GOx and tetrathiafulvalene. When glucose in sweat came into contact with GOx on the surface of the flexible mask-based bioanode, the bioanode generated electrons. These electrons produced by glucose oxidation on the bioanode then drove the self-sustainable energy system. Finally, the electrons flowed to the cathode modified with PEDOT/CNTs and completed the circuit. This mask-based flexible device could hold potential for the next generation of smart textiles and wearable devices.

Smartwatches are interesting devices for integrating with glucose biosensors. In this regard, a fully integrated and self-powered smartwatch was developed for continuous monitoring of glucose in sweat [98]. The smartwatch included flexible silicon-based photovoltaic cells for energy harvesting and conversion, flexible Zn-MnO_2

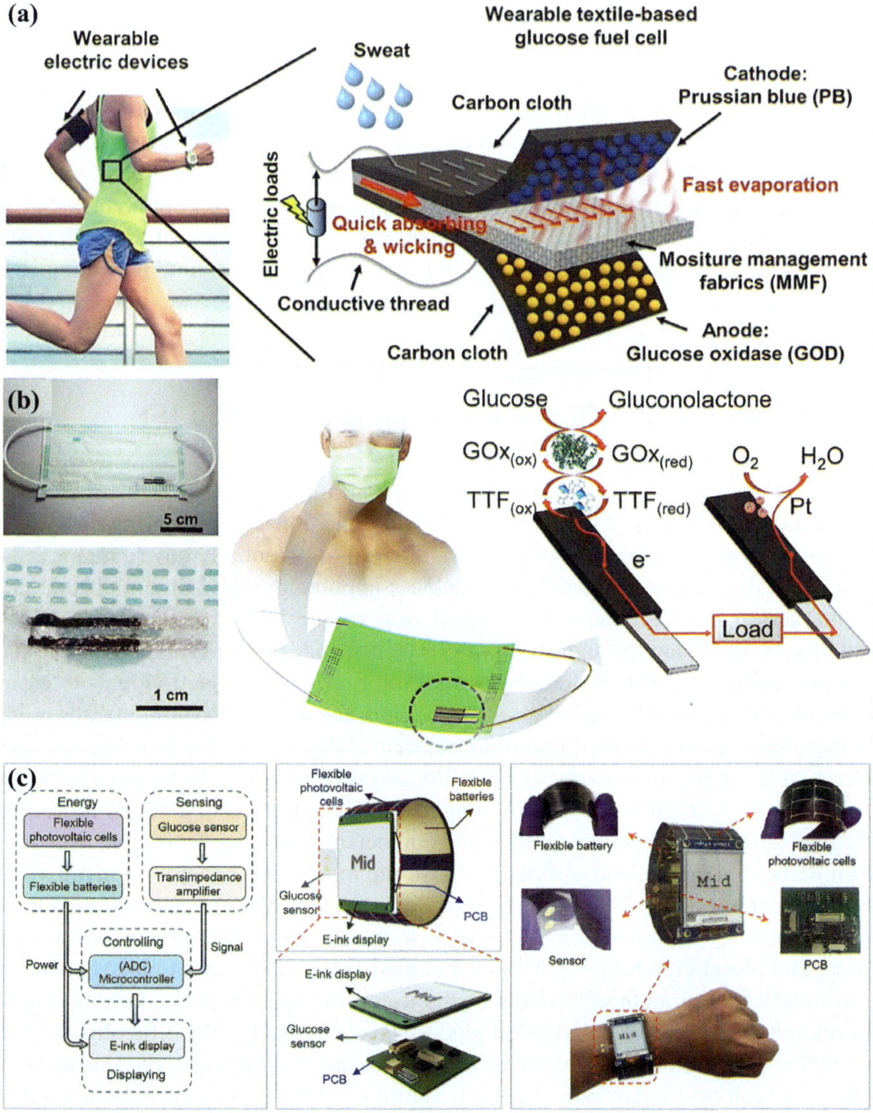

Fig. 6 **a** A wearable textile-based glucose biofuel cell employed moisture control fabrics used in sportswear. Reprinted from Ref. [96] with permission; **b** A schematic representation of a mask-based bioelectronic system for energy capturing from glucose and self-powered measurement of glucose. Reprinted from Ref. [97] with permission; **c** A schematic illustration of the totally combined self-powered smartwatch for continuous determination of glucose; from left to right respectively: System-level block layout of the self-powered smartwatch; schematic representation of the self-powered smartwatch; and pictures of the smartwatch on an individual's wrist and separate parts. Reprinted from Ref. [98] with permission

rechargeable batteries as intermediate energy storage devices, an electrochemical glucose sensor, a printed circuit board (PCB) as a controlling module, and electronic ink (E-ink) display to facilitate direct and real-time monitoring in the form of a "watch strap". In the sensing part, two electrodes including working and reference/counter electrodes were fabricated on flexible PET polymer using photolithography and thermally evaporated nano-layers of Cr/Au. The electrolyte made of polyvinyl alcohol (PVA) was located between MnO_2@ nitrogen-doped carbon foams (N-CF) cathode and Zn@N-CF anode (Fig. 6c). In this system, the sweat extraction was based on the iontophoresis method. This smartwatch showed a good performance in glucose detection.

4 Comparison of Glucose Detection in Sweat and ISF

Recent efforts have focused on epidermal biomonitoring devices of two easily accessible biofluids including ISF and sweat. Reliable and convenient non-invasive monitoring of desired biomarkers in these biofluids could potentially have a major influence on a wide range of health and wellness applications. Sweat contains several biomarkers that reflect the physiological state of the blood, as they are released from the bloodstream through sweat [88]. The presence of blood-related biomarkers in sweat, non-invasiveness, continuous monitoring, easy collection and its easier accessibility than ISF makes it an ideal medium for non-invasive biomarker monitoring. However, one of the limitations of sweat is that it can only be evaluated once it is excreted to the outer surface of the skin and several available methods for sweat generation (e.g. thermal heating, prolonged exercise, stress, and iontophoretic stimulation) can be both time-consuming, and cumbersome and discomfort when dealing with diabetic patients, and ultimately cause the loss of the benefits of using sweat as a matrix biomarker detection. Moreover, skin surface glucose can confound these measurements. On the other hand, despite some reports on the correlation between sweat and blood glucose [99], it's not specified yet whether sweat glucose concentration reflects the metabolic status of the users. In addition to the lack of a good correlation between sweat and blood glucose levels, as well as the difficulty of sweat collection, other limitations of sweat include variation in GOx activity due to lactic acid secretion and environmental temperature changes and enzyme delamination when exposed to mechanical friction and skin deformation. Relatively low sensitivity due to lower glucose concentration in sweat and distinct hysteresis compared to blood glucose level are also the main aspects limiting its use [77]. ISF is of particular importance because of high glucose content in ISF (the glucose concentration in blood ranges from 2 ~ 40 mmol/L, in ISF is 1.99 ~ 22.2 mmol/L, in sweat is 0.01 ~ 1.11 mmol/L) [100], good stability (compared to the discontinuous and unstable secretion of sweat), good linear range, high sensitivity, and high correlation coefficient with blood glucose levels due to its proximity to blood capillaries, which allows for rapid diffusion from the blood into the ISF. Moreover, the detection of glucose levels in ISF possesses mature technologies since several commercial devices have been developed [101–103]. Thus, accuracy of ISF analysis in much better than other biofluids (i.e. tear, saliva and sweat). However, a suitable element to pierce the stratum corneum and reach the epidermis and dermis is required for accurate ISF measurement [66]. In this regard, MNs can be used for both ISF extraction and ISF analysis. A less invasive or even non-invasive alternative approach to avoid piercing the skin is the employing reverse iontophoresis method to extract ISF. However, the latter method may cause skin irritation [66]. The other challenge with ISF application is that since the material exchange between blood and ISF exists for a certain period of time, the measured results often lag behind, and may not reflect the real-time value of blood glucose [77]. Regarding the comparison of the sensitivity of wearable sensors based on sweat or interstitial fluid, it should be mentioned that it is not possible to say with certainty which one is more sensitive, since the sensitivity depends on various factors, including the elements used in the sensor's construction. However, we have to notice to the presence of quite low levels of sweat glucose and, thus, the need to develop devices with higher sensitivity. In this context, a combination of some specific materials has shown that the increased surface area and porosity provided, for example, by nanofibers or nanoparticle-modified electrodes, can improve sensor performance with high sensitivity, low LOD, and wide linear range. For commercial development of sweat-based sensors, in vivo validation tests are still required for meaningful medical applications to investigate the correlation of sweat with blood readings.

5 Conclusions

Continuous monitoring of glucose using wearable devices is considered as an ideal approach to control glucose levels in diabetic individuals. Most commercially available CGM devices that have received FDA approval are based on the principle of electrochemical glucose sensing due to high accuracy and sensitivity. Two groups of wearable devices including minimally invasive and non-invasive sensors based on glucose in tears, sweat, saliva and urine have become of great interest to researchers. In this regard, the possibility of using different biofluids as detection samples and also innovative monitoring devices including microneedle arrays, flexible electrode-based sensors and data transmission systems including wireless and self-powered have been studied in recent years. However, despite recent advances in wearable biosensors, commercialized wearable devices still face considerable challenges. Several limitations such as high production costs and inflexible and uncomfortable tools with the human body have prevented the widespread use of these health care monitoring devices in the daily life of diabetic people. Development of flexible materials with high surface area, special mechanical characteristics, and adaptability to the human body has opened new avenues in innovative glucose monitoring devices. Therefore, it is still necessary to understand as much as possible the characteristics of these materials in order to use them properly and use appropriate technology to fabricate wearable glucose monitoring devices. In addition, considerable attention is being paid to microfluidic and self-powered/sustainable devices based on flexible materials to reduce manufacturing costs and increase device compatibility, stability, and applicability. Another interesting topic regarding wearable glucose monitoring devices is the development of multiplexed monitoring systems that are able to measure other diabetes-related biomarkers to provide a better assessment of diabetes treatment and improve the life quality of diabetic patients. It is also necessary to mention that not only CGM devices can be used for monitoring purposes but also they have the potential for therapeutic applications with a combination of closed-loop systems for insulin infusion in individuals with type I diabetes, i.e., an artificial pancreas. The next generation of wearable CGM devices can certainly provide more accurate and unique glucose monitoring capabilities for diabetic patients.

References

1. Wang, J.: Electrochemical glucose biosensors. Chem. Rev. **108**(2), 814–825 (2008). https://doi.org/10.1021/cr068123a
2. Yoo, E.-H., Lee, S.-Y.: Glucose biosensors: an overview of use in clinical practice. Sensors **10**(5), 4558–4576 (2010). https://doi.org/10.3390/s100504558
3. Sehit, E., Altintas, Z.: Biosensors for glucose detection. In: Advanced Sensor Technology, pp. 235–259. Elsevier (2023)
4. Sanati, A., Esmaeili, Y., Bidram, E., Shariati, L., Rafienia, M., Mahshid, S., Parlak, O.: Recent advancement in electrode materials and fabrication, microfluidic designs, and self-powered systems for wearable non-invasive electrochemical glucose monitoring. Appl. Mater. Today **26**, 101350 (2022). https://doi.org/10.1016/j.apmt.2021.101350
5. Teymourian, H., Parrilla, M., Sempionatto, J.R., Montiel, N.F., Barfidokht, A., Van Echelpoel, R., De Wael, K., Wang, J.: Wearable electrochemical sensors for the monitoring and screening of drugs. ACS Sens. **5**(9), 2679–2700 (2020). https://doi.org/10.1021/acssensors.0c01318
6. Zhang, S., Zhao, W., Zeng, J., He, Z., Wang, X., Zhu, Z., Hu, R., Liu, C., Wang, Q.: Wearable non-invasive glucose sensors based on metallic nanomaterials. Mater. Today Bio. 100638 (2023). https://doi.org/10.1016/j.mtbio.2023.100638
7. Zafar, H., Channa, A., Jeoti, V., Stojanović, G.M.: Comprehensive review on wearable sweat-glucose sensors for continuous glucose monitoring. Sensors **22**(2), 638 (2022). https://doi.org/10.3390/s22020638
8. Heifler, O., Borberg, E., Harpak, N., Zverzhinetsky, M., Krivitsky, V., Gabriel, I., Fourman, V., Sherman, D., Patolsky, F.: Clinic-on-a-needle array toward future minimally invasive wearable artificial pancreas applications. ACS Nano **15**(7), 12019–12033 (2021). https://doi.org/10.1021/acsnano.1c03310
9. Mohammadpour-Haratbar, A., Mohammadpour-Haratbar, S., Zare, Y., Rhee, K.Y., Park, S.-J.: A review on non-enzymatic electrochemical biosensors of glucose using carbon nanofiber nanocomposites. Biosensors **12**(11), 1004 (2022). https://doi.org/10.3390/bios12111004
10. Naikoo, G.A., Salim, H., Hassan, I.U., Awan, T., Arshad, F., Pedram, M.Z., Ahmed, W., Qurashi, A.: Recent advances in non-enzymatic glucose sensors based on metal and metal oxide nanostructures for diabetes management-a review. Front. Chem. **9**, 748957 (2021). https://doi.org/10.3389/fchem.2021.748957
11. Price, C.P.: Point-of-care testing in diabetes mellitus. Clin. Chem. Lab. Med. **41**, 1213–1219 (2003). https://doi.org/10.1515/CCLM.2003.185
12. Heller, A., Feldman, B.: Electrochemical glucose sensors and their applications in diabetes management. Chem. Rev. **108**(7), 2482–2505 (2008). https://doi.org/10.1021/cr068069y

13. Bankar, S.B., Bule, M.V., Singhal, R.S., Ananthanarayan, L.: Glucose oxidase—an overview. Biotechnol. Adv. **27**(4), 489–501 (2009). https://doi.org/10.1016/j.biotechadv.2009.04.003
14. Jeon, W.-Y., Kim, H.-H., Choi, Y.-B.: Development of a glucose sensor based on glucose dehydrogenase using polydopamine-functionalized nanotubes. Membranes **11**(6), 384 (2021). https://doi.org/10.3390/membranes11060384
15. Ferri, S., Kojima, K., Sode, K.: Review of glucose oxidases and glucose dehydrogenases: a bird's eye view of glucose sensing enzymes. J. Diabetes Sci. Technol. **5**(5), 1068–1076 (2011). https://doi.org/10.1177/193229681100500507
16. Jeon, W.-Y., Hyug-Han Kim, C.-J., Choi, Y.-B., Lee, B.-H., Jo, H.-J., Jeon, S.-Y.: Glucose detection via Ru-mediated catalytic reaction of glucose dehydrogenase. Adv. Mater. Lett. **9**(3), 220–224 (2018). https://doi.org/10.5185/amlett.2018.1947
17. Lee, I., Probst, D., Klonoff, D., Sode, K.: Continuous glucose monitoring systems-current status and future perspectives of the flagship technologies in biosensor research. Biosens. Bioelectron. **181**, 113054 (2021). https://doi.org/10.1016/j.bios.2021.113054
18. Karyakin, A.A., Gitelmacher, O.V., Karyakina, E.E.: Prussian blue-based first-generation biosensor. A sensitive amperometric electrode for glucose. Anal. Chem. **67**(14), 2419–2423 (1995). https://doi.org/10.1021/ac00110a016
19. Wang, J.: Glucose biosensors: 40 years of advances and challenges. Electroynalysis (N.Y.N.Y.) **13**(12), 983–988 (2001). https://doi.org/10.1002/1521-4109(200108)13:12<983::AID-ELAN983>3.0.CO;2-%23
20. Hassan, M.H., Vyas, C., Grieve, B., Bartolo, P.: Recent advances in enzymatic and non-enzymatic electrochemical glucose sensing. Sensors **21**(14), 4672 (2021). https://doi.org/10.3390/s21144672
21. Teymourian, H., Moonla, C., Tehrani, F., Vargas, E., Aghavali, R., Barfidokht, A., Tangkuaram, T., Mercier, P.P., Dassau, E., Wang, J.: Microneedle-based detection of ketone bodies along with glucose and lactate: toward real-time continuous interstitial fluid monitoring of diabetic ketosis and ketoacidosis. Anal. Chem. **92**(2), 2291–2300 (2019). https://doi.org/10.1021/acs.analchem.9b05109
22. Ahmed, M.U., Hossain, M.M., Tamiya, E.: Electrochemical biosensors for medical and food applications. Electroynalysis (N.Y.N.Y.) **20**(6), 616–626 (2008). https://doi.org/10.1002/elan.200704121
23. Teymourian, H., Barfidokht, A., Wang, J.: Electrochemical glucose sensors in diabetes management: an updated review (2010–2020). Chem. Soc. Rev. **49**(21), 7671–7709 (2020). https://doi.org/10.1039/D0CS00304B
24. Suzuki, N., Lee, J., Loew, N., Takahashi-Inose, Y., Okuda-Shimazaki, J., Kojima, K., Mori, K., Tsugawa, W., Sode, K.: Engineered glucose oxidase capable of quasi-direct electron transfer after a quick-and-easy modification with a mediator. Int. J. Mol. Sci. **21**(3), 1137 (2020). https://doi.org/10.3390/ijms21031137
25. Dong, S., Wang, B., Liu, B.: Amperometric glucose sensor with ferrocene as an electron transfer mediator. Biosens. Bioelectron. **7**(3), 215–222 (1992). https://doi.org/10.1016/0956-5663(92)87018-K
26. Cass, A.E., Davis, G., Francis, G.D., Hill, H.A.O., Aston, W.J., Higgins, I.J., Plotkin, E.V., Scott, L.D., Turner, A.P.: Ferrocene-mediated enzyme electrode for amperometric determination of glucose. Anal. Chem. **56**(4), 667–671 (1984). https://doi.org/10.1021/ac00268a018
27. Bruen, D., Delaney, C., Florea, L., Diamond, D.: Glucose sensing for diabetes monitoring: recent developments. Sensors **17**(8), 1866 (2017). https://doi.org/10.3390/s17081866
28. Goodnight, L., Butler, D., Xia, T., Ebrahimi, A.: Non-enzymatic detection of glucose in neutral solution using PBS-treated electrodeposited copper-nickel electrodes. Biosensors **11**(11), 409 (2021). https://doi.org/10.3390/bios11110409
29. Niu, X., Shi, L., Zhao, H., Lan, M.: Advanced strategies for improving the analytical performance of Pt-based nonenzymatic electrochemical glucose sensors: a minireview. Anal. Methods **8**(8), 1755–1764 (2016). https://doi.org/10.1039/C5AY03181H

30. Shim, K., Lee, W.-C., Park, M.-S., Shahabuddin, M., Yamauchi, Y., Hossain, M.S.A., Shim, Y.-B., Kim, J.H.: Au decorated core-shell structured Au@ Pt for the glucose oxidation reaction. Sens. Actuators B: Chem. **278**, 88–96 (2019). https://doi.org/10.1016/j.snb.2018.09.048
31. Abunahla, H., Mohammad, B., Alazzam, A., Jaoude, M.A., Al-Qutayri, M., Abdul Hadi, S., Al-Sarawi, S.F.: MOMSense: metal-oxide-metal elementary glucose sensor. Sci. Rep. **9**(1), 5524 (2019). https://doi.org/10.1038/s41598-019-41892-w
32. Zhu, H., Li, L., Zhou, W., Shao, Z., Chen, X.: Advances in non-enzymatic glucose sensors based on metal oxides. J. Mater. Chem. B **4**(46), 7333–7349 (2016). https://doi.org/10.1039/C6TB02037B
33. Rahman, M.M., Saleh Ahammad, A., Jin, J.-H., Ahn, S.J., Lee, J.-J.: A comprehensive review of glucose biosensors based on nanostructured metal-oxides. Sensors **10**(5), 4855–4886 (2010). https://doi.org/10.3390/s100504855
34. Zhang, Y., Su, L., Manuzzi, D., de los Monteros, H.V.E., Jia, W., Huo, D., Hou, C., Lei, Y.: Ultrasensitive and selective non-enzymatic glucose detection using copper nanowires. Biosens. Bioelectron. **31**(1), 426–432 (2012). https://doi.org/10.1016/j.bios.2011.11.006
35. Yi, W., Liu, J., Chen, H., Gao, Y., Li, H.: Copper/nickel nanoparticle decorated carbon nanotubes for nonenzymatic glucose biosensor. J. Solid State Electrochem. **19**, 1511–1521 (2015). https://doi.org/10.1007/s10008-015-2766-2
36. Phetsang, S., Kidkhunthod, P., Chanlek, N., Jakmunee, J., Mungkornasawakul, P., Ounnunkad, K.: Copper/reduced graphene oxide film modified electrode for non-enzymatic glucose sensing application. Sci. Rep. **11**(1), 9302 (2021). https://doi.org/10.1038/s41598-021-88747-x
37. Fall, B., Sall, D.D., Hémadi, M., Diaw, A.K.D., Fall, M., Randriamahazaka, H., Thomas, S.: Highly efficient non-enzymatic electrochemical glucose sensor based on carbon nanotubes functionalized by molybdenum disulfide and decorated with nickel nanoparticles (GCE/CNT/MoS2/NiNPs). Sens. Actuators Rep. **5**, 100136 (2023). https://doi.org/10.1016/j.snr.2022.100136
38. Fang, L., Wang, F., Chen, Z., Qiu, Y., Zhai, T., Hu, M., Zhang, C., Huang, K.: Flower-like MoS_2 decorated with Cu_2O nanoparticles for non-enzymatic amperometric sensing of glucose. Talanta **167**, 593–599 (2017). https://doi.org/10.1016/j.talanta.2017.03.008
39. Khan, A., Khan, A.A.P., Marwani, H.M., Alotaibi, M.M., Asiri, A.M., Manikandan, A., Siengchin, S., Rangappa, S.M.: Sensitive non-enzymatic glucose electrochemical sensor based on electrochemically synthesized PANI/bimetallic oxide composite. Polymers **14**(15), 3047 (2022). https://doi.org/10.3390/polym14153047
40. Osuna, V., Vega-Rios, A., Zaragoza-Contreras, E.A., Estrada-Moreno, I.A., Dominguez, R.B.: Progress of polyaniline glucose sensors for diabetes mellitus management utilizing enzymatic and non-enzymatic detection. Biosensors **12**(3), 137 (2022). https://doi.org/10.3390/bios12030137
41. Cappon, G., Acciaroli, G., Vettoretti, M., Facchinetti, A., Sparacino, G.: Wearable continuous glucose monitoring sensors: a revolution in diabetes treatment. Electronics **6**(3), 65 (2017). https://doi.org/10.3390/electronics6030065
42. Vaddiraju, S., Burgess, D.J., Tomazos, I., Jain, F.C., Papadimitrakopoulos, F.: Technologies for continuous glucose monitoring: current problems and future promises. J. Diabetes Sci. Technol. **4**(6), 1540–1562 (2010). https://doi.org/10.1177/193229681000400632
43. Srivastava, A., Chowdhury, M.K., Sharma, S., Sharma, N.: Blood glucose monitoring using non invasive optical method: Design limitations and challenges. Int. J. Adv. Res. Electr. Electron. Instrum. Eng. **2**(1) (2013)
44. Wang, G., He, X., Wang, L., Gu, A., Huang, Y., Fang, B., Geng, B., Zhang, X.: Non-enzymatic electrochemical sensing of glucose. Microchim. Acta **180**, 161–186 (2013). https://doi.org/10.1007/s00604-012-0923-1
45. Zhao, H., Guo, X., Wang, Y., Duan, X., Qu, H., Zhang, H., Zhang, D., Pang, W.: Microchip based electrochemical-piezoelectric integrated multi-mode sensing system for continuous glucose monitoring. Sens. Actuators B: Chem. **223**, 83–88 (2016). https://doi.org/10.1016/j.snb.2015.09.022

46. Garg, S.K., Potts, R.O., Ackerman, N.R., Fermi, S.J., Tamada, J.A., Chase, H.P.: Correlation of fingerstick blood glucose measurements with GlucoWatch biographer glucose results in young subjects with type 1 diabetes. Diabetes Care **22**(10), 1708–1714 (1999). https://doi.org/10.2337/diacare.22.10.1708
47. Johnston, L., Wang, G., Hu, K., Qian, C., Liu, G.: Advances in biosensors for continuous glucose monitoring towards wearables. Front. Bioeng. Biotechnol. **9**, 733810 (2021). https://doi.org/10.3389/fbioe.2021.733810
48. Zou, Y., Chu, Z., Guo, J., Liu, S., Ma, X., Guo, J.: Minimally invasive electrochemical continuous glucose monitoring sensors: Recent progress and perspective. Biosens. Bioelectron. 115103 (2023). https://doi.org/10.1016/j.bios.2023.115103
49. Todaro, B., Begarani, F., Sartori, F., Luin, S.: Is Raman the best strategy towards the development of non-invasive continuous glucose monitoring devices for diabetes management? Front. Chem. **10**, 1139 (2022). https://doi.org/10.3389/fchem.2022.994272
50. Acciaroli, G., Vettoretti, M., Facchinetti, A., Sparacino, G.: Calibration of minimally invasive continuous glucose monitoring sensors: state-of-the-art and current perspectives. Biosensors **8**(1), 24 (2018). https://doi.org/10.3390/bios8010024
51. Liu, Y., Yu, Q., Luo, X., Yang, L., Cui, Y.: Continuous monitoring of diabetes with an integrated microneedle biosensing device through 3D printing. Microsyst. Nanoeng. **7**(1), 75 (2021). https://doi.org/10.1038/s41378-021-00302-w
52. Xie, Z., Zhang, X., Chen, G., Che, J., Zhang, D.: Wearable microneedle-integrated sensors for household health monitoring. Eng. Regen. (2022). https://doi.org/10.1016/j.engreg.2022.09.002
53. Zhang, B.L., Zhang, X.P., Chen, B.Z., Fei, W.M., Cui, Y., Guo, X.D.: Microneedle-assisted technology for minimally invasive medical sensing. Microchem. J. **162**, 105830 (2021). https://doi.org/10.1016/j.microc.2020.105830
54. García-Guzmán, J.J., Pérez-Ràfols, C., Cuartero, M., Crespo, G.A.: Microneedle based electrochemical (Bio) Sensing: Towards decentralized and continuous health status monitoring. TrAC Trends Anal. Chem. **135**, 116148 (2021). https://doi.org/10.1016/j.trac.2020.116148
55. Sharma, S., Huang, Z., Rogers, M., Boutelle, M., Cass, A.E.: Evaluation of a minimally invasive glucose biosensor for continuous tissue monitoring. Anal. Bioanal. Chem. **408**, 8427–8435 (2016). https://doi.org/10.1007/s00216-016-9961-6
56. Samavat, S., Lloyd, J., O'Dea, L., Zhang, W., Preedy, E., Luzio, S., Teng, K.S.: Uniform sensing layer of immiscible enzyme-mediator compounds developed via a spray aerosol mixing technique towards low cost minimally invasive microneedle continuous glucose monitoring devices. Biosens. Bioelectron. **118**, 224–230 (2018). https://doi.org/10.1016/j.bios.2018.07.054
57. Mishra, R.K., Mohan, A.V., Soto, F., Chrostowski, R., Wang, J.: A microneedle biosensor for minimally-invasive transdermal detection of nerve agents. Analyst **142**(6), 918–924 (2017). https://doi.org/10.1039/C6AN02625G
58. Valdés-Ramírez, G., Li, Y.-C., Kim, J., Jia, W., Bandodkar, A.J., Nuñez-Flores, R., Miller, P.R., Wu, S.-Y., Narayan, R., Windmiller, J.R.: Microneedle-based self-powered glucose sensor. Electrochem. Commun. **47**, 58-62 (2014). https://doi.org/10.1016/j.elecom.2014.07.014
59. Bollella, P., Sharma, S., Cass, A.E., Tasca, F., Antiochia, R.: Minimally invasive glucose monitoring using a highly porous gold microneedles-based biosensor: Characterization and application in artificial interstitial fluid. Catalysts **9**(7), 580 (2019). https://doi.org/10.3390/catal9070580
60. Sharma, S., Saeed, A., Johnson, C., Gadegaard, N., Cass, A.E.: Rapid, low cost prototyping of transdermal devices for personal healthcare monitoring. Sens. Bio-Sens. Res. **13**, 104–108 (2017). https://doi.org/10.1016/j.sbsr.2016.10.004
61. Invernale, M.A., Tang, B.C., York, R.L., Le, L., Hou, D.Y., Anderson, D.G.: Microneedle electrodes toward an amperometric glucose-sensing smart patch. Adv. Healthc. Mater. **3**(3), 338–342 (2014). https://doi.org/10.1002/adhm.201300142
62. Bollella, P., Sharma, S., Cass, A.E.G., Antiochia, R.: Minimally-invasive microneedle-based biosensor array for simultaneous lactate and glucose monitoring in artificial interstitial fluid. Electroanalysis **31**(2), 374–382 (2019). https://doi.org/10.1002/elan.201800630

63. Bao, L., Park, J., Bonfante, G., Kim, B.: Recent advances in porous microneedles: materials, fabrication, and transdermal applications. Drug Deliv. Transl. Res. **12**(2), 395–414 (2022). https://doi.org/10.1007/s13346-021-01045-x
64. Ju, J., Li, L., Regmi, S., Zhang, X., Tang, S.: Microneedle-based glucose sensor platform: from vitro to wearable point-of-care testing systems. Biosensors **12**(8), 606 (2022). https://doi.org/10.3390/bios12080606
65. Ribet, F., Stemme, G., Roxhed, N.: Real-time intradermal continuous glucose monitoring using a minimally invasive microneedle-based system. Biomed. Microdevices **20**, 1–10 (2018). https://doi.org/10.1007/s10544-018-0349-6
66. Parrilla, M., Detamornrat, U., Domínguez-Robles, J., Donnelly, R.F., De Wael, K.: Wearable hollow microneedle sensing patches for the transdermal electrochemical monitoring of glucose. Talanta **249**, 123695 (2022). https://doi.org/10.1016/j.talanta.2022.123695
67. Xie, L., Zeng, H., Sun, J., Qian, W.: Engineering microneedles for therapy and diagnosis: a survey. Micromachines **11**(3), 271 (2020). https://doi.org/10.3390/mi11030271
68. Zheng, M., Zhang, Y., Hu, T., Xu, C.: A skin patch integrating swellable microneedles and electrochemical test strips for glucose and alcohol measurement in skin interstitial fluid. Bioeng. Transl. Med. e10413 (2022). https://doi.org/10.1002/btm2.10413
69. Zheng, L., Zhu, D., Wang, W., Liu, J., Thng, S.T.G., Chen, P.: A silk-microneedle patch to detect glucose in the interstitial fluid of skin or plant tissue. Sens. Actuators B: Chem. **372**, 132626 (2022). https://doi.org/10.1016/j.snb.2022.132626
70. Kai, H., Kumatani, A.: A porous microneedle electrochemical glucose sensor fabricated on a scaffold of a polymer monolith. J. Phys. Energy **3**(2), 024006 (2021). https://doi.org/10.1088/2515-7655/abe4a1
71. Yun, S.M., Kim, M., Kwon, Y.W., Kim, H., Kim, M.J., Park, Y.-G., Park, J.-U.: Recent advances in wearable devices for non-invasive sensing. Appl. Sci. **11**(3), 1235 (2021). https://doi.org/10.3390/app11031235
72. Chen, X., Song, Y., Su, Z., Chen, H., Cheng, X., Zhang, J., Han, M., Zhang, H.: Flexible fiber-based hybrid nanogenerator for biomechanical energy harvesting and physiological monitoring. Nano Energy **38**, 43–50 (2017). https://doi.org/10.1016/j.nanoen.2017.05.047
73. Sun, L., Chen, S., Guo, Y., Song, J., Zhang, L., Xiao, L., Guan, Q., You, Z.: Ionogel-based, highly stretchable, transparent, durable triboelectric nanogenerators for energy harvesting and motion sensing over a wide temperature range. Nano Energy **63**, 103847 (2019). https://doi.org/10.1016/j.nanoen.2019.06.043
74. Promphet, N., Ummartyotin, S., Ngeontae, W., Puthongkham, P., Rodthongkum, N.: Non-invasive wearable chemical sensors in real-life applications. Anal. Chim. Acta **1179**, 338643 (2021). https://doi.org/10.1016/j.aca.2021.338643
75. Dervisevic, M., Alba, M., Prieto-Simon, B., Voelcker, N.H.: Skin in the diagnostics game: Wearable biosensor nano-and microsystems for medical diagnostics. Nano Today **30**, 100828 (2020). https://doi.org/10.1016/j.nantod.2019.100828
76. Bandodkar, A.J., Wang, J.: Non-invasive wearable electrochemical sensors: a review. Trends Biotechnol. **32**(7), 363–371 (2014). https://doi.org/10.1016/j.tibtech.2014.04.005
77. Tang, L., Chang, S.J., Chen, C.-J., Liu, J.-T.: Non-invasive blood glucose monitoring technology: a review. Sensors **20**(23), 6925 (2020). https://doi.org/10.3390/s20236925
78. Kudo, H., Sawada, T., Kazawa, E., Yoshida, H., Iwasaki, Y., Mitsubayashi, K.: A flexible and wearable glucose sensor based on functional polymers with soft-MEMS techniques. Biosens. Bioelectron. **22**(4), 558–562 (2006). https://doi.org/10.1016/j.bios.2006.05.006
79. Chu, M.X., Miyajima, K., Takahashi, D., Arakawa, T., Sano, K., Sawada, S.-I., Kudo, H., Iwasaki, Y., Akiyoshi, K., Mochizuki, M.: Soft contact lens biosensor for in situ monitoring of tear glucose as non-invasive blood sugar assessment. Talanta **83**(3), 960–965 (2011). https://doi.org/10.1016/j.talanta.2010.10.055
80. Zheng, H., Chen, H., Pu, Z., Li, D.: A breathable flexible glucose biosensor with embedded electrodes for long-term and accurate wearable monitoring. Microchem. J. **181**, 107707 (2022). https://doi.org/10.1016/j.microc.2022.107707

81. Wang, Y., Guo, H., Yuan, M., Yu, J., Wang, Z., Chen, X.: One-step laser synthesis platinum nanostructured 3D porous graphene: a flexible dual-functional electrochemical biosensor for glucose and pH detection in human perspiration. Talanta **257**, 124362 (2023). https://doi.org/10.1016/j.talanta.2023.124362
82. Zhang, L., Du, W., Nautiyal, A., Liu, Z., Zhang, X.: Recent progress on nanostructured conducting polymers and composites: synthesis, application and future aspects. Sci. China Mater. **61**(3), 303–352 (2018). https://doi.org/10.1007/s40843-017-9206-4
83. Xu, C., Jiang, D., Ge, Y., Huang, L., Xiao, Y., Ren, X., Liu, X., Zhang, Q., Wang, Y.: A PEDOT: PSS conductive hydrogel incorporated with Prussian blue nanoparticles for wearable and noninvasive monitoring of glucose. Chem. Eng. J. **431**, 134109 (2022). https://doi.org/10.1016/j.cej.2021.134109
84. Ajmal Mokhtar, S.M., Alvarez de Eulate, E., Sethumadhavan, V., Yamada, M., Prow, T.W., Evans, D.R.: Electrochemical stability of PEDOT for wearable on-skin application. J. Appl. Polym. Sci. **138**(44), 51314 (2021). https://doi.org/10.1002/app.51314
85. Lin, Y.-C., Rinawati, M., Chang, L.-Y., Wang, Y.-X., Wu, Y.-T., Yen, Y.-H., Chen, K.-J., Ho, K.-C., Yeh, M.-H.: A non-invasive wearable sweat biosensor with a flexible N-GQDs/PANI nanocomposite layer for glucose monitoring. Sens. Actuators B: Chem. **383**, 133617 (2023). https://doi.org/10.1016/j.snb.2023.133617
86. Deroco, P.B., Wachholz Junior, D., Kubota, L.T.: Paper-based wearable electrochemical sensors: a new generation of analytical devices. Electroanalysis **35**(1), e202200177 (2023). https://doi.org/10.1002/elan.202200177
87. Li, M., Wang, L., Liu, R., Li, J., Zhang, Q., Shi, G., Li, Y., Hou, C., Wang, H.: A highly integrated sensing paper for wearable electrochemical sweat analysis. Biosens. Bioelectron. **174**, 112828 (2021). https://doi.org/10.1016/j.bios.2020.112828
88. Kim, J., Sempionatto, J.R., Imani, S., Hartel, M.C., Barfidokht, A., Tang, G., Campbell, A.S., Mercier, P.P., Wang, J.: Simultaneous monitoring of sweat and interstitial fluid using a single wearable biosensor platform. Adv. Sci. **5**(10), 1800880 (2018). https://doi.org/10.1002/advs.201800880
89. Sun, J.-Y., Zhao, X., Illeperuma, W.R., Chaudhuri, O., Oh, K.H., Mooney, D.J., Vlassak, J.J., Suo, Z.: Highly stretchable and tough hydrogels. Nature **489**(7414), 133–136 (2012). https://doi.org/10.1038/nature11409
90. Kim, G.J., Yoon, K.J., Kim, K.O.: Glucose-responsive poly (vinyl alcohol)/β-cyclodextrin hydrogel with glucose oxidase immobilization. J. Mater. Sci. **54**(19), 12806–12817 (2019). https://doi.org/10.1007/s10853-019-03805-0
91. Kim, K.O., Kim, G.J., Kim, J.H.: A cellulose/β-cyclodextrin nanofiber patch as a wearable epidermal glucose sensor. RSC Adv. **9**(40), 22790–22794 (2019). https://doi.org/10.1039/C9RA03887F
92. Li, T., Liang, B., Ye, Z., Zhang, L., Xu, S., Tu, T., Zhang, Y., Cai, Y., Zhang, B., Fang, L.: An integrated and conductive hydrogel-paper patch for simultaneous sensing of Chemical-Electrophysiological signals. Biosens. Bioelectron. **198**, 113855 (2022). https://doi.org/10.1016/j.bios.2021.113855
93. Sanati, A., Jalali, M., Raeissi, K., Karimzadeh, F., Kharaziha, M., Mahshid, S.S., Mahshid, S.: A review on recent advancements in electrochemical biosensing using carbonaceous nanomaterials. Microchim. Acta **186**, 1–22 (2019). https://doi.org/10.1007/s00604-019-3854-2
94. Rayegani, A., Saberian, M., Delshad, Z., Liang, J., Sadiq, M., Nazar, A.M., Mohsan, S.A.H., Khan, M.A.: Recent advances in self-powered wearable sensors based on piezoelectric and triboelectric nanogenerators. Biosensors **13**(1), 37 (2022). https://doi.org/10.3390/bios13010037
95. Zeng, X., Peng, R., Fan, Z., Lin, Y.: Self-powered and wearable biosensors for healthcare. Mater. Today Energy **23**, 100900 (2022). https://doi.org/10.1016/j.mtener.2021.100900
96. Wang, C., Shim, E., Chang, H.-K., Lee, N., Kim, H.R., Park, J.: Sustainable and high-power wearable glucose biofuel cell using long-term and high-speed flow in sportswear fabrics. Biosens. Bioelectron. **169**, 112652 (2020). https://doi.org/10.1016/j.bios.2020.112652

97. Jeerapan, I., Sangsudcha, W., Phokhonwong, P.: Wearable energy devices on mask-based printed electrodes for self-powered glucose biosensors. Sens. Bio-Sens. Res. **38**, 100525 (2022). https://doi.org/10.1016/j.sbsr.2022.100525
98. Zhao, J., Lin, Y., Wu, J., Nyein, H.Y.Y., Bariya, M., Tai, L.-C., Chao, M., Ji, W., Zhang, G., Fan, Z.: A fully integrated and self-powered smartwatch for continuous sweat glucose monitoring. ACS Sens. **4**(7), 1925–1933 (2019). https://doi.org/10.1021/acssensors.9b00891
99. Moyer, J., Wilson, D., Finkelshtein, I., Wong, B., Potts, R.: Correlation between sweat glucose and blood glucose in subjects with diabetes. Diabetes Technol. Ther. **14**(5), 398–402 (2012). https://doi.org/10.1089/dia.2011.0262
100. Peng, Z., Xie, X., Tan, Q., Kang, H., Cui, J., Zhang, X., Li, W., Feng, G.: Blood glucose sensors and recent advances: a review. J. Innov. Opt. Health Sci. **15**(02), 2230003 (2022). https://doi.org/10.1142/S1793545822300038
101. Mastrototaro, J.J.: The MiniMed continuous glucose monitoring system. Diabetes Technol. Ther. **2**(1, Supplement 1), 13–18 (2000). https://doi.org/10.1089/15209150050214078
102. Roze, S., Isitt, J., Smith-Palmer, J., Javanbakht, M., Lynch, P.: Long-term cost-effectiveness of Dexcom G6 real-time continuous glucose monitoring versus self-monitoring of blood glucose in patients with type 1 diabetes in the UK. Diabetes Care **43**(10), 2411–2417 (2020). https://doi.org/10.1007/s13300-020-00959-y
103. Feldman, B., McGarraugh, G., Heller, A., Bohannon, N., Skyler, J., DeLeeuw, E., Clarke, D.: FreeStyle™: a small-volume electrochemical glucose sensor for home blood glucose testing. Diabetes Technol. Ther. **2**(2), 221–229 (2000). https://doi.org/10.1089/15209150050025177
104. Wang, Y., Wang, X., Lu, W., Yuan, Q., Zheng, Y., Yao, B.: A thin film polyethylene terephthalate (PET) electrochemical sensor for detection of glucose in sweat. Talanta **198**, 86–92 (2019). https://doi.org/10.1016/j.talanta.2019.01.104
105. Cao, Q., Liang, B., Tu, T., Wei, J., Fang, L., Ye, X.: Three-dimensional paper-based microfluidic electrochemical integrated devices (3D-PMED) for wearable electrochemical glucose detection. RSC Adv. **9**(10), 5674–5681 (2019). https://doi.org/10.1039/c8ra09157a
106. Xuan, X., Yoon, H.S., Park, J.Y.: A wearable electrochemical glucose sensor based on simple and low-cost fabrication supported micro-patterned reduced graphene oxide nanocomposite electrode on flexible substrate. Biosens. Bioelectron. **109**, 75–82 (2018). https://doi.org/10.1016/j.bios.2018.02.054
107. Yokus, M.A., Songkakul, T., Pozdin, V.A., Bozkurt, A., Daniele, M.A.: Wearable multiplexed biosensor system toward continuous monitoring of metabolites. Biosens. Bioelectron. **153**, 112038 (2020). https://doi.org/10.1016/j.bios.2020.112038
108. Zhai, Q., Gong, S., Wang, Y., Lyu, Q., Liu, Y., Ling, Y., Wang, J., Simon, G.P., Cheng, W.: Enokitake mushroom-like standing gold nanowires toward wearable noninvasive bimodal glucose and strain sensing. ACS Appl. Mater. Interfaces **11**(10), 9724–9729 (2019). https://doi.org/10.1021/acsami.8b19383
109. Feng, R., Chu, Y., Wang, X., Wu, Q., Tang, F.: A long-term stable and flexible glucose sensor coated with poly (ethylene glycol)-modified polyurethane. J. Electroanal. Chem. **895**, 115518 (2021). https://doi.org/10.1016/j.jelechem.2021.115518
110. Shu, Y., Shang, Z., Su, T., Zhang, S., Lu, Q., Xu, Q., Hu, X.: A highly flexible Ni–Co MOF nanosheet coated Au/PDMS film based wearable electrochemical sensor for continuous human sweat glucose monitoring. Analyst **147**(7), 1440–1448 (2022). https://doi.org/10.1039/D1AN02214H
111. Lin, P.-H., Sheu, S.-C., Chen, C.-W., Huang, S.-C., Li, B.-R.: Wearable hydrogel patch with noninvasive, electrochemical glucose sensor for natural sweat detection. Talanta **241**, 123187 (2022). https://doi.org/10.1016/j.talanta.2021.123187
112. Zhao, Y., Zhai, Q., Dong, D., An, T., Gong, S., Shi, Q., Cheng, W.: Highly stretchable and strain-insensitive fiber-based wearable electrochemical biosensor to monitor glucose in the sweat. Anal. Chem. **91**(10), 6569–6576 (2019). https://doi.org/10.1021/acs.analchem.9b00152
113. Kim, J., Kim, M., Lee, M.-S., Kim, K., Ji, S., Kim, Y.-T., Park, J., Na, K., Bae, K.-H., Kyun Kim, H.: Wearable smart sensor systems integrated on soft contact lenses for wireless ocular diagnostics. Nat. Commun. **8**(1), 14997 (2017). https://doi.org/10.1038/ncomms14997

ECG and EMG Monitoring with Smart Textile hitoe™

Masumi Yamaguchi and Hiroyoshi Togo

1 Introduction

Ischemic heart disease and stroke were the top two global killers from 2000 to 2019. In 2019, they killed more than 15 million people worldwide [1]. It is known that stroke is often caused by atrial fibrillation (AF); however, heart disease is an indirect but significant factor in death by stroke. In view of these alarming statistics, there is growing interest in monitoring heart rates (HRs) and measuring electrocardiograms (ECGs) on a daily basis in order to reduce the risk of health problems. Furthermore, there is a growing need for individuals to be able to monitor their own biometric information during daily activities, not just for medical purposes but to better understand their health status, improve their lifestyle, and enhance their performance in sports and other physical endeavors. Because recent wearable technology makes it simpler to measure one's own biosignals, it is now possible to easily monitor and use this information in daily life.

The functional material hitoe™, developed by NTT Corporation and Toray Industries, Inc., is a conductive fabric that functions as an electrode in contact with our skin surface for measuring ECGs and electromyograms (EMGs). It is made of conductive polymer PEDOT-PSS (polyethylene dioxythiophene-polystyrene sulfonate) coated onto a polyester nanofiber fabric [2, 3].

The diameter of the nanofiber fabric used in hitoe™ is about 700 nm, which is smaller than that of polyester fibers used in general clothing, which is about 15 μm. This ultrasmall fiber diameter results in a larger contact area with the skin and a

M. Yamaguchi (✉)
NTT Basic Research Laboratories and Bio-Medical Informatics Research Center, NTT Corporation, 3-1, Morinosato Wkamiya, Atsugi-Shi, Kanagawa Pref 243-0198, Japan
e-mail: msm.yamaguchi@ntt.com

H. Togo
NTT Device Technology Laboratories, NTT Corporation, 3-1, Morinosato Wkamiya, Atsugi-Shi, Kanagawa Pref 243-0198, Japan

K. Mitsubayashi (ed.), *Wearable Biosensing in Medicine and Healthcare*,
https://doi.org/10.1007/978-981-99-8122-9_4

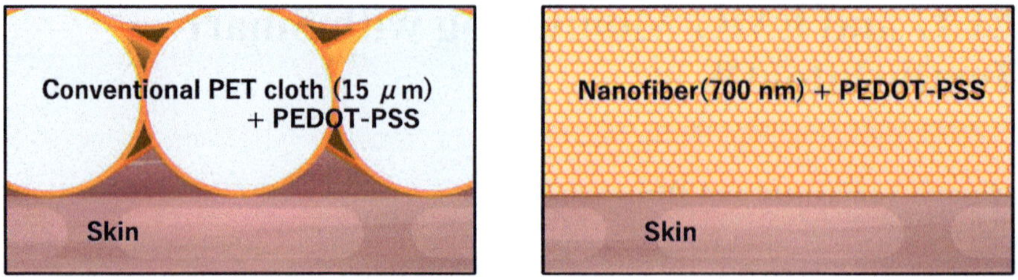

Fig. 1 Schematics of PEDOT-PSS-coated fabric on conventional PET cloth and on nanofiber (hitoe)

higher degree of adhesion compared to polyester fibers (Fig. 1). Additionally, the fabric is more comfortable to the touch than traditional polyester fibers owing to its softer texture. Also, unlike conventional metallic electrodes, hitoe™ electrodes absorb water well, as shown in Figs. 1 and 2; therefore, they absorb perspiration and blend well with the skin. The hitoe™ electrodes rarely cause rashes even when adhered to the skin for long periods of time and are gentle to the skin without any of the allergic reactions to metal.

Wearable sensing devices with hitoe™ electrodes have been commercialized as functional wear for measuring HRs during sports and for monitoring workers' vital signs, as well as for medical use of Holter ECGs. In these clothes, the hitoe™ electrodes are embedded inside shirts or belts. In developing this wear, related technologies were also important in terms of the intended use, such as electrode placement in the shirts and suitable garment materials that, for example, allow stretching with low tension with high fabric elasticity. Various hitoe™ weares have been developed considering the intended use case.

Section 2 introduces commercialized hitoe™ wear for ECG monitoring. Monitoring ECG on a daily basis requires not only sensing clothing but also wearable measuring devices that are lightweight and have long battery life. Section 3 describes how these requirements are met with transmitter technology saves power and reduces transmitter weight for daily measurements with hitoe™. Sections 4 and 5 describe examples of hitoe™ use cases for daily monitoring.

2 hitoe™ Electrode and ECG Sensing Wear

ECG measurements with shirts or belts are often performed with dry electrodes made of metal or conductive rubber. The most significant difference between hitoe™ and these materials is its hydrophilic properties, as shown in Fig. 2. Since 2014, various ECG measurement wear with hitoe™ have been developed and released in Japan. Here, we will introduce the basic properties of hitoe™ as a biological electrode and the lineup of commercialized garments.

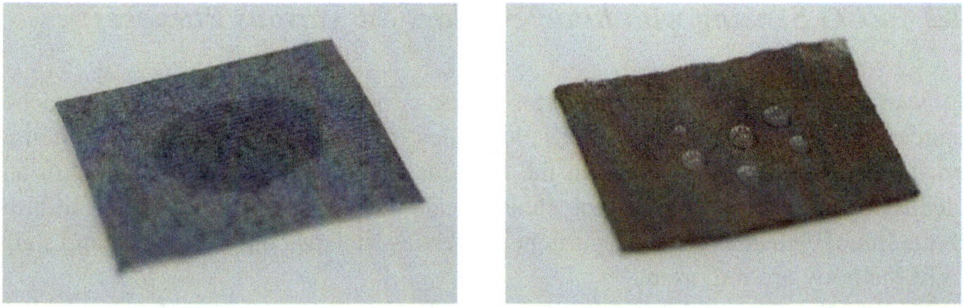

Fig. 2 Photographs showing how, **a** hitoe™ absorbs water droplets and **b** silver-coated fabric repels water droplets

2.1 Basic Properties of hitoe™ Electrode

The hitoe™ electrode is a fabric made of conductive polymer PEDOT-PSS coated onto a polyester nanofiber cloth. The electrical properties of hitoe™ fulfill the requirements of disposal medical electrodes, which were evaluated based on ANSI standards for disposable electrodes for ECG (Table 1).

For an electrode material for daily use, besides the electrical properties for measuring biological signals, comfortableness for users and washing durability are very important. The hitoe™ electrode has a soft texture due to its nanofiber construction and hydrophilic properties. This allows for a good feel on the skin and absorption of perspiration, resulting in a comfortable fit. Regarding durability, the conductive polymer is less likely to peel off because it is firmly impregnated in the small gaps between the nanofibers. Tests have confirmed that ECGs can be measured even after 100 washes in a home washing machine.

Table 1 Electrode characteristics and ANSI/AMI requirements for disposable ECG electrode. After Ref. [4]

	Textile	ANSI/AAMI
DC resistance	<0.1k Ω	<2k Ω
AC impedance	1.26 ± 0.18K Ω	<3K Ω
Internal noise	1–3 μV	<150μV
Defibrillation discharge at 200 V		
Recovery polarization potential	1.95 ± 0.80 mV	<100 mV
Rate of change of polarization potential	No change	<1 mV/s
AC impedance after test	1.14 ± 0.07 KΩ	<3 KΩ
DC offset voltage	0.0028 ± 0.0020 mV	<100 mV
	Data are reported as means ± SD	

2.2 ECG Sensing with hitoe™ Wears for Various Purposes

In wearable sensing devices with hitoe™, the conductive hitoe™ fabric is attached inside shirts or on belts and used as electrodes for detecting the electrical potential on the surface of human skin. In the system, good contact between the skin and electrode is essential for acquiring clear signals without noise; therefore, in addition to the hitoe™ electrode itself, garment properties such as compressibility and shape to fit the body are important.

"C3fit IN-pulse" (GOLDWIN), a garment with hitoe™ electrodes and a dedicated transmitter (hitoe transmitter 01, NTT DOCOMO) that can measure HRs, was released as sports shirts in 2014 and then as sports bras for women in 2015 in Japan (Fig. 2a and b). These compression-type shirts/bras fit the contours of the torso, and thus the electrode has good contact to the skin during even rather hard sports activity. The basic properties of C3fit IN-pulse for measuring ECGs were evaluated for supine and seated users and for various body motions, such as upper body twisting, and foot stomping exercises, assuming movements in daily life (Fig. 4). The results showed that when the electrode was moistened with glycerin solution in advance to inhibit drying and maintain continuous contact with the skin, the signals obtained with this wear were almost comparable to those measured by a Holter ECG recorder (Kenz Cardy 303 Pico+®, SUZUKEN Co., Ltd., Nagoya, Japan) with medical gel electrodes patched directly to the skin [4].

Since the release of C3fit IN-pulse, it has been utilized for ECG monitoring in various sports with hard body motion such as in baseball, skiing, snowboard jumping [5], and badminton. Although the first version of C3fit IN-pulse was released as sportswear, it was also used in medical research to evaluate the feasibility of a T-shirt-type wearable ECG monitor. This research included ECG monitoring during a full marathon race to create novel preventive strategies against sudden cardiac events [6], as well as clinical research for the detection of covert atrial fibrillation [7] and analysis of autonomic nervous system functions during natural defecation in patients with irritable bowel syndrome [8].

Though C3fit IN-pulse has been accepted by most athletes who regularly wear compression-type inner shirts, it is a little tight for everyday wear. With a focus on wearing comfort in daily life, less compressive wear was developed with the introduction of new materials and the design of the compression of the body fabric. Comfortable all-day wear is expected to expand the utilization scenarios of ECG monitoring for various purposes. After several attempts, a new type of C3fit IN-pulse was released in 2020 (Fig. 3c) along with a new transmitter TX02 (details of the TX02 are described in Sect. 3.1). The new wear significantly improved comfort as innerwear without compromising HR measurement accuracy. This opened the possibility of widespread use among field workers who require a garment that can be worn all-day (this wear is suitable for worker monitoring, as described in Sect. 4.1). HR detection from ECG signal was evaluated with this innerwear especially for worker-specific movements such as applying external vibration and lifting and lowering loads, assuming real-world cases of physical load in light to moderate intensity tasks

[9]. The accuracy of the obtained HR values was competitive with that of conventional gel-type electrodes (Fig. 5).

Daily monitoring in rehabilitation wards is desired as well (see Sect. 4.2). For stroke rehabilitation, users will mostly be elderly or paraplegic, unlike the users of sportswear or work clothes. Improvements have been made for practical use in rehabilitation wards, such as an opening at the front to make it easier for caregivers to put it on and allowing only the top of the electrodes to be fastened with a belt (Fig. 3d) or by wear shape (Fig. 3e) so that both comfortableness with a loose-fit and clear ECG can be taken.

In addition to the several types of sensing wear shown above, belt-type wear that can be used in a variety of situations has also been developed and marketed. To reduce

Fig. 3 hitoe™ wear for various purposes

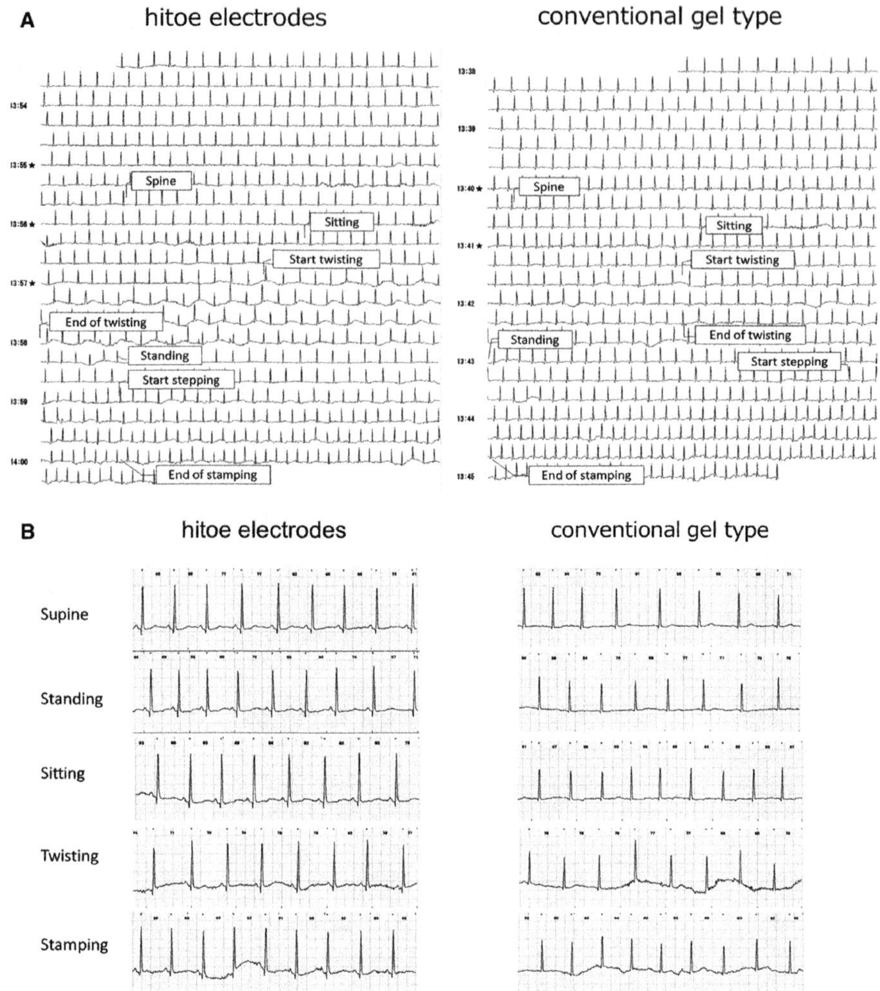

Fig. 4 ECG signals detected with hitoe™ electrodes (C3fit IN-pulse and transmitter 01) and conventional gel-type electrodes in various body motions: supine, seated, upper body twisting, and foot stomping exercises. After Ref. [4]

the influence of electrode displacement caused by body movement, the position of electrodes is designed to be less susceptible to belt expansion and contraction, and the belt itself is designed to be less prone to twisting (Fig. 3f).

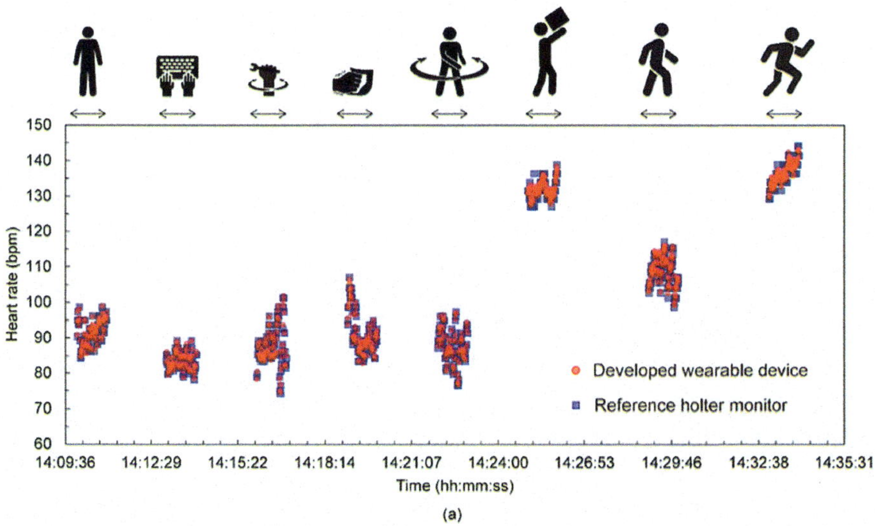

Fig. 5 Example of time-series of HR data obtained from the wearable device, C3fit IN-pulse 2020 with TX02, and gel-type Holter monitor at each movement. After Ref. [9]

2.3 hitoe™ Medical Wear

The hitoe™ medical electrodes and lead wires have been approved for use as general medical devices in Japan. Toray Medical Co. Ltd. has been distributing the hitoe™ Wearable ECG Measurement System since 2018 (Fig. 3f).

This system can be used for Holter ECG examination with the bipolar lead CC5 condition. The dedicated ECG logger is capable of continuous ECG measurement for up to 14 days without recharging. The hitoe™ medical electrodes are attached to the dedicated hitoe™ medical leads with a snap. The wearable Holter ECG system is suitable for long-term measurement over the course of days because there is no need for a specialist to place the electrode. The medical electrodes are highly electrically conductive with snap hook buttons, and the medical lead wires are made of silver-plated nylon thread to integrate them with the wear. The system is especially useful for monitoring the occurrence of AF, which often occurs intermittently in the early stages of a disease and is often undetectable with short-term measurements. The AF detection rate can be improved by measuring the ECG over a longer period. In tests comparing the hitoe™ Holter ECG medical system and conventional gel electrode system, the wearable ECG with hitoe™ system showed a slight increase in noise signal episodes with associated deterioration in R-wave amplitudes (Fig. 6). However, this increased noise signal count caused a negligible interruption in a continuous AF episode in a patient with intermittent AF [10]. A clinical study reported that the ECG acquisition rate was higher for gel-type Holter ECG than for the hitoe™ Wearable ECG Measurement System, but the hitoe™ system provided longer total analysis

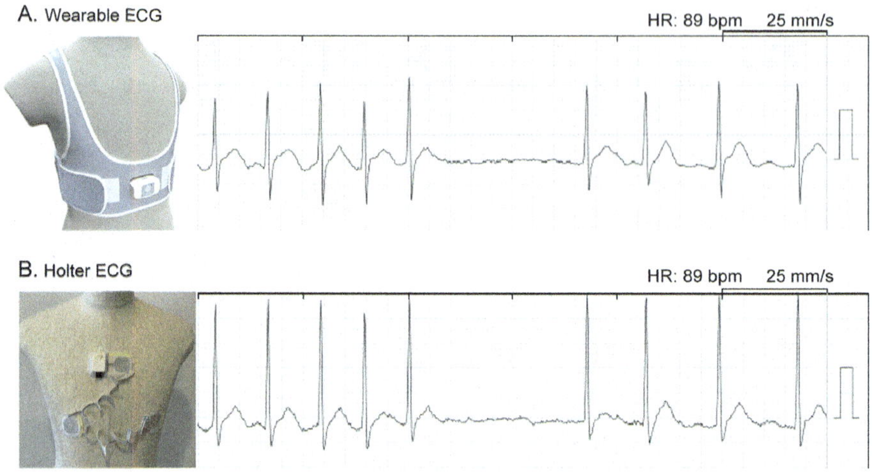

Fig. 6 Simultaneous tracing from wearable ECG (**a**) and Holter ECG (**b**). Electrodes were positioned to compose a bipolar lead CC5 in both ECG devices. After Ref. [10]

time. Despite its lower ECG acquisition rate, the two-week ECG with the hitoe™ Wearable ECG Measurement System has revealed instances of AF recurrence after ablation in patients who were underdiagnosed by 24-h Holter ECG [11].

Later, the hitoe™ Wearable ECG Measurement System was modified into disposable garments, and the hitoe™ Wearable ECG Measurement System II was released in 2020 in order to avoid hospital laundering (Fig. 3h).

3 New Wearable Biological/Environmental Sensor

Wearable devices equipped with biological sensors have emerged as promising tools for monitoring an individual's lifestyle and providing personalized advice for maintaining a healthful life. These devices have also found applications in managing the physical well-being of workers, particularly during the summer season when temperatures are high. However, for wearable devices to make a meaningful and sustainable contribution to healthful living, they need to be seamlessly integrated into an individual's daily life and exhibit a high degree of naturalness and intuitiveness.

To this end, NTT has collaborated with Toray to develop the conductive textile hitoe™, which takes the form of shirt-type clothing and enables daily biological measurements. In this section, we present a novel wearable sensor that, in conjunction with hitoe™, enables the simultaneous detection of biological and environmental information while featuring a small form factor and low-power consumption.

3.1 *Wearable Biological/Environmental Sensing with hitoe™*

For the purpose of daily biological monitoring, hitoe™ electrodes are incorporated into garments such as innerwear and connected to a sensor comprising analog front-end circuits, an analog–digital converter, accelerometer, central processing unit (CPU), and radio-frequency circuits for Bluetooth Low Energy communication [12]. The measured data are transmitted to a smartphone for visualization and subsequently sent to cloud servers for data storage and further signal processing. To enhance the naturalness of wearable sensing technology in our daily lives, miniaturization and long battery life of the sensor are imperative to minimize interference with user movement and reduce the frequency of battery charging. Moreover, for smart healthcare applications, such as the physical condition management of workers in hot environments, environmental information must also be evaluated to assess the thermal stress on individuals [13].

To address the aforementioned requirements, NTT has developed a wearable sensor (hitoe™ Transmitter TX02), as depicted in Fig. 7, that is compact (weighing 12 g), has low-power consumption, and enables multi-sensing of biological and environmental information. The sensor can be affixed to hitoe™ innerwear to measure biological information, such as HR and ECG. Furthermore, it can measure the environment between clothing layers. When outerwear is worn over hitoe™ innerwear, the sensor can monitor the temperature and humidity of the space between the two garments. These data are instrumental in assessing a wearer's heat stress and comfort levels. Additionally, acceleration and angular velocity can be measured to track body movements. From these measurement data, the sensor can derive various feature values, including the HR, R-R interval (RRI), number of steps, amount of body movement, and angle. The measurement data and extracted feature values are transmitted to a smartphone or Internet-of-Things gateway via Bluetooth Low Energy. The extracted feature values are stored in the sensor's internal memory, enabling data collection in cases where the wearer does not possess a smartphone during measurement. Thanks to the ergonomically designed housing, the sensor fits snugly in the gap between the chest and abdomen, allowing it to be worn comfortably without hindering movement, such as bending and lying down.

3.2 *Multi-sensing Technique for Low-Power Operation*

Continuous measurement, analysis, and transmission of diverse data lead to an increase in CPU load and power consumption, necessitating frequent recharging or the use of a large battery to maintain operational time. This challenge is addressed by the deployment of a novel biological/environmental sensor that employs a multi-sensing technique developed to mitigate power consumption [12]. The circuit configuration and flow of this technique are illustrated in Fig. 8. Through the efficient utilization of a direct memory access controller, the sensor measures and stores

Appearance	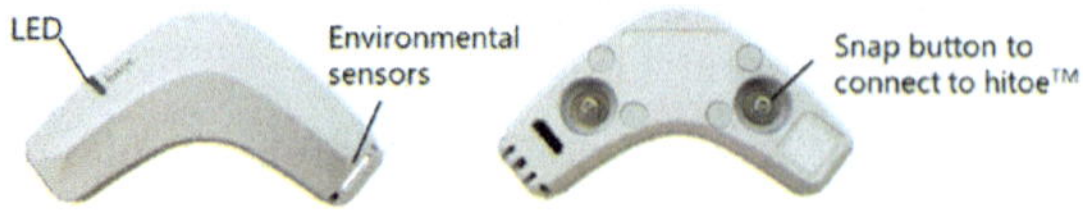
Size	64 x 36 x 9 mm
Weight	12 g
Measurement data	Cardiac potential, Temperature and humidity, Acceleration, and Angular velocity
Extracted feature values	Heart rate, R-R interval Number of Steps, Amount of body movement, and Posture
Memory	Storage of measurement data and extracted feature values

Fig. 7 Specifications of new wearable biological/environmental sensor

diverse data in a buffer while the CPU is in sleep mode. Once a considerable amount of data has been collected, the CPU is briefly activated for waveform processing and feature-value extraction.

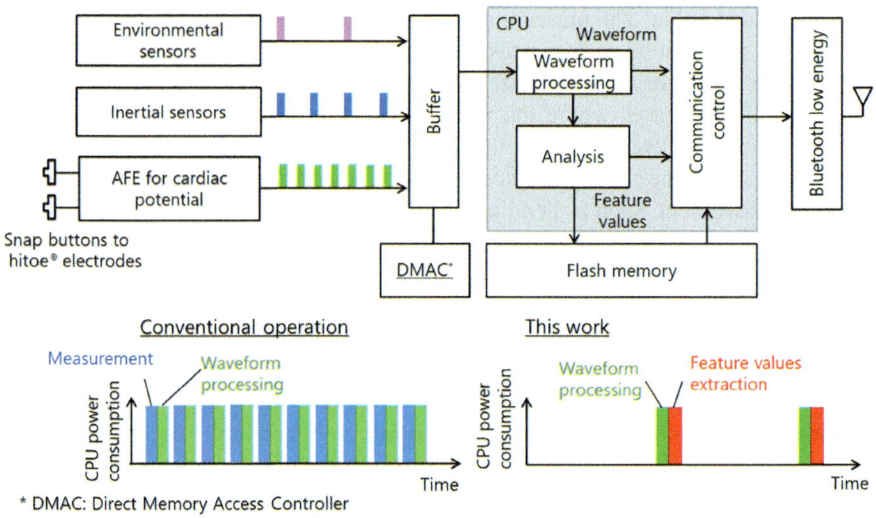

Fig. 8 Low-power multi-sensor processing. After Ref. [12]

The three operating modes of this new biological/environmental sensor are designed to meet different requirements depending on the application. The standard mode allows for the transmission of the cardiac potential waveform at a sampling rate of 200 Hz, acceleration waveform, and extracted feature values. The high-performance mode offers an improved sampling rate of the cardiac potential waveform at 1 kHz and enables angular velocity measurement. The low-power mode extends battery life by stopping the transmission of the waveform. By reducing the CPU load, this sensor is approximately half the size of the conventional one and uses a smaller battery. In the low-power mode, the sensor can operate for more than 100 h. These operating modes provide flexibility in data acquisition and power management to suit different application requirements.

Moreover, NTT proposed a simple method for RRI correction from ECG at a low sampling rate to mitigate the trade-off between high resolution and energy consumption for the calculation of the RRI [14]. The resolution of the RRI is generally dependent on the ECG sampling rate [15]. However, in wearable devices, a high sampling rate like that required for standard ECG testing such as 500 or 1000 sps results in heavy calculation loads, which increase power consumption. The proposed method is suitable for such algorithms and involves combining QRS enhancement using the first derivatives of the filtered ECG signal and QRS detection with adaptive thresholding [16].

Figure 9 shows a schematic of the proposed method. The method uses zero crossing points (Z[n]) in the filtered ECG signal. The points are estimated by $Z[n] = Y[n] - Y[n]\Delta t/(Y[n] - Y[n-1])$, where Y[n] is a sampling point corresponding to the detecting point obtained by adaptive thresholding, Y[n-1] is the previous sampling point of Y[n], and Δt is the sampling interval. Each estimated zero crossing point is used for RRI calculation.

To demonstrate the efficacy of the proposed method, an ideal ECG signal (120 bpm) was generated using an ECG generator (AX-301D, Nihon Kohden). The resulting analog signal was digitized under two sampling conditions: 1 ksps, which

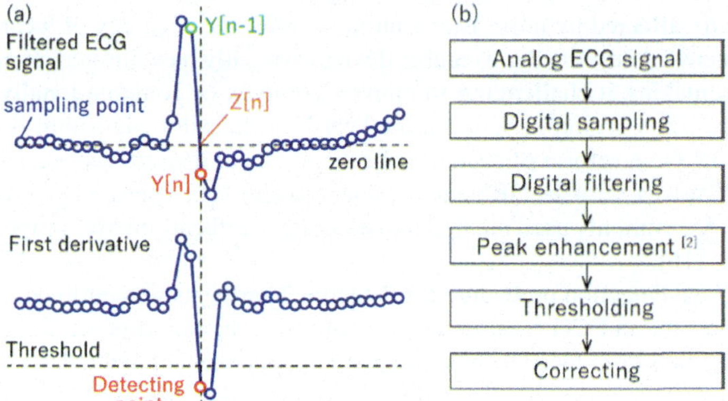

Fig. 9 **a** Schematic of RRI correction. **b** QRS detection and RRI correction flow chart

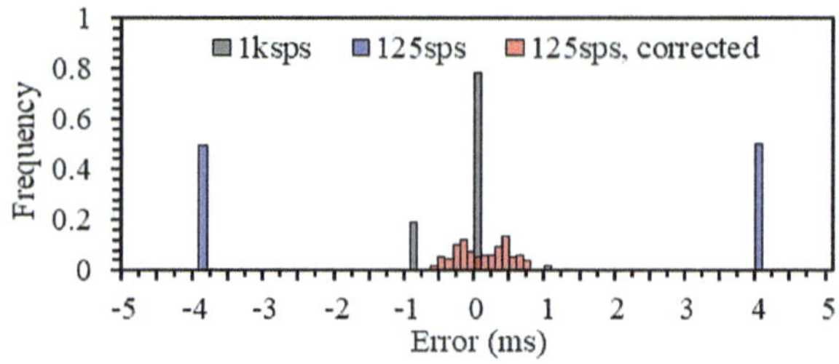

Fig. 10 Comparison of RRI error histogram

is equivalent to standard ECG testing, and 125 sps. RRI detection was performed based on the flow chart presented in Fig. 9b. Specifically, the digitized ECG signal was filtered using a digital filter appropriate for the sampling rate, and peak enhancement was conducted following QRS detection with adaptive thresholding and RRI calculation. In the case of 125 sps, the raw RRI was corrected using the proposed method.

The resulting RRI error histogram is shown in Fig. 10, which was calculated based on the difference between the calculated RRI for each condition and the reference value (500 ms). The RRI accuracy at the low sampling rate of 125 sps was ±4 ms, which was larger than that at the high sampling rate of 1 ksps (±1 ms). However, the proposed correction method improved the accuracy at the low sampling rate to ±0.8 ms. These results demonstrate the effectiveness of the proposed method in achieving the RRI accuracy required for standard ECG testing under low sampling conditions.

3.3 Lightweight Heartbeat Detection Algorithm and Its Implementation

In wearable measurement environments, ECG waveforms are often corrupted by body movement, resulting in significant noise. Therefore, developing a technique that is minimally affected by noise is crucial to improve the accuracy of wearable ECG measurements. Additionally, wearable devices typically have limited computational resources, making it challenging to employ complex or computationally intensive techniques. Thus, a simple yet accurate heartbeat detection algorithm is necessary [17]. The ECG waveform of each heartbeat consists of distinct waves, namely P, Q, R, S, and T waves, with the QRS complex being a region of steep changes in electrical potential. A commonly used method for detecting heartbeats involves capturing QRS complexes [16].

A practical detection procedure for detecting heartbeats was outlined in an application report for an ECG front-end processor [18]. The report presented a reference design for a mass-produced semiconductor chip, and the technique described can be regarded as a general approach. In this algorithm, a threshold-based peak search is applied to the time differential of the ECG signal. A threshold value is set for the time differential values, and the time point at which the value exceeds the threshold is

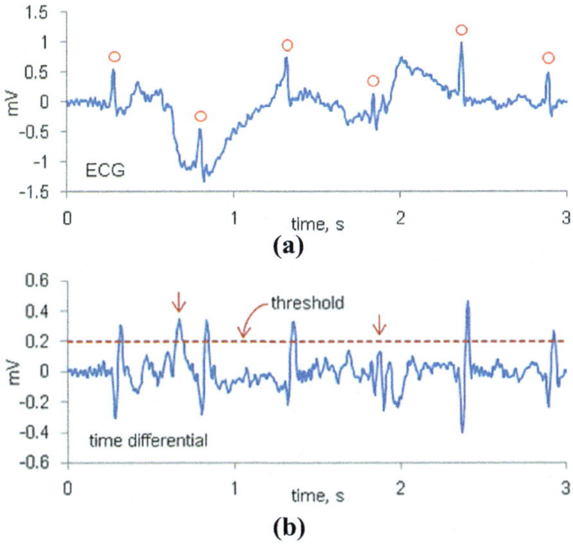

Fig. 11 Example of ECG waveforms and heartbeat detection. **a** Raw ECG. **b** Inverted time differential of ECG. After Ref. [17]

identified as the time of the heartbeat. Figure 11a and b provide an example of ECG waveforms captured by the C3fit IN-pulse and its corresponding time differential waveform. In Fig. 11b, the time differential of the time-series data x(n) is obtained by subtracting x(n − 1) from x(n + 1) for each sampling time. The small circles on the ECG waveform represent the QRS complexes. The time differential value shows slightly shifted peak positions and is inverted.

As shown in Fig. 12, using the time differential instead of the raw ECG signal allows for the cancellation of baseline slopes and swings. It should be noted that depending on the individual's physique, some individuals may have ECG waveforms with a low top R and a deep bottom S, caused by the positional relationship between the electrodes and the heart. However, even in such cases, using the inverted time differential can treat the amount of rapid decrease in electrical potential from R to S as positive amplitude. To adapt to the waveform amplitude trend, which is subject to change with wearable sensing, the threshold value can be set adaptively according to the latest peak values [19]. Nevertheless, during measurements with a wearable device, the electrodes may rub against or peel off the skin, introducing sharp fluctuations in the form of noise that inevitably disturbs the ECG waveform. Consequently, the threshold-based logic described above may lead to errors, such as detecting noise as a QRS unnecessarily (around 0.7 s) or missing a true QRS buried in noise (around 1.9 s), as shown in Fig. 12b.

To address the issues described above, NTT has developed a new lightweight algorithm for detecting heartbeats in wearable ECG measurements [17]. Based on the mechanism of ECG, the QRS complex is generated by a displacement current that occurs during the depolarization of the ventricular muscle propagating from the endocardium to the epicardium. Therefore, the duration of the QRS complex is expected to be almost constant among individuals without cardiorespiratory diseases. Thus,

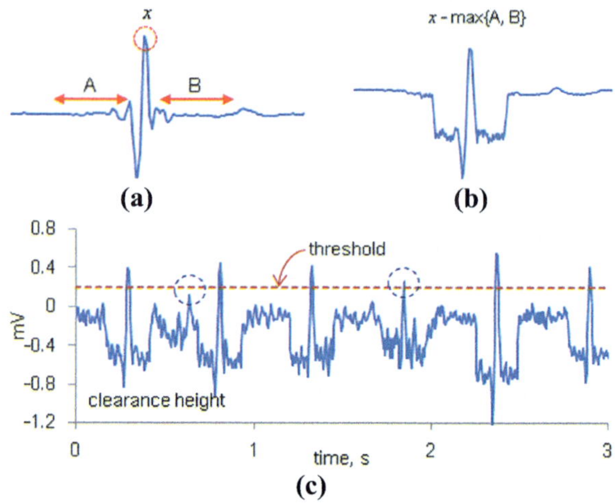

Fig. 12 Schematic of the principle of the algorithm. **a** Inverted time differential of ECG. **b** Calculated clearance height. **c** Heartbeat detection using the proposed algorithm. After Ref. [17]

it is appropriate to consider sensitivity to time width in waveforms. Furthermore, a pinching approach, where a protruding portion is detected from above, is more effective than catching something exceeding a threshold from below. Additionally, it is desirable to consider the area before and after a point, rather than using a one-way scan that focuses only on one point, as in primitive threshold logic.

The concept described above is implemented as an algorithm as follows. First, a peak in the inverted time differential of an ECG, shown in Fig. 12a, is identified. A typical QRS waveform is used to illustrate the principle. If the peak has a certain time width and sufficient clearance around it, it is most likely attributable to QRS. Let x be the value of the inverted time differential at a specific point in time, and calculate 'x—max {A, B}', where A and B correspond to specific time ranges before and after x. Max {A, B} corresponds to the floor level in the area surrounding x, A, and B. Figure 12b displays the time sequence of the value 'x—max {A, B}' calculated for each time point. This value represents the height of the clearance around each peak and, as a result, selectively enhances QRS-derived peaks. By using the clearance height, a normal threshold-based peak search can be redefined as an accurate heartbeat detection method. The width of time is determined based on the nature of an ECG. For example, the duration of A and B is 100 ms, and the interval between them is 50 ms. Figure 12c displays a case where this method was applied to the ECG in Fig. 11a. The threshold for clearance height can also be set adaptively. The results show that the QRSs that were not handled appropriately by the conventional method are now processed correctly by our algorithm, as indicated by the dotted circles.

4 Application for Smart Healthcare

The feature of the hitoe™ is its ability to acquire and transmit real-time information on HR, ECG, acceleration, temperature, and humidity just by wearing it, and its strength lies in its applications that capture global needs by computing and extracting features from the information. In this section, we will introduce two applications that contribute to smart healthcare with the aim of extending human life toward the realization of a sustainable society.

4.1 Novel Health Risk Alert System for Occupational Safety in Hot Environments

The global increase in temperature over the last century, commonly referred to as global warming, has been recognized as a significant issue by the World Meteorological Organization. This organization has reported that 2020 was one of the warmest years on record, with temperatures about 1.2 °C above preindustrial levels [20]. The adverse effects of global warming on human health are a crucial concern, particularly as extreme heat can lead to health risks [21]. In fact, this warming has been shown to severely limit human activity in tropical and mid-latitude regions [22]. Such warming has also been shown to limit human activity, especially for outdoor and manual workers who are susceptible to increased health risks from exposure to ambient heat during working hours.

To address this issue, workers should pay attention to their own physical conditions and take breaks when they feel uncomfortable. Supervisors must also manage worker conditions and schedule regular breaks. Additionally, a new integrated system has been developed to notify individuals at risk based on their thermal physiology. This method uses wearable sensors to obtain biological and environmental information, and core body temperature is estimated on-ground, as it cannot be measured wirelessly and noninvasively [23]. Overall, it is essential to take measures to mitigate the risks associated with global warming and protect the health and well-being of workers.

The core body temperature, HR, and subjective symptoms of dizziness and nausea are listed as indicators of the risk of poor physical conditions in hot environments [24]. The new method developed in this trial aims to estimate core body temperature variation in real-time using wearable sensors and a calculation model that considers various factors, including heat exchange between the core and skin layers of the human body, heat produced by human activities, the thermoregulatory function due to sweating, and clothing (Fig. 13) [25]. The model takes into account personal information, such as age, height, weight, gender, and clothing, along with heart rate and temperature/humidity inside clothes, to estimate core temperature variation. The calculation model is designed to be simple yet effective and can be performed using limited computation resources, such as smartphones, in real-time. This new method

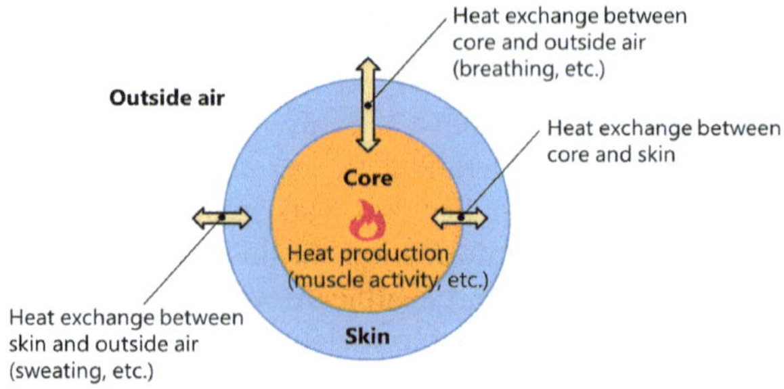

Fig. 13 Calculation human body model with core and skin layer for estimating core body temperature. After Ref. [26]

can be used to notify individuals at risk based on their thermal physiology, allowing them to take proactive measures to prevent health risks associated with extreme heat.

To verify the effectiveness of the proposed technique, a clinical experiment was conducted at Shigakkan University in an artificial weather room to estimate body temperature variations using the model. Rectal thermometer measurements were taken, as illustrated in Fig. 14, and the estimation accuracy was confirmed to be 0.15 °C [26, 27]. Criteria and algorithms were established to determine the high risk of poor physical condition for each worker. Three indicators, including HR, estimated core temperature fluctuation, and subjective information based on joint experiments and thermal physiology, were used to set criteria. An algorithm was also developed to determine the overall risk of poor physical condition, based on the state of the three indicators. The system issues an alert in the case of high risk. The health risk alert system was applied to 49 construction workers employed during the summer of 2020 for the world's largest sports events in Tokyo (a total of 834 person-days from August 5 to September 30), as illustrated in Fig. 15.

The implemented technologies for core temperature variation estimation, poor physical condition risk determination, and workload estimation were integrated into a smartphone app as a health risk alert tool for occupational safety. A system was also built to remotely monitor the physical condition of each worker wearing wearable biological or environmental sensors via the cloud. Alerts were sent to workers, supervisors, and remote monitoring systems for those identified as being at high risk. Comprehensive detection of risk cases was achieved during the construction period by supplementing each case with the three indicators of the poor physical condition risk determination technology, including HR, core temperature increase, and subjective information. Workers who received the alert paid attention to their own physical condition and took proactive measures to rest and avoid heat exposure. Supervisors assessed workers' physical conditions and directed breaks accordingly. No serious cases of poor physical condition, such as heat stroke, were reported during the construction period, indicating the usefulness of this health risk alert system.

The health management of workers and volunteers is crucial for event operations, in addition to athlete safety. The Nippon Telegraph and Telephone East Corporation applied this health risk alert system to construction and maintenance workers for the world's biggest sports events held in Tokyo during the summer of 2021. Future

Fig. 14 Experiment in the artificial weather room. Subject is using an exercise bike in a room with temperature control to simulate outdoor activity in hot weather (photo courtesy of Shigakkan University)

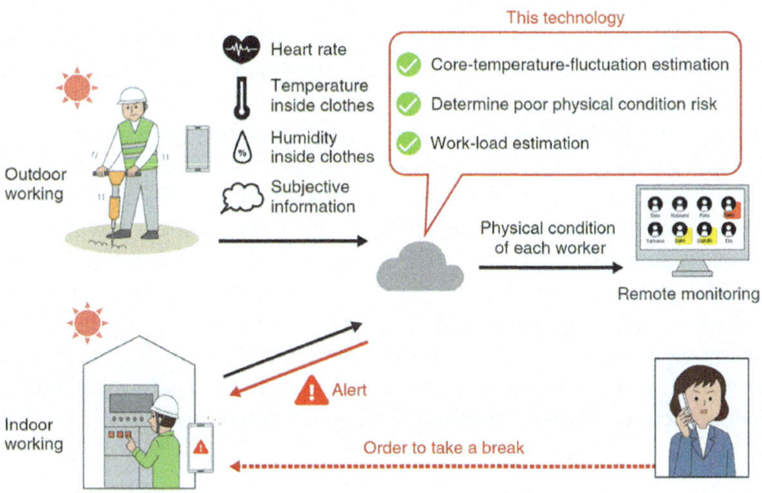

Fig. 15 Schematic diagram of the health risk alert system for workers and their supervisors. After Ref. [26]

studies may involve the application of the system to athletes, as well as other outdoor and manual workers, such as firefighters and outdoor police officers.

4.2 Application for Rehabilitation Medicine

The second application is rehabilitation support using HRs and acceleration data obtained by hitoe™. Since 2017, a joint research project between Fujita Health University and Toray Industries, Inc. has been underway. In stroke rehabilitation, the more training opportunities a patient has after stroke onset, the better progress can be expected [28]. However, in actual rehabilitation settings, patients should be as active as possible in their daily lives through getting out of bed and doing voluntary training, as training opportunities with therapists are limited. However, it is not easy for patients to vigorously load their hemiplegic bodies on their own, and support is needed to achieve a better recovery. As a tool to provide this support, a system that can precisely monitor the activity data of stroke patients on a 24-h scale using a wearable device has been developed (Fig. 16). In this system, a patient wears a garment that even a hemiplegic patient can put on and take off easily, which was introduced in Sect. 2.2, and the data is automatically collected on a server via a relay such as a smartphone or gateway, providing a report for the patient and family and detailed measurement results for the medical team. These reports provide concrete information about the recovery process, allowing the patient to have a conversation with the therapist about progress and the next training session. Detailed measurement results provide information on changes in activities of daily living for each patient, and can be shared with the entire healthcare team for collaboration. The technology that makes this possible is characterized by algorithms that process the data, and it features advanced downsampling that avoids network load at hospital facilities while preserving activity characteristics and compensation processing for partial data loss. Such processing is enabling new clinical approaches, such as using machine learning to predict indicators of recovery in stroke patients [29–31]. The technology that allows patients to move around hospital facilities without having to carry a smartphone features a gateway relay for data collection, handover to reduce data loss between adjacent gateways, and time-stamp correction to match the time difference in data collected at each gateway [32]. Furthermore, as a feature for medical validation, the system supports tests widely used in rehabilitation medicine, such as the six-minute walk test, allowing new comparative studies by easily matching existing testing approaches with the results of 24-h activity monitoring.

5 Improving Performance of Athletes by Using hitoe™ EMG Sensing Data

The hitoe™ electrode can also be used as an electrode for EMG measurements. Although EMG hitoe™ clothing has not yet been commercialized, it has been used for specialized tasks, such as training elite athletes or racing drivers, and have evaluated them [33, 34]. As an example, here we show how top cyclists utilized hitoe™ for monitoring their muscle condition.

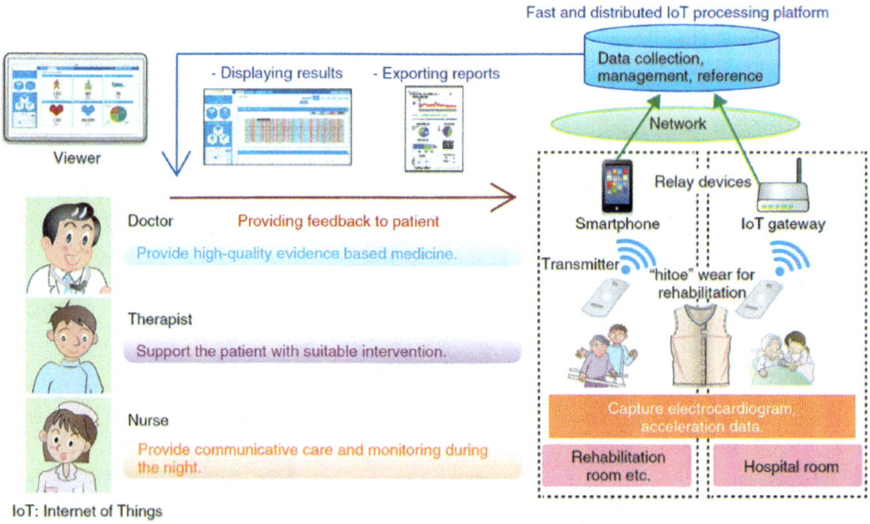

Fig. 16 Rehabilitation support system using hitoe™

For athletes, knowing the state of their own muscle activities is very important for conditioning and improving performance. Surface EMG is an effective way to ascertain muscle activity with little physical burden.

Conventional EMG sensors had practical problems when utilized in real cycling fields, such as electrodes falling off due to perspiration, problems with preparation time to fix electrodes one by one, and hampering of athletes' movements. In contrast, with the measurement using hitoe™, the electrodes can be easily fixed to the skin by simply putting on the training wear with hitoe™ mounted on the lining, and there is no need to worry about them falling off.

To evaluate their EMG during cycling on a real cycling bank, cyclists wore special spats, with the hitoe™ electrodes attached at the position of the targeted muscles inside (Fig. 17).

Figure 18 shows an overview of the system that the cyclists used for their training. The real field data collected with the hitoe EMG wear and a smartphone were analyzed in the server. The analyzed data combined with other sensor data, such as GPS, cycling power, and speed were presented via the dashboard interface to the analyst. A comprehensive report was returned to the athlete and coach. Figure 18b shows the surface myopotential data acquired during a time trial. It is known that the frequency components of the surface EMG signal change as muscle fatigue progresses. The data shows the reduction of the average frequency component of the surface EMG signal with time. Furthermore, the differences in pedaling styles on the basis of muscle fatigue and muscle activity of cyclists were evaluated from those data. Such data helped the cyclists realize the training issue and improve their performance [33].

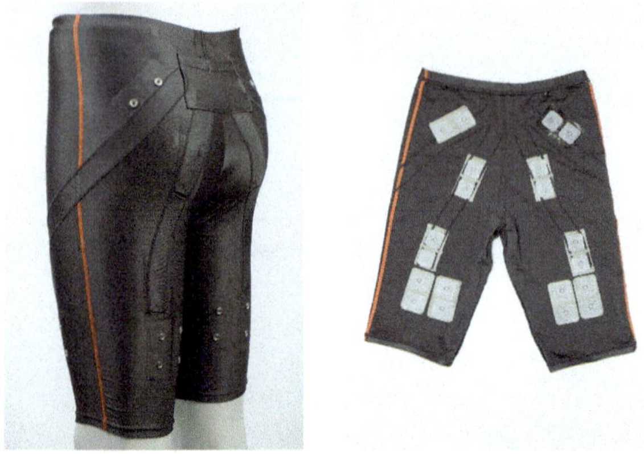

Fig. 17 Specially designed spats for cyclists. The hitoe™ electrodes are attached at the position of targeted muscles

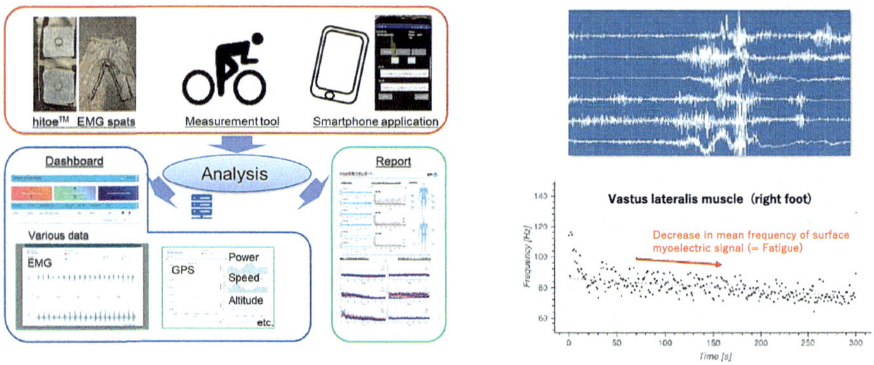

(a) System overview for cyclist training

(b) EMG signal (top) and mean frequency of surface EMG on vastus lateralis during the time trial(bottom)

Fig. 18 System using hitoe™ for measuring surface myoelectric potential. After Ref. [33]

6 Summary

As a material for skin electrodes suitable for detecting our biological signals in everyday life, hitoe™ is expanding its applications not only in the medical field but also in the healthcare field, including sports, monitoring of workers, and rehabilitation recovery support. In addition to the material's features of good adhesion to the skin and the ability to obtain clean signals, we believe that new applications will expand in the future as data analysis technologies using transmitters, signal processing, and machine learning, as well as effective intervention methods using these technologies are developed.

References

1. World health Organization: https://www.who.int/news-room/fact-sheets/detail/the-top-10-causes-of-death (2020). Accessed 28 Jul 2023
2. Tsukada, S., Nakashima, H., Torimitsu, K.: Conductive polymer combined silk fiber bundle for bioelectrical signal recording. PLoS ONE **7**(4), e33689 (2012). https://doi.org/10.1371/journal.pone.0033689
3. TORAY INDUSTRIES, INC., hitoe. https://www.hitoe.toray/en/. Accessed 28 Jul 2023
4. Tsukada, Y.T., Tokita, M., Murata, H., Hirasawa, Y., Yodogawa, K., Iwasaki, Y., Asai1, K., Shimizu, W., Kasai, N., Nakashima, H., Tsukada, S.: Validation of wearable textile electrodes for ECG monitoring. Heart Vessels **34**, 1203–1211 (2019). https://doi.org/10.1007/s00380-019-01347-8
5. Matsumura, S., Watanabe, K., Saijo, N., Ooishi, Y., Kimura, T., Kashino, M.: Positive relationship between precompetitive sympathetic predominance and competitive performance in elite extreme sports athletes. Front. Sports Act. Living **3**, 712439 (2021). https://doi.org/10.3389/fspor.2021.712439
6. Ousaka, D., Hirai, K., Sakano, N., Morita, M., Haruna, M., Hirano, K., Yamane, T., Teraoka, A., Sanou, K., Oozawa, S., Kasahara, S.: Initial evaluation of a novel electrocardiography sensor-embedded fabric wear during a full marathon. Heart Vessels **37**, 443–450 (2021). https://doi.org/10.1007/s00380-021-01939-3
7. Fukuma, N., Hasumi, E., Fujiu, K., Waki, K., Toyooka, T., Komuro, I., Ohe, K.: Feasibility of a T-shirt-type wearable electrocardiography monitor for detection of covert atrial fibrillation in young healthy adults. Sci. Rep. **9**, 11768 (2019)
8. Nakata, R., Tanaka, F., Sugawara, N., Kojima, Y., Takeuchi, T., Shiba, M., Higuchi, K., Fujiwara, Y.: Analysis of autonomic function during natural defecation in patients with irritable bowel syndrome using real-time recording with a wearable device. PLoS ONE **17**(12), e0278922 (2022). https://doi.org/10.1371/journal.pone.0278922
9. Hashimoto, Y., Sato, R., Takagahara, K., Ishihara, T., Watanabe, K., Togo, H.: Validation of wearable device consisting of a smart shirt with built-in bioelectrodes and a wireless transmitter for heart rate monitoring in light to moderate physical work. Sensors **22**(23), 9241 (2022). https://doi.org/10.3390/s22239241
10. Machino, T., Aonuma, K., Komatsu, Y., et al.: Dry textile electrode for ambulatory monitoring after catheter ablation of atrial fibrillation: a pilot study of simultaneous comparison to the Holter electrocardiogram [version 2; peer review: 2 approved]. F1000Research **11**, 97 (2022). https://doi.org/10.12688/f1000research.75712.2
11. Machino, T., Aonuma, K., Maruo, K., Komatsu, Y., Yamasaki, H., Igarashi, M., et al.: Randomized crossover trial of 2-week Garment electrocardiogram with dry textile electrode to reveal instances of post-ablation recurrence of atrial fibrillation underdiagnosed during 24-hour Holter monitoring. PLoS ONE **18**(2), e0281818 (2023). https://doi.org/10.1371/journal.pone.0281818
12. Kuwabara, K., Tokura, A., Hashimoto, Y., Higuchi, Y., Togo, H.: Wearable biological/environmental sensor and its application for smart healthcare services. NTT Tech. Rev. **18**(10), 46–51 (2020). https://www.ntt-review.jp/archive/ntttechnical.php?contents=ntr202010ra2.pdf&mode=show_pdf
13. Hirata, A., Nomura, T., Laakso, I.: Computational estimation of body temperature and sweating in the aged during passive heat exposure. Int. J. Therm. Sci. **89**, 154–163 (2015). https://doi.org/10.1016/j.ijthermalsci.2014.11.001
14. Hashimoto, Y., Matsuura, N., Kuwabara, K., Togo, H.: Simple and accurate RRI correction method from ECG signal with low sampling rate. In: 2020 42th Annual International Conference of the IEEE Engineering in Medicine and Biology Society, MoAT14.56 (2020)
15. Moller, M., et al.: Standard ECG versus 24-hour holter monitoring in the detection of ventricular arrhythmias. Clin. Cardiol. **4**, 322–324 (1981)
16. Elgendi, M., et al.: Revisiting QRS detection methodologies for portable, wearable, battery-operated, and wireless ECG systems. PLoS ONE **9**, 1 (2014)

17. Matsuura, N., Kuwabara, K., Ogasawara, T.: Lightweight heartbeat detection algorithm for consumer grade wearable ECG measurement devices and its implementation. In: 2022 44th Annual International Conference of the IEEE Engineering in Medicine and Biology Society, ThEP-21.4 (2022). https://doi.org/10.1109/EMBC48229.2022.9871514
18. Texas Instruments Incorporated : ECG Implementation on the TMS320C5515 DSP Medical Development Kit (MDK) with the ADS1298 ECG-FE
19. Luo, J., et al.: A dual-mode ECG processor with difference-insensitive QRS detection and lossless compression. IEICE Electron. Express **14**, 12 (2017)
20. World Meteorological Organization: Climate change indicators and impacts worsened in 2020 (2020). https://public.wmo.int/en/media/press-release/climate-change-indicators-and-impacts-worsened-2020. Accessed 3 Aug 2023
21. International Labor Organization: Working on a Warmer Planet: The Effect of Heat Stress on Productivity and Decent Work. ILO, Geneva (2019)
22. Dunne, J.P., Stouffer, R.J., John, J.G.: Reductions in labor capacity from heat stress under climate warming. Nat. Clim. Chang. **3**, 563–566 (2013)
23. Wong, L.: Temperature of a healthy human (body temperature). The Physics Factbook. https://hypertextbook.com/facts/1997/LenaWong.shtml. Accessed 3 Aug 2023
24. ACGIH: Threshold Limit Values and Biological Exposure Indices (2019)
25. Hirata, A., et al.: Body core temperature estimation using new compartment model with vital data from wearable devices. IEEE Access **9**, 124452–124462 (2021)
26. Takagahara, K., et al.: Physical condition management technology for making a more comfortable work site. NTT Tech. Rev. **19**(6), 48–54 (2021)
27. Hashimoto, Y., Takagahara, K., Togo, H., Uematsu, R., Miyazawa, T., Hirata, A., Kawahara, T., Tanaka, H.: Body core temperature estimation using biometric and environmental data measured by integrated wearable device. In: 2021 43th Annual International Conference of the IEEE Engineering in Medicine and Biology Society, ThDT2.14, November 2021
28. Ogasawara, T., Matsunaga, K., Ito, H., Mukaino, M.: Application for rehabilitation medicine using wearable textile "hitoe." NTT Tech. Rev. **16**(9), 6–12 (2018)
29. Ogasawara, T., Mukaino, M., Otaka, Y., Matsuura, H., Aoshima, Y., Suzuki, T., Togo, H., Nakashima, H., Yamaguchi, M., Tsukada, S., Saito, E.: Validation of data imputation by ensemble averaging to quantify 24-h behavior using heart rate of stroke rehabilitation inpatients. J. Med. Biol. Eng. **41**, 322–330 (2021)
30. Ogasawara, T., Mukaino, M., Matsuura, H., Aoshima, Y., Suzuki, T., Togo, H., Nakashima, H., Saitoh, E., Yamaguchi, M., Otaka, Y., Tsukada, S.: Ensemble averaging for categorical variables: validation study of imputing lost data in 24-h recorded postures of inpatients. Front. Physiol. **14**, 1094946 (2023). https://doi.org/10.3389/fphys.2023.1094946
31. Matsuura, H., Mukaino, M., Ogasawara, T., Aoshima, Y., Suzuki, T., Inukai, A., Hattori, E., Saitoh, E.: Preliminary study on activity monitoring for over 24 hours among stroke patients in a rehabilitation ward. Jpn. J. Compr. Rehabil. Sci. **10**, 37–41 (2019)
32. Matsunaga, K., Ogasawara, T., Kodate, J., Mukaino, M., Saitoh, E.: On-site evaluation of rehabilitation patients monitoring system using distributed wireless gateways. In: Proceeding on 2019 41st Annual International Conference of the IEEE Engineering in Medicine and Biology Society (EMBC), Berlin, July 2019
33. Tanaka, K., Tsukada, S., Yamaguchi, M.: Cycling × hitoe™: dialogue with the body via surface myoelectric potentials. NTT Tech. Rev. **20**(2), 39–43 (2022). https://doi.org/10.53829/ntr202202fa4
34. NTT STORY, NTT: Supporting the world's top racing car drivers at speeds of more than 300 km/h, NTT's biometric information solutions are opening up the way to new sporting possibilities (2019). https://group.ntt/en/magazine/blog/indycar/. Accessed 3 Aug 2023

Wearable Transdermal Biosensors

Govind Rao, Venkatesh Srinivasan, Zach Sheffield, Preety Ahuja, Sanjeev Kumar, Xudong Ge, Ketan Dighe, and Chad Sundberg

1 Introduction

The twenty-first century is revolutionizing the biomedical industry with myriad innovations from point-of-care diagnosis to telehealth solutions. Among the various fields of biomedical innovations, wearable biosensors have shown noteworthy advancements. The first few generations of wearable transdermal biosensors (WTBs) are capable of monitoring various physiological parameters such as temperature, blood pressure, and electrocardiogram and are restricted to non-invasive physiological parameters that are indicative of overall health status, but cannot provide disease/condition-specific diagnosis or continuous status monitoring. The current generation of WTBs have improved capabilities owing to both technological and biochemical sensing developments. This chapter discusses the concepts and components of WTBs. In general, WTBs have three major components, namely (i) Biorecognition elements (sensing element)—chosen sensing element specific to target analytes; (ii) Detection methods—methods to observe the interactions between biofluids and biorecognition elements; (iii) Biofluids, which are sampled by the wearable device to detect the target analyte(s).

G. Rao (✉) · V. Srinivasan · Z. Sheffield · P. Ahuja · S. Kumar · X. Ge · C. Sundberg
Department of Biochemical, Chemical, and Environmental Engineering, Center for Advanced Sensing TechnologyUniversity of Maryland Baltimore County, Maryland, USA
e-mail: grao@umbc.edu

Z. Sheffield
Department of Nuclear Engineering, The Pennsylvania State University, University Park, USA

K. Dighe
Department of Biomedical Engineering, The Pennsylvania State University, University Park, USA

K. Mitsubayashi (ed.), *Wearable Biosensing in Medicine and Healthcare*,
https://doi.org/10.1007/978-981-99-8122-9_5

2 Biorecognition Elements

Biorecognition elements are the component of WTBs responsible for the selective detection of the target analyte(s) by the sensor. Their necessity as a component of the final sensor is dependent on the target analyte. Some small, electroactive compounds can be detected using methods like differential scanning voltammetry or potentiometry purely based on their unique electrochemical properties. When the target analyte(s) are chemically complex, biorecognition elements are necessary for distinguishing the analytes from non-targets. Biorecognition elements rely on steric and intermolecular interactions to specifically bind with a complementary substrate. The binding event is then transduced to a measurable signal that is dependent on the chosen detection method. They can be either native biological macromolecules or biomimetic. Enzymes, antibodies, and aptamers represent the three commonly used biorecognition elements used in WTBs, while molecularly imprinted polymers are an emerging approach.

2.1 Enzymes

Enzymes, as biological catalysts, accelerate chemical reactions. Through billions of years of evolution, enzymes have become highly specific, efficient, and robust. Their active binding sites fit and lock specifically to particular substrates, aided by functional groups that bind or lower activation energy. This specificity makes enzymes ideal for biosensors. Enzymes can also repeatedly catalyze reactions without being consumed. These properties make enzymes suitable for biosensing, including WTBs. Biosensors combine biological components, like enzymes, with transducers to detect and quantify analytes. However, challenges such as enzyme stability, reproducibility, and calibration must be addressed. Ongoing research and technological advancements aim to enhance the performance of enzyme-based biosensors for health care and personal monitoring.

2.2 Antibodies

Antibodies are immune system-produced proteins that target foreign substances called antigens. They identify, neutralize, and eliminate harmful pathogens. Antibodies consist of two identical heavy chains and two identical light chains, forming a Y shape. Their variable regions, known as antigen-binding sites, bind specifically to antigens. Like enzymes, this binding is highly specific, with antibody sites matching antigen structures precisely. Antibodies are classified into different isotypes (IgG, IgM, IgA, IgE, IgD) that function in various immune response stages. The immune system generates new antibodies through V(D)J recombination, in which gene segments (variable (V), diversity (D), and joining (J) segments) are rearranged, recombined, and imprecisely joined. This process allows for an essentially random and highly diverse array of antibodies that contain antibodies that successfully recognize foreign antigens, which can be separated from self-recognizing antibodies. Antibodies have applications in diagnostic tests but are limited to binding biological materials, not smaller chemical compounds or analytes. Nonetheless, their specificity and engineering potential make them valuable biosensing recognition elements.

2.3 Aptamers

Aptamers are slowly gaining popularity as alternatives to enzymes and antibodies as the chosen biorecognition element in WTBs. Aptamers, also known as oligonucleotides, are short, single-stranded DNA or RNA molecules that exhibit high affinity and selectivity for specific target molecules. They can be chemically modified to enhance stability and prevent degradation. Aptamers, often called "chemical antibodies", have a broad target range beyond biological entities. Their production involves SELEX [1] (Systematic Evolution of Ligands by Exponential Enrichment), a process similar to antibody generation. SELEX includes iterative rounds of selection and amplification to isolate aptamers that bind to the desired target. Aptamers can be designed to bind small molecules, proteins, peptides, nucleic acids, viruses, and cells. Offering advantages such as easy synthesis, chemical modification, and high specificity, aptamers are valuable tools for biomedical research, diagnostics, and therapeutics. Their small size compared to enzymes and antibodies enables greater densities of aptamer probes at the active layer. Consequently, devices that employ aptamers can generally achieve greater analytical sensitivity.

2.4 Molecularly Imprinted Polymers

Molecularly imprinted polymers (MIPs) are synthetic biorecognition elements that aim to emulate the natural formation of biorecognition patterns between antibodies and antigens. They are formed from the polymerization of a chosen monomer and subsequent crosslinking of the resulting polymers in the presence of a target analyte. The target analyte acts like a template which, after removal from the MIP, leaves a cavity that acts as the biorecognition pattern between the analyte and the MIP [2]. Advantages of MIPs over other biorecognition elements are they are generally more cost-effective and can be produced for nearly any target analyte [3]. A disadvantage is the general decrease in selectivity with analyte size. MIPs have thus far been successfully implemented in wearable sweat sensors for detecting non-electroactive species such as cortisol [4–6].

3 Detection Methods

3.1 Electrochemical Detection

Electrochemical detection is widely used in analytical chemistry, biochemistry, and environmental science for its high sensitivity, specificity, simplicity, and affordability. Techniques such as amperometry, voltammetry, potentiometry, and impedance spectroscopy are the most employed detection strategies in WTBs.

Amperometry measures current for measuring analyte concentration. Amperometry is most commonly used in WTBs to measure the enzymatically catalyzed oxidation or reduction of a target molecule. One of the key advantages of amperometry is its high sensitivity, which allows for the detection of very low concentrations of analytes. One of the main limitations is that it is a single-point measurement, hence only provides information about the concentration of the analyte at a specific location. Amperometry is most used in continuous glucose monitors (CGMs) [7] to measure the amount of glucose oxidation occurring at the working electrode.

Voltammetry measures current as a function of applied potential. One of the key advantages of voltammetry is its ability to provide information about the redox behavior of analytes, including the number of electrons involved in the reaction and the standard reduction potential that can be used to identify and quantify the analyte in complex matrices. However, the complexity of the voltammogram can make interpretation difficult, requiring careful analysis of the data.

Potentiometry measures potential difference between electrodes, with high selectivity for ions and small molecules when combined with ion-selective membranes, making it useful in clinical applications. It is relatively simple and inexpensive. It is also a non-destructive technique, which means that the sample can be reused for further analysis. Potentiometry is generally used with WTBs in conjunction with ion-selective membranes for the detection of specific electrolytes in sweat or ISF.

Electrical impedance spectroscopy (EIS) is a method for measuring interfacial changes in capacitance, resistance, and conductance at electroactive surfaces. EIS can be either Faradaic or non-Faradaic. In the former, redox reactions govern changes in the interfacial electrical properties. For non-Faradaic EIS, disruptions of the electric double layer at the electrode-fluid interface is most responsible for measured changes in electrical properties. The latter is particularly useful for detecting non-electroactive analytes and electroactive analytes that cannot be easily distinguished from non-target entities. As such, EIS has been employed in wearable transdermal sensors (WTS) alongside affinity-based biorecognition to detect hydrophobic molecules like cortisol [8–10].

While electrochemical detection in sweat differs from that in ISF, the underlying principles of the various electrochemical approaches remain the same. Amperometry is primarily employed for detecting enzymatically catalyzed redox reactions while potentiometry is mainly used for ion detection. The disparity between electrochemical sensing in sweat and ISF arises from the contrasting compositions, properties, sampling methods, and sensing environments of the two fluids. Electrochemical approaches involving a transistor as the biosensor or relying on electrochemical impedance spectroscopy (EIS) for analyte detection are not utilized for in situ ISF sampling. However, they find applications in in situ sweat sampling. Transdermal gases are usually detected based on gas-specific optical sensors. For example, transdermal CO_2 is detected using an IR detector that observes the change in the IR absorbance in the sampling chamber specific to CO_2 [11].

3.2 Optical Detection

Optical detection methods comprise a variety of techniques used to detect and quantify molecules based on their optical properties. They generally involve the use of light to interact with the analyte of interest, which can then be measured and analyzed to determine the concentration of the molecule. There are several types of optical detection techniques that have been utilized in WTSs, including colorimetry, surface plasmon resonance, fluorescence spectroscopy, and surface-enhanced Raman scattering.

Colorimetry involves measuring the color change of a sample solution in response to the presence of an analyte. Certain chemical reactions or interactions between molecules can cause a change in the absorption or reflection of light in the visible spectrum, which are detected using a colorimetric assay. Colorimetric detection is commonly used in various applications, such as clinical diagnostics, environmental monitoring, and food safety testing. It is often preferred due to its simplicity, low cost, and ease of use compared to more complex analytical techniques.

Plasmonic detection utilizes the optical properties of metallic nanoparticles to detect the presence of an analyte [12]. Noble metal nanomaterials show collective oscillations of surface electrons that can interact with the surrounding molecules causing change in their optical properties [13]. These changes are measured using spectrophotometers as absorbance, scattering, or SPR to detect the presence or concentration of the analyte. Plasmonic detection is used in a variety of applications, including biosensing, medical diagnostics, and environmental monitoring. One of the main challenges is the potential for non-specific binding, which can lead to false positives or inaccurate readings.

Fluorometric detection relies on measuring the fluorescence emission of a sample in response to the presence of an analyte. Certain molecules, known as fluorophores, can absorb light at a specific wavelength and emit the absorbed energy as radiative emission at a longer wavelength, which is measured using a spectrophotometer [14]. It is also a versatile technique, allowing for the detection of a wide range of molecules and analytes. Additionally, fluorescence measurements can be affected by factors such as pH, temperature, and sample preparation.

Surface-Enhanced Raman Scattering (SERS) is a detection technique that utilizes the enhanced Raman signal of molecules adsorbed onto a metal surface, typically silver or gold, to detect the presence and concentration of analytes. The enhancement is achieved by the plasmonic resonance of the metal surface, which greatly amplifies the Raman signal of the analyte. Advantages of SERS-based detection are high sensitivity, label-free technique, simplified assay, and cost reduction. Limitations of SERS are signal interference and variability due to the complex nature of the metal surface and sensitivity to temperature and pH.

Detection in interstitial fluid (ISF) in situ is currently performed exclusively with electrochemical methods. Optical detection methods, such as colorimetry and fluorometry, are difficult to implement because of the low penetration depth of visible light into skin. There are a few exceptions, including the commercially available continuous glucose monitor (CGM), Eversense® [15, 16], developed by Senseonics that uses fluorescence for analyte detection in ISF. However, in such cases, the fluorescence is measured subcutaneously. Sweat, on the other hand, is not limited by the poor penetration depth of light because analysis requires secretion outside the body. Therefore, optical detection methods can be more easily integrated into wearable sweat sensors (WSSs) for in situ analysis than wearable sensors that sample ISF. Optical detection methods have the advantage of not requiring a power supply for their operation, which sets them apart from electrochemical methods. This advantage is most noticeable in WSSs that use colorimetry. When combined with smartphone technology [17], this detection method has shown to be an effective, point-of-care approach to measuring sweat rate and simple analytes like electrolytes and metabolites [18–20] in secreted eccrine sweat.

4 Biofluids

4.1 Interstitial Fluid

Interstitial fluid (ISF) is the aqueous solution that bathes the extracellular space acting as the transport medium for exchange of nutrients and waste products between the capillaries and surrounding tissues; signaling molecules between cells and larger solutes such as protein macromolecules, lipoproteins, antigens, and endogenous vesicles are delivered from the interstitium to the lymphatic system [21, 22]. Ex situ approaches for analyzing ISF involve minimally invasive methods like wicking, microdialysis, suction blisters, etc., which require puncturing the skin, or non-invasive methods like reverse-iontophoresis or sonification to extract ISF for off-body analysis. In situ WTBs provide us with a means to expand the utility of ISF beyond laboratory-based diagnostics and proteomics to healthcare applications that would benefit from rapid, continuous sampling of a target analyte.

4.1.1 Sourcing Interstitial fluid—anatomical Understanding

The skin consists of three layers: the epidermis, dermis, and hypodermis in that order from external to internal. The epidermis (50–150 μm thick), the outermost layer, includes the stratum corneum and viable epidermis (Fig. 1a). The viable epidermis is avascular and is mainly comprised of special antigen-presenting cells, known as Langerhans cells, and keratinocytes [23]. ISF comprises 15–35% [24] of the volume of the viable epidermis. This ISF is the target of microneedle platforms that are designed to be minimally invasive because of the absence of blood vessels and nerves in the epidermis. The dermis is divided into two essential layers: the papillary dermis (~15 μm thick) and the reticular dermis (~1–4 mm thick) containing the lowest cellular content of the layers, but the highest ISF content [24]. The extracellular matrix (ECM) of the dermis which consists of glycoaminoglycans, collagen, and proteoglycans [25] plays a crucial role in the extraction and sampling of ISF. The hypodermis, composed of adipose tissue, connects the dermis to muscle and is relevant for continuous glucose monitors (CGMs). While WTBs focus on the dermis and epidermis, the hypodermis provides nutrient-rich ISF and allows secure sensor insertion without flexible electronics. Ions and small, non-lipophilic molecules (molecular weight < 3 kDa) freely partition between the blood and dermal ISF via paracellular diffusion through intercellular junctions of the capillary endothelium. Small lipophilic solutes passively diffuse through the cells' lipid bilayers. Solutes with sizes in the range of 3–70 kDa are transported by paracellular and transcellular routes. Which transport mechanism dominates depends on the size [25, 26] of the solute as is illustrated in Fig. 1b.

4.1.2 Sampling Interstitial Fluid

ISF and sweat need different sampling methodologies. One of the early methods used for ISF extraction that is relevant to WTSs is reverse-iontophoresis. It requires the application of a small current across the skin to initiate the electroosmotic flow of ISF from the epidermis and papillary dermis (Fig. 2). Electroosmosis [27, 28] is bulk fluid flow in the direction of counter ion flow due to the application of an electric field across a membrane. Since ISF is an aqueous solution, it is incapable of passing through the lipophilic stratum corneum, and, hence, exits the body through sweat ducts or hair follicles before collecting at the cathode [29]. Concerns that prolonged use or increased current density can cause skin irritation hinder the wider use of reverse-iontophoresis in WTSs. In addition, there is a general lack of validation of correlations between measured concentrations of analytes in iontophoretically extracted ISF and circulating concentrations. Nonetheless, proof of concepts demonstrating its feasibility for ISF extraction has been presented [30–32].

Microneedles constitute an approach to sampling ISF that has the advantage of direct sampling, while being less invasive than the probes used with commercial glucose sensors. Generally, microneedles refer to needles with lengths of 50–2000 μm, base widths of 25–500 μm, and tip diameters of 1–100 μm [33, 34]. Microneedle platforms can be categorized as either ex situ or in situ [35]—referring to off-body and on-body detection, respectively. Most advances in microneedle technology for transdermal biosensing can be classified as in situ. Microneedles for

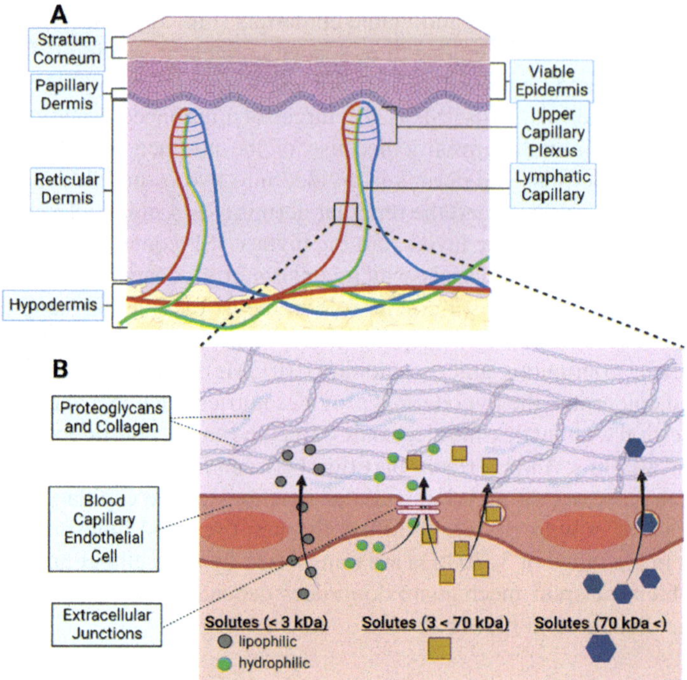

Fig. 1 **a** An illustration showing the layers of skin. **b** Illustration of the partitioning of solutes between the blood and dermal ISF

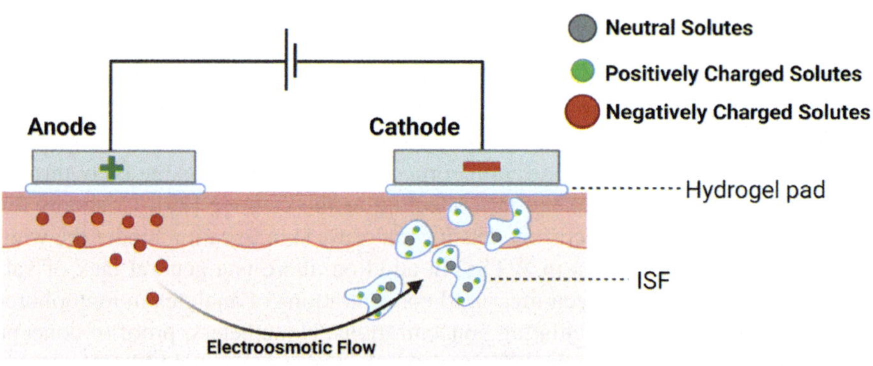

Fig. 2 An illustration of the principle of reverse-iontophoresis

in situ, in vivo sensing come in four basic configurations (Fig. 3): hollow microneedles with the sensor situated ex vivo, hollow microneedles where the sensor is inserted or packed into the tip, solid-tip microneedles where the tips are dry electrodes, and solid-tip microneedles where the sensor is fabricated on the tip [36]. The hollow microneedles, not interfaced with the sensor element, aid in extracting ISF to the

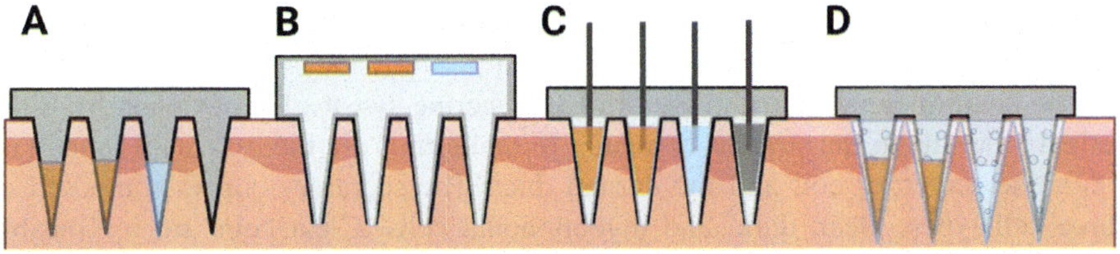

Fig. 3 Illustrations of conventional microneedle configurations for transdermal sensing. Orange denotes the working and counter electrodes and blue denotes the reference electrode of a typical three-electrode electrochemical cell. **a** Solid-tip microneedles. **b** Hollow microneedles with an ex vivo sensor. **c** Packed-hollow-tip microneedles. Wires are imbedded from the backside of the sensor

biosensor. The high fluidic resistance of the dermis makes this method impractical for in situ monitoring; however, the configuration of hollow microneedles packed with the biosensor has been demonstrated [37] as a possible way to circumvent the complexity of the dermal interstitium. Thus far, microneedles for in situ monitoring have been demonstrated for metabolites [37–40], drugs [41–44], and electrolytes [45–47]. However, few have been clinically validated.

4.2 *Biofluids*

4.2.1 Sweat

Sweat is our bodies' natural way to regulate internal temperature for maintaining thermal homeostasis. Sweat secretion is controlled by the sympathetic nervous system and is initiated through either thermal or psychological stimulation. Sweat is easy to access, naturally secreted, and contains a cornucopia of biomarkers also found in ISF and blood. Above all, sweat can be sampled non-invasively. However, wearable sweat sensor devices still have not seen the same level of commercial or clinical success as their ISF sampling counterparts due to operational challenges such as variations in composition and rate of secreted sweat. Rate of secretion due to environmental and physiological factors requires consideration of not only the biosensor, but also how the sweat is being sampled. Ultimately, development of a wearable sweat sensor amounts to a uniquely interdisciplinary problem.

4.2.2 Sourcing sweat—anatomy of Eccrine Sweat Gland

Eccrine sweat glands are the smallest, present in whole body, most numerous, and contribute the most to the total volume of sweat secreted [48]. The other two sweat gland types, apocrine and apoeccrine, are primarily found in regions less accessible for sweat sampling. Sweat secreted from apocrine and apoeccrine glands is also comparatively more complex [49]. For these reasons, eccrine sweat is mainly chosen for sweat analysis.

The anatomical structure (Fig. 4) of the eccrine sweat gland extends from the reticular dermis to the stratum corneum. In the dermis are the two main functional units, the secretory coil and the dermal duct. The secretory coil is composed of three cell types: clear, dark, and myoepithelial cells. Clear cells are responsible for primary sweat secretion as is evidenced by the cells' system of intercellular canaliculi, glycogen, large amount of mitochondria, and high Na–K-ATPase activity [50]. The function of the dark cells is not entirely understood; however, they are known to secrete glycoproteins, dermicidin, and sialomucin and are necessary for proper sweat secretion [51, 52]. The function of myoepithelial cells is primarily as structural support for the secretory coil [50]. The dermal duct is composed of basal and luminal epithelial cells. The function of the duct is reabsorption of Na^+ and Cl^- as secreted sweat flows through the duct [49, 50].

Solutes that comprise secreted sweat's composition include metabolites like glucose, lactate, urea, amino acids, and creatinine; micronutrients like Mg^{2+}, Ca^{2+}, iron, zinc, and vitamins like ascorbic acid; adrenal steroids like cortisol and dehydroepiandrosterone (DHEA); neurotransmitters like serotonin and dopamine; and cytokines like IL-6, IL-8, IL-10, IL-31, and TNF-α [49, 53, 54]. Mechanisms and correlations with blood plasma concentrations have not been fully established for most of these solutes. Complicating these deficits in data are physiological factors like sweat rate and sweat pH [52, 55], which can vary with sex, age, diet, weight, temperature, etc. If they are not properly accounted for, these variabilities can negatively affect the accuracy of measurements depending on the analyte. As such, WSSs may require sensors for measuring sweat rate, pH, skin temperature, and ion concentration alongside the biosensor for the analyte.

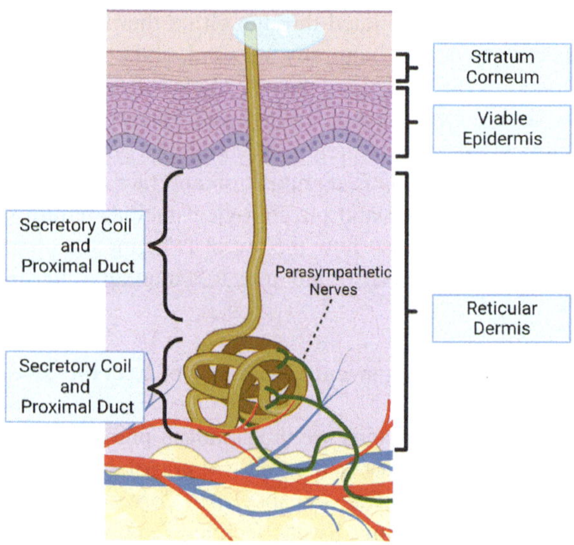

Fig. 4 An illustration showing the basic anatomy of an eccrine sweat gland

4.2.3 Sampling Sweat

Traditional sweat sampling methods include the whole body-washdown technique, patches, polymer bags/films, and macroducts. The limited-use whole body-washdown technique collects all sweat runoff to determine whole body-sweat electrolyte loss [56]. Using absorbent patches is flexible and can be applied to specific sites for sweat accumulation, but they may overestimate analyte concentration and have time-consuming application processes [57]. Polymer bags/films and macroducts [58, 59] offer alternative collection methods but are also associated with limitations such as sweat composition changes and inhibiting sweat evaporation. Current approaches to sweat sampling for wearable transdermal biosensing aim to actively collect secreted sweat for on-body analyte monitoring. These methods aim to enable real-time, continuous sweat sensing with improved reliability and sensitivity. Absorbent materials integrated with wearable sensors, microfluidic devices with superhydrophobic/superhydrophilic surfaces, fabrics with moisture-wicking properties, and controlled sweat secretion via the iontophoretic delivery of sweat-stimulating drugs are emerging techniques that address some of the limitations of traditional sampling methods [33].

4.3 Transcutaneous Gases

Despite the benefits offered by sweat-based analysis, the successful diagnosis of the related problem is still a big challenge, sometimes resulting in misleading information if a standalone sensor shows a signal change. For example, sensing potassium levels in the sweat may provide incomplete status as they are invariant with sweat rate and with normal physiological changes in the body. While change in sodium, lactate, or glucose signals and the potassium signal holds steady, multiple sensor could provide a solution for monitoring a real physiological event [60]. Alternative solutions involve the monitoring of transcutaneous gases that include volatile organics and gases emanating through human skin (skin gas). Skin emanation gas was firstly proved by Hirabayashi et al. with the releasing hydrogen gas and acetone vapor from skin [61].

4.3.1 Volatile Organic Compounds (Vocs)

The commonly studied VOCs include acetone, ethanol, and ethylene having insight into the health state of the human body. Diabetic patients have shown high acetone concentration in their breath that can be used for diagnosis. Prompted by this, Kondo et al. measured skin acetone concentration and its correlation with blood glucose for regular examination of diabetes to avoid later complications [62]. However, this can only be applied to patients with insulin therapy and not in regular patients on oral medication or with controlled diet. Although the results have large deviations, they proved that the acetone concentration varied with the diabetic condition of the patients diagnosed with diabetes. In other recent study, Misra et al. fabricated a wrist-worn system to monitor skin acetone with output represented by the resistance change [63]. Their sensing system comprised commercial off the shelf components along with a microcontroller and a rechargeable Li ion battery. They made a wristband form factor constituting an array from four sensors (n-type metal oxide semiconductor-based gas sensor, BME680) operated at different temperatures as the sensing modality. Due to the reducing effect of acetone on the metal oxide, a decrement in the resistance was detected. The acetone reacts with oxygen anions of metal oxide and increases the carrier concentration thereby declining the resistance of the sensor. Besides the instrumentation evaluation of the fabricated wrist-worn system, they provided real-time analysis showing larger changes in the subjects with intermittent fasting than the normal subjects.

Other volatile organic compounds (VOCs) such as ethanol and acetic acid are also analyzed in recent studies and correlated with different health conditions for example psoriasis [64]. One such study reported wearable e-nose constituting an array of chemical sensors integrated in a wireless communication system [65]. The array comprised eight elements formed by varying the printing patterns and sensing materials (MWCNT/polymer films) to produce the cross responses toward different analytes. The applicability was demonstrated with fabricated armband and corresponding monitoring of ammonia, acetic acid, acetone, and ethanol. The electrical responses of the fabricated system yielded discrete patterns to ammonia and acetic acid unlike the similar patterns for acetone and ethanol. Another study reported the skin gas sensor and named bio-sniffer for measurement of ethanol by using alcohol dehydrogenase (ADH) [66]. It measured the change in fluorescence intensity of nicotinamide adenine dinucleotide (NADH), as a result of oxidation of ethanol by ADH. Later, it was modified as sniff-cam or skin gas cam extending the bio-sniffer toward the measurement of the time-dependent spatial distribution of ethanol [67].

4.3.2 Trace Gases

Apart from the VOCs, skin gas also contains important trace gases such as NH_3, NO_2, and CO_2 which have particular benefits in healthcare systems [68]. For example, measurement of skin CO_2 can provide ventilation status of newborn infants along with respiratory and cardiac status [69]. They constructed a wearable prototype for monitoring transcutaneous CO_2 by employing polymer films (Poly(propyl methacrylate), PPMA) with pH-sensitive fluorescent dye (8-hydroxy-1,3,6-pyrenetrisulfonic acid trisodium salt or HPTS). Their sensor is based on the fluorescence changes of a pH indicator while protonation or deprotonation. The fabricated device comprised a microcontroller along with other electronics. The fluorescence response of the film is monitored via intensity of the sensing film on interaction with CO_2 skin gas and was found to be highly sensitive in the physiological CO_2 range. They proposed that their future work will aim to make a CO_2 permeable scattering layer which will not require heating for transcutaneous applications. Another study reported a new approach based on the initial pseudo steady state diffusion rate monitoring for non-invasive measurement of transcutaneous partial pressure of carbon dioxide ($tcpCO_2$) [70]. They provided additional reliable data to compare their sensing device with commercially available equilibrium-based transcutaneous monitor, Radiometer TCM4, used for CO_2 monitoring in hospitals [71]. Radiometer TCM4 requires the start-up at 37 °C and took 85 min before to provide meaningful results. Even at a higher temperature of 42 °C, it requires 20 min to stabilize. Unlike this, their device can provide immediate reliable data, almost independent of temperature by using a rate-based approach.

Recently, Ahuja et al. reported the room temperature measurement of skin CO_2 gas with an air permeable transcutaneous sensor comprising of polyaniline (PANI)-based sensing material as shown in Fig. 5 [72]. Porous sugar-templated polydimethylsiloxane (PDMS) elastomer sponge (PP) was used as a matrix for PANI-based ink (oxSWCNT-PANI) and fabricated using a dip coating method (Fig. 5a). The dispersant (Zn-Al) used for ink fabrication offered uniform active sites for the analyte (transdermal CO_2 gas) as shown by TEM analysis (Fig. 5b). The interaction of CO_2 with the sensor (partial extraction of electrons from PANI) resulted in an augmented charge transfer interaction leading to increased response of the sensor. Besides the sensor's fast response and recovery (Fig. 5c), it can selectively detect transcutaneous CO_2 at room temperature, unlike the commercial devices operated at a high temperature. Noteworthily, the deformability tolerance capability of the sensor opens up novel opportunities for skin-compliant wearable systems.

5 The Case of Continuous Glucose Monitors

The Center for Disease Control and Prevention (CDC) reported [73] in 2022, that, in the United States, 37.3 million people have diabetes and 96 million people aged 18 or older are prediabetic. In the late twentieth century until the beginning of the twenty-first century, personal management of one's glycemic level exclusively required pricking a finger to draw blood, then testing the glucose concentration of the blood sample with a disposable electrochemical sensor integrated with a digital-readout device. In the year 1999, Medtronic's Minimed Gold became the first continuous glucose sensor (CGM) to be approved by the FDA for market distribution [74]. The device still required daily calibration via finger pricking; however, the decrease in frequency of the pricking was a dramatic improvement to patients' quality of life and demonstrated the future healthcare potential of wearable biosensors.

CGM devices that sample ISF are currently the most commercially successful type of WTB available on the market—being capable of long-term continuous monitoring of their target analyte with analytical sensitivity and selectivity great enough [75] for the devices to be used as either adjuncts to fingerstick blood glucose monitors (BGMs) or replacements of fingerstick BGMs. A CGM [76] is any device that can automatically track glucose levels throughout the day and night. Current, commercially available CGMs use a small probe (Fig. 6), either inserted or implanted

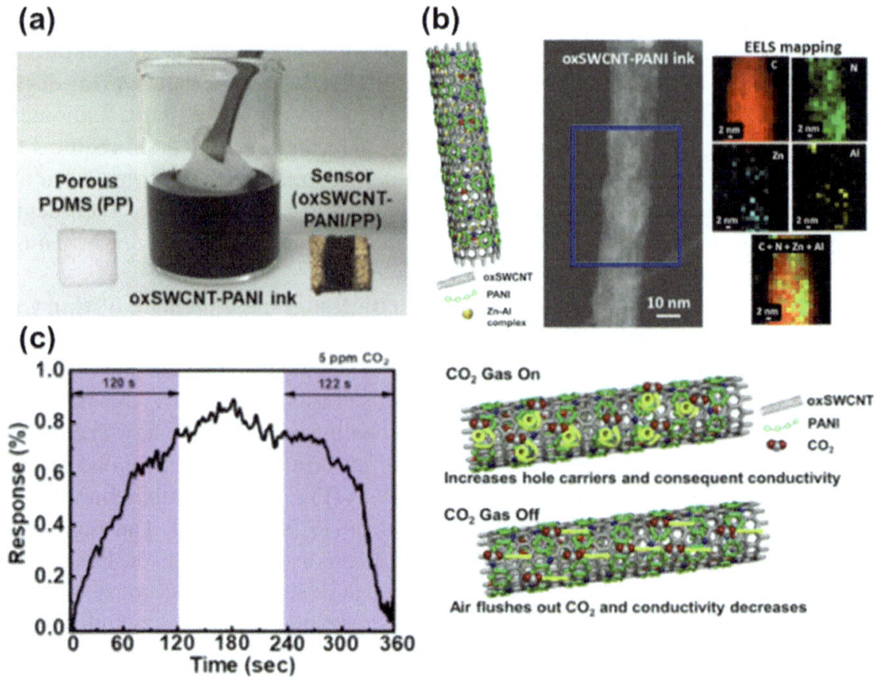

Fig. 5 **a** Photograph of the porous PDMS sponge (PP), oxSWCNT-PANI ink and fabricated sensor (oxSWCNT-PANI/PP); **b** ADF-STEM micrograph of the oxSWCNT-PANI ink with schematic of the fabricated sensor. EELS mapping of particular region in blue box. **c** Response and recovery time of the oxSWCNT-PANI/PP sensor in CO_2 gas sensing (5 ppm). **d** Schematic representation of the probable CO_2 sensing mechanism of the sensor

deep in the dermis or into the hypodermis. The typical glucose sensor relies on the oxidation of glucose catalyzed by glucose-oxidase after the molecule binds with the enzyme. First-generation glucose monitors use oxygen as an electron acceptor/redox mediator by using the electron from the oxidation of glucose to reduce oxygen to hydrogen peroxide. Then the electron obtained from the oxidation of hydrogen peroxide to water by the working electrode is detected by the change in current at the working electrode. This change in current is subsequently correlated to the blood-glucose concentration. These data are wirelessly transmitted to a handheld device, thus providing the patient with real-time changes of their blood glucose levels throughout the wear period. Some commercial CGMs still rely on first-generation glucose sensing. Second-generation CGMs, that use artificial electron acceptors or redox mediators, and third-generation CGMs, that use modified enzymes instead of mediators, are becoming more prevalent in CGM technology [7].

CGMs share four key components: an enzymatic biorecognition element, an electrochemical transduction element, a glucose limiting membrane, and interference rejection [7]. Most CGMs use glucose oxidase (GOx) as the enzyme, which catalyzes the oxidation of glucose. The generated electron participates in a redox reaction

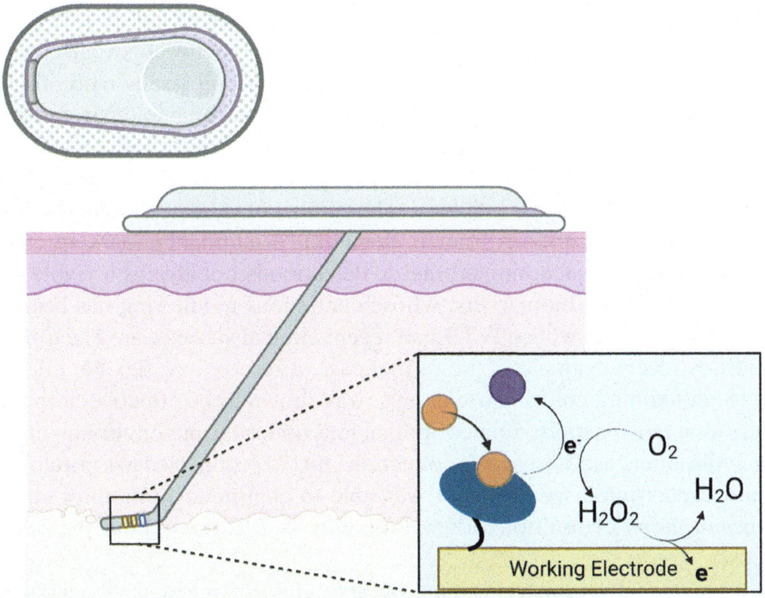

Fig. 6 Illustration of a continuous glucose monitor and first-generation detection of glucose. The orange circle represents glucose, which is oxidized by glucose oxidase into glucono-δ-lactone, represented by the purple circle. The electron removed from glucose is ultimately used to reduce oxygen dissolved in ISF to hydrogen peroxide, which is subsequently oxidized by the working electrode to water

with a mediator molecule near the transduction element, detected amperometrically. First-generation sensors relied on the H_2O_2–O_2 redox reaction but had issues with unintended oxidation of other molecules. Second-generation sensors used artificial mediators or glucose dehydrogenases (GDHs) as replacements but face stability and toxicity challenges. Recent advancements aim to improve enzyme stability, electron transfer rate, or eliminate redox mediators in the reaction. Incorporating GDHs into commercial CGMs is still pending.

6 Discussion and Concluding Remarks

Within the past 5 years, WTB technology has slowly branched away from glucose monitoring for diabetes management toward other analytes and applications. The success of CGMs using ISF as the sampling fluid punctuates the viability of the biofluid for other transdermal sensing applications and sets them as the gold standards which other WTBs aim to emulate regarding sensitivity, selectivity, and longevity. However, many aspects of their designs do not necessarily integrate well with alternative sampling methods or can be easily repurposed for the detection of biomarkers dissimilar to glucose. Sweat has been extensively investigated as a non-invasive alternative to ISF for glucose monitoring. Correlations between sweat-glucose and blood-glucose have been validated; however, they are still issues with reliable and continuous sampling throughout the day due to uncontrollable environmental and physiological factors—issues which extend to other wearable sweat sensing applications. To this point, monitoring transcutaneous gases has a clear advantage; however, the technology is still in a nascent phase. The number of validated biomarkers that can be detected as gases is currently limited. Regarding dissimilar biomarkers, enzymatic recognition and subsequent amperometric detection is not always a viable strategy. Two examples of such biomarkers, whose continuous monitoring has been repeatedly demonstrated and whose WTB had recent clinical success, are electrolytes and hormones. A recent example of the former case, a zero-power, flexible microfluidic patch [18] containing colorimetric assays, was shown to be effective at monitoring Cl^- and sweat rate for sports-related applications under various environmental conditions. For the latter case, a wearable collection unit [77] connected to a portable liquid-chromatographer/mass spectrometer was able to continuously monitor changes in the concentrations of multiple adrenal steroids in ISF throughout patients' daily activities.

WTBs represent an emerging and variable class of medical devices that have viable potential for improving point-of-care medicine. Their successful development requires careful consideration of the interactions between the sensor and the biofluid containing the target analyte. WTBs most often target sweat and ISF because they contain a plethora of biomarkers and can be minimally or non-invasively sampled. On-body detection of the target analyte(s) is normally done in situ using electrochemical detection methods. Access to these biofluids for continuous, on-body monitoring, however, is not trivial. WTBs for transcutaneous gases are an emerging technology that, in comparison to ISF and sweat WTBs, have the advantage of continuous monitoring with minimal sensor degradation. However, there are few transcutaneous gases that have been identified as viable biomarkers. So far, CGMs have been the most successful WTBs, but that same success has yet to be replicated. However, recent successes highlight the prospective benefits of the WTBs and herald its continuing evolution into a staple of modern medical technology.

References

1. Zhuo, Z., Yu, Y., Wang, M., Li, J., Zhang, Z., Liu, J., Wu, X., Lu, A., Zhang, G., Zhang, B.: Recent advances in selex technology and aptamer applications in biomedicine. IJMS **18**, 2142 (2017). https://doi.org/10.3390/ijms18102142
2. Morales, M.A., Halpern, J.M.: Guide to selecting a biorecognition element for biosensors. Bioconjugate Chem. **29**, 3231–3239 (2018). https://doi.org/10.1021/acs.bioconjchem.8b00592
3. BelBruno, J.J.: Molecularly imprinted polymers. Chem. Rev. **119**, 94–119 (2019). https://doi.org/10.1021/acs.chemrev.8b00171

4. Parlak, O., Keene, S.T., Marais, A., Curto, V.F., Salleo, A.: Molecularly selective nanoporous membrane-based wearable organic electrochemical device for noninvasive cortisol sensing. Sci. Adv. 4, eaar2904 (2018). https://doi.org/10.1126/sciadv.aar2904
5. Mugo, S.M., Robertson, S.V., Lu, W.: A molecularly imprinted electrochemical microneedle sensor for multiplexed metabolites detection in human sweat. Talanta **259**, 124531 (2023). https://doi.org/10.1016/j.talanta.2023.124531
6. Mugo, S.M., Lu, W., Robertson, S.: A wearable, textile-based polyacrylate imprinted electrochemical sensor for cortisol detection in sweat. Biosensors **12**, 854 (2022). https://doi.org/10.3390/bios12100854
7. Lee, I., Probst, D., Klonoff, D., Sode, K.: Continuous glucose monitoring systems – Current status and future perspectives of the flagship technologies in biosensor research. Biosens. Bioelectron. **181**, 113054 (2021). https://doi.org/10.1016/j.bios.2021.113054
8. Ganguly, A., Lin, K.C., Muthukumar, S., Prasad, S.: Autonomous, real-time monitoring electrochemical aptasensor for circadian tracking of cortisol hormone in sub-microliter volumes of passively eluted human sweat. ACS Sens. **6**, 63–72 (2021). https://doi.org/10.1021/acssensors.0c01754
9. Pali, M., Jagannath, B., Lin, K.-C., Upasham, S., Sankhalab, D., Upashama, S., Muthukumar, S., Prasad, S.: CATCH (Cortisol Apta WATCH): 'Bio-mimic alarm' to track anxiety, stress immunity in human sweat. Electrochim. Acta. **390**, 138834 (2021). https://doi.org/10.1016/j.electacta.2021.138834
10. Upasham, S., Osborne, O., Prasad, S.: Demonstration of sweat-based circadian diagnostic capability of SLOCK using electrochemical detection modalities. RSC Adv. **11**, 7750–7765 (2021). https://doi.org/10.1039/D0RA10561A
11. Grieshaber, D., MacKenzie, R., Vörös, J., Reimhult, E.: Electrochemical biosensors - sensor principles and architectures. Sensors **8**, 1400–1458 (2008)
12. Rai, B., Malmberg, R., Srinivasan, V., Ganesh, K.M., Kambhampati, N.S.V., Andar, A., Rao, G., Sanjeevi, C.B., Venkatesan, K., Ramamurthy, S.S.: Surface plasmon-coupled dual emission platform for ultrafast oxygen monitoring after SARS-CoV-2 infection. ACS Sens. **6**, 4360–4368 (2021). https://doi.org/10.1021/acssensors.1c01665
13. Srinivasan, V., Ramamurthy, S.S.: Purcell factor: a tunable metric for plasmon-coupled fluorescence emission enhancements in cermet nanocavities. J. Phys. Chem. C **120**, 2908–2913 (2016). https://doi.org/10.1021/acs.jpcc.5b11311
14. Srinivasan, V., Badiya, P.K., Ramamurthy, S.S.: Low-dimensional carbon spacers in surface plasmon-coupled emission with femtomolar sensitivity and 1000-fold fluorescence enhancements. Chem. Commun. **51**, 7809–7811 (2015). https://doi.org/10.1039/C5CC01040C
15. Boscari, F., Vettoretti, M., Cavallin, F., Amato, A.M.L., Uliana, A., Vallone, V., Avogaro, A., Facchinetti, A., Bruttomesso, D.: Implantable and transcutaneous continuous glucose monitoring system: a randomized cross over trial comparing accuracy, efficacy and acceptance. J. Endocrinol. Invest. **45**, 115–124 (2022). https://doi.org/10.1007/s40618-021-01624-2
16. Joseph, J.I.: Review of the long-term implantable senseonics continuous glucose monitoring system and other continuous glucose monitoring systems. J. Diabetes Sci. Technol. **15**, 167–173 (2020). https://doi.org/10.1177/1932296820911919
17. Ghaffari, R., Rogers, J.A., Ray, T.R.: Recent progress, challenges, and opportunities for wearable biochemical sensors for sweat analysis. Sens. Actuators, B Chem. **332**, 129447 (2021). https://doi.org/10.1016/j.snb.2021.129447
18. Baker, L.B., Model, J.B., Barnes, K.A., Anderson, M.L., Lee, S.P., Lee, K.A., Brown, S.D., Reimel, A.J., Roberts, T.J., Nuccio, R.P., Bonsignore, J.L., Ungaro, C.T., Carter, J.M., Li, W., Seib, M.S., Reeder, J.T., Aranyosi, A.J., Rogers, J.A., Ghaffari, R.: Skin-interfaced microfluidic system with personalized sweating rate and sweat chloride analytics for sports science applications. Sci. Adv. 6, eabe3929 (2020). https://doi.org/10.1126/sciadv.abe3929
19. Koh, A., Kang, D., Xue, Y., Lee, S., Pielak, R.M., Kim, J., Hwang, T., Min, S., Banks, A., Bastien, P., Manco, M.C., Wang, L., Ammann, K.R., Jang, K.-I., Won, P., Han, S., Ghaffari, R., Paik, U., Slepian, M.J., Balooch, G., Huang, Y., Rogers, J.A.: A Soft, Wearable Microfluidic Device for the Capture, Storage, and Colorimetric Sensing of Sweat. Sci Transl Med. 8, 366ra165 (2016). https://doi.org/10.1126/scitranslmed.aaf2593

20. Choi, J., Bandodkar, A.J., Reeder, J.T., Ray, T.R., Turnquist, A., Kim, S.B., Nyberg, N., Hourlier-Fargette, A., Model, J.B., Aranyosi, A.J., Xu, S., Ghaffari, R., Rogers, J.A.: Soft, skin-integrated multifunctional microfluidic systems for accurate colorimetric analysis of sweat biomarkers and temperature. ACS Sens. **4**, 379–388 (2019). https://doi.org/10.1021/acssensors.8b01218
21. Trevaskis, N.L., Kaminskas, L.M., Porter, C.J.H.: From sewer to saviour—targeting the lymphatic system to promote drug exposure and activity. Nat. Rev. Drug Discov. **14**, 781–803 (2015). https://doi.org/10.1038/nrd4608
22. Wiig, H., Swartz, M.A.: Interstitial fluid and lymph formation and transport: physiological regulation and roles in inflammation and cancer. Physiol. Rev. **92**, 1005–1060 (2012). https://doi.org/10.1152/physrev.00037.2011
23. Kabashima, K., Honda, T., Ginhoux, F., Egawa, G.: The immunological anatomy of the skin. Nat. Rev. Immunol. **19**, 19–30 (2019). https://doi.org/10.1038/s41577-018-0084-5
24. Groenendaal, W., von Basum, G., Schmidt, K.A., Hilbers, P.A.J., van Riel, N.A.W.: Quantifying the composition of human skin for glucose sensor development. J. Diabetes Sci. Technol. **4**, 1032–1040 (2010)
25. Friedel, M., Thompson, I.A.P., Kasting, G., Polsky, R., Cunningham, D., Soh, H.T., Heikenfeld, J.: Opportunities and challenges in the diagnostic utility of dermal interstitial fluid. Nat. Biomed. Eng. (2023). https://doi.org/10.1038/s41551-022-00998-9
26. Heikenfeld, J., Jajack, A., Feldman, B., Granger, S.W., Gaitonde, S., Begtrup, G., Katchman, B.A.: Accessing analytes in biofluids for peripheral biochemical monitoring. Nat. Biotechnol. **37**, 407–419 (2019). https://doi.org/10.1038/s41587-019-0040-3
27. Pikal, M.J.: The role of electroosmotic flow in transdermal iontophoresisq. Adv. Drug Delivery Rev. 25 (2001)
28. Giri, T.K., Chakrabarty, S., Ghosh, B.: Transdermal reverse iontophoresis: a novel technique for therapeutic drug monitoring. J. Control. Release **246**, 30–38 (2017). https://doi.org/10.1016/j.jconrel.2016.12.007
29. Bakshi, P., Vora, D., Hemmady, K., Banga, A.K.: Iontophoretic skin delivery systems: success and failures. Int. J. Pharm. **586**, 119584 (2020). https://doi.org/10.1016/j.ijpharm.2020.119584
30. Cheng, Y., Gong, X., Yang, J., Zheng, G., Zheng, Y., Li, Y., Xu, Y., Nie, G., Xie, X., Chen, M., Yi, C., Jiang, L.: A touch-actuated glucose sensor fully integrated with microneedle array and reverse iontophoresis for diabetes monitoring. Biosens. Bioelectron. **203**, 114026 (2022). https://doi.org/10.1016/j.bios.2022.114026
31. Yang, B., Fang, X., Kong, J.: Engineered microneedles for interstitial fluid cell-free DNA capture and sensing using iontophoretic dual-extraction wearable patch. Adv. Func. Mater. **30**, 2000591 (2020). https://doi.org/10.1002/adfm.202000591
32. Kim, J., Sempionatto, J.R., Imani, S., Hartel, M.C., Barfidokht, A., Tang, G., Campbell, A.S., Mercier, P.P., Wang, J.: Simultaneous monitoring of sweat and interstitial fluid using a single wearable biosensor platform. Adv. Sci. **5**, 1800880 (2018). https://doi.org/10.1002/advs.201800880
33. Liu, C., Xu, T., Wang, D., Zhang, X.: The role of sampling in wearable sweat sensors. Talanta **212**, 120801 (2020). https://doi.org/10.1016/j.talanta.2020.120801
34. Le, Z., Yu, J., Quek, Y.J., Bai, B., Li, X., Shou, Y., Myint, B., Xu, C., Tay, A.: Design principles of microneedles for drug delivery and sampling applications. Mater. Today. S1369702122002772 (2022). https://doi.org/10.1016/j.mattod.2022.10.025
35. Iitani, K., Ramamurthy, S.S., Ge, X., Rao, G.: Transdermal sensing: in-situ non-invasive techniques for monitoring of human biochemical status. Curr. Opin. Biotechnol. **71**, 198–205 (2021). https://doi.org/10.1016/j.copbio.2021.08.005
36. Teymourian, H., Tehrani, F., Mahato, K., Wang, J.: Lab under the skin: microneedle based wearable devices. Adv. Healthcare Mater. **10**, 2002255 (2021). https://doi.org/10.1002/adhm.202002255
37. Jina, A., Tierney, M.J., Tamada, J.A., McGill, S., Desai, S., Chua, B., Chang, A., Christiansen, M.: Design, development, and evaluation of a novel microneedle array-based continuous glucose monitor. J. Diabetes Sci. Technol. **8**, 483–487 (2014). https://doi.org/10.1177/1932296814526191

38. Miller, P.R., Skoog, S.A., Edwards, T.L., Lopez, D.M., Wheeler, D.R., Arango, D.C., Xiao, X., Brozik, S.M., Wang, J., Polsky, R., Narayan, R.J.: Multiplexed microneedle-based biosensor array for characterization of metabolic acidosis. Talanta **88**, 739–742 (2012). https://doi.org/10.1016/j.talanta.2011.11.046
39. Freeman, D.M.E., Ming, D.K., Wilson, R., Herzog, P.L., Schulz, C., Felice, A.K.G., Chen, Y.-C., O'Hare, D., Holmes, A.H., Cass, A.E.G.: Continuous measurement of lactate concentration in human subjects through direct electron transfer from enzymes to microneedle electrodes. ACS Sens. **8**, 1639–1647 (2023). https://doi.org/10.1021/acssensors.2c02780
40. Yang, J., Gong, X., Chen, S., Zheng, Y., Peng, L., Liu, B., Chen, Z., Xie, X., Yi, C., Jiang, L.: Development of smartphone-controlled and microneedle-based wearable continuous glucose monitoring system for home-care diabetes management. ACS Sens. **8**, 1241–1251 (2023). https://doi.org/10.1021/acssensors.2c02635
41. Mishra, R.K., Goud, K.Y., Li, Z., Moonla, C., Mohamed, M.A., Tehrani, F., Teymourian, H., Wang, J.: Continuous opioid monitoring along with nerve agents on a wearable microneedle sensor array. J. Am. Chem. Soc. **142**, 5991–5995 (2020). https://doi.org/10.1021/jacs.0c01883
42. Vinu Mohan, A.M., Windmiller, J.R., Mishra, R.K., Wang, J.: Continuous minimally-invasive alcohol monitoring using microneedle sensor arrays. Biosens. Bioelectron. **91**, 574–579 (2017). https://doi.org/10.1016/j.bios.2017.01.016
43. Rawson, T.M., Gowers, S.A.N., Freeman, D.M.E., Wilson, R.C., Sharma, S., Gilchrist, M., MacGowan, A., Lovering, A., Bayliss, M., Kyriakides, M., Georgiou, P., Cass, A.E.G., O'Hare, D., Holmes, A.H.: Microneedle biosensors for real-time, minimally invasive drug monitoring of phenoxymethylpenicillin: a first-in-human evaluation in healthy volunteers. Lancet Digital Health **1**, e335–e343 (2019). https://doi.org/10.1016/S2589-7500(19)30131-1
44. Goud, K.Y., Moonla, C., Mishra, R.K., Yu, C., Narayan, R., Litvan, I., Wang, J.: Wearable electrochemical microneedle sensor for continuous monitoring of levodopa: toward Parkinson management. ACS Sens. **4**, 2196–2204 (2019). https://doi.org/10.1021/acssensors.9b01127
45. Yuan, D., Anthis, A.H.C., Ghahraman Afshar, M., Pankratova, N., Cuartero, M., Crespo, G.A., Bakker, E.: All-solid-state potentiometric sensors with a multiwalled carbon nanotube inner transducing layer for anion detection in environmental samples. Anal. Chem. 87, 8640–8645 (2015). https://doi.org/10.1021/acs.analchem.5b01941
46. Parrilla, M., Cuartero, M., Padrell Sánchez, S., Rajabi, M., Roxhed, N., Niklaus, F., Crespo, G.A.: Wearable all-solid-state potentiometric microneedle patch for intradermal potassium detection. Anal. Chem. **91**, 1578–1586 (2019). https://doi.org/10.1021/acs.analchem.8b04877
47. Molinero-Fernández, Á., Casanova, A., Wang, Q., Cuartero, M., Crespo, G.A..: In vivo transdermal multi-ion monitoring with a potentiometric microneedle-based sensor patch. ACS Sens. **8**, 158–166 (2023). https://doi.org/10.1021/acssensors.2c01907
48. Taylor, N.A., Machado-Moreira, C.A.: Regional variations in transepidermal water loss, eccrine sweat gland density, sweat secretion rates and electrolyte composition in resting and exercising humans. Extrem. Physiol. Med. **2**, 4 (2013). https://doi.org/10.1186/2046-7648-2-4
49. Baker, L.B., Wolfe, A.S.: Physiological mechanisms determining eccrine sweat composition. Eur. J. Appl. Physiol. **120**, 719–752 (2020). https://doi.org/10.1007/s00421-020-04323-7
50. Baker, L.B.: Physiology of sweat gland function: The roles of sweating and sweat composition in human health. Temperature **6**, 211–259 (2019). https://doi.org/10.1080/23328940.2019.1632145
51. Cui, C.-Y., Schlessinger, D.: Eccrine sweat gland development and sweat secretion. Exp. Dermatol. **24**, 644–650 (2015). https://doi.org/10.1111/exd.12773
52. Min, J., Tu, J., Xu, C., Lukas, H., Shin, S., Yang, Y., Solomon, S.A., Mukasa, D., Gao, W.: Skin-interfaced wearable sweat sensors for precision medicine. Chem. Rev. acs.chemrev.2c00823 (2023). https://doi.org/10.1021/acs.chemrev.2c00823
53. Steckl, A.J., Ray, P.: stress biomarkers in biological fluids and their point-of-use detection. ACS Sens. 32 (2018)
54. Upasham, S., Churcher, N.K.M., Rice, P., Prasad, S.: Sweating out the circadian rhythm: a technical review. ACS Sens. **6**, 659–672 (2021). https://doi.org/10.1021/acssensors.0c02622

55. Sonner, Z., Wilder, E., Heikenfeld, J., Kasting, G., Beyette, F., Swaile, D., Sherman, F., Joyce, J., Hagen, J., Kelley-Loughnane, N., Naik, R.: The microfluidics of the eccrine sweat gland, including biomarker partitioning, transport, and biosensing implications. Biomicrofluidics **9**, 031301 (2015). https://doi.org/10.1063/1.4921039
56. Shirreffs, S.M., Maughan, R.J.: Whole body sweat collection in humans: an improved method with preliminary data on electrolyte content. J. Appl. Physiol. **82**, 336–341 (1997). https://doi.org/10.1152/jappl.1997.82.1.336
57. Baker, L.B., Stofan, J.R., Hamilton, A.A., Horswill, C.A.: Comparison of regional patch collection vs. whole body washdown for measuring sweat sodium and potassium loss during exercise. J. Appl. Physiol. 107 (2009)
58. van Heyningen, R., Weiner, J.S.: A comparison of arm-bag sweat and body sweat. J. Physiol. **116**, 395–403 (1952)
59. Ladell, W.S.S.: The measurement of chloride losses in the sweat. J. Physiol. **107**, 465–471 (1948)
60. Heikenfeld, J.: Non-invasive analyte access and sensing through eccrine sweat: challenges and outlook circa 2016. Electroanalysis **28**, 1242–1249 (2016). https://doi.org/10.1002/elan.201600018
61. Naitoh, K., Tsuda, T., Nose, K., Kondo, T., Takasu, A., Hirabayashi, T.: New measurement of hydrogen gas and acetone vapor in gases emanating from human skin. Instrum. Sci. Technol. **30**, 267–280 (2002). https://doi.org/10.1081/CI-120013506
62. Yamane, N., Tsuda, T., Nose, K., Yamamoto, A., Ishiguro, H., Kondo, T.: Relationship between skin acetone and blood β-hydroxybutyrate concentrations in diabetes. Clin. Chim. Acta **365**, 325–329 (2006). https://doi.org/10.1016/j.cca.2005.09.016
63. Lee, B., Lim, M., Misra, V.: Wearable skin vapor sensing system for continuous monitoring of various health and lifestyles. In: 2021 IEEE Sensors. pp. 1–4 (2021)
64. Farkas, RpD, Kemny, L., Szll, M., Dobozy, A., Bata-Csrgo, Z.: Ethanol and acetone stimulate the proliferation of HaCaT keratinocytes. Arch. Dermatol. Res. 295, 56–62 (2003). https://doi.org/10.1007/s00403-003-0399-2
65. Lorwongtragool, P., Sowade, E., Watthanawisuth, N., Baumann, R., Kerdcharoen, T.: A novel wearable electronic nose for healthcare based on flexible printed chemical sensor array. Sensors **14**, 19700–19712 (2014). https://doi.org/10.3390/s141019700
66. Arakawa, T., Sato, T., Iitani, K., Toma, K., Mitsubayashi, K.: Fluorometric biosniffer camera "Sniff-Cam" for direct imaging of gaseous ethanol in breath and transdermal vapor. Anal. Chem. **89**, 4495–4501 (2017). https://doi.org/10.1021/acs.analchem.6b04676
67. Iitani, K., Toma, K., Arakawa, T., Mitsubayashi, K.: Transcutaneous blood VOC imaging system (Skin-Gas Cam) with real-time bio-fluorometric device on rounded skin surface. ACS Sens. **5**, 338–345 (2020). https://doi.org/10.1021/acssensors.9b01658
68. Schmidt, F.M., Vaittinen, O., Metsälä, M., Lehto, M., Forsblom, C., Groop, P.-H., Halonen, L.: Ammonia in breath and emitted from skin. J. Breath Res. **7**, 017109 (2013). https://doi.org/10.1088/1752-7155/7/1/017109
69. Cascales, J.P., Li, X., Roussakis, E., Evans, C.L.: A Patient-ready wearable transcutaneous CO_2 sensor. Biosensors **12**, 333 (2022). https://doi.org/10.3390/bios12050333
70. Chatterjee, M., Ge, X., Kostov, Y., Tolosa, L., Rao, G.: A novel approach towards noninvasive monitoring of transcutaneous CO_2. Med. Eng. Phys. **36**, 136–139 (2014). https://doi.org/10.1016/j.medengphy.2013.07.001
71. Ge, X., Adangwa, P., Lim, J.Y., Kostov, Y., Tolosa, L., Pierson, R., Herr, D., Rao, G.: Development and characterization of a point-of care rate-based transcutaneous respiratory status monitor. Med. Eng. Phys. **56**, 36–41 (2018). https://doi.org/10.1016/j.medengphy.2018.03.009
72. Ahuja, P., Ujjain, S.K., Kukobat, R., Urita, K., Moriguchi, I., Furuse, A., Hattori, Y., Fujimoto, K., Rao, G., Ge, X., Wright, T., Kaneko, K.: Air-permeable redox mediated transcutaneous CO2 sensor. Chem. Eng. J. **457**, 141260 (2023). https://doi.org/10.1016/j.cej.2022.141260
73. National Diabetes Statistics Report | Diabetes | CDC, https://www.cdc.gov/diabetes/data/statistics-report/index.html

74. Block, L., MS, RD, CDE: Continuous Glucose Monitoring: CGMS System Gold, https://www.diabetesselfmanagement.com/diabetes-resources/tools-tech/continuous-glucose-monitoring/
75. Bailey, T.S., Alva, S.: Landscape of continuous glucose monitoring (CGM) and integrated CGM: accuracy considerations. Diab. Technol. Therap. 23, S-5-S-11 (2021). https://doi.org/10.1089/dia.2021.0236
76. Continuous Glucose Monitoring—NIDDK, https://www.niddk.nih.gov/health-information/diabetes/overview/managing-diabetes/continuous-glucose-monitoring
77. Upton, T.J., Zavala, E., Methlie, P., Kämpe, O., Tsagarakis, S., Øksnes, M., Bensing, S., Vassiliadi, D.A., Grytaas, M.A., Botusan, I.R., Ueland, G., Berinder, K., Simunkova, K., Balomenaki, M., Margaritopoulos, D., Henne, N., Crossley, R., Russell, G., Husebye, E.S., Lightman, S.L.: High-resolution daily profiles of tissue adrenal steroids by portable automated collection. Sci. Transl. Med. 15, eadg8464 (2023). https://doi.org/10.1126/scitranslmed.adg8464

Wearable Physical Sensors for Non-invasive Health Monitoring

Cong Thanh Nguyen, Khoa Tuan Nguyen, Toan Dinh, Van Thanh Dau, and Dzung Viet Dao

1 Introduction

Advances in sensing technology have revolutionized the way we approach health care, with a particular focus on non-invasive methods for monitoring and managing our health. Among the remarkable breakthroughs in this field, wearable sensors have emerged as a promising solution, enabling continuous and real-time health monitoring. These sensors use various sensing mechanisms to gather physiological data and provide valuable insights into our well-being.

This chapter presents wearable physical sensors for non-invasive health monitoring, exploring the sensing mechanisms used, the concepts and capabilities of human motion monitoring sensors, the advancements in heart rate and blood pressure sensors, and the respiratory sensors. By understanding the principles behind these sensors and their applications, we can unlock new possibilities for personalized health care, early disease detection, and improved well-being.

The chapter is structured into four sections, each focusing on its key aspects as follows.

Section 2 introduces a wide range of sensing mechanisms or sensing effects which are used in wearable sensors for non-invasive health monitoring.

Section 3 explores the structure and working principle of sensors used for human motion monitoring, with a specific focus on Micro Electromechanical Systems (MEMS) sensors. One of fundamental aspects of health monitoring is understanding human motion and activity. Human motion monitoring sensors, such as accelerometers, gyroscopes, and magnetometers, enable the assessment of physical activity levels, gait analysis, and the detection of abnormal movements. These sensors

C. T. Nguyen · K. T. Nguyen · T. Dinh · V. T. Dau · D. V. Dao (✉)
School of Engineering and Built Environment, Griffith University, Parklands Drive, Southport, QLD 4222, Australia
e-mail: d.dao@griffith.edu.au

K. Mitsubayashi (ed.), *Wearable Biosensing in Medicine and Healthcare*,
https://doi.org/10.1007/978-981-99-8122-9_6

integrated into wearable devices can accurately track and analyze an individual's movements throughout the day.

Section 4 introduces heart rate and blood pressure sensors. The cardiovascular system plays a vital role in our overall health, making heart rate and blood pressure crucial parameters to monitor. Wearable sensors equipped with heart rate and blood pressure monitoring capabilities offer a non-invasive means of tracking these vital signs in real time. This section examines the different sensor technologies used for heart rate and blood pressure monitoring and explores their potential applications in non-invasive health monitoring.

Section 5 discusses the advancements in respiratory sensing technologies and their applications in wearable devices for non-invasive health monitoring. Respiratory function is a key indicator of overall health and plays a pivotal role in diagnosing and managing various conditions. Wearable respiratory sensors have the potential to monitor respiratory patterns and detect abnormalities without the need for invasive procedures. These sensors, such as strain gauges, chest bands, and respiratory sensors, offer a non-invasive approach to respiratory monitoring.

2 Physical Sensing Mechanisms for Health Monitoring

Different functional activities of the human body can be considered as stimuli for transducing into electrical signals, such as heart rate, blood pressure, breath, joint movement, and so on. A large number of wearable sensors have been developed to monitor these stimuli based on different sensing mechanisms, including piezoresistive, thermoresistive, capacitive, piezoelectric, triboelectric, and optoelectronic effects. The working principles of these sensing effects are presented in this section.

2.1 Piezoresistive Effect

The electrical resistance of a resistor is computed as a function of the resistivity (ρ), length (l), width (w), and thickness (t) as below

$$R = \rho \frac{l}{wt} \tag{1}$$

The change in the resistivity and dimensions of the resistor causes the alteration in the electrical resistance as below

$$\frac{\Delta R}{R} = \frac{\Delta \rho}{\rho} + \frac{\Delta l}{l} - \frac{\Delta w}{w} - \frac{\Delta t}{t} = \frac{\Delta \rho}{\rho} + (1 + 2\gamma)\varepsilon \tag{2}$$

Table 1 Longitudinal gauge factor of typical metals and semiconductors

Materials	Longitudinal gauge factors	References
Aluminum	3.1	[1]
Copper	2.9	[1]
Gold	4.48	[1]
p-Si [110]	120	[2]
n-Si [100]	−133	[2]
p-Ge [111]	105	[2]
n-Ge [111]	−155	[2]
n-GaAs [111]	−8.9	[2]

where γ is Poisson's ratio of the material and $\varepsilon = \Delta l / l$ is the strain in the longitudinal direction.

Piezoresistive effect is the change in the electrical resistance under the applied mechanical stress or strain. Gauge factor (GF) is commonly used to quantify the piezoresistive effect, which is defined as below

$$GF = \frac{\Delta R/R}{\varepsilon} = \frac{\Delta\rho/\rho}{\varepsilon} + (1 + 2\gamma) \tag{3}$$

As the resistivity of metals changes slightly under the applied strain, the gauge factor of metals is quite small and mainly depends on the mechanical/geometric properties. In contrast, the charge carrier mobility of semiconductors changes significantly under the applied stress or strain, resulting in a significant change in the resistivity or conductivity. Table 1 shows the GF of typical metals and semiconductors.

The piezoresistive effect in semiconductors has been investigated since the first experiment on the effect of strain on the electrical conductance of germanium and silicon [3], followed by other theoretical studies. So far, the theoretical models of the charge carrier mobility change under the applied stress or strain have been used to explain the piezoresistive effect in semiconductors [4]. The general idea is that the stress-induced modification of energy band diagram of semiconductors causes the repopulation of charge carriers (holes or electrons) in the valence band and conduction band, leading to the change in charge carrier effective mass and mobility, consequently changing the conductivity [5].

2.2 *Thermoresistive Effect*

Thermoresistive effect is the change in electrical resistance with the temperature variation. Therefore, the resistance is calculated as a function of temperature as below

$$R = R_0\left(1 + \alpha\Delta T + \beta\Delta T^2 + \ldots\right) \tag{4}$$

where α is the linear thermoresistive coefficient and β is the quadratic thermoresistive coefficient.

The resistance of metals increases with temperature ($\alpha > 0$), which is so-called positive temperature coefficient of resistance (TCR). Low-doped semiconductors have negative TCR since a large number of generated electrons and holes at elevated temperature reduce the resistivity. Nevertheless, highly doped semiconductors could have a positive TCR while they behave like metals and show small positive TCR at very high doping (metallic degeneration).

Thermistors are created for extremely high TCR. Metal oxides, which are used to make negative TCR thermistors, have considerably greater TCR than metals but also highly nonlinear characteristics. These metal oxides are used to develop small resistive temperature sensors. Some ceramic materials have high positive TCR, so these materials change from low resistivity to high resistivity within a few degrees. These ceramic materials are employed as temperature switches to protect electronic circuits from overcurrent.

2.3 *Capacitive Effect*

The capacitance between two parallel conductive plates is calculated below

$$C = \frac{q}{V} = \epsilon \frac{A}{d} \tag{5}$$

where q, V, A, d are the charge, voltage, overlap area, and distance between the two plates, and ϵ is the permittivity of the material between the two plates, defined as

$$\epsilon = \epsilon_0 \kappa \tag{6}$$

where ϵ_0 is the electric constant and κ the dielectric constant.

The capacitive effect is the change in the capacitance between two conductive plates, caused by the change in the distance (Δd), overlapped width (Δw), overlapped length (Δl), and permittivity ($\Delta \epsilon$) between the plates. To maintain simplicity, both w and l are generally kept constant ($\Delta w = 0$, $\Delta l = 0$). Therefore, the capacitance is computed as below

$$C = (\epsilon + \Delta\epsilon) \frac{wl}{d - \Delta d} \tag{7}$$

where $\Delta\epsilon$ is the change in permittivity of the dielectric material and Δd represents the change in the distance between the parallel plates.

The capacitance relies not only on the physical deformation (Δd) but also the permittivity ($\Delta\epsilon$). Depending on the chosen dielectric material, the capacitance between two plates can be greatly affected by temperature or humidity which alters the permittivity significantly [6–8].

2.4 Piezoelectric Effect

Piezoelectric effect is the creation of mechanical deformation in some non-conductive materials in reaction to the application of an electric field, or the creation of electric charges on the surface of such materials when they experience a mechanical stress.

The piezoelectric effect can also be interpreted as electromechanical interactions between mechanical states and electrical states. Here, piezoelectric constants are defined to represent these linear and proportional relations. With the piezoelectric constants being in the form of third-rank tensors, the electric field and displacement as first-rank tensors, and the stress and strain as second-rank tensors, the piezoelectric relations are formulated as below ($i, j, k = 1, 2, 3$)

$$D_k = d_{kij} T_{ij} \tag{8}$$

$$S_{ij} = d^*_{kij} E_k \tag{9}$$

where D_k and E_k are the electric displacement and electric field, S_{ij} and T_{ij} are the strain and stress components, and d_{kij} and d^*_{kij} are the piezoelectric charge and strain constants.

The indexes of piezoelectric constants are generally represented by the reduced Voigt notation d_{km} with the index k designated for the electric displacement or electric field component in a reference Cartesian coordinate system (x_1, x_2, x_3), and $m(1, \ldots, 6)$ for specifying the stress or strain. Specifically, $m = 1, 2, 3$ stand for the normal stresses along axes x_1, x_2, x_3, while $m = 4, 5, 6$ are for the shear stresses T_{23}, T_{13}, T_{12}, respectively. Both d_{km} and d^*_{km} are the piezoelectric constants, however their units are different.

2.5 Triboelectric Effect

Triboelectric effect is a type of contact electrification happening when certain materials are electrically charged after they contact and then separate from other materials. Depending on different characteristics, i.e., the selected material, surface roughness, exerted stress, the polarities, and strengths of the created electric charges are different. For instance, amber can acquire an electric charge by the process of contact and separation with other materials like wool. Glass and silk, or hard rubber and fur, are other examples of materials obtaining a significant charge when rubbed together.

Triboelectric materials are commonly listed according to the polarity of charge separation when in contact with another material [9]. In this list, a material near the bottom will acquire larger negative charge when it touches another one close to the top, and vice versa. In other words, the greater the charge is transferred, the further away two materials are from each other on the list. On the other hand, materials near each other on the list may not exchange any charge or may exchange the opposite of what is implied by the list. This depends more on other factors than the type of material, such as rubbing, contaminants, and other properties.

2.6 *Optoelectronic Effects*

2.6.1 Photoconductive effect

Photoconductive effect is the change in the conductivity or resistivity of semiconductors under the light illumination. This is explained by the generation of electron–hole pairs when incident photons are absorbed into the materials. Here, the photoconductivity is defined as below

$$\sigma = q(\mu_e n + \mu_h p) \tag{10}$$

where q is the electric charge, μ_e and μ_h are the mobilities of electrons and holes, $n = n_0 + \Delta n$ and $p = p_0 + \Delta p$, where n_0 and p_0 are the concentrations of electrons and holes in the dark condition, and Δn and Δp are the created electrons and holes under the light excitation.

The number of photogenerated charge carriers ($\Delta n = \Delta p$) can be expressed as

$$\Delta n = \eta \frac{\alpha \cdot I \cdot \tau}{h\nu} \tag{11}$$

where $\eta, h\nu, \alpha, I,$ and τ are the quantum efficiency, incident photon energy, absorption coefficient, illumination intensity, and carrier lifetime, respectively.

2.6.2 Photovoltaic effect

Photovoltaic effect is the generation of output voltage across p–n junctions under the light illumination. The phenomenon is explained by the creation of electron–hole pairs in the p–n junctions when illuminated by certain wavelengths. The created electrons then move to the n-side while the generated holes transport oppositely toward the p-side of the p–n junctions. The accumulation of photogenerated holes and electrons in two sides of the p–n junction generates the output voltage [5, 10].

3 Human Motion Monitoring Sensors

3.1 Importance of Human Motion Monitoring

Human motion is a vital physical capacity which is crucial for a large number of daily activities. However, human mobility may be compromised by injury, aging, or illness progression. In healthcare applications, the accuracy of characterizing human motion is vital to diagnose, categorize, and manage a wide range of movements and physiological disorders. Recently, the change in human movement has been considered as a key factor not only in common and chronic diseases, such as heart failure, diabetes, and stroke, but also in some motion disorders, such as Parkinson's disease and Huntington's disease [11]. Therefore, maintaining and rehabilitating human motion have been considered as an essential part of disease management. An essential factor for enhancing motion rehabilitation in health care is to accurately characterize comprehensive human motion. Thus, the function of human motion monitoring sensors is to acquire real-time data, and then process to characterize related movements.

In the past, human motion monitoring mostly relied on image processing, which is known as camera-based human motion capture [12]. In this case, the movement of multiple sensing elements attached to the body is used to determine the body motion. This technique is effective; however, incorporating it into contemporary medical systems, such as portable medical devices, is expensive and obtrusive. Other widely used human mobility management sensors were wrist-worn systems which tracked simple activities and provide data on aspects of body translation, for instance the total step number and travel distance. Nevertheless, detailed analyses of the movements of local body parts, such as arms or legs as well as their velocity, acceleration, or rotational parameters, are necessary to properly characterize and comprehend biomechanics and other complicated movements. Alterations in movements of these components may be associated with disease, atrophy, and injury, while tracking the recovery progress is crucial for effective rehabilitation.

In recent years, substantial advancements in microscale motion sensing technology have greatly aided the development of human balance prostheses, sports medicine, radiotherapy, and biomechanical research [13–15]. Specifically, strong toolkits for body motion detection have been made possibly by the fast development of MEMS sensors with high precision, high reliability, and various functionalities. Over the past two decades, MEMS sensors for motion monitoring have drawn a lot of attention and are still a thriving research field.

3.2 Human Motion Monitoring Sensors and Methods

The most prevalent uses of human motion monitoring sensors are in the field of health care. These applications include gait monitoring and motion capture. The most frequent use is for gait assessment because of its connection to cognitive impairments. Besides, motion capture is used to gather data on human kinematics for motion analysis or to look for potential illnesses.

3.2.1 Gait Monitoring

Gait analysis is currently a critical task in therapeutic and rehabilitation programs because it supplies extensive knowledge about the gait quality, the behavior of the gait pattern, and other dynamic aspects. The results of a gait study might also provide information on a specific gait disease or impairment. Based on that analysis, customized therapies can be suggested. Currently, the most common wearable sensors developed for monitoring human motion consist of inertial sensors, ultrasonic sensors, and laser rangefinder systems.

Inertial sensors are becoming prevalent in studies of human mobility. An inertial measurement unit (IMU) combining accelerometers, gyroscopes, and magnetometers is the most common inertial sensors. Attaching to body parts such as chest, arms, or legs, an IMU is capable of measuring their acceleration, rotation, and gravitational force. Estimates of the gait phases and spatiotemporal characteristics can be made using these parameters. IMUs are extremely resilient sensors since they integrate multiple sources of data, yet they frequently need sophisticated fusion algorithms to produce better estimations.

Accelerometers are the most popular choice when an outpatient gait analysis is needed. They have significant benefits including compact size, high mobility, low cost, and low energy consumption. Accelerometers can be used for both translation and rotational motions either with uniaxial or multi-axial configurations. These sensors work based on Newton's second law of motion and Hooke's law. Different sensing effects have been applied to develop accelerometers, such as piezoresistive effect, capacitive effect, and piezoelectric effect. For example, a piezoresistive accelerometer consists of a proof mass hung on surrounding beams which are fixed at the middle to the frame, Fig. 1(a) [16, 17]. Nanowire p-type silicon piezoresistors are embedded on the sensing beams. In human gait analyses, this piezoresistive accelerometer is attached to a specific position of a human part. Depending on the human motion, the gravitational force deforms the surrounding beams of the accelerometer differently. This results in the resistance change of the piezoresistors, which is converted to the change in output voltage by using the Wheatstone bridge circuit.

Gyroscopes are sensors for measuring the angular velocity, using the Coriolis effect to convert the rotational velocity to a measurable linear motion. This velocity is not affected by the gravitational force. The structure of a symmetric and decoupled gyroscope consisting of a proof mass hung by two sets of springs placed in perpendicular directions is presented in Fig. 1(b) [18]. This creates two mass-spring systems in perpendicular directions, working as linear accelerometers. However, these systems have different functions of driving mode and sensing mode. Upon the rotational movement of the frame around the third axis, the created Coriolis acceleration in

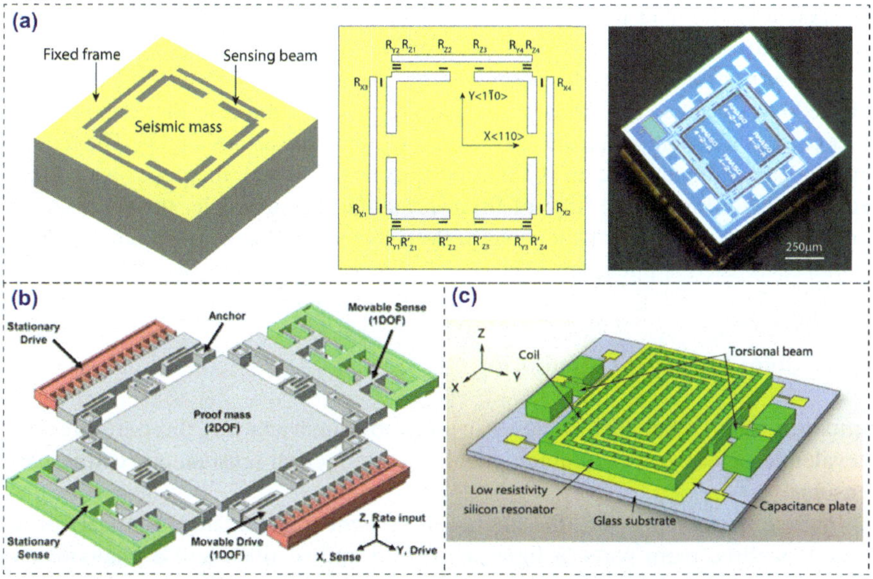

Fig. 1 **a** A three degrees of freedom MEMS piezoresistive accelerometer, **b** Design of a vibratory gyroscope; and **c** Design of a Lorentz-force MEMS magnetometer. Reproduced with permission from [17–19]

the orthogonal orientation of the driving mode provides the input which is measured by the sensing mode accelerometer. In gait analysis applications, gyroscopes can be applied to measure the periodic and repeated rotational movement of different body parts. Current commercial inertial sensors with accelerometer and gyroscope can measure six degrees of freedom of both linear accelerations and rotational speeds.

Magnetometers, which measure changes in magnetic fields, are typically the third sensing component of an IMU. These instruments measure the air magnetic flux density and identify variations in the Earth's magnetic field. For example, the design of a torsional resonant magnetometer is illustrated in Fig. 1(c), working based on the Lorentz-force principle [19]. This magnetometer is fabricated on a glass substrate, consisting of a coil, torsional beams, capacitance plates, and silicon resistors. When the electrical current in the coil is placed in an external magnetic field, the generated Lorentz force causes the displacement of the torsional structure. This displacement is measured by two sensing capacitors fabricated on the torsional structure, then the external magnetic field can be determined. Magnetometers are often employed to improve the accuracy of IMUs by determining the direction toward Earth's magnetic North. Commercial inertial sensors with accelerometer, gyroscope, and magnetometer are considered as nine degrees of freedom IMUs.

Ultrasonic range sensors determine the distance to a target object by converting the reflected sound of ultrasonic waves to a measurable electrical signal. For example, gait parameters are estimated by determining the distance between human feet and the ground [20]. Other parameters that can be measured consist of the step length, stride length, and distance between two steps.

Laser range finders are optical sensors that measure the distance of an object using infrared laser beams. These systems typically comprise a transmitter that sends out laser pulses on a rotational frame to measure distance at various angles. The required duration since the laser beam is sent out reaches the object, then goes back, and is recorded according to the time-of-flight principle. For example, these sensors are positioned in the user's front at about his knee level to track the position of user legs in walker-assisted applications [21].

3.2.2 Joint Movement Monitoring

Among different wearable techniques in joint movement analysis, there are three most prevalent used sensors, including optical sensors, inertial sensors, and textile-based sensors.

Optical sensors measure the alteration in light transmittance. Optical sensors consist of three main parts: a light source, an optical fiber, and a photodetector, Fig. 2(b) [22]. A light beam is generated by the light source, then its intensity after traveling through the optical fiber is measured by the photodetector. To monitor the joint movement, the light source and photodetector are attached on the upper half and lower half of a joint. By measuring the attenuation of beam intensity, the optical sensor can determine how the optical fiber is bent. The immunity to electromagnetic noises is a significant advantage of optical sensors in comparison with other methods.

Inertial sensors including accelerometers, gyroscopes, and magnetometers are normally integrated in an IMU. In this case, the joint alignment using two IMUs must be applied to calculate a joint angle accurately, Fig. 2(a) [22]. First, typical joint movements are based on to define a collection of postures for aligning these IMUs. The calculation of the joint angle is then obtained by fusion algorithms. Later, based on the data achieved by these algorithms, errors are assessed by comparing with those of a reference system. Finally, calibration functions are proposed to compensate

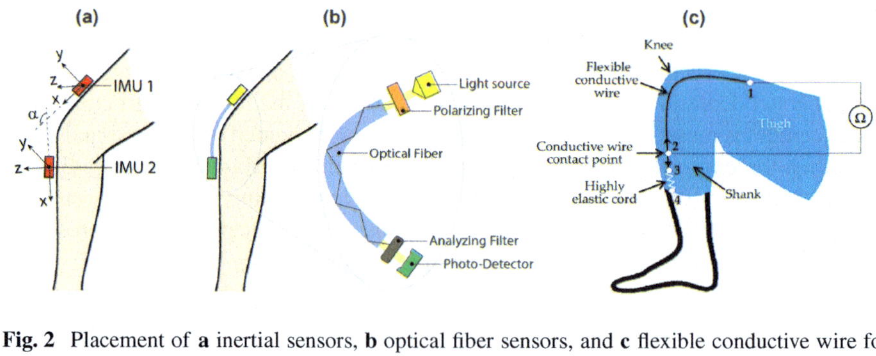

Fig. 2 Placement of **a** inertial sensors, **b** optical fiber sensors, and **c** flexible conductive wire for joint movement monitoring. Reproduced with permission from [22, 23]

the errors. This technique can be used to track complicated movements of different joints, such as elbow, ankle, and knee.

Textile-based sensors for example flexible conductive wire and strain sensors are ideal for making wearable joint monitoring systems. These sensors work based on the monitor of resistance change which corresponds to the joint angle. Textile-based sensors are a great option to create wearable monitoring devices because of their flexibility, simple structure, and easy integration into stretchable skin-tight fabrics. For instance, a flexible conductive wire sensor was integrated into stretchable garments for joint bending measurement, Fig. 2(c) [23]. In this case, a single conductive wire was fabricated in the fabrics around the knee, then the resistance change of this wire was measured to determine the knee bending. Another example is a textile-based wearable strain sensor that employed conductive yarns as a sensing element. A variety of textile-based materials with preferred conformability and elasticity were selected for developing wearable systems.

4 Wearable Heart Pulse and Blood Pressure Sensors

4.1 Importance of Heart Rate and Blood Pressure Monitoring

As the central organ of the circulatory system, the heart pumps blood that carries oxygen and nutrients to all parts of the body. Therefore, measuring heart rate and blood pressure can provide valuable data to predict potential cardiovascular diseases such as atrial fibrillation, coronary heart disease, heart attack, heart failure, and stroke. With rapid technological developments in materials, devices, and integration, wearable heart rate sensors have been becoming mainstream devices for continuous and real-time monitoring of heart conditions. Currently, heart rate monitoring is a must-have function in commercial wearable devices, such as Apple Watch, Fitbit, and smart bands [24, 25]. Most of these commercial devices utilize either optical-based—photoplethysmography (PPG)), or electrical-based—electrocardiography (ECG) transducers. The former technology monitors the heart rate by the reflection of light beams penetrating the blood flow through arteries, whereas the latter measures the body electrical signals to estimate the heart rate. Despite the tremendous commercialization progress of these two technologies, there are considerable obstacles that hinder their accuracy and reliability. For instance, due to employing optical-based methods, PPG is greatly affected by ambient lights, skin tones/colors, relative distance between devices, and the artery; meanwhile, ECG strictly requires multiple electrodes attached to the skin that typically cause discomforts and reduced performance with wearing durations due to sweat and the varying skin contact. To break through the current limits, tremendous efforts in device development and integration have been devoted toward high precision, and miniaturization wearable heart rate monitoring devices. For instance, strain-based MEMS heart rate sensors [26,

27] were introduced that can accurately measure blood pressure and pulse waves without the skin effect. Beyond the widely available information regarding commercially available PPG- and ECG-based devices, this section will focus on the latest research on micro-electromechanical system (MEMS)-based wearable sensors for heart pulse monitoring (HPM) or blood pressure monitoring (BPM).

4.2 Heart Rate and Blood Pressure Sensing Methods

Monitoring blood pressure and heart pulse can provide a range of quantitative evaluations and diagnostics for cardiovascular diseases and health conditions. Compared with inflatable brachial arm cuffs, wearable sensors yield real-time, continuous data for the variation of blood pressure during daily activities or exercise so that more insightful quantification of heart conditions can be gathered [30]. Blood flow is cycled around the body by a periodic pump pressure from the heart. A BPM sensor typically acts as a strain transducer unit placed on a skin site just above a blood artery, Fig. 3(a). When the heart pumps a blood flow, blood vessels expand and lead to the deformation of skin and the sensor. The sensor then translates the mechanical deformation into electrical signals (i.e., resistance, capacitance, voltage, triboelectric, and optical) recorded by electronic circuit and data logging units. This artery pulse waveform can be extracted as non-invasive medical diagnosis information containing the heart rate, arterial blood pressure, and blood vessel stiffness [31, 32]. These data are useful for the prediction of cardiovascular diseases such as atherosclerosis, coronary heart disease, and peripheral arterial disease [29].

There are two stages of a pulse waveform: the ascending and descending stages. The ascending stage corresponds to the expansion of the artery during the ventricular contraction, whereas the descending stage occurs with ventricular diastole, Fig. 3(b). Corresponding to different states, the blood pressure varies as a waveform with three distinct "waves": percussion (P) wave, tidal (T) wave, and diastolic (D) wave, Fig. 3(c). As such, P-wave and T-wave are the early and late stages of the systolic

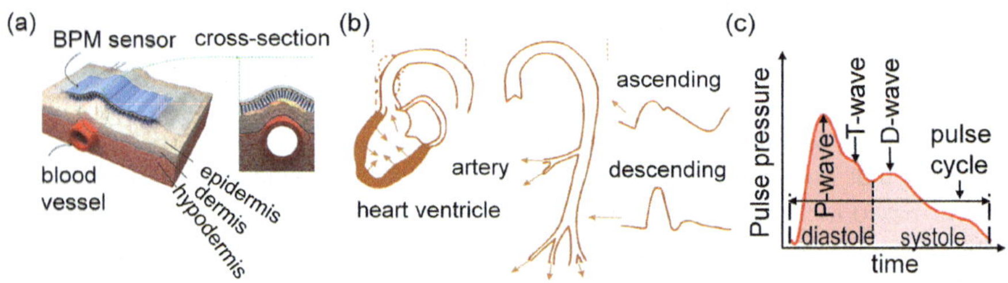

Fig. 3 **a** Working principle of blood pressure sensing: planar sensor placed just above a blood artery to sense its deformation upon blood pump cycle from the heart. **b** Ventricular contraction and diastole during a heart pulse. **c** Diastolic and systolic stages incorporating P-, T-, and D-wave. Reproduced with permission from [28, 29]

peak pressure, respectively; and D-wave refers to the diastolic pulse waveform in the diastole stage [33]. These pulse wave peaks are important parameters to clinically validate cardiovascular conditions. For instance, quantitative indicators can be derived to assess the arterial stiffness corresponding to health status and age, including radial artery index, Air, defined as the amplitude ratio of P-wave and T-wave, and the time delay, ΔT_{DVP}, between the P-wave and T-wave peaks.

4.3 Heart Rate and Blood Pressure Monitoring Sensors

There are various types of sensing systems that are used in wearable heart pulse and blood pressure monitoring as summarized in Fig. 4. Therefore, a system classification can be applied based on the strain or displacement transducing mechanisms to record heart pulses and blood pressure. Wearable BPM devices adopt various physical sensing effects, aforementioned in Section 2 of this chapter, showing difference advantages/disadvantages that are applicable for variety of case scenarios.

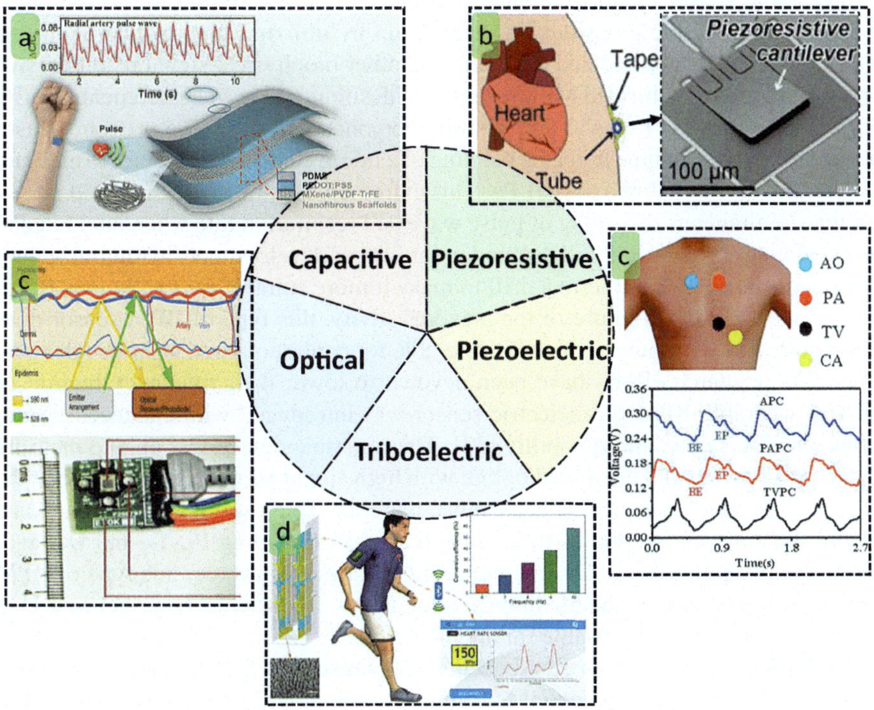

Fig. 4 Wearable heart pulse and blood pressure monitoring sensors with different sensing mechanisms: **a** Capacitive, **b** Piezoresistive, **c**, Piezoelectric, **d** Triboelectric/Nanogenerators, and **e** Optical

Capacitive effect-based BPM sensors are among the most commonly used owing to their low power operation and immunity to thermal noises from capacitive sensing elements. The capacitive readout and configuration can also be applied to a wide range of conductive and semiconductive materials making it widely adoptable in wearable BPM devices. As such, a MEMS capacitive pressure sensor element was used to detect arterial pulse waveform to predict atrial fibrillation [29]. The rated power consumption was reported as low as 5 μW at 3.3 V power supply and 100 kHz readout frequency, indicating the suitability for the long-term and continuous recording. Flexible nanocomposites were also preferable choices for capacitive BPM. A composite nanofibrous scaffold structure was reported with low-pressure detection limit and high stability across a wide pressure range [34]. Due to the soft nature of the composite, it exhibited good mechanical strength and resilience, high dielectric constant, and low hysteresis in radial artery pulse wave measurements.

Piezoresistive effect-based BPM sensors have been widely adopted for pulse wave monitoring due to the simple readout, wide range of material compatibility, and miniaturization. As such, a flexible multiplexed strain-gauge sensor fabricated from Pt-coated polymeric nanofibers was presented that can sense the heartbeat pattern [35]. Due to its highly directional selectivity and low detection range down to 5 Pa, the wearable sensor could accurately measure pressure variation in blood vessels while minimizing the cross-talk of shear strain by utilizing fiber-to-fiber contacts. This effect can also be utilized together with other biophysical signal recordings. A pressure-based miniaturized MEMS sensor with simultaneous measurement of pulse wave and respiration rate was successfully demonstrated [27]. The pressure sensor utilizes resistance change upon the bending of a micro cantilever structure to measure differential pressure down to 0.01 Pa. This extremely low-pressure threshold allows for the simultaneous recording of pulse wave and respiration rate when integrated in an eyeglass. However, it is still demanded for further developments to minimize their power consumption and thermal drift to make it more suitable for continuous BPM. Due to the transducer nature of the piezoresistivity, this type of BPM sensor does not require a power source which is preferable for continuous and wearable devices. Therefore, research efforts have been devoted to lower their frequency regime. A passive wearable BPM piezoelectric sensor was introduced with enhanced wearability, electric safety, and reliability [33]. The wearable device was able to measure radial pulsilogram and apex cardiogram with high signal-to-noise (SNR) ratio at the applicable frequency range of heart pulse patterns. The kind of BPM sensors can record the consecutive phases to identify the peaks including P-, T-, and D-waves of the cardiac pulse wave. Despite not requiring a power source, this type of BPM sensors is typically not capable in measuring at low frequencies (e.g., below 1 Hz), making it not universally usable for all BPM ranges.

Triboelectric nanogenerator-based BPM sensors is a new class of wearable devices that shows great potential for human health monitoring and specially BPM sensors. These devices can convert human motions during daily activities or exercise into electric power that are sufficient for powering sensing devices. This unique feature positions it as a promising candidate for wireless and self-power BPM devices. A triboelectric nanogenerator driven by wireless body sensor network real-time human heart-rate monitoring [36]. The device could convert up to a maximum power of 2.28 mW from inertia energy when walking with a high conversion efficiency at 58%. However, the sensitivity and detection of this BPM type are still limited and require further research and development for new material and device integration.

Optical based BPM is the most widely adopted transducing mechanism in commercial products due to its ease of use. Researchers have been continuously optimizing system design and readout schemes to overcome these limitations. A 590 nm (yellow-orange) wavelength-based optical system was introduced for BPM devices [37]. At this chosen wavelength, there was improvement in the signal quality of PPG from varied skin tones, indicating the effective utilization of longer wavelength as optical emitters in the wrist-based wearable BPM. Due to significant influences by skin variations of the wearer such as colors and thickness, more efforts are still required to minimize the discrepancies in developing optical based BPM systems.

5 Respiratory Sensors

5.1 Importance of Respiratory Monitoring

Respiratory monitoring is a critical task in hospital settings, particularly for patients with respiratory issues, such as chronic obstructive pulmonary disease (COPD), asthma, and sleep apnea. Continuous monitoring of respiration will provide useful information about the respiratory rate, depth, pattern, and overall lung function of patients. Data collected from respiration assists healthcare professionals analyze and deal with respiratory disorders and assess treatment effectiveness, as well as identify potential problems.

Respiratory sensors are also used as an effective tool for early detection of respiratory issues. Their rapid and accurate monitoring of respiratory signals is essential for timely intervention and ensuring the safety of patients. Respiration sensors can detect abnormal breathing depth and rate and their deviations, patterns, or signs of respiratory distress. This early warning sign enables healthcare providers to act quickly and prevent severe problems, especially in anesthesia and in high-risk settings such as intensive care units [38].

In the optimization of sports performance, respiration monitoring will help athletes and coaches understand the impact of breathing activities on performance. Data is used to determine optimal breathing patterns during training and exercises and identify areas for improvement. Respiration sensors provide real-time signals and information about breathing rate and volume, allowing athletes to optimize their breathing duration for enhancing the endurance and maximizing their performance.

5.2 Design of Respiration Sensors

To accurately measure respiration signals, it is crucial to design wearable sensors capable of monitoring respiration with a high signal-to-noise ratio. These sensors operate based on the sensing mechanisms discussed in previous sections. They are capable of detecting various factors, including temperature, humidity, airflow velocity, as well as the movement of the chest and abdomen during breathing. The primary objectives in designing wearable respiration sensors are to ensure wearable functionality with high sensitivity and fast response. This necessitates careful selection of materials and structural designs to determine the sensors' sensitivity.

There are various types of systems that include wearable respiratory sensors available for monitoring respiration. These include the following.

Chest bands are a wearable device that is worn around the chest to monitor respiratory activity, Fig. 5. Respiratory sensors are integrated into the belts to detect the expansion and contraction of the chest during breathing. These sensors, often based on piezoelectric or strain gauge technologies, can measure the changes in the circumference of the chest as the user inhales and exhales. The detected movements are converted into electrical signals that can be analyzed to obtain information about respiration.

Sensor patches are wearable devices that are designed in the form of adhesive patches or sticker sensors that are applied directly to the body to monitor respiration, Fig. 5. These patches contain adhesive sensors and electronic components that can detect and measure various aspects of respiratory activity. These sensors typically work based on piezoelectric or capacitive effects, which are embedded within the patch and are in direct contact with the skin to capture the relevant signals. The special case of the sensor patches is a sensor that is attached to the philtrum and directly measure temperature or pressure, or flow induced by exhalation or inhalation.

In recent years, respiration sensors are integrated into a mask, Fig. 5. The sensors embedded in the mask can detect the airflow during breathing. They utilize various

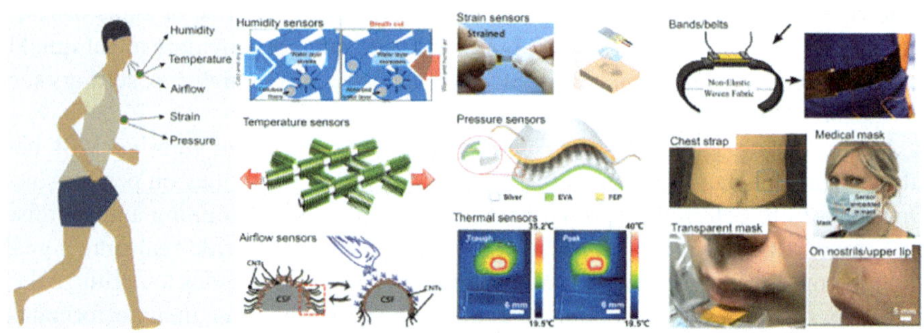

Fig. 5 Wearable respiratory sensors which are fixed in mask, incorporated into belts and patches to detect respiration activities (humidity, temperature, airflow, strain, and pressure). Reproduced with permission from [39]

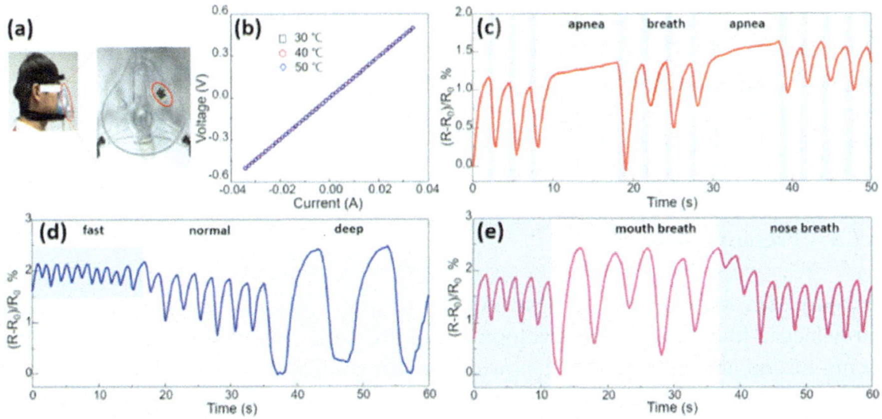

Fig. 6 A graphene respiration sensor that detects humidity changes. **a** The photograph of a graphene sensor attached in a medical breathing mask, **b** the current–voltage (I–V) characteristics of porous graphene network at different temperatures, **c** the resistance changes with breath and apnea, **d** the resistance changes to fast, normal, and deep breathing, **e** the resistance changes to breathing when fixed to mouth and nose. Reproduced with permission from [40]

sensing effect such as triboelectric, allowing the relevant respiratory signals to be captured with high sensitivity.

5.2.1 Humidity Sensors

The sensitivity of humidity-based respiration sensors depends on the selection of suitable materials. These sensors measure changes in humidity caused by breathing and require materials that are highly responsive to moisture variations. Designers consider materials with excellent hygroscopic properties, such as certain polymers or hydrogels, to enhance the sensitivity of the sensors to humidity changes associated with respiration. The typical structure of humidity sensors include a thin film or a porous three-dimensional structure [40]. Fig. 6 shows an example of a porous graphene-based humidity sensor which is incorporated into a medical breathing mask for monitoring of respiration [40] (also see Fig. 5 for medical masks used for humidity measurement). The sensor can detect normal, deep, fast and apnea [40].

5.2.2 Strain Sensors

The strain sensors are typically integrated into a belt or band (Fig. 5). Strain-based respiration sensors rely on detecting the deformation or stretching of the sensor caused by chest and abdominal movements during breathing [41]. Various structural designs are employed to improve the sensitivity of these sensors. This can include incorporating specialized patterns or microstructures that respond to strain variations, allowing for more accurate and sensitive measurement of respiratory movements. Additionally, the use of advanced fabrication techniques, such as microelectromechanical systems (MEMS) or nanomaterials, can enhance the sensitivity of strain-based respiration sensors.

5.2.3 Pressure Sensors

Pressure sensors measure the flow-induced pressure on a sensing structure such as triboelectric fibers [42]. The development of pressure sensors for respiration monitoring has advanced toward self-powered sensing capability. Self-powered generation capability is a desirable feature in wearable respiration sensors. It allows the sensors to generate or harvest energy from the body movements or the surrounding environment, eliminating the need for external power sources or frequent battery replacements. This energy harvesting capability can be achieved through the integration of piezoelectric materials or triboelectric nanogenerators into the sensor design. Fig. 7 shows the recent design of respiration sensors based on the triboelectric effect induced between two fibers [42]. The sensors are typically integrated into a mask which converts the flow-induced pressure into electricity. These energy conversion mechanisms enable the sensors to convert mechanical motion or vibrations into electrical energy, ensuring long-term and sustainable operation.

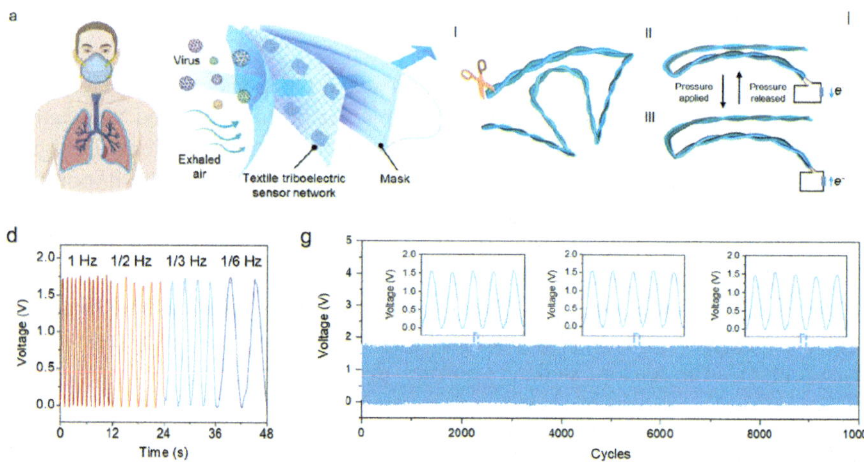

Fig. 7 Respiratory sensors embedded on mask based on triboelectric effect of fibers. **a** Wearable triboelectric sensors embedded on mask, **b** working principle of fiber-based triboelectric sensor, **c** response of sensor under different frequency, and **d** long-term stability of the sensor. Reproduced with permission from [42]

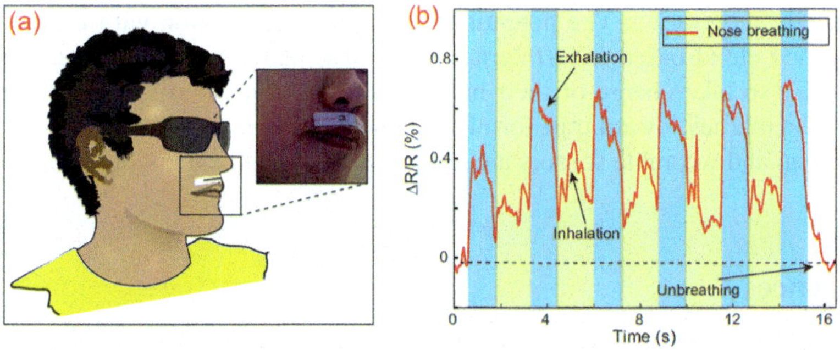

Fig. 8 Carbon nanotube fiber-based thermal sensor for respiration monitoring. **a** Sketch of a wearable thermal flow sensor affixed to human upper lip; **b** The resistance changes corresponding to nose breathing conditions. Reproduced with permission from [43]

5.2.4 Thermal and Flow Sensors

Exhalation leads to an increase in the temperature of the surrounding environments and increase in the airflow. To detect the temperature changes, temperature sensors can be utilized as respiration sensors. In order to achieve high-sensitivity measurement of respiration, hot-wire and hot-film thermal flow sensors can be employed. Hot-wire or hot-film flow sensors are supplied by a large electric current that causes a temperature rise due to the Joule heating effect. The breathing airflow cools the sensing elements (i.e., hot wire or hot film) down, and its resistance change (according to the piezoresistive effect) can be detected. Wearable thermal flow sensors are typically attached to human nostrils to enable the detection of these temperature changes. Fig. 8 shows a wearable thermal sensor using carbon nanotube yarn as a hot wire for measuring the direct airflow [43].

6 Conclusion

This chapter has provided an in-depth exploration of various physical sensing mechanisms and sensor technologies utilized in wearable devices for non-invasive health monitoring, including human motion monitoring, heart rate and blood pressure sensing, and respiratory monitoring. It also highlighted the significance of human motion monitoring, heart rate and blood pressure sensing, and respiratory monitoring in achieving continuous and accurate tracking of human physiological data. These advancements open up new possibilities for personalized health care, early disease detection, and improved overall health and life quality. As research and technology continue to evolve, wearable sensors are expected to play an increasingly vital role in transforming the way we monitor and manage our health. Further advancements in miniaturization, accuracy, and integration with other healthcare systems will enhance the capabilities of these sensors, enabling even more comprehensive and precise monitoring. The potential for seamless integration with mobile applications and cloud-based IoT (Internet of Things) platforms will further empower individuals to take charge of their health, fostering a proactive approach to well-being. The journey toward transforming the way we monitor and manage our health is ongoing, and wearable sensors are at the forefront of this transformative wave.

References

1. Lide, D.R.: CRC Handbook of Chemistry and Physics. CRC press (2004)
2. Fiorillo, A.S., Critello, C.D., Pullano, S.A.: Theory, technology and applications of piezoresistive sensors: A review. Sens.S Actuators A: Phys. 281, 156–175 (2018). doi:https://doi.org/10.1016/j.sna.2018.07.006
3. Smith, C.S.: Piezoresistance effect in germanium and silicon. Phys. Rev. **94**(1), 42 (1954)
4. Bardeen, J., Shockley, W.: Deformation potentials and mobilities in non-polar crystals. Phys. Rev. **80**(1), 72 (1950)
5. Nguyen, C.T., et al.: Vertical piezo-optoelectronic coupling in 3C-SiC/Si heterostructure for self-powered and highly-sensitive mechanical sensing. ACS Appl. Mater. Interfaces (2023). doi:https://doi.org/10.1021/acsami.3c03045
6. Lazarus, N., Bedair, S.S., Lo, C.-C., Fedder, G.K.: CMOS-MEMS capacitive humidity sensor. J. Microelectromech. Syst. **19**(1), 183–191 (2009)
7. Riddle, B., Baker-Jarvis, J., Krupka, J.: Complex permittivity measurements of common plastics over variable temperatures. IEEE Trans. Microw. Theory Tech. **51**(3), 727–733 (2003)
8. Fastier-Wooller, J.W., et al.: Multimodal fibrous static and dynamic tactile sensor. ACS Appl. Mater. Interfaces (2022). doi:https://doi.org/10.1021/acsami.2c08195
9. Drobny, J.G.: Polymers for Electricity and Electronics: Materials, Properties, and Applications. John Wiley & Sons, (2012)
10. Md Foisal, A.R., et al.: Self-powered broadband (UV-NIR) photodetector based on 3C-SiC/Si heterojunction. IEEE Trans. Electron Devices 66(4), 1804–1809 (2019). doi:https://doi.org/10.1109/ted.2019.2899742
11. Wada, O., Nagai, K., Hiyama, Y., Nitta, S., Maruno, H., Mizuno, K.: Diabetes is a risk factor for restricted range of motion and poor clinical outcome after total knee arthroplasty. J. Arthroplasty **31**(9), 1933–1937 (2016)
12. Moeslund, T.B., Hilton, A., Krüger, V.: A survey of advances in vision-based human motion capture and analysis. Comput. Vis. Image Underst. **104**(2–3), 90–126 (2006)
13. Roriz, P., Ribeiro, A.B.L.: Fiber Optical Sensors in Biomechanics. In: Opto-Mechanical Fiber Optic Sensors, pp. 263–300. Elsevier (2018)
14. Kassanos, P., Rosa, B.G., Keshavarz, M., Yang, G.-Z.: From wearables to implantables-clinical drive and technical challenges." In: Wearable Sensors, pp. 29–84. Elsevier (2021)
15. Pizzolato, C., et al.: Bioinspired technologies to connect musculoskeletal mechanobiology to the person for training and rehabilitation. Front. Comput. Neurosci. **11**, 96 (2017)
16. Dao, D.V., Toriyama, T., Sugiyama, S.: Noise and frequency analyses of a miniaturized 3-DOF accelerometer utilizing silicon nanowire piezoresistors. SENSORS (2004)
17. Dao, D.V., Okada, S., Dau, V., Toriyama, T., Sugiyama, S.: Development of a 3-DOF silicon piezoresistive micro accelerometer. In: Micro-Nanomechatronics and Human Science and The Fourth Symposium Micro-Nanomechatronics for Information-Based Society (2004)
18. Alper, S.E., Azgin, K., Akin, T.: A high-performance silicon-on-insulator MEMS gyroscope operating at atmospheric pressure. Sens. Actuators A: Phys. 135(1), 34–42 (2007). doi:https://doi.org/10.1016/j.sna.2006.06.043
19. Wu, L., Tian, Z., Ren, D., You, Z.: A Miniature Resonant and Torsional Magnetometer Based on Lorentz Force. Micromachines (Basel) 9(2), (2018). doi:https://doi.org/10.3390/mi9120666

20. Muro-De-La-Herran, A., Garcia-Zapirain, B., Mendez-Zorrilla, A.: Gait analysis methods: An overview of wearable and non-wearable systems, highlighting clinical applications. Sensors 14(2), 3362–3394 (2014)
21. Aguirre, A., Sierra M, S.D., Munera, M., Cifuentes, C.A.: Online system for gait parameters estimation using a LRF sensor for assistive devices. IEEE Sensors J. 21(13), 14272–14280 (2021). doi:https://doi.org/10.1109/jsen.2020.3028279
22. Homayounfar, S.Z., Andrew, T.L., Wearable sensors for monitoring human motion: A review on mechanisms, materials, and challenges. SLAS Technol. 25(1), 9–24 (2020). doi:https://doi.org/10.1177/2472630319891128
23. Faisal, A.I., Majumder, S., Mondal, T., Cowan, D., Naseh, S., Deen, M.J.: Monitoring methods of human body joints: State-of-the-art and research challenges. Sensors (Basel) 19(11) (2019). doi:https://doi.org/10.3390/s19112629
24. Jeong, H., Rogers, J.A., Xu, S.: Continuous on-body sensing for the COVID-19 pandemic: Gaps and opportunities. Sci. Adv. 6(36) (2020)
25. Thomson, E.A., et al.: Heart rate measures from the Apple Watch, Fitbit Charge HR 2, and electrocardiogram across different exercise intensities. J. Sports Sci. **37**(12), 1411–1419 (2019)
26. Mizuki, Y., Nguyen, T.-V., Takahata, T., Shimoyama, I.: Highly sensitive pulse wave sensor with a piezoresistive cantilever inside an air chamber. In IEEE 32nd International Conference on Micro Electro Mechanical Systems (MEMS), 2019
27. Nguyen, T.-V., Ichiki, M.: MEMS-based sensor for simultaneous measurement of pulse wave and respiration rate. Sensors **19**(22), 4942 (2019)
28. Meng, K., et al.: Wearable pressure sensors for pulse wave monitoring. Adv. Mater. **34**(21), 2109357 (2022)
29. Kaisti, M., et al.: Clinical assessment of a non-invasive wearable MEMS pressure sensor array for monitoring of arterial pulse waveform, heart rate and detection of atrial fibrillation. NPJ Digit. Med. **2**(1), 39 (2019)
30. Lin, Q., et al.: Highly sensitive flexible iontronic pressure sensor for fingertip pulse monitoring. Adv. Healthcare Mater. 9(17), (2020). doi:https://doi.org/10.1002/adhm.202001023
31. Wang, X., Liu, Z., Zhang, T.: Flexible sensing electronics for wearable/attachable health monitoring. Small **13**(25), 1602790 (2017)
32. Lin, Q., et al.: Highly sensitive flexible iontronic pressure sensor for fingertip pulse monitoring. Adv. Healthcare Mater. **9**(17), 2001023 (2020)
33. Cai, F., et al.: Ultrasensitive, passive and wearable sensors for monitoring human muscle motion and physiological signals. Biosens. Bioelectron. **77**, 907–913 (2016)
34. Sharma, S., Chhetry, A., Sharifuzzaman, M., Yoon, H., Park, J.Y.: Wearable capacitive pressure sensor based on MXene composite nanofibrous scaffolds for reliable human physiological signal acquisition. ACS Appl. Mater. Interfaces **12**(19), 22212–22224 (2020)
35. Pang, C., et al.: A flexible and highly sensitive strain-gauge sensor using reversible interlocking of nanofibres. Nat. Mater. **11**(9), 795–801 (2012)
36. Lin, Z., et al.: Triboelectric nanogenerator enabled body sensor network for self-powered human heart-rate monitoring. ACS Nano 11(9), 8830–8837 (2017). doi:https://doi.org/10.1021/acsnano.7b02975
37. Mohapatra, P., Preejith, S., Sivaprakasam, M.: A novel sensor for wrist based optical heart rate monitor. In: IEEE International Instrumentation and Measurement Technology Conference (I2MTC) (2017)
38. Mirjalali, S., Peng, S., Fang, Z., Wang, C.H., Wu, S.: Wearable sensors for remote health monitoring: Potential applications for early diagnosis of covid-19. Adv. Mater. Technol. 7(1), 2100545 (2022). doi:https://doi.org/10.1002/admt.202100545
39. Dinh, T., Nguyen, T., Phan, H.-P., Nguyen, N.-T., Dao, D.V., Bell, J.: Stretchable respiration sensors: Advanced designs and multifunctional platforms for wearable physiological monitoring. Biosens. Bioelectron. 112460 (2020)
40. Pang, Y., et al.: Wearable humidity sensor based on porous graphene network for respiration monitoring. Biosens. Bioelectron. **116**, 123–129 (2018)

41. Ho, M.D., et al.: Percolating network of ultrathin gold nanowires and silver nanowires toward "invisible" wearable sensors for detecting emotional expression and apexcardiogram. Adv. Func. Mater. **27**(25), 1700845 (2017)
42. Fang, Y., et al.: A deep-learning-assisted on-mask sensor network for adaptive respiratory monitoring. Adv. Mater. **34**(24), 2200252 (2022)
43. Dinh, T., et al.: Environment-friendly carbon nanotube based flexible electronics for noninvasive and wearable healthcare. J. Mater. Chem. C **4**(42), 10061–10068 (2016)

Novel Approaches for Wearable Biosensing

Recent Progress in Wearable Microneedle Sensor Devices for Continuous Screening of Interstitial Fluid: A Journey Toward Lab Under the Skin

Lakshmi R. Panicker, M. R. Keerthanaa, and Kotagiri Yugender Goud

1 Introduction

Wearable sensors have attained enormous interest recently due to their plethora of applications. Today's healthcare system focusing on proactive care, prevention, and early diagnostics of diseases requires decentralized, portable, miniaturized, user-friendly, sensitive, and reliable sensors which cannot be achieved by traditional healthcare systems that rely on blood and urine [1]. Blood being the common source of biomarkers has limitations with accessibility, while biofluids like saliva and urine have issues with concentration and biomarker numbers. On the other hand, ISF is a unique source of biomarkers, also carrying almost all the analytes found in the blood [2]. The sensors can be integrated into watches, rings, bracelets, contact lenses, textiles, mouthguards, etc., depending upon their applications. Wearable sensors can be either in non-invasive or minimally invasive fashion [3, 4]. Among the various wearable platforms, microneedle (MNs) sensors have attained greater interest recently since it has greater advantages such as being minimally invasive in nature, able to access the biofluid (ISF) continuously, and, moreover, the ISF is a structurally less complex biofluid than the others. ISF was previously accessed with blister approaches, planted wicks, and micropipette insertion, but in recent days this ISF extraction is mostly done with the MNs [5].

Microneedles can potentially be used for continuous monitoring and detection of different targets such as therapeutic drugs, alcohol, lactate, physiological ions, amino acids, urea, and most significantly in the monitoring of glucose. In the near future, these microneedle sensors will be able to monitor several types of analytes; that's the reason we are calling these MN sensing technologies "Lab Under the Skin".

L. R. Panicker · M. R. Keerthanaa · K. Y. Goud (✉)
Department of Chemistry, Indian Institute of Technology Palakkad, Palakkad, Kerala 678 557, India
e-mail: yugenderkotagiri@iitpkd.ac.in

K. Mitsubayashi (ed.), *Wearable Biosensing in Medicine and Healthcare*,
https://doi.org/10.1007/978-981-99-8122-9_7

Analytes in ISF can be either detected directly by penetration of MNs through the skin or ISF can be extracted with MNs and further detection can be carried out in-vitro, considering geometry (length, diameter, size, and shape) and biocompatibility of the material used in MN making. The fabrication of MNs has been done in various ways such as hollow, solid, and hydrogel-based or porous forms. Fabrication of hollow MNs is complicated compared to solid and porous. Solid types of MNs are generally used in sensors with integrated electronics, owing to their advantages like ease of fabrications, and sharp tips that are helpful in case of drug delivery solid MNs [6]. Abnormalities like inflammations and burn injuries can reduce the ISF flow and, thereby, can be detected by a drop in ISF concentration. In the real-time applications of biomarker monitoring, MNs themselves may not provide satisfying results; hence, it has to be surface-modified with bio/chemical recognition element coatings which can be used based on the analyte [7].

This chapter has been constructed with discussions including the importance of ISF biofluid, fabrication protocols of microneedles, the delivery performance of microneedles, and sensing application of microneedles toward the continuous monitoring of physiological ions, biomarkers, and therapeutic drugs. Finally, we discussed the future prospects of microneedle devices in sensing and deliver.

2 Importance of Interstitial Fluid (ISF) in Sensing

Interstitial fluid formed by trans capillary filtration is considered to be an alternative to blood, as a source of biological information for continuous monitoring of various biomarkers and therapeutic drugs. ISF is rich in biological components that are found in blood, 79% similar components, while ISF has only smaller protein molecules and misses out larger protein structures present in plasma and some biomarkers that are unique to ISF and not available in serum [8–10]. In terms of volume, ISF present in the human body is at least three times more than blood [11]. The analytes present in ISF come from blood through continuous capillaries [12]. It also proves to be more advantageous than blood in the aspect of coagulation-free characteristics [8]. Initially, implanted probes were used for ISF-based sensors to investigate limited biomarkers. However, in order to make wearable sensors less painful, minimally invasive microneedle-based sensors were developed.

Microneedles can be defined as an array of microscale cones or pyramid-shaped protuberances. Microneedles are less than 1 mm in length and width and can penetrate the epidermis of the skin and reach the dermis layer (rupture of stratum corneum takes places), thereby preventing blood capillaries or nerve endings from damage [5, 13, 14]. Microneedle sensors are now used to simultaneously monitor physiological ions such as sodium (Na^+), potassium (K^+), and calcium (Ca^{2+}) in plants and animals. This electrolyte balance can provide personalized health care and help in smart agriculture [9]. Similarly in the recent study by Xinshuo and coworkers, the fluctuations in these ions were monitored using 3D assembled microneedle arrays [10].

Ex vivo studies with chicken and porcine skin, K^+ was detected with microneedle-based sensor, where a robust potassium sensor was developed with stainless-steel microneedle arrays covered with different coatings and reference electrode fixed with a flexible polymer substrate [15]. Lithium (Li^+) levels in ISF can also be monitored with spectroscopic methods after ISF is extracted from the porcine skin samples using microneedle as a tool for extraction; this is a kind of therapeutic drug monitoring since patients with bipolar disorder (chronic in nature) are first prescribed with Li-based drugs [16]. Chronic diseases like Parkinson's disease, Alzheimer's, etc. need continuous monitoring of therapeutic drugs in the patients. ISF-based microneedle sensors can be employed in such applications. Researchers have developed sensors for the continuous monitoring of apomorphine and Levodopa using ISF as a biofluid. The advancement in these sensors is believed to guide dose regulation and personalized drug dosing [17, 18].

Among various biomarkers present in the ISF, glucose detection and monitoring have been most attractive due to the prevalence of diabetes all over the world. From a medical perspective, biofluids such as saliva, tear, and sweat are interesting but may not be very much suitable for glucose monitoring; here, ISF comes into play offering a good correlation with blood glucose levels. Recently, continuous monitoring of glucose has been achieved through the self-powered wearable sensor where a colorimetric glucose paper integrated microneedle patch [19]. Other biomarkers like lactate, alcohol, etc., which have significant roles in the body, are detected with microneedle sensors in ISF.

3 Fabrication of Microneedles

The performance of the microneedle-based sensor is highly dependent on the geometry, shape, and size of the microneedles used. So, fabrication is an important step in developing a microneedle sensor. Microneedle fabrications consist of two major steps: one is the development of three-dimensional (3D) computer-aided designs and the other one is the printing of the microneedle electrodes. Usually, these 3D model designs are prepared with software like CAD, Fusion 360, SOLIDWORKS, etc. Generally, microneedles are fabricated using 3D printing technology, and this technique offers rapid designing of customizable microneedles with the desired shape and size [20]. Lithography is also one of the important fabrication strategies in the preparation of microneedles. Based on the structure, microneedles are broadly classified into two types—hollow and solid. Figure 1 displays the tree diagram of hollow and solid microneedles sensors including their making materials.

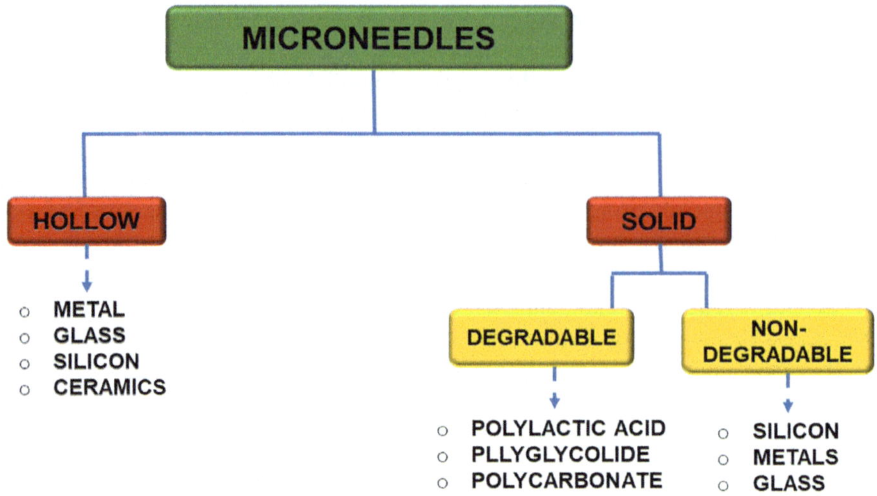

Fig. 1 Classification of microneedles

3.1 Hollow Microneedles

Hollow microneedles have an empty cavity inside each needle and a bore on the needle tip. In early times, hollow MN was created in 2D arrays with in-plane fabrication. These 2D arrays faced many limitations like the limited number of needles in an array. Hence, 3D hollow MN arrays replaced them. The general methods of fabricating 3D hollow MN are lithography, laser cutting and ablation, metal electroplating, isotropic and anisotropic etching, and micro-molding, and the main substrates for hollow MN fabrication include ceramics, metal, silicon, and glass [21]. Silicon-based hollow MN is created using plasma etching and photolithography methods. In plasma-based etching, also called reactive ion etching (RIE) [22], a gas is excited into a reactive state, facilitating reactions between the gas and the substrate. The silicon substrate can be either monocrystalline or polycrystalline. Hollow MN can be developed using 3D laser cutting with metal substrates like stainless steel, palladium, titanium, etc. *Vinay Kumar* et al. fabricated a hollow stainless-steel MN using electron discharge machining (EDM) followed by the laser micromachining technique [23]. The fabricated MN array showed high mechanical strength and could obtain good fluid flow characteristics. They have got desired geometry and shape using computer-aided design (CAD) software. *Davis P.* et al. created a nickel hollow microneedle array by a modified LIGA process using polyethylene terephthalate as molds [24]. The microneedle array displayed proper insertion into the skin without breaking and also showed excellent insulin delivery performance.

Hollow MNs can be used to monitor various biomarkers from the ISF by piercing the skin's stratum corneum. This gives a better alternative for blood-based diagnosis. *Ribet* et al. focused on real-time glucose monitoring by placing an electrochemical sensing probe in the lumen of a single hollow silicone MN, which relies on capillary filling action rather than conventional invasive blood extraction [25]. The smaller size of the device ensures that the MN is in direct contact with the ISF, and it is flexible to wear on various body parts. *Nicholas* et al. [26] reported a device for the detection of glucose in ISF using an enzyme-based colorimetric glucose sensor attached to the backplate of polymeric hollow MN. *Li* et al. [27] developed a hollow MN sensor for detecting both glucose and cholesterol by coupling the MN with a paper-based sensor.

The application of hollow MN can be extended to drug delivery as well since the hollow needle can act as a carrier for the specific drug. *Ryan* et al. [28] developed a hollow MN array from polycarbonate via the micro-molding technique for the transdermal delivery of *Escherichia coli*-specific bacteriophages for the potential treatment of infectious diseases. The study confirmed that bacteriophages can be successfully transferred to the blood via hollow MN. *Davis* et al. [29] explored the usage of hollow MN for insulin delivery to diabetes-induced rats. Due to their smaller sizes, hollow MN can be used for drug delivery in the case of ocular diseases. *Morales* et al. [30, 31] used hollow MN to deliver sulprostone, a prostaglandin antagonist, to assess the decrease of intraocular pressure (IOP). Hollow MN delivery enabled significant dose sparing to that of topical administration. *Jiang* et al. investigated the intrascleral delivery of solutions containing soluble molecules, microparticles, and nanoparticles to overcome conventional invasive procedures.

Researchers have deeply studied the applicability of hollow MN for vaccine delivery also. *Van der Maden* et al. [32] developed a silica-based hollow MN produced by HF etching for the intradermal delivery of inactivated Poliovirus vaccine (IPV) in rats. The group compared the immune response of the MN-delivered vaccine and the conventional hypodermic needle-delivered vaccine. The studies got similar results for the above two methods. *Du* et al. [33] investigated the delivery of diphtheria toxoid (DT) encapsulated in silica nanoparticles in mice using hollow MNs. *Ogai* et al. [34] studied intradermal (ID) and subcutaneous (SC) delivery of vaccines using a hollow MN array and 27-G stainless needles, respectively, for the administration of mumps and varicella vaccines in rats. *Pamornpathomkul* et al. [35] explored the use of hollow MN for the administration of plasmid DNA-encoding ovalbumin and found that there was an increased IgG antibody production upon intradermal vaccine delivery.

3.2 *Solid Microneedles*

Solid microneedles are considered as second-generation microneedles, developed after the hollow ones. Like the hollow MNs, solid MNs can also be used for sensing purposes as well as for drug delivery. Solid microneedles have no passage [36, 37]; hence, they are used as a pre-treatment to create channels in the epidermis for collecting various biomarkers for the detection or for delivering a desired drug. In the case of drug delivery, small patches [37] containing the drug are introduced to

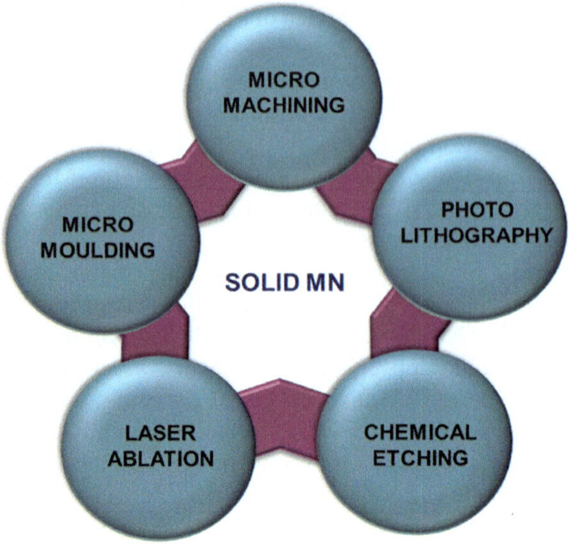

Fig. 2 Different fabrication methods for solid microneedles [38]

the skin surface where channels are created. As in the case of hollow MNs, substrate selection is an important process while developing a solid MN too. Degradable and non-degradable substances are used to fabricate solid MNs. Non-degradable substrates consist of silicon, metals, ceramics, etc. Out of these, metals are considered the best. They have adequate mechanical strength to pierce the skin to create channels. Biodegradable materials have the added advantage of biocompatibility over non-biodegradable substrates. Many polymers like polyglycolide, poly(l-lactic acid), polydimethylsiloxane (PDMS), polycarbonate, poly(methyl methacrylate), etc. are used for fabricating solid MNs. The most popular fabrication methods of solid MNs are Surface micromachining, Micro-molding, Photolithography, Chemical etching, Laser ablation, etc. Figure 2 illustrates the schematic representation of solid microneedle fabrication methods.

Solid MN arrays are mainly used for drug delivery purposes. *Wang* et al. [39] developed a SU-8 polymer-based solid MN by lithography. He combined the fabricated MN with CNT nano filters for delivering three different biomolecules: glucose, insulin, and hemagglutinin. The transdermal delivery of small hydrophilic molecules using solid MN was studied by Li et al. [40]. They have developed a super-short silicon solid MN using a wet etching technique.

4 Application of Microneedles as Sensors

Various clinically significant analytes can be continuously monitored with microneedle sensors [41]. Currently, for therapeutic drug monitoring, we rely on blood-based measurements. This is invasive (painful in nature), and there is a limit to the temporal resolution of results obtained. The microneedle patch comprises needles that enable access to the interstitial fluid (ISF) with minimal invasiveness through a direct interface with the skin. The sensor platform has the option of being directly integrated onto the needles or externally fixed [42]. Other laboratory-based methods are labor-intensive and need to be substituted with a more user-friendly and faster approach. Microneedle's minimally invasive nature is less painful and efficient in continuous monitoring. Microneedle sensor's painless monitoring and simple design enable various applications such as in situ analysis of therapeutic drug monitoring [43], continuous monitoring of biomarkers like glucose, lactate, proteins, etc., biologically significant ions (sodium, potassium, and calcium) [44], and drug/vaccine delivery in small concentrations [45].

4.1 Physiological Ion Monitoring

Maintaining the concentrations of different ions in the body is essential for perfect health management. Human health relies on the concentration of various ions like Ca^{2+}, K^+, Na^+, etc. Any fluctuations in the ion concentrations can be considered as an indicator of illnesses. Diseases like hypernatremia and hypokalemia are related to concentrations of sodium and calcium in the body. Hence, monitoring the concentration of ions is significant. ISF carrying these ions in close concentration to blood can be used as a biofluid for continuous monitoring. Microneedles are minimally invasive and can extract the ISF or monitor the ions in situ [9].

Miller et al. coined the initial efforts of ion sensing using a microneedle platform. He developed a hollow microneedle with a microfluidic chip to extract fluid through a channel toward a downstream solid-state ion selective electrode (ISE). 3D porous carbon and 3D porous graphene were selected as the ISE for K^+ sensing. The developed sensor showed capable sensing of potassium ions within the physiological pH in the presence of interfering molecules, with greater stability. Parilla et al. [15] carried out the potassium ion sensing from interstitial fluid (ISF) using an all-solid potentiometric sensor. They found that the potassium responses obtained using the sensor were resilient even after many insertions through the animal skin. Ex vivo tests based on the intradermal detection of potassium in chicken and porcine skin demonstrate that the microneedle patch is suitable for monitoring potassium changes inside the skin. Zhu et al. [9] designed a flexible polymeric microneedle patch coupled electrode array (MNP-EA) for the in situ multiplexed detection of

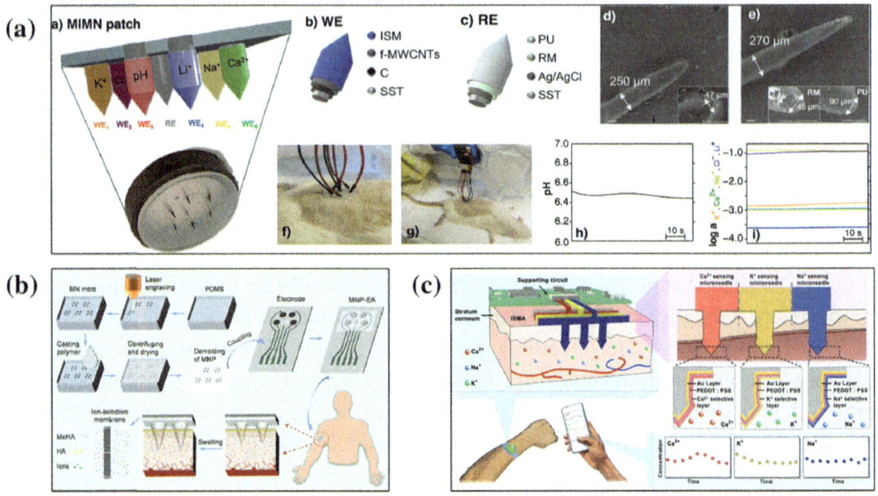

Fig. 3 Different microneedle sensors for physiological ion monitoring. **a** Microneedle sensor for monitoring Na^+, K^+, Ca^{2+}, and H^+, Cl^-, and Li^+ [46]. **b** Microneedle sensor for Na^+, K^+, Ca^{2+}, and H^+ monitoring [9]. **c** 3D microneedle sensor for various physiological ion monitoring [47]

ion species (Na^+, K^+, Ca^{2+}, and H^+) in tissue interstitial fluid (ISF). The potentiometric sensors were highly efficient in the simultaneous measurement of the above-described ions within the clinical range of determination. The response ranges were (Na^+: 0.75 – 200 mM; K^+: 1 – 128 mM; Ca^{2+}: 0.25 – 4.25 mM; pH: 5.5 – 8.5). Fernandez et al. [46] employed the applicability of membrane-based microneedles for real-time rapid detection of multi-ions simultaneously. The group have developed a potentiometric sensor consisting of a wearable microneedle patch with seven electrodes (6 working electrodes and 1 reference electrode). The 6 working electrodes are specific to each of the ions, i.e., pH, Na^+, K^+, Ca^{2+}, Li^+, and Cl^-. In vitro studies revealed the reversibility, reproducibility, and selectivity of the sensor. Huang et al. [10] fabricated a planar microneedle sheet containing 3D microneedles with multiple electrodes. The sensor was sufficiently able to detect the changes in ions like Ca^{2+}, Na^+, and K^+. By combining the sensor with an integrated circuit, the group developed a wearable sensor for real-time multiple ion monitoring. Figure 3 illustrates the various microneedle sensor applications toward the monitoring of several physiological ions.

4.2 Biomarkers Monitoring

Wearable devices, especially microneedle sensors, have been extended for biomarker sensing. Various biomarkers have been extensively detected using microneedle sensors. Biomarkers found in blood play a significant role in bodily activities.

Abnormalities in their concentrations are found to be responsible for numerous diseases. Monitoring of the biomarkers like glucose, protein, lactate, alcohol, etc. is essential in order to track human health. ISF which is a promising biofluid that is recently explored enormously, as a substitute for blood, can be easily accessed using microneedles [19]. MNs are used to collect ISF through their hollow needles or can be integrated with an electronics part and punctured through the skin and detect the biomarkers underneath (wearable sensors). Glucose, a biomarker responsible for diabetes, continuous monitoring can improve diabetes management [48].

Ciui et al. [49] developed a minimally invasive microneedle sensor capable of detecting tyrosinase enzyme cancer biomarkers in the presence of its catechol substrate, immobilized on the transducer surface. In the presence of the surface TYR biomarker, the immobilized catechol is rapidly converted to benzoquinone, which is detected amperometrically, with a current signal proportional to the TYR level. The hollow microneedle device is filled with catechol-coated carbon paste for assessing TYR levels in melanoma tissues. Dervisevic et al. [50] fabricated a gold-coated silicon microneedle for sensing epidermal growth factor receptor 2 (ErbB2), a key breast cancer biomarker. The developed microneedle acts as the biomarker extraction platform and the electrochemical transducer. The sensor was able to detect the desired biomarker from artificial ISF even in a low range than present in breast cancer cells. Wang et al. [51] formulated a microneedle patch for the ultrasensitive quantification of protein biomarkers in interstitial fluid. A highly efficient fluorescent probe was incorporated with the sensor for enhancing the limit of detection. The group under J. Wang fabricated an integrated wearable microneedle array for the continuous monitoring of multiple biomarkers in interstitial fluid [20]. The sensor was designed for measuring glucose, lactate, and alcohol by enzymatic reduction. The corresponding microneedle array was coupled with a CWS board to make a wearable device that was connected to a newly developed customized application via Bluetooth. Molecularly imprinted polymers (MIP) are becoming a commonly used analytical sensing platform due to their increased selectivity. Mugo et al. [52] recently reported MIP-based electrochemical microneedle sensor for multiplexed metabolite detection in human sweat. They have carried out electrochemical detection of biomarkers including pH, epinephrine, lactate, and dopamine from sweat using a PDMS microneedle platform coated with a conductive PDMS/carbon nanotube (CNT)/cellulose nanocrystal (CNC) composite (consider revising with reference to the citation). Figure 4 displays the schematic representations of microneedle sensing devices toward the monitoring of several biomarkers.

4.3 Therapeutic Drug Monitoring

Another wide application of microneedle sensors is therapeutic drug monitoring. Efficient management of chronic diseases demands continuous monitoring of therapeutic drugs. Accurate monitoring of these long-term used drugs will provide personalized patient care. Another major concern affecting the healthcare community is

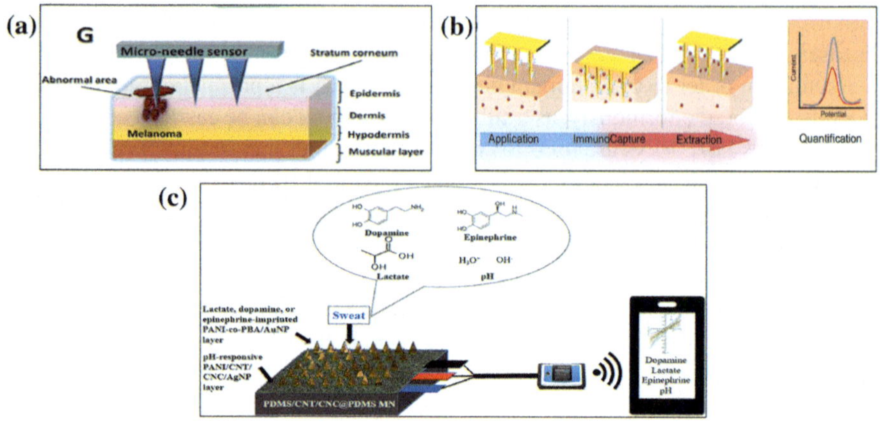

Fig. 4 Continuous monitoring of various biomarkers using microneedle sensors. **a** Wearable microneedle sensor for melanoma screening [46]. **c** Antibody-based microneedle biosensor for breast cancer biomarker detection [9]

the increased prevalence of antibiotic resistance. The difference in individual pharmacokinetics resulted in variable numbers and concentrations of antibiotics. So to address this issue, personalized monitoring is needed. Conventional methods of monitoring are based on blood extraction and this method lacks the quality of continuous monitoring and also, these methods are invasive and painful too. Wearable sensors can be used as a better alternative. In the class of wearable sensors, microneedles are the best candidates for the continuous monitoring of antibiotics, therapeutic drugs, or nerve agents. Here, we are going to discuss different kinds of microneedle sensors for the detection and monitoring of various classes of antibiotics and therapeutic drugs.

Ranamukhaarachchi et al. [53] fabricated a hollow microneedle optofluidic biosensor for therapeutic drug monitoring in sub-nanolitre volume. The hollow microneedle can extract a volume <nL of ISF to detect a commonly used vancomycin. The developed biosensor could obtain a LOD of <100 nM. Rawson et al. [54] developed a microneedle biosensor for real-time detection of the antibiotic phenoxymethylpenicillin. The group conducted studies on ten healthy volunteers using a β-lactam microneedle sensor for up to 6 h, after which the volunteers were allowed to take a constant dosage of phenoxymethylpenicillin. They compared the data obtained with the developed sensor with the data obtained from gold-standard methods, which was shown to have good agreement with the former. The monitoring of antibiotics using a microneedle wearable sensor was continued by Gowers et al. [43] The group came up with a minimally invasive potentiometric microneedle sensor for real-time monitoring of β-lactam antibiotics. The biosensor consists of an iridium layer which is pH-sensitive. This layer monitors the changes in pH occurring via the enzymatic hydrolysis of β-lactam by β-lactamase immobilized on the electrode surface. The biosensor was found to be appreciably stable and could obtain significant results.

Recently, microneedle sensing platforms have been explored enormously toward the detection of various analytes including pharmaceutical drugs, biomarkers, nerve agents, and illicit drugs. Goud et al. [55] fabricated a wearable electrochemical microneedle sensor for continuous monitoring of levodopa, which is a gold-standard drug used for the treatment of Parkinson's disease. They have introduced an orthogonal detection for L-dopa using an enzymatic amperometric and non-enzymatic square-wave voltametric method independently. The studies were performed on phantom-gel skin-mimicking model and rat skin, and excellent analytical results were obtained. Goud et al. [41] developed a wearable microneedle sensor array for continuous opioid monitoring along with nerve agents. The sensor relies on unmodified and organophosphorus hydrolase (OPH) enzyme-modified carbon paste (CP) microneedle electrodes for the detection of the fentanyl and nerve agent targets, respectively. The same group under K. Y. Goud came up with a wearable electrochemical microneedle [56] sensing platform for continuous monitoring of apomorphine, a drug used for treating Parkinson's disease. They used Rh nanoparticles as the electrode modification, and the detection was done using square-wave voltammetry and chronoamperometry techniques. Yao Wu et al. [57] used aptamers to design a microneedle sensor for continuous monitoring of therapeutic drugs. The system works on biorecognition and thus the applicability can be extended for sensing non-redox active drugs. Figure 5 displays the schematic representations of microneedle sensing devices toward the monitoring of several therapeutic drugs.

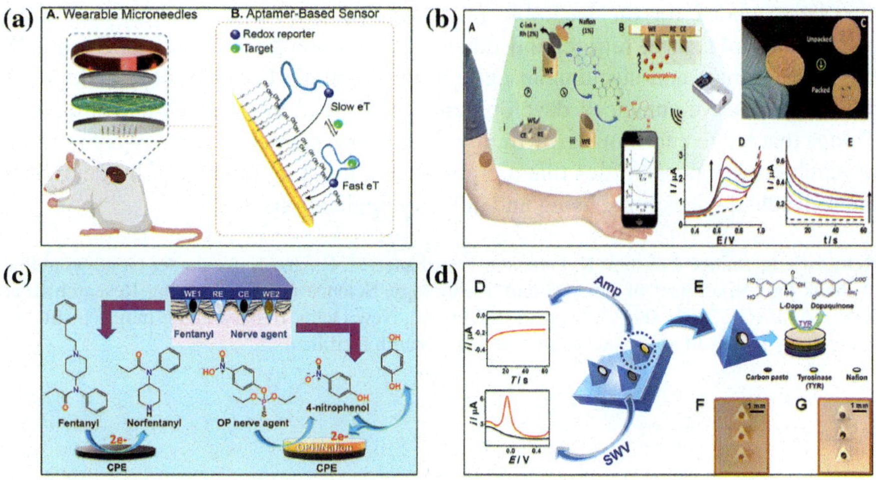

Fig. 5 Therapeutic drug monitoring using microneedle sensors. **a** Aptamer-based microneedle biosensor for multiple drug detection [52]. **b** Microneedle sensor for continuous monitoring of apomorphine [5]. **c** Opioid and nerve agent detection using microneedles [36]. **d** Continuous monitoring of Parkinson's drug levodopa by hollow Microneedles [51]

5 Summary and Conclusions

Recent developments in wearable microneedle sensing methodologies are clearly described in this book chapter including the importance/advantages of ISF, fabrication protocols of MNs, continuous monitoring of biomarkers, physiological ions, therapeutic drugs, and delivery of drugs/vaccines. Critical discussions have been included related to the sensing strategies, electrode fabrications, and sensor analytical performances. We have added a discussion on the importance of ISF biofluid over the other ones and the feasibility of the continuous monitoring under the skin. The rich biomarker/drug sources of ISF and advanced developments of micro/nano fabrications and semiconductor/electronics industries would definitely help in the growth of advancements in microneedle fabrications in the near future. Moreover, multiplex analysis would be easily feasible in microneedle sensing array platforms. Simultaneously, we can determine the various biomarkers that would help in diagnosing the various chronic diseases in their early stages. This technology enables personalized monitoring devices for individual people who are suffering from chronic diseases. Currently, microneedle developments majorly focus on diabetic and neurological disease management. These platforms can be further extended to other chronic diseases like cardiac, cancer, renal, etc. One more major possible advantage of these MNs are simultaneously we can monitor the respective biomarker or drug levels of the chronic disease, and corresponding antidotes will be delivered through the same microneedle patch. So, these MNs will be made for the "Sens and Act", which means the same microneedle patch can detect the biomarker levels of the chronic disease and also deliver the respective antidotes. These microneedle platforms could also be extended for the wearable plant applications for continuous monitoring of moisture, pH, urea, nitrate, and other plant physiological elements. Eventually, this led to smart and sustainable agriculture forming. The current decade would definitely observe enhanced developments in artificial intelligence and Internet of Things that led to enormous growth in wearable sensing devices. There are potential research growth possibilities that are possible in the development of biodegradable microneedle devices for sensing and delivery applications.

Acknowledgements Lakshmi R. Panicker, Keerthanaa M. R., and K. Yugender Goud would like to thank the Department of Science and Technology, Science and Engineering Research Board (DST-SERB/RJF/2021/000113, Ramanujan Award), New Delhi, India, and the Indian Institute of Technology Palakkad for financial support and research facilities.

References

1. García-Guzmán, J.J., Pérez-Ràfols, C., Cuartero, M., Crespo, G.A.: Microneedle based electrochemical (Bio)Sensing: Towards decentralized and continuous health status monitoring. TrAC-Trends Anal. Chem. **135** (2021) Preprint at https://doi.org/10.1016/j.trac.2020.116148
2. Leanpolchareanchai, J., Nuchtavorn, N. Wearable microneedle-based colorimetric and fluorescence sensing for transdermal diagnostics. Talanta Open **8** (2023)
3. Bariya, M., Nyein, H.Y.Y., Javey, A.: Wearable sweat sensors. Nat. Electron. **1**, 160–171 (2018) Preprint at https://doi.org/10.1038/s41928-018-0043-y
4. Iqbal, S.M.A., Mahgoub, I., Du, E., Leavitt, M.A., Asghar, W.: Advances in healthcare wearable devices. NPJ Flex. Electron. **5** (2021) Preprint at https://doi.org/10.1038/s41528-021-00107-x
5. Miranda, B. et al.: Hollow microneedle-based plasmonic sensor for on patch detection of molecules in dermal interstitial fluid. Adv. Mater. Technol. (2023). doi:https://doi.org/10.1002/admt.202300037
6. Tariq, N., Ashraf, M.W., Tayyaba, S.A.: Review on solid microneedles for biomedical applications. J. Pharm. Innov. 17, 1464–1483 (2022) Preprint at https://doi.org/10.1007/s12247-021-09586-x
7. Kashaninejad, N. et al.: Microneedle arrays for sampling and sensing skin interstitial fluid. Chemosensors **9** (2021) Preprint at https://doi.org/10.3390/chemosensors9040083
8. Li, J., et al.: Interstitial fluid biomarkers' minimally invasive monitoring using microneedle sensor arrays. Anal. Chem. **94**, 968–974 (2022)
9. Zhu, D. D. et al.: Microneedle-coupled epidermal sensors for in-situ-multiplexed ion detection in interstitial fluids. ACS Appl. Mater. Interfaces (2023) doi:https://doi.org/10.1021/acsami.3c00573
10. Huang, X. et al.: 3D-assembled microneedle ion sensor-based wearable system for the transdermal monitoring of physiological ion fluctuations. Microsyst. Nanoeng. **9** (2023)
11. Friedel, M., et al.: Opportunities and challenges in the diagnostic utility of dermal interstitial fluid. Nat. Biomed. Eng. (2023). https://doi.org/10.1038/s41551-022-00998-9
12. Heikenfeld, J. et al.: Accessing analytes in biofluids for peripheral biochemical monitoring. Nat. Biotechnol. **37**, 407–419 (2019) Preprint at https://doi.org/10.1038/s41587-019-0040-3
13. Xie, Z., Zhang, X., Chen, G., Che, J., Zhang, D.: Wearable microneedle-integrated sensors for household health monitoring. Eng. Regen. **3**, 420–426 (2022) Preprint at https://doi.org/10.1016/j.engreg.2022.09.002
14. Zhang, X., Lu, M., Cao, X., Zhao, Y.: Functional microneedles for wearable electronics. Smart Med. **2**, (2023)
15. Parrilla, M. et al.: Wearable all-solid-state potentiometric microneedle patch for intradermal potassium detection. Anal. Chem. **91**, 1578–1586 (2019)
16. Sheikh, M., Qassem, M., Kyriacou, P.A.: Optical determination of lithium therapeutic levels in micro-volumes of interstitial fluid. Bipolar Disord. (2023). https://doi.org/10.1111/bdi.13291
17. Goud, K.Y., et al.: Wearable electrochemical microneedle sensor for continuous monitoring of levodopa: Toward Parkinson management. ACS Sens. **4**, 2196–2204 (2019)
18. Goud, K.Y. et al.: Wearable electrochemical microneedle sensing platform for real-time continuous interstitial fluid monitoring of apomorphine: Toward Parkinson management. Sens. Actuators B Chem. **354**, (2022)
19. You, X.-Q. et al.: Multi-groove microneedles based wearable colorimetric sensor for simple and facile glucose detection. Microchem. J.**190**, 108570 (2023)
20. Tehrani, F. et al.: An integrated wearable microneedle array for the continuous monitoring of multiple biomarkers in interstitial fluid. Nat. Biomed. Eng. **6**, 1214–1224 (2022)
21. Ji, J., Tay, F.E.H., Miao, J.: Microfabricated hollow microneedle array using ICP etcher. J. Phys. Conf. Ser. **34**, 1132–1136 (2006)
22. Liu, Y., et al.: Advanced deep reactive-ion etching technology for hollow microneedles for transdermal blood sampling and drug delivery. IET Nanobiotechnol. **7**, 59–62 (2013)
23. Vinayakumar, K.B. et al.: A hollow stainless steel microneedle array to deliver insulin to a diabetic rat. J. Micromech. Microeng. **26**, (2016)

24. Davis, S.P., Landis, B.J., Adams, Z.H., Allen, M.G., Prausnitz, M.R.: Insertion of microneedles into skin: Measurement and prediction of insertion force and needle fracture force. J. Biomech. **37**, 1155–1163 (2004)
25. Ribet, F., Stemme, G., Roxhed, N.: Real-time intradermal continuous glucose monitoring using a minimally invasive microneedle-based system. Biomed. Microdevices **20** (2018)
26. Nicholas, D., et al.: Rapid paper based colorimetric detection of glucose using a hollow microneedle device. Int. J. Pharm. **547**, 244–249 (2018)
27. Li, C.G., et al.: One-touch-activated blood multidiagnostic system using a minimally invasive hollow microneedle integrated with a paper-based sensor. Lab Chip **15**, 3286–3292 (2015)
28. Ryan, E., et al.: Microneedle-mediated transdermal bacteriophage delivery. Eur. J. Pharm. Sci. **47**, 297–304 (2012)
29. Davis, S.P., Martanto, W., Allen, M.G., Prausnitz, M.R.: Hollow metal microneedles for insulin delivery to diabetic rats. IEEE Trans. Biomed. Eng. **52**, 909–915 (2005)
30. Morales, M., Gómez-Cabrero, A., Peral, A., Gasull, X., Pintor, J.: Hypotensive effect of profilin on rabbit intraocular pressure. Eur. J. Pharmacol. **567**, 145–148 (2007)
31. Jiang, J., Moore, J.S., Edelhauser, H.F., Prausnitz, M.R.: Intrascleral drug delivery to the eye using hollow microneedles. Pharm. Res. **26**, 395–403 (2009)
32. Van Der Maaden, K., et al.: Novel hollow microneedle technology for depth-controlled microinjection-mediated dermal vaccination: A study with polio vaccine in rats. Pharm. Res. **31**, 1846–1854 (2014)
33. Du, G. et al.: Coated and hollow microneedle-mediated intradermal immunization in mice with diphtheria toxoid loaded mesoporous silica nanoparticles. Pharm. Res. **35** (2018)
34. Ogai, N., et al.: Enhanced immunity in intradermal vaccination by novel hollow microneedles. Skin Res. Technol. **24**, 630–635 (2018)
35. Pamornpathomkul, B., et al.: A combined approach of hollow microneedles and nanocarriers for skin immunization with plasmid DNA encoding ovalbumin. Int. J. Nanomed. **12**, 885–898 (2017)
36. Mcallister, D.V et al.: Microfabricated needles for transdermal delivery of macromolecules and nanoparticles: Fabr. Methods Transp. Stud. **100** (2003) www.pnas.orgcgidoi10.1073pnas.2331316100
37. Lin, W. et al.: Transdermal Delivery of Antisense Oligonucleotides with Microprojection Patch (Macroflux) Technology. (2001)
38. Tariq, N., Ashraf, M.W., Tayyaba, S. A review on solid microneedles for biomedical applications. J. Pharm. Innov. **17**, 1464–1483 (2022) Preprint at https://doi.org/10.1007/s12247-021-09586-x
39. Wang, H., et al.: Microneedle array integrated with CNT nanofilters for controlled and selective drug delivery. J. Microelectromech. Syst. **23**, 1036–1044 (2014)
40. Wei-Ze, L., et al.: Super-short solid silicon microneedles for transdermal drug delivery applications. Int. J. Pharm. **389**, 122–129 (2010)
41. Mishra, R.K., et al.: Continuous opioid monitoring along with nerve agents on a wearable microneedle sensor array. J. Am. Chem. Soc. **142**, 5991–5995 (2020)
42. Saha, T., et al.: Wearable electrochemical glucose sensors in diabetes management: A comprehensive review. Chem. Rev. **123**, 7854–7889 (2023)
43. Gowers, S. A. N. et al.: Development of a minimally invasive microneedle-based sensor for continuous monitoring of β-lactam antibiotic concentrations in vivo. ACS Sens. **4**, 1072–1080 (2019)
44. Li, H., et al.: Microneedle-based potentiometric sensing system for continuous monitoring of multiple electrolytes in skin interstitial fluids. ACS Sens. **6**, 2181–2190 (2021)
45. Bollella, P., Sharma, S., Cass, A.E.G., Antiochia, R.: Microneedle-based biosensor for minimally-invasive lactate detection. Biosens. Bioelectron. **123**, 152–159 (2019)
46. Molinero-Fernández, Á., Casanova, A., Wang, Q., Cuartero, M., Crespo, G.A.: In vivo transdermal multi-ion monitoring with a potentiometric microneedle-based sensor patch. ACS Sens. **8**, 158–166 (2023)

47. Huang, X. et al.: 3D-assembled microneedle ion sensor-based wearable system for the transdermal monitoring of physiological ion fluctuations. Microsyst. Nanoeng. **9**, (2023)
48. Liu, Y., Yu, Q., Luo, X., Yang, L., Cui, Y.: Continuous monitoring of diabetes with an integrated microneedle biosensing device through 3D printing. Microsyst. Nanoeng. **7**, (2021)
49. Ciui, B. et al.: Wearable wireless tyrosinase bandage and microneedle sensors: toward melanoma screening. Adv. Healthcare Mater. **7**, (2018)
50. Dervisevic, M., Alba, M., Adams, T. E., Prieto-Simon, B., Voelcker, N.H.: Electrochemical immunosensor for breast cancer biomarker detection using high-density silicon microneedle array. Biosens. Bioelectron. **192** (2021)
51. Wang, Z., et al.: Microneedle patch for the ultrasensitive quantification of protein biomarkers in interstitial fluid. Nat. Biomed. Eng. **5**, 64–76 (2021)
52. Mugo, S.M., Robertson, S.V., Lu, W.: A molecularly imprinted electrochemical microneedle sensor for multiplexed metabolites detection in human sweat. Talanta **259**, 124531 (2023)
53. Ranamukhaarachchi, S.A. et al.: Integrated hollow microneedle-optofluidic biosensor for therapeutic drug monitoring in sub-nanoliter volumes. Sci. Rep. **6** (2016)
54. Rawson, T.M., et al.: Microneedle biosensors for real-time, minimally invasive drug monitoring of phenoxymethylpenicillin: a first-in-human evaluation in healthy volunteers. Lancet Digit Health **1**, e335–e343 (2019)
55. Goud, K.Y., et al.: Wearable electrochemical microneedle sensor for continuous monitoring of levodopa: Toward Parkinson management. ACS Sens **4**, 2196–2204 (2019)
56. Goud, K.Y. et al.: Wearable electrochemical microneedle sensing platform for real-time continuous interstitial fluid monitoring of apomorphine: Toward Parkinson management. Sens. Actuators B Chem. **354** (2022)
57. Wu, Y., et al.: Microneedle aptamer-based sensors for continuous real-time therapeutic drug monitoring. Anal. Chem. **94**, 8335–8345 (2022)

Wireless Biosensors for Healthcare: Smart Contact Lenses and Microbial Devices

Saman Azhari, Gábor Méhes, and Takeo Miyake

1 Introduction to Wireless Power Transmission

Wireless power transmission (WPT) refers to the transfer of electrical energy without physical connections, which enhances the safety and convenience of electronic devices by removing the need for power storage devices [1]. The roots of WPT can be traced back to James Clerk Maxwell's merging of Ampère's circuital law and Faraday's law of induction in 1865, which laid the foundation for the conventional electromagnetic theory [2]. John Henry Poynting introduced a mathematical equation for contactless energy transfer in 1884 [3], and Heinrich Rudolf Hertz discovered radio waves in 1888, confirming the existence of electromagnetic propagation [4]. In 1894, Hutin and Leblanc's patented WPT system was employed to supply power to an electric railway system [5]. Nikola Tesla made the first breakthrough in WPT technology between 1891 and 1904, using inductive coupling WPT with Tesla coils that produced high AC voltage, demonstrating wireless lighting of phosphorescent bulbs [6–8]. Half a century later, in 1960, W. C. Brown accomplished long-distance energy transfer using advancements in microwave technology, and he used a rectenna to convert electromagnetic waves into DC power in 1964. In 1975, Brown achieved a short-distance transfer of 475 W using microwaves at 54% DC to DC efficiency [9]. In recent years, the Nuclear and Quantum Engineering team at KAIST University utilized inductive coupling WPT to transfer energy to distances of 3, 4, and 5 m with efficiencies of 29%, 16%, and 8% in 2015 [10]. Such experiments began with physicists from the Massachusetts Institute of Technology lighting a 60 W light bulb 2 m away from a transmitting coil in 2007 [11].

S. Azhari—equal contributions
G. Méhes—equal contributions

S. Azhari · G. Méhes · T. Miyake (✉)
Graduate School of Information, Production and Systems, Waseda University, 2-7 Hibikino, Wakamatsu, Kitakyushu, Fukuoka 808-0135, Japan
e-mail: miyake@waseda.jp

K. Mitsubayashi (ed.), *Wearable Biosensing in Medicine and Healthcare*,
https://doi.org/10.1007/978-981-99-8122-9_8

WPT technology is divided into two primary categories: near-field and far-field. Near-field WPT is designed for short-range power transmission at a low-frequency range [12]. These systems are further subdivided into capacitive and inductive coupling categories, discussed later [13]. In contrast, the far-field is used for long-distance WPT at high frequency, and there are two primary approaches: microwaves and lasers. However, the use of far-field WPT technology for biosensors, particularly contact lenses, is not ideal due to the frequency range and power of these systems [14]. Based on the Federal Communication Commission (FCC), the specific absorption rate (SAR) limit for public exposure from cellular telephones is 1.6 W/kg [15], which is lower than the power provided by microwave and laser WPT. Furthermore, these systems operate in the GHz and THz frequency ranges, respectively, which can rapidly generate heat and therefore harm the human body and tissues. For this reason, near-field WPT is preferred for biosensing applications due to its lower frequency range and lack of harmful effects on the body and tissues. The use of far-field WPT technology is also limited due to factors such as high cost, complex fabrication processes, and insufficient sensitivity [16, 17]. The summary of different coupling technologies is shown in Table 1.

Table 1 A comparison of wireless power transmission technologies

Coupling type	Principle	Advantages	Disadvantages	Distinguishing features	Applications
Inductive	Magnetic field coupling between two coils	Simple, robust, widely used	Low efficiency, short range, sensitive to misalignment	Power transfer depends on mutual inductance and load impedance	Smartphone battery charging, biomedical implants, electric vehicles
Capacitive	Electric field coupling between two plates	High power density, low electromagnetic interference	High voltage, complex circuitry, safety issues	Power transfer depends on electric displacement and capacitance	Mobile devices, low-power biomedical devices
Resonant inductive	Magnetic field coupling between two resonant coils	High efficiency, longer range, multiple devices charging	Frequency tuning, coil design, heat generation	Power transfer depends on resonance frequency and quality factor	WiTricity, Rezence, eCoupled, WREL, RFID tags, contactless smart cards
Far-field	Electromagnetic wave propagation between transmitter and receiver	Long distance, high power	Beam alignment, atmospheric attenuation, health and safety concerns	Power transfer depends on beam intensity and aperture size	Solar power satellites, wireless sensor networks, directed-energy weapons

The following sections will describe the principle of operation and the advantages and disadvantages of near-field WPT technologies. We will also discuss recent advancements in integrating WPT in biomedical applications and devices, particularly contact lenses, and explore the possibilities and prospects of such devices.

1.1 Near-Field Wireless Power Transmission

The term "near-field" pertains to the region surrounding the antenna conductors, where electromagnetic wave emission and propagation can be hampered. Consequently, the absorption of radiation in this area by nearby conductive objects detectably affects the load on the signal generator, commonly referred to as the transmitter. Near-field WPT is a wireless power transfer mechanism that uses the electromagnetic field to transmit power from a transmitter to a receiver without necessitating physical contacts, cables, or wires. This technology utilizes either magnetic or electric fields to establish a coupling between the transmitter and the receiver. The transmitter is usually connected to a power source that generates an alternating current (AC) in a wire coil. On the other hand, the receiver is another coil commonly attached to an electronic device that requires power, such as a smartphone, laptop, or smart contact lens, which receives the AC signal from the transmitter. The AC is generally converted to direct current (DC) by the receiver's circuitry and then used to charge a battery or directly power up the device [18].

Various factors, including the distance between the transmitter and the receiver, the size and shape of the coils, and the frequency of the AC, influence the strength of the coupling in near-field WPT. Typically, near-field WPT is the most effective when the distance between the transmitter and the receiver is within approximately one wavelength (λ) of the antenna length, which refers to the area in close proximity to the antenna where the electromagnetic field is the most powerful [11]. Near-field wireless power transfer can be classified into two types: inductive coupling and capacitive coupling. Both types of near-field WPT have unique advantages and disadvantages that are specific to various applications.

1.2 Inductive Coupling (Non-Resonant)

Inductive coupling is a technique used for wireless power transfer between coils of wire, which involves transferring power through a magnetic field. This method is based on the principle of a transformer, where the transmitter and receiver coils form the primary and secondary coils, respectively.

The transmitter coil (L_1) is supplied with an alternating current (AC), which generates an oscillating magnetic field (B) according to Ampere's law. This oscillating magnetic field then passes through the receiver coil (L_2), which is placed in close proximity to the transmitter coil, inducing an alternating voltage, according

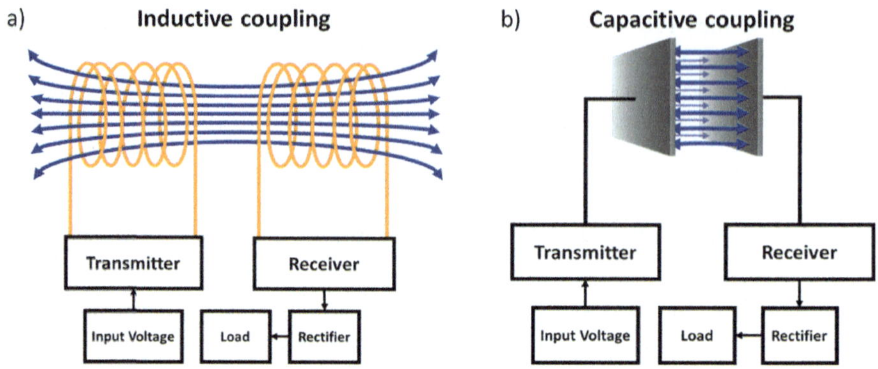

Fig. 1 The principal schematic of wireless power transmission via, **a** inductive/resonant inductive coupling [21] and **b** two-plate capacitive coupling [22]. The transmitter and receiver in each figure contain a compensation network that provides impedance-matching circuitry. The main components of impedance-matching circuitry are usually passive components such as resistor (R) and capacitor, **c** in the case of inductive/resonant inductive coupling and R and inductor (L) in the case of capacitive coupling. Such configurations are also referred to as LCR circuits

to Faraday's law of induction. The induced voltage in the receiver coil produces an alternating current, which can either directly drive the load or be converted to direct current (DC) using a rectifier in the receiver, which in turn drives the load (Fig. 1a). This enables wireless electrical power transfer from the transmitter to the receiver over a short distance without physical contact between the two coils [19, 20].

Inductive coupling has been widely used in charging stands for cordless appliances in wet environments, such as electric toothbrushes and shavers, to reduce the risk of electric shock. Recently, inductive coupling has also been rapidly adopted as wireless charging pads for mobile and handheld wireless devices, including laptop and tablet computers and cell phones. In December 2017, the FCC certified the first wireless transmission charging system in the United States [23].

The power transferred between the coils in inductive coupling depends on the frequency and the mutual inductance M between the coils. The geometry of the coils and the distance between them determines the mutual inductance M. As the frequency and mutual inductance increase, the power transferred also increases. In inductive coupling, the coupling coefficient $k = M/\sqrt{L_1 L_2}$ is of great importance. This dimensionless parameter represents the fraction of the magnetic flux through L_1 that passes through L_2 when L_2 is open-circuited.

When the two coils are on the same axis and are positioned close to each other, all the magnetic flux from L_1 passes through L_2, resulting in a coupling coefficient of $k = 1$ and a 100% link efficiency. As the separation between the coils increases, more of the magnetic field from the first coil misses the second, resulting in a lower k and lower link efficiency, which approaches zero at large separations. The link efficiency and power transferred are approximately proportional to k^2. Therefore, the coils must be positioned close to each other to achieve high efficiency, typically within a fraction of the coil diameter, usually within centimeters, with the coils' axes aligned.

Flat coil shapes are often used to increase coupling, and ferrite “flux confinement” cores can improve coupling and reduce interference with nearby electronics, although are heavy and bulky. Air-core coils are commonly used in small wireless devices.

Most modern inductive systems use resonant inductive coupling to achieve high efficiency at greater distances, increasing efficiency by using resonant circuits. This method enables higher efficiencies over greater distances compared to non-resonant inductive coupling.

1.3 Resonant Inductive Coupling

The transfer of power via magnetic fields (B) between two resonant circuits, namely the transmitter and the receiver, is referred to as resonant inductive coupling. In each resonant circuit, a wire coil is connected to a capacitor, a self-resonant coil, or other resonators having internal capacitance. Both circuits are adjusted to resonate at the same frequency, and the resonance between the coils can considerably boost coupling and power transfer. Similarly, a vibrating tuning fork can generate sympathetic vibration in a distant tuning fork tuned to the same pitch [20].

Resonant inductive coupling systems are based on the idea that high-Q factor resonators exchange energy quicker than they lose energy owing to internal damping. The same amount of power can be transported across longer distances by exploiting resonance. The resonant inductive coupling can attain excellent efficiency at ranges of 4 to 10 times the coil diameter. This is known as “mid-range” transfer, in contrast to non-resonant inductive coupling’s “short-range” transfer, which may attain equal efficiency only when the coils are adjacent. Another benefit of resonant coupling is that resonant circuits interact far more strongly with each other compared to non-resonant ones, resulting in low-power losses.

Nikola Tesla discovered resonant coupling during his pioneering studies in wireless power transfer at the start of the twentieth century. However, the prospects of exploiting resonant coupling to extend transmission range have only recently been explored. A team led by Marin Soljačić at MIT used two coupled tuned circuits, each made of a 25 cm self-resonant coil of wire at 10 MHz, to transmit 60 W of power over a distance of 2 m (8 times the coil diameter) with roughly 40% efficiency in 2007[11].

A noteworthy disadvantage of resonant coupling is the tight coupling of the two resonant circuits at close ranges. In such cases, the system’s resonant frequency “splits” into two resonant peaks, which means that the maximum power transfer no longer happens at the original resonant frequency. As a result, the oscillator frequency must be adjusted to the new resonance peak [24, 25].

Modern inductive wireless power systems increasingly utilize resonant technology while expanding their applications by integrating with advanced materials, such as carbon nanotubes, for physical sensing [26]. An example of combining WPT with nanomaterials for pressure sensing Area wireless power coverage is one of the applications envisaged for this technology. A coil installed in a room’s wall or ceiling may wirelessly power lights and mobile devices everywhere in the room with reasonable efficiency. Wirelessly powering small gadgets such as clocks, radios, music players, and remote controls has the potential to significantly reduce the number of batteries discarded each year [27], which is a major source of toxic waste and groundwater contamination.

1.4 Capacitive Coupling

Capacitive coupling, also called electric coupling, is a technique that employs electric fields to transfer power between two electrodes. The power transfer is achieved by forming a capacitance between the electrodes. The electric fields between the electrodes, typically metal plates, transmit energy through capacitive coupling. The gap between the electrodes serves as the dielectric, forming a capacitor between the transmitter and receiver electrodes.

To transfer power, the transmitter applies an alternating voltage to the transmitting plate, and the resulting oscillating electric field induces an alternating potential on the receiver plate through electrostatic induction. This, in turn, causes an alternating current to flow in the load circuit (Fig. 1b). The amount of power transferred is directly proportional to the frequency, the square of the voltage, and the capacitance between the plates, which is proportional to the area of the smaller plate and inversely proportional to the separation [28].

However, capacitive coupling poses several risks and has limitations. High voltages are required to transfer significant power, making this technology potentially hazardous, while it can also generate harmful ozone. Thus, practical applications have been limited to low-power applications. Additionally, unlike magnetic fields, electric fields interact strongly with most materials, including those making up the human body, leading to dielectric polarization and energy absorption in the affected materials. This may cause excessive electromagnetic field exposure, making capacitive coupling less suitable for some applications [29].

Despite these limitations, capacitive coupling has several advantages over inductive coupling. The electric field is confined mainly between the capacitor plates, reducing interference and eliminating the need for heavy ferrite "flux confinement" cores required in inductive coupling. Moreover, alignment requirements between the transmitter and receiver are less critical. The capacitive coupling has been used to charge battery-powered portable devices and for charging or continuous wireless power transmission in biomedical implants. It is also being explored in transferring power across substrate layers in integrated circuits.

2 Wireless Biosensors for Smart Contact Lenses

2.1 Introduction: A Historical Outlook

The eyes are a crucial sensing organ for human survival, and their sensitivity and importance have earned them the moniker "the window to the soul." However, in the US alone, three out of four people require corrective lenses to improve their vision [30]. As shown in Fig. 2, the first known lens dates back to 2600 BC in ancient Egypt, where they polished stones and crystals such as quartz for decorative purposes [31, 32]. The Nimrud lens is one of the most well-known examples of such lenses (750 BC).

The first record of eyeglasses dates back to 1290 in central Italy [46], although the exact story behind is a controversial topic. In 1608 Hans Lippershey, a German-Dutch eyeglasses maker invented the first two-lens telescope [47], which Galileo Galilei modified to 3X magnification in the following year. Later, Galileo modified the telescope further to 30X magnification [48]. In 1887, Louis J. Girard invented the first scleral contact lens, which covered the entire visible surface of the eye, but with no vision correction. German ophthalmologist Adolf Gaston Eugen Fick developed the first prototype of afocal scleral contact lenses in 1888 [49].

The first corneal lenses were developed in 1949 by using polymethyl methacrylate (PMMA) [50, 51]. These lenses were much smaller and only worn on the cornea, allowing wearers to use them for up to 16 h a day. With advances in manufacturing technology, PMMA corneal lenses became increasingly popular during the 1960s due to their improved lens designs. However, PMMA-based lenses have the drawback of low oxygen permeability, causing insufficient oxygen concentrations in the conjunctiva and cornea [52]. Various rigid materials with oxygen permeability were developed in the late 1970s, 1980s, and 1990s to address this issue. Chemist Norman Gaylord played a key role in creating these new contact lenses that allowed oxygen to pass through [53].

In contrast to hard lens development using PMMA materials, Czech chemists Otto Wichterle and Drahoslav Lím achieved a significant breakthrough in soft contact lenses, reported in their publication "Hydrophilic Gels for biological use" in Nature in 1960 [54]. This pioneering work led to the introduction of the first soft (hydrogel) contact lenses in several countries during the 1960s, and the US Food and Drug Administration (FDA) approved the soft lens material in 1971 [55]. Soft lenses quickly became more popular than rigid lenses due to their immediate and greater comfort. Over the next 25 years, the materials used to make soft lenses were refined by altering the ingredients to enhance oxygen permeability. In 1998, Ciba Vision launched the first silicone hydrogel contact lenses in Mexico, combining the exceptional oxygen permeability of silicone with the comfort and clinical performance of traditional hydrogels [56].

In 2008, the Babak Parviz group at the University of Washington developed the first smart contact lens with a built-in circuit that could power a single micro-LED device [57]. In 2009 Babak Parviz and Brian P. Otis reported on an RF-powered

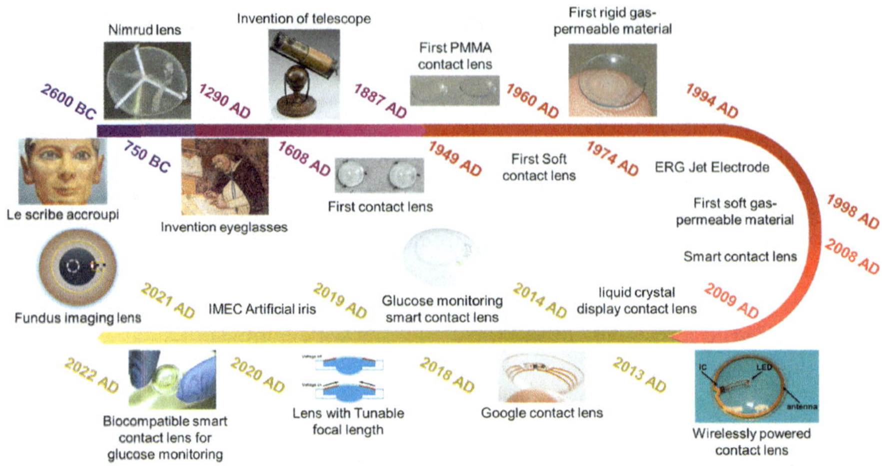

Fig. 2 The evolution of lens throughout history. Images used in this timeline, starting from top left to right, are le scribe accroupi, courtesy of Manuel Parada López de Corselas in Louvre-Lens [33], Nimrud lens in the British Museum [34], portrait of the Dominican Cardinal and renowned biblical scholar Hugh of Saint-Cher, painted by Tommaso da Modena in 1352 [35], a replica of Newton's second reflecting telescope which was presented to the Royal Society in 1672 [36], scleral lenses made by Otto Himmler now in the Deutsches Museum, Munich, constructed in 1887 and donated by San. Rat Dr. A Müller, München Gladbach (Copyright 2007 by Richard M Pearson. Reprint with permission.), [37], A pair of contact lenses positioned with the concave side facing upward [38], Rigid gas-permeable (RGP) lens [39]. Images used starting from bottom right to left are photographs of a wirelessly powered contact lens system (Copyright 2011 by IOP Publishing, Ltd. Reprint with permission.) [40], Google's smart lens (Copyright 2014 by Springer Nature. Reprint with permission.) [41], Schematic illustration of the soft, smart contact lens. The soft, smart contact lens is composed of a hybrid substrate, functional devices (rectifier, LED, and glucose sensor), and a transparent, stretchable conductor (for antenna and interconnects); (Copyright 2018 by American Association for the Advancement of Science. Reprint with permission.) [42], The expansion of the annular dielectric elastomer film in the lens could reduce the curvature of the film on both sides and thus increase the focal length of the lens (Copyright 2019 by John Wiley and Sons. Reprint with permission.) [43], Photograph of the sensor layer transferred onto a dome-shaped PDMS substrate (Copyright 2021 by Elsevier. Reprint with permission.) [44], Top view of contact lens placed on pig eye; the gold circle is the receiver coil, the red square is the infrared LED, wavelength: 870 nm, and white dotted circle is the observation area (Copyright 2022 by IOP Publishing, Ltd. Reprint with permission.) [45]

contact lens with a single-element display [58, 59]. In 2014, Google announced its smart contact lens project, which aimed to measure glucose levels in tears using a tiny sensor [60]. In 2018, researchers at the UNIST developed a contact lens that could monitor glucose levels and other biomarkers [42]. The University of California, San Diego, developed a lens in 2019 that could zoom in and out on objects by responding to the wearer's eye movements [43]. The latest development in smart contact lenses by the University of Surrey in 2021 is a biocompatible, multifunctional ultrathin sensor system incorporated into soft contact lenses with a high detection sensitivity for optical signals, glucose levels, and corneal temperature [44].

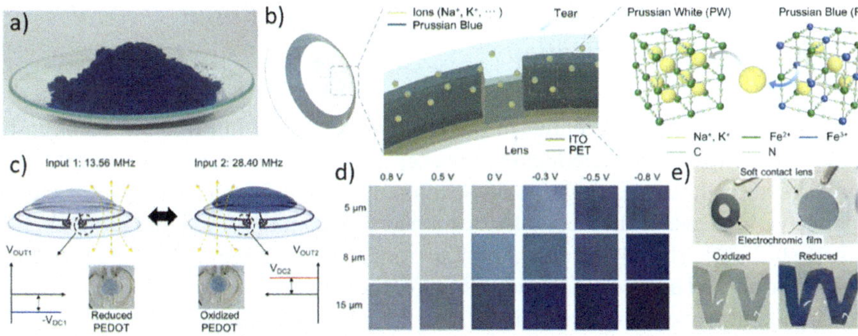

Fig. 3 **a** Prussian blue as the first known electrochromic material; **b** application of Prussian blue on a contact lens for the first time (Copyright 2020 by Elsevier. Reprint with permission.) [66]; **c** color change mechanism of an electrochromic soft contact lens with PEDOT as the electrochromic material with **d** variable colors because of a change in potential and **e** the final fabricated electrochromic soft contact lens. **c**–**e** Copyright 2022 by Elsevier. Reprint with permission [67]

At the same time, work on electrochromic lenses began to expand the functionality of lenses. As the name suggests, electrochromic technology allows the material to change its color or opacity in response to an electrical stimulus. The origin of electrochromic (EC) technology can be traced back to the discovery of Prussian blue (hexacyanoferrate), shown in Fig. 3a, by Diesbach in 1704 [61]. The transition of Prussian blue from a transparent state to a blue state because of iron oxidation marked the beginning of electro-coloration. In the 1930s, Kobosew and Nekrassow discovered electrochemical coloration in bulk tungsten oxide, paving the way for future advancements in EC technology [62, 63]. In 1953, T. Kraus provided a comprehensive description of electrochemical coloration in a thin film of tungsten trioxide (WO_3) while working at Balzers in Lichtenstein [62, 64]. Later in 1969, S. K. Deb demonstrated electrochromic coloration in WO_3 thin films by applying an electric field across the film [65].

Deb's seminal paper of 1973, describing the coloration mechanism in WO_3, is widely regarded as the true birth of EC technology [68]. Through his work, Deb introduced the fundamental principles of EC technology and laid the foundation for developing various EC devices that we use today. Such devices were initially (1960–1970) developed and used for automotive applications, but in the late 1980s and early 1990s, researchers focused on developing electrochromic materials that could be used in eyewear. In 2020, a research group at NTU reported the fabrication of an electrochromic contact lens using Prussian blue that can change color using an electric signal with 28.8 μW power consumption (Fig. 3b). They suggested using such lenses as an alarm system to monitor vital health signals [66].

Overall, contact lenses have come a long way since their early days. From scleral lenses that cover the entire visible surface of the eye to modern soft contact lenses and smart contact lenses, these devices have the potential to revolutionize the way we monitor and interact with our bodies and the world around us.

2.2 Near-Field WPT for Smart Contact Lens-Based Biosensing Applications

The human eye has the potential to provide physiological information that can be employed as physical and chemical biomarkers for diagnostic purposes. Over the past few years, there have been significant developments in using smart contact lenses to monitor various eye-related conditions such as glaucoma [69–71] and ailments like diabetes [42, 72]. The advancement in micro and nano fabrications and the growth in nanomaterials and nanotechnologies allow for the development of electronic devices and circuits that can fit onto a small contact lens without obstructing vision.

The eyes are among the body parts with great potential for non-invasive biosensing due to their continuous contact with tear fluid. The importance of tear fluid for non-invasive biosensing arises from the plasma leaking into the tear fluid through the blood-tear barrier, as illustrated in Fig. 4 for the canine eye. This leakage is due to the palpebral conjunctiva, supplied by the ophthalmic artery, diverging from the internal carotid artery, which is the primary blood supplier to the brain. The smart contact lens can detect various chemical biomarkers, such as lactic acidosis, cancer, keratoconus, AIDS, dry eye syndrome, and Blepharitis, in the tear fluid [73, 74] as well as physical biomarkers, such as intraocular pressure (IOP) for detecting glaucoma or eye movement tracking for determining concussion. Overall, smart contact lens technology has immense potential in biosensing.

A significant progress has been made in the field of biosensing using smart contact lenses. One example is the work by Lam's group, who developed a contact lens with an inductor-capacitor sensor that is sensitive to curvature and can non-invasively and continuously measure the eye's intraocular pressure (IOP). The curvature of the lens correlates mechanically with the IOP in the eye, and the sensor's resonance frequency can be used to determine the IOP. Their sensor exhibited high sensitivity (>200 ppm/

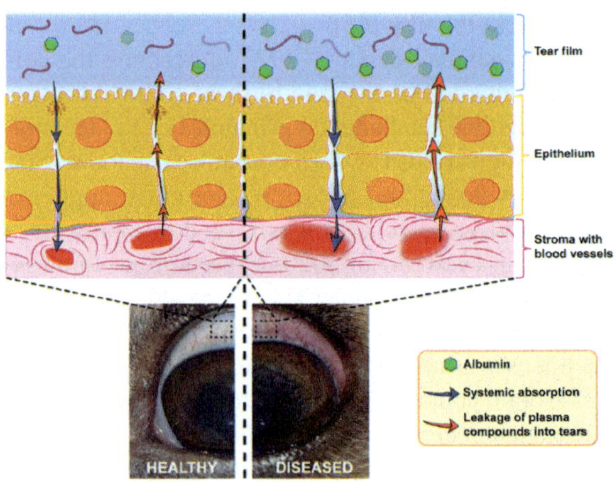

Fig. 4 Graphical representation of the blood-tear barrier in the canine eye, as a representative mammalian model; (Copyright 2020 by John Wiley and Sons. Reprint with permission.) [75]

mmHg) and good linearity (R > 0.997) while tracking pressure changes over time [70].

Another wireless smart contact lens system was developed by Chiou et al., which included a reconfigurable capacitive sensor wirelessly powered at 920 MHz. The sensor was embedded into a soft contact lens of 200 μm thickness using biocompatible materials. The sensor interface achieved high sensitivity and baseline tuning up to 120 pF while consuming only 110 μW of power. The on-lens system was capable of detecting changes in capacitance due to pressure variations within the range of 2.25 to 30 mmHg, as well as changes in hydration level from 1 cm. This detection was accomplished using incident power from an RFID reader at 26.5 dBm [76].

Next, Park's research group developed a multifunctional contact lens sensor fabricated on an ocular contact lens, which uses the resistance and capacitance of the device to monitor glucose levels in tears and IOP. The researchers achieved selective binding of target biomarkers onto graphene by using pyrene-chemistry, which could be adjusted for various analytes. The contact lens sensor's hybrid structure of 1D and 2D nanomaterials provided high conductivity, flexibility, and transparency, making it reliable and robust [71]. To test their sensor, the team performed ex vivo experiments on a bovine eyeball, able to control IOP in the 5–50 mmHg range. The frequency response of the contact lens sensor exhibited a linear decrease according to the pressure (2.64 MHz/mmHg) with negligible hysteresis. Their approach allows for simultaneous and independent monitoring of IOP and glucose levels in tear fluid, making it a promising method for the non-invasive monitoring of ocular biomarkers [71].

In 2022, the same team reported the development of a wireless and soft smart contact lens that can record cholesterol levels in real-time in tear fluids using a smartphone. The contact lens incorporated an electrochemical biosensor for continuous detection of cholesterol concentrations, a stretchable antenna, and integrated circuits for wireless communication. This makes a smartphone the only device required to operate the lens remotely without obstructing the wearer's vision. The electrochemical cholesterol biosensor on this contact lens platform showed selective, sensitive, and reliable sensing performance of tear cholesterol with wide dynamic ranges, allowing continuous and repetitive detections. The smart contact lens showed a promising correlation between the cholesterol levels in both blood and tears, highlighting its potential as a diagnostic device for hyperlipidemia by measuring cholesterol levels in tear fluids [77].

Miyake and colleagues developed a multifunctional power generation device that combines a wireless power transfer system with a bioabsorbable metal-air primary battery. The system can provide DC and/or AC current output. The DC power is generated from artificial tears using a Zn loop anode and a bilirubin oxidase (BOD) biocathode. The Zn loop anode also functions as the antenna of a wireless power transfer system, resulting in high power transfer efficiency of 17.6% at 13.56 MHz. A DC offset boosted the wireless-powered AC voltage from 1.5 V to 1.5 V +0.5 Vpp, allowing red light emission from a light-emitting diode (LED). The team also demonstrated using their smart contact lens to determine eye movement [78]. Furthermore, they recently developed a wireless-powered electrochromic soft contact lens with a dual-band wireless power transfer system resonating at 13.56 and 28.4 MHz to modulate color. The system also transmitted Morse code signals by modulating the wireless power supply voltage and duration to change the duration of color [67].

In another report by the Miyake group in 2022, they utilized the wireless power supply to power up a NIR LED with a wavelength of 870 nm to perform fundus imaging (Fig. 5e). They suggested that the reported imaging system can detect choroidal vasculatures and a few retinal blood vessels at high resolutions, which opens up the possibility of observing choroidal abnormalities [45]. Moreover, their latest work (Fig. 5f) demonstrated the use of parity-time symmetry wireless power transmission technology in combination with biosensors to detect the small concentration of glucose and lactate as low as 0.1 mM and 0.5 mM, respectively. The reported system exhibits a 2000-fold higher sensitivity for linear detection, although a correction mechanism is required for the possible misalignment between transmitter and receiver coils [79].

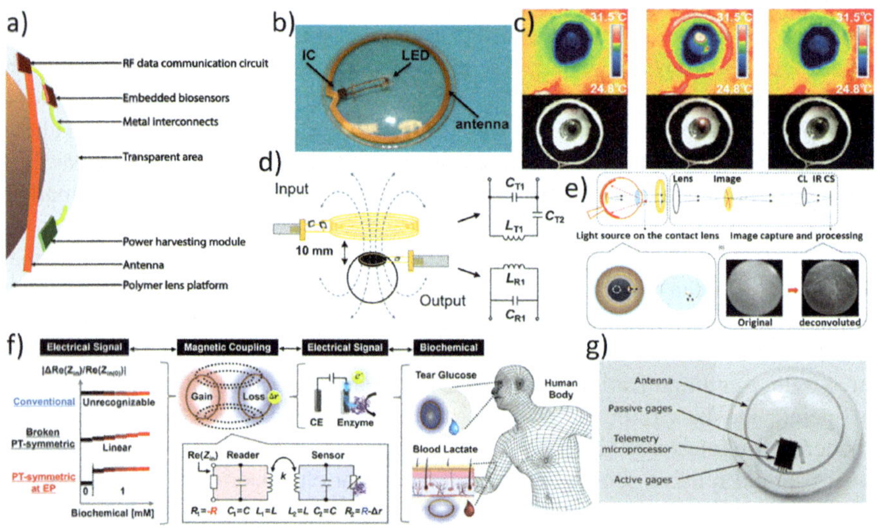

Fig. 5 **a** Schematic of a smart contact lens inspired by [58] while, **b** shows the fabricated device (Copyright 2011 by IOP Publishing, Ltd. Reprint with permission.) [40]; to assess the safety of wireless power transmission, **c** temperature variation on the surface of the eye due to the wireless power transmission is measured using the **d** wireless power transmission setup (Copyright 2019 by John Wiley and Sons. Reprint with permission.) [18]; **e** shows the use of wireless power transmission for fundus imaging using NIR (Copyright 2022 by IOP Publishing, Ltd. Reprint with permission.) [45] while **f** displays the mechanism and principle behind using parity-time symmetry on a contact lens for high sensitivity biosensing application (Copyright 2023 by John Wiley and Sons. Reprint with permission.) [79]; **g** exhibits the diagram of the contact lens sensor, showing the location of the sensor-active strain gages (which are placed circumferentially to measure changes in the corneal curvature caused by intraocular pressure variations) and the sensor-passive strain gages for thermal compensation (which are placed radially, where no strain is measured). For wireless powering and communication, a microprocessor and an antenna have been integrated into the soft contact lens (Copyright 2009 by John Wiley and Sons. Reprint with permission.) [80]

2.3 Future Directions in Smart Contact Lenses

Smart contact lenses—contact lenses with built-in electronics—are a next-generation wearable product with capabilities beyond simple vision correction. Since the electrical lenses are in continuous contact with the eyeball surface, they have three main applications: (i) biomedical sensing of tears to monitor health conditions, (ii) wearable displays for augmented reality (AR), and (iii) actively regulating eye accommodation to ensure perfect vision. Thus, a smart contact lens has substantially greater functionality than an electrical eyeglass. However, technology is still in the early stages of development, and it may be several years before we see widespread adoption. Power consumption is a significant challenge for these devices, which has led to the development of wireless-powered smart contact lenses that function through wireless power transmission. With continued research and development, we may see more innovations in contact lens technology that improve comfort, vision correction, and monitoring of health conditions.

3 Microbial Devices for Wearable Biosensing in Medicine and Healthcare

3.1 Introduction

Microbes, including bacteria, in medicine and healthcare, have been commonly associated with harmful, pathogenic intruders, sources of diseases that need to be eradicated. Analogously, for industry, detrimental effects of microbes manifest in the spoilage of foodstuff, infections of livestock, biofouling of underwater surfaces, such as on ships and pipes [81], as well as pipeline corrosion by, for example, sulfate-reducing bacteria [82]. While efforts to understand and treat microbial diseases, as well as to enhance antimicrobial, antifouling, and anticorrosion properties of surface coatings and chemicals, remain the main driving forces behind applied microbial research, there is a yet underexploited but promising way of employing microbes on the opposite side of the beneficiary spectrum. A typical representative technology on this 'opposite side' is microbial fuel cells (MFCs) which can generate electrical power by breaking down chemicals in industrial water, often producing only carbon dioxide and water as byproducts and products, respectively [83]. MFCs, as a subset of microbial electrochemical systems (MES), have the advantage of scalability and tunability toward specific target chemicals by device and material engineering and advanced analysis techniques. For medical and healthcare applications, MESs are designed as hybrid bio-electrochemical sensors of analytes where bacteria are integrated into devices to react to the presence of an analyte, in turn triggering an electrical response that can be measured [84]. On the other hand, the detection of pathogens or their chemical constituents is typically done by abiotic electrochemical biosensors not containing microbes as signal transducers [85].

The skin, oral cavity, and gut are the main places of the human body that could benefit from deploying advanced microbial sensing devices. The skin constitutes an easy-to-access and oxygenated surface for mostly non-invasive monitoring. The continually wet oral cavity is the host for several pathogens; the composition varies based on the presence of oxygen, with anoxygenic areas located typically in the interdental space and in the grooves of the tongue. The gut is a very important place for the transfer, absorption, and metabolism of nutrients and drugs, immune system maturation, and host protection from pathogens. The concentration of microbes in the gut ranges from 10^3–10^4 cells/g in the jejunum to 10^{11}–10^{12} cells/g in the colon, affected by many factors, including the amount of oxygen and pH. A growing number of intestinal and extraintestinal diseases and health conditions have been linked to the activity of gut microbes, including irritable bowel syndrome, colorectal cancer, obesity, type 2 diabetes, depression, and more [86, 87]. Intriguingly, evidence shows that microbes in the gut can influence the central nervous system by altering gut-brain axis signaling, causing or at least contributing to neuroinflammatory conditions such as multiple sclerosis, Alzheimer's, and Parkinson's diseases through secreting neuroactive metabolites and products [88]. Because microbes exert most of their influence on health through biochemical conversions, the possibility of monitoring such chemical conversions directly inside the human body through bioelectrical and bio-electrochemical devices opens.

The main mechanism of electrical signal transduction between bacteria and devices is charge transfer to an electrode by extracellular electron transfer (EET), a process linked to the metabolic and respiration cycles of microbes. Typically, EET is activated under anoxygenic conditions, whereby oxygen is substituted by solid terminal-end electron acceptors, in this case, an electrode polarized at a sufficient potential. Indeed, EET-capable pathogens have been found in the mammalian gut [89]. EET may proceed directly through physical contact between the electrode and multiheme outer membrane-bound cytochrome complexes, such as in the MTR pathway of the well-studied model organism *Shewanella oneidensis* MR-1 (*S. oneidensis*), as well as through nanowires, pili and indirectly through endogenously synthesized flavin shuttles. It is to be noted that EET can proceed in both ways: from the microbe outward to the electrode, in this case, cells will be loaded into the anodic chamber of an MFC, and from the electrode into the microbe, called bio-electrosynthesis, in which case the microbes will be in the cathodic compartment of an MFC. An example of the latter is sulfur-reducing bacteria that extract electrons from iron, often causing the oxidation of steel pipes. Traditional device architectures are based on dual or single-chamber fuel cells or electrochemical cells, while typical electrodes include cheap carbon-based porous anodes (or working electrodes in the case of MESs) and Pt-based cathodes (or counter electrodes in MESs) to fulfill counter-reactions.

In this part of the chapter, first, we will look at the major hurdles that prevent the efficient use of microbes as chemical-to-electrical signal converters, as well as novel approaches from the field of materials science and fabrication techniques to address them. This is followed by introducing non-traditional device architectures and detection principles to monitor bacterial activities, and then highlight a few innovative examples of using MESs in combination with wireless signal transmission and hybrid electronics in the medical and healthcare sector, especially on the skin, in the gut, and in the mouth. Finally, we outline our vision for future microbial medical diagnostics by integrating recent engineering advancements across several fields, namely synthetic biology, applied chemistry, soft materials, and hybrid electronics.

3.2 Advanced Materials and Fabrication Processes Create Biohybrid Electrodes to Enhance Signal Output

Efficient biosensing with bacteria relies on electron transfer from cells to the electrode. Major hurdles in this regard are (1) insufficient number of cells in direct contact with the electrode; (2) thin biofilms or cells within biofilms not connected electrically to charge collectors; and (3) high charge transfer resistance at the bioelectrical interface. To alleviate these issues, several conducting materials were proposed for integration with biofilms on electrodes, including metals and conducting polymers. Such approaches usually lead to electrically interconnected biohybrid electrodes, where cells, even from inside the biofilm, can contribute charges to the current running through the external circuit. For example, gold nanoparticles were integrated with *Geobacter sulfurreducens*, another model organism for EET, to yield 1.4 times improved current density outputs [90]. However, gold may not be the best material for wide-scale implementation due to its high price. Conducting polymers have also been investigated due to their biocompatibility, chemical tunability, versatile coating methods, high surface area, high electrochemical activity, and in some cases, mixed ionic-electronic coupling. In the first example of electro-polymerization in the presence of bacteria, *S. oneidensis*, *Escherichia coli* (*E. coli*), *Ochrobacterium anthropic,* or *Streptococcus thermophilus* were coated by the conducting polymer polypyrrole (PPy). MFCs using anodes prepared by such a method achieved a 14.1-fold increase in power output compared to native *S. oneidensis* [91]. This method allows for excellent control of the thickness and morphology of the coating, as well as yielding nearly 100% attachment of the polymer to the base conducting material.

On the other hand, the thickness of the biofilm, which in the case of *S. oneidensis* is only a monolayer, did not significantly increase, ultimately limiting the total signal output. To improve on this method, in our work, we succeeded in the creation of an 80 μm thick multi-layer conductive bacterial-composite film (MCBF) by the electro-polymerization of the conducting polymer poly(3,4-ethylenedioxythiophene):polystyrene sulfonate (PEDOT:PSS) around *S. oneidensis* (Fig. 6a) [92]. Notably, our approach led to a high viability of over 90%, as assessed by live-dead fluorescent staying. Key factors to keep the cells alive even while mixed with the toxic EDOT solution before electro-polymerization were the decrease of the contact time of cells with EDOT by controlling the flow speed and the separation of individual constituents of the electro-polymerization solution and the bacteria (Fig. 6a). In addition, by keeping the bacteria in a dormant state by performing all the processes at 4 °C, we further decreased the chance of the monomer interacting with bacterial bioprocesses. As a result of this process, we embedded a large number of bacteria into a 3D composite of base carbon felt with PEDOT:PSS, as evidenced by fluorescent confocal microscopy and scanning electron microscopy (SEM), shown in Fig. 6b. These images were in stark contrast to the typical monolayers of cells forming by *S. oneidensis* in anoxygenic, electron-respiring conditions. [92] By employing the biohybrid electrode as a working electrode in MESs, we improved the electrical signal output by 20 times compared to native biofilms on carbon felt. Importantly, this method can be used for many different bacteria when integrated with electrodes.

An interesting approach for the efficient fabrication of biohybrid electrodes and microbial devices is the printing of bio-inks containing live bacteria. Freyman, Kou et al. recently demonstrated the printing of *S. oneidensis* by dispensing a mixture containing bacterial cells suspended in a growth medium, as well as carbon black for conductivity, cellulose for mechanical support, and alginate for entrapment of bacteria within the electrode (Fig. 6c) [93]. To entrap bacteria in this electrode, carboxylic acid salts in alginate strands were cross-linked by Ca^{2+}. The resulting electrode was successfully employed as the anode in MFCs with at least 93 h of stable operation. Another promising example of printing is inkjet printing of photosynthetic cyanobacteria *Synechocystis*, where digital addressing of the inkjet printer enabled the creation of complex patterns, allowing to tailor the device shape to specific applications (Fig. 6d) [94].

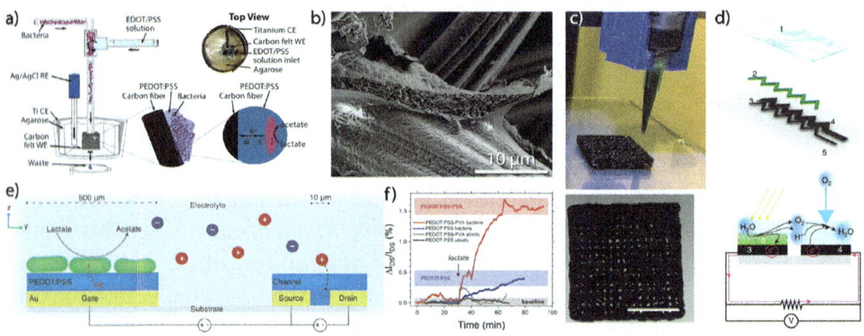

Fig. 6 Advanced fabrication methods and sensing principles of microbial devices. **a** Electropolymerization process of PEDOT:PSS-based multi-layer conductive bacterial-composite films (MCBFs) and **b** scanning electron micrograph of the resulting electrode [92]. **c** 3D printing of living bacterial electrodes by dispensing carbon black, nano-cellulose, alginate, and *S. oneidensis* [93]. **d** Photosynthetic energy harvesting device made by inkjet printing hydrogel (1), cyanobacteria (2), and carbon electrodes (3, 4) onto paper support (5) [94]. e) Operation principle and f) normalized current response curves of the organic microbial electrochemical transistor (OMECT) [95]. **a, b** Reproduced with permission from Zajdel et al. [92] Copyright 2018 Springer Nature Limited under open access CC BY-NC-ND 4.0 http://creativecommons.org/licenses/by/4.0/. **c** Reproduced with permission from Freyman et al. [93]. **d** Reprinted with permission from Sawa et al. [94] Copyright 2017 Springer Nature Limited under open access CC BY-NC-ND 4.0 http://creativecommons.org/licenses/by/4.0/. **e, f** Reprinted with permission from Méhes et al. [95] Copyright 2020 Wiley–VCH under CC BY 4.0 http://creativecommons.org/licenses/by/4.0/

3.3 Novel Device Architectures for Monitoring Bacterial Activities

Typical microbial electrochemical devices are based on fuel cell or half-cell architecture, where the detected current or voltage difference is directly related to the electrochemical conversion process at the electrode surfaces. Due to the low current harvested per bacterial cell, in the case of *S. oneidensis,* amounting to 10–100 fA [96], these systems rely on recruiting a high number of microbes to produce substantial signals above the noise limit. For this reason, detecting low numbers of bacteria is difficult, even though the presence of pathogens should be identified in the early phase of infection when cells are present still in low numbers. In this regard, we recently demonstrated the first organic microbial electrochemical transistor (OMECT) to amplify EET [95]. In this device, *S. oneidensis* metabolizes lactate and transfers charges to a mixed PEDOT:PSS-polyvinyl alcohol gate electrode with a surface area of only 0.25 mm^2, hosting ~ 18 000 cells (Fig. 6e). The device is based on the technology of organic electrochemical transistors (OECTs), which has been successfully used in many bio-electronics applications, including stimulation and amplification of signals from brain activity in transcranial recordings [97], or real-time monitoring of glucose export from chloroplasts [98]. OECTs are highly sensitive to charging at their gate terminal compared to field-effect transistors, including electrolyte-gated variants. This is thanks to the fact that ionic species can penetrate into the gate material (usually a conducting polymer with mixed ionic-electronic coupling), thereby inducing capacitive changes on the volumetric scale, as compared to a 2D interfacial capacitive charging in the case of field-effect transistors (FET)s. For this reason, OECTs can successfully amplify and transduce small changes in charges received by the gate into larger changes measured in the drain-source loop with a reduced footprint (gate size) compared to FETs. In the case of the OMECT, first bacteria transfer charges to the gate resulting from lactate conversion. At the same time, to keep charge neutrality, ions of opposite polarity enter the PEDOT:PSS source-drain channel from the electrolyte, thereby decreasing the conductivity of the polymer, finally leading to a decrease in source-drain current flow (Fig. 6e). In addition to the high sensitivity, the response time on lactate using the OMECT was one order of magnitude faster compared to typical centimeter-size electrodes in MESs based on *S. oneidensis* (Fig. 6f).

Biobattery is a relatively newly introduced device concept relevant to medical and healthcare applications [99]. This device is very similar to MFCs, with the distinction of having a semi-closed shell, pre-fed with a set amount of nutrients, as compared to the continuously fed MFCs. It is envisioned that biobatteries will be useful for powering miniature sensor systems deployed on the human skin [100] and in the human gut [101] for health monitoring and disease prevention while harvesting energy directly from sweat on the skin and from food remains in the gut, respectively.

A promising novel device platform for studying the functioning of the gut is Gut-on-Chip (GoC), which is based on Organ-on-chip (OoC), a new type of microfluidic and micromachined device to study the fundamental mechanisms relevant to cell biology, diseases, and drug action [102, 103]. In contrast to OoC, GoC, with its layered form, tries to mimic the multi-layer structure of the gut and the complex interactions within it. 3D hydrogel scaffolds for cells in GoC systems can be fabricated from either non-conducting or conducting materials, such as polydimethylsiloxane (PDMS) and PEDOT:PSS, respectively. GoCs are typically combined with sensors to measure physical, chemical, and biochemical signals during cell growth and interactions. Equipping GoC platforms with microbial devices will allow the study and monitoring of both the mammalian and the microbial cells, which is important due to the intricate interplay between these two domains of life.

3.4 Applications: On Skin, in Gut, and in Oral Microbial Sensing

Human sweat is a rich source for biochemical conversions. Biobatteries based on several different skin bacteria attached to the epidermis were able to generate energy from human sweat, especially from the presence of lactate and glucose (Fig. 7a). To increase the low output voltage (0.5 V) to meet practical applications, the biobattery was connected to a DC-DC voltage booster circuit, and together with a charge-storing supercapacitor, the biobattery-based system was able to operate a thermometer requiring 3.0 V, input voltage (Fig. 7b) [100]. In a similar manner, in another study, glucose served as the analyte, and the resulting bio-electrochemical conversion produced a current in 16 connected MFCs built on a paper platform that powered up a light-emitting diode (LED) [104]. The LED intensity was proportional to the glucose concentration. In this case, exoelectrogens (EET-capable bacteria) were pre-inoculated into a PEDOT:PSS-coated conducting paper substrate in lyophilized (freeze-dried) form (Fig. 7c) and activated by saliva.

Operation of devices inside the gut, differently from sensing on the skin and in the oral space, poses much higher challenges in terms of delivery of the device, its stable operation, and biosafety. Nevertheless, considering its influence on health and well-being, the gut is a very important arena for future biosensing. In 2018, Mimee, Nadeau, and coworkers published their work about an ingestible capsule bacterial-electronic system to monitor gastrointestinal health [101]. The authors genetically engineered *E. coli* to produce bioluminescence in the presence of *heme* groups in blood. These cellular biosensors were enclosed into a cell carrier permeable to chemicals and together with a battery-powered photo-sensing circuit on a printed circuit board (PCB) were encapsulated by soft PDMS. The device could successfully detect porcine gastric bleeding 52 min after the administration of blood and simultaneously transmit the signal to a receiver unit through the antenna of the device. To avoid the detrimental effect of acidic gastric fluid on the biosensor, prior to delivering the device, the researchers neutralized the gastric fluid by administering a bicarbonate-glucose solution [101].

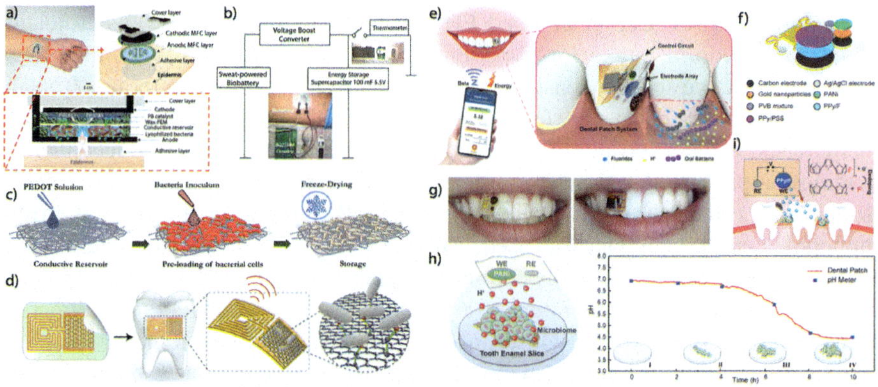

Fig. 7 Applications of microbial devices for medical/ healthcare purposes. **a** Biobattery skin-patch based on lyophilized electroactive bacteria and **b** voltage-boosting and energy-storing circuit to operate a thermometer based on skin-biobattery [100]; **c** preparation of the lyophilized conducting bacterial layer of biobattery on filter paper using PEDOT solution [104]. **d** Oral pathogen detection based on resistance change of a functionalized graphene sensing sheet and wireless transfer via inductive coupling [105]. **e**–**i**: Two-layer FHE dental patch device for monitoring pH changes over time and injection of fluorides when reaching the signal threshold [106], where **e** shows the overall principle and application of the device, **f** the functional layers, **g** photographs of the sensing part only (left) and the whole device (right) as dental patches, **h** sensing principle of pH (left) and changes in pH over time-based on the presence of microbes (right), and **i** principle of fluoride delivery from the device to the teeth. **a, b** Reprinted from Mohammadifar et al. [100] with permission from Elsevier. **c** Reprinted with permission from Mohammadifar and Choi [104] Copyright 2020 Wiley–VCH under CC BY 4.0 http://creativecommons.org/licenses/by/4.0/. **d** Reproduced with permission from Mannoor et al. [105] **e**–**i** Reproduced with permission from Shi et al. [106] Copyright 2022 Springer Nature Limited under open access CC BY-NC-ND 4.0 http://creativecommons.org/licenses/by/4.0/

For oral and dental applications, we highlight two methods of wireless pathogen detection: (1) utilizing the above-explained WPT by inductive coupling method and based on (2) flexible hybrid electronics (FHE). Mannoor et al. wirelessly detected oral pathogenic bacteria based on the change in the resistance of a graphene sensing sheet upon the attachment of cells (Fig. 7d).[105] To ensure selectivity, the authors prepared antimicrobial peptides specific to capturing Gram-negative bacteria *E. coli* and *Helicobacter pylori* and Gram-positive bacteria *Staphylococcus aureus*. The change in the resistance of the graphene sheet ultimately led to a change in its impedance, which, when incorporated into a receiver circuit as an impedance element, the bacterial attachment could be monitored wirelessly based on the inductive coupling method. The graphene-based sensing element was connected to a planar meander line inductor and interdigitated capacitive electrodes, printed onto a bioresorbable silk film, allowing the device to be transported and attached to a human molar (Fig. 7d). The characteristics of FHE technology are the seamless combination of soft interconnection circuitry and rigid surface mount devices (SMDs), such as IC chips and components, on a flexible substrate, allowing for advanced sensing, processing, and data transmission on a thin, small-size, mechanically flexible, and often soft platform. Lee et al. used FHE technology for real-time quantification of intraoral sodium intake for hypertension management [107]. The microfabricated ion-selective sodium sensor with functionalized polymer membrane, ICs, and stretchable interconnects were embedded in breathable and bio-friendly elastomeric membranes, while a soft microporous dental retainer allowed for a reduced thermal loading compared with typical elastomers. For wireless communication to an external device, e.g., a smartphone, Bluetooth low energy (BLE) combined with a low-profile monopole ceramic antenna was used. Another interesting recent example of using FHE for oral health includes the continuous quantification of oral chemicals simultaneously with pH and temperature [106]. In this device, the conducting polymers of PEDOT and polyaniline (PANI) were used for the transduction of ionic-to-electronic signals. In addition to intraoral wireless pH sensing, a dental patch based on FHE was also equipped with electrically controlled on-demand drug delivery of fluorides, offering local treatment coupled with detection (Fig. 7e) [108]. This versatile device consisted of a multi-layer thin sensing-and actuating circuit (Fig. 7f) and a similarly thin control and wireless circuit that could be attached to a dental surface (Fig. 7g). Real-time pH monitoring was realized through the collection of H^+ by a PANI film, while fluoride drug was released by a polypyrrole/fluoride (PPy/F) layer (Fig. 7f, h, i). In addition, to slow down the spontaneous leaking of fluorides, a PPy/PSS layer was prepared on top of the PPy/F layer (Fig. 7f).

3.5 Future Prospects in Microbial Electronic Devices for Healthcare

Due to better stability compared to enzymatic sensors, bacterial sensors are expected to be a viable and cheap technology for wearable biosensors after improving specificity by genetic engineering. Future research efforts are expected to make progress not only in improving the selectivity, sensitivity, biocompatibility, and integrability into humans but also in developing fully biodegradable devices. To avoid the detrimental effect of the stomach's acidic environment, devices relying on living cellular detectors could be preloaded by lyophilized (freeze-dried) bacteria that can be reactivated in the gut by a specific chemical or physical trigger. Recently, EET-capable pathogens have been identified in the human gut [109–111], and therefore future gut-based electrochemical detection methods can leverage EET as the detection method. This is because EET is directly linked to the respiratory cycle and thus the metabolic activity of bacterial cells, differently from passively capturing bacteria based on chemical signatures on the surface of the bacterial envelope, offering richer information compared to passive methods. In conclusion, as we have seen above, microbial sensing with high selectivity toward the desired bio-analytes and an advanced supporting electronics system capable of data analysis and wireless transmission to external devices are key in achieving functional microbial devices that can aid healthcare applications. Gene-engineered bacteria integrated with soft and hybrid electronics, presumably co-deposited by printing technologies, will allow highly selective biosensing coupled with powerful data processing based on SMD IC chips and wireless transfer. The role of such microbial technologies will not only be a real-time monitoring and augmenting of the chemical and metabolic state on and in the human body for the improvement of health and prevention of diseases but also the uncovering of yet hidden interrelationships between our own cells and the microbes living in us, the latter constituting 90% of cells in our body.

4 Concluding Remarks

In this chapter, we described current wireless biosensor technologies for smart contact lenses and microbial devices, especially focusing on their potential applications in healthcare. To create wireless biosensors, the principles of wireless systems using inductor, capacitor, and resistor (LCR) circuits between the transmitter and the receiver are important, so we showed several examples to detect physical and chemical bio-signals with the LCR circuits. The wireless system is as simple as the detected resonant frequency shifts by the changes of L and C values in the sensors, while the amplitude at the resonant frequency varies by the resistance change. For smart contact lenses, the main targets relate to (1) vision enhancement and (2) diagnosis: (1) by using wearable displays and holographical systems and (2) by analyzing biomarkers in tears and intraocular pressure in eyes. In practical use, bacterial detection on the lenses is crucially important, so we described several microbial devices that can detect wirelessly on the surfaces of tissues as well as in the oral cavity and soon also in the gut to wirelessly perform the early detection of disease markers and monitor health-related (bio)chemical signals.

References

1. Ibrahim, F.N., Jamail, N.A.M., Othman, N.A.: Development of wireless electricity transmission through resonant coupling. In: 4th IET Clean Energy and Technology Conference (CEAT 2016), p. 33-5. Institution of Engineering and Technology (2016)
2. Maxwell, J.C.: VIII. A dynamical theory of the electromagnetic field. Philos. Trans. R. Soc. Lond. **155**, 459–512 (1865). https://doi.org/10.1098/rstl.1865.0008
3. Poynting, J.H.: XV. On the transfer of energy in the electromagnetic field. Philos. Trans. R. Soc. Lond. **175**, 343–361 (1884). https://doi.org/10.1098/rstl.1884.0016
4. Huurdeman, A.A.: The Worldwide History of Telecommunications. Wiley, Hoboken (2003)
5. Hutin, M., Leblance, M.: Transformer system for electric railways (1894)
6. Tesila, N.: System of transmission of electrical energy (1900)
7. Jawad, A.M., Nordin, R., Gharghan, S.K., Jawad, H.M., Ismail, M.: Opportunities and challenges for near-field wireless power transfer: a review. Energies (Basel) **10**, 1022 (2017). https://doi.org/10.3390/en10071022
8. Zhong, W., Xu, D., Hui, R.S.Y.: Wireless Power Transfer. Springer, Singapore (2020)
9. Brown, W.C.: The history of power transmission by radio waves. IEEE Trans. Microw. Theory Tech. **32**, 1230–1242 (1984). https://doi.org/10.1109/TMTT.1984.1132833

10. Park, C., Lee, S., Cho, G.-H., Rim, C.T.: Innovative 5-m-off-distance inductive power transfer systems with optimally shaped dipole coils. IEEE Trans. Power Electron. **30**, 817–827 (2015). https://doi.org/10.1109/TPEL.2014.2310232
11. Kurs, A., Karalis, A., Moffatt, R., Joannopoulos, J.D., Fisher, P., Soljačić, M.: Wireless power transfer via strongly coupled magnetic resonances. Science (1979). **317**, 83–86 (2007). https://doi.org/10.1126/science.1143254
12. Garnica, J., Chinga, R.A., Lin, J.: Wireless power transmission: from far field to near field. Proc. IEEE **101**, 1321–1331 (2013). https://doi.org/10.1109/JPROC.2013.2251411
13. Popovic, Z.: Near- and far-field wireless power transfer. In: 2017 13th International Conference on Advanced Technologies, Systems and Services in Telecommunications (TELSIKS), pp. 3–6. IEEE (2017)
14. El Rayes, M., Nagib, G., Ali Abdelaal, W.: A review on wireless power transfer. Int. J. Eng. Trends Technol. **40**, 272–280 (2016). https://doi.org/10.14445/22315381/IJETT-V40P244
15. Federal Communication Commission: Specific Absorption Rate (SAR) for Cellular Telephones. https://www.fcc.gov/general/specific-absorption-rate-sar-cellular-telephones
16. Mahmood, A.I., Gharghan, S.K., Eldosoky, M.A., Soliman, A.M.: Near-field wireless power transfer used in biomedical implants: a comprehensive review. IET Power Electron. **15**, 1936–1955 (2022). https://doi.org/10.1049/pel2.12351
17. Kim, H.-J., Hirayama, H., Kim, S., Han, K.J., Zhang, R., Choi, J.-W.: Review of near-field wireless power and communication for biomedical applications. IEEE Access **5**, 21264–21285 (2017). https://doi.org/10.1109/ACCESS.2017.2757267
18. Takamatsu, T., Chen, Y., Yoshimasu, T., Nishizawa, M., Miyake, T.: Highly efficient, flexible wireless-powered circuit printed on a moist. Soft Contact Lens. Adv Mater Technol. **4**, 1800671 (2019). https://doi.org/10.1002/admt.201800671
19. Okasili, I., Elkhateb, A., Littler, T.: A review of wireless power transfer systems for electric vehicle battery charging with a focus on inductive coupling. Electronics (Basel) **11**, 1355 (2022). https://doi.org/10.3390/electronics11091355
20. Shinohara, N.: The wireless power transmission: inductive coupling, radio wave, and resonance coupling. Wiley Interdiscip. Rev. Energy Environ. **1**, 337–346 (2012). https://doi.org/10.1002/wene.43
21. Maulana, E., Abidin, Z., Djuriatno, W.: Wireless power transfer characterization based on inductive coupling method. In: 2018 Electrical Power, Electronics, Communications, Controls and Informatics Seminar (EECCIS), pp. 164–168. IEEE (2018)
22. Erel, M.Z., Bayindir, K.C., Aydemir, M.T., Chaudhary, S.K., Guerrero, J.M.: A comprehensive review on wireless capacitive power transfer technology: fundamentals and applications. IEEE Access **10**, 3116–3143 (2022). https://doi.org/10.1109/ACCESS.2021.3139761
23. Federal Register, United States Government Publishing Office: Code of Federal Regulations. https://www.govinfo.gov/app/collection/cfr/2022/
24. Huang, S., Li, Z., Lu, K.: Frequency splitting suppression method for four-coil wireless power transfer system. IET Power Electron. **9**, 2859–2864 (2016). https://doi.org/10.1049/iet-pel.2015.0376
25. Huang, R., Zhang, B., Qiu, D., Zhang, Y.: Frequency splitting phenomena of magnetic resonant coupling wireless power transfer. IEEE Trans. Magn. **50**, 1–4 (2014). https://doi.org/10.1109/TMAG.2014.2331143
26. Azhari, S., Kimizuka, K., Méhes, G., Usami, Y., Hayashi, Y., Tanaka, H., Miyake, T.: Integration of wireless power transfer technology With hierarchical multiwalled carbon nanotubes-polydimethylsiloxane piezo-responsive pressure sensor for remote force measurement. IEEE Sens. J. **23**, 7902–7909 (2023). https://doi.org/10.1109/JSEN.2023.3248021
27. U.S. EPA Office of Solid Waste: Batteries. https://guides.library.illinois.edu/battery-recycling
28. Sun, S.W.: Understanding the capacitive coupling with influence factors and applications. J. Phys. Conf. Ser. **1087**, 042011 (2018). https://doi.org/10.1088/1742-6596/1087/4/042011
29. Liu, Q., Sun, X.-B.: Indirect electrical injuries from capacitive coupling: a rarely mentioned electrosurgical complication in monopolar laparoscopy. Acta Obstet. Gynecol. Scand. **92**, 238–241 (2013). https://doi.org/10.1111/aogs.12049

30. Consumer Action: The Contact Lens Rule and the Eyeglass Rule. https://www.consumer-action.org/english/articles/contact_lens_rule
31. Enoch, J.: First known lenses originating in Egypt about 4600 years ago. Doc. Ophthalmol. **99**, 303–314 (1999). https://doi.org/10.1023/A:1002747025372
32. Enoch, J.M., Lakshminarayanan, V.: Duplication of unique optical effects of ancient Egyptian lenses from the IV/V Dynasties: lenses fabricated ca 2620±2400 BC or roughly 4600 years ago. Ophthalmic Physiol. Opt. **20**, 126–130 (2000). https://doi.org/10.1046/j.1475-1313.2000.00496.x
33. Wikipédia: Le Scribe accroupi—wikipédia, l'encyclopédie libre. http://fr.wikipedia.org/w/index.php?title=Le_Scribe_accroupi&oldid=202725852
34. Wikipedia contributors: Nimrud lens—wikipedia, the free encyclopedia. https://en.wikipedia.org/w/index.php?title=Nimrud_lens&oldid=1134659653
35. Wikipedia contributors: Glasses—wikipedia, the free encyclopedia. https://en.wikipedia.org/w/index.php?title=Glasses&oldid=1165948579
36. Wikipedia contributors: History of the telescope—wikipedia, the free encyclopedia, https://en.wikipedia.org/w/index.php?title=History_of_the_telescope&oldid=1161734303
37. Pearson, R.M.: Karl Otto Himmler, manufacturer of the first contact lens. Cont. Lens Anterior Eye **30**, 11–16 (2007). https://doi.org/10.1016/j.clae.2006.10.003
38. Wikipedia contributors: Contact lens—wikipedia, the free encyclopedia. https://en.wikipedia.org/w/index.php?title=Contact_lens&oldid=1163860896
39. Wikipedia contributors: Rigid gas permeable lens—wikipedia, the free encyclopedia. https://en.wikipedia.org/w/index.php?title=Rigid_gas_permeable_lens&oldid=1150558911
40. Lingley, A.R., Ali, M., Liao, Y., Mirjalili, R., Klonner, M., Sopanen, M., Suihkonen, S., Shen, T., Otis, B.P., Lipsanen, H., Parviz, B.A.: A single-pixel wireless contact lens display. J. Micromech. Microeng. **21**, 125014 (2011). https://doi.org/10.1088/0960-1317/21/12/125014
41. Senior, M.: Novartis signs up for Google smart lens. Nat. Biotechnol. **32**, 856–856 (2014). https://doi.org/10.1038/nbt0914-856
42. Park, J., Kim, J., Kim, S.-Y., Cheong, W.H., Jang, J., Park, Y.-G., Na, K., Kim, Y.-T., Heo, J.H., Lee, C.Y., Lee, J.H., Bien, F., Park, J.-U.: Soft, smart contact lenses with integrations of wireless circuits, glucose sensors, and displays. Sci. Adv. **4** (2018). https://doi.org/10.1126/sciadv.aap9841
43. Li, J., Wang, Y., Liu, L., Xu, S., Liu, Y., Leng, J., Cai, S.: A biomimetic soft lens controlled by electrooculographic signal. Adv. Funct. Mater. **29** (2019). https://doi.org/10.1002/adfm.201903762
44. Guo, S., Wu, K., Li, C., Wang, H., Sun, Z., Xi, D., Zhang, S., Ding, W., Zaghloul, M.E., Wang, C., Castro, F.A., Yang, D., Zhao, Y.: Integrated contact lens sensor system based on multifunctional ultrathin MoS2 transistors. Matter **4**, 969–985 (2021). https://doi.org/10.1016/j.matt.2020.12.002
45. Cui, Y., Takamatsu, T., Shimizu, K., Miyake, T.: Near-infrared fundus imaging system with light illumination from an electronic contact lens. Appl. Phys. Express **15**, 027001 (2022). https://doi.org/10.35848/1882-0786/ac4675
46. Ilardi, V.: Renaissance Vision from Spectacles to Telescopes. American Philosophical Society (2007)
47. King, H.C.: The History of the Telescope. Dover Publications Inc. (2003)
48. Drake, S.: Galileo: Pioneer Scientist. University of Toronto Press (1990)
49. Hellemans, A., Bunch, B.: The Timetables of Science: A Chronology of the Most Important People and Events in the History of Science. Simon & Schuster (1988)
50. Butterfield, G.H.: Corneal contact lens (1950)
51. Tuohy, K.M.: Contact lens (1948)
52. Musgrave, C.S.A., Fang, F.: Contact lens materials: a materials science perspective. Materials. **12**, 261 (2019). https://doi.org/10.3390/ma12020261
53. Key, J.E.: Development of contact lenses and their worldwide use. Eye Contact Lens: Sci. Clin. Pract. **33**, 343–345 (2007). https://doi.org/10.1097/ICL.0b013e318157c230

54. Wichterle, O., Lím, D.: Hydrophilic gels for biological use. Nature **185**, 117–118 (1960). https://doi.org/10.1038/185117a0
55. Akerman, D.: Our greatest opportunity. Cont. Lens Anterior Eye **41**, 319–320 (2018). https://doi.org/10.1016/j.clae.2018.05.007
56. Efron, N.: Twenty years of silicone hydrogel contact lenses: a personal perspective. Clin. Exp. Optom. **103**, 251–253 (2020). https://doi.org/10.1111/cxo.13062
57. Ho, H., Saeedi, E., Kim, S.S., Shen, T.T., Parviz, B.A.: Contact lens with integrated inorganic semiconductor devices. In: Proceedings of the IEEE International Conference on Micro Electro Mechanical Systems (MEMS), pp. 403–406 (2008)
58. Pandey, J., Liao, Y.-T., Lingley, A., Mirjalili, R., Parviz, B., Otis, B.P.: A fully integrated RF-powered contact lens with a single element display. IEEE Trans. Biomed. Circuits Syst. **4**, 454–461 (2010). https://doi.org/10.1109/TBCAS.2010.2081989
59. Pandey, J., Liao, Y.-T., Lingley, A., Parviz, B., Otis, B.: Toward an active contact lens: Integration of a wireless power harvesting IC. In: 2009 IEEE Biomedical Circuits and Systems Conference, pp. 125–128. IEEE (2009)
60. Otis, B., Parviz, B.: Introducing our smart contact lens project. https://blog.google/alphabet/introducing-our-smart-contact-lens/
61. Kraft, A.: What a chemistry student should know about the history of Prussian blue. ChemTexts **4**, 16 (2018). https://doi.org/10.1007/s40828-018-0071-2
62. Granqvist, C.G.: Handbook of Inorganic Electrochromic Materials. Elsevier (1995)
63. Hjelm, A., Granqvist, C., Wills, J.: Electronic structure and optical properties of WO_3, $LiWO_3$, $NaWO_3$, and HWO_3. Phys. Rev. B Condens. Matter **54**, 2436–2445 (1996). https://doi.org/10.1103/physrevb.54.2436
64. Knittlmayer, C., Muffler, H.-J., Fischer, C.-H., Weppner, W.: Investigation of electrochromic tungsten trioxide thin films prepared by the ILGAR method. Ionics (Kiel) **12**, 127–130 (2006). https://doi.org/10.1007/s11581-006-0022-6
65. Deb, S.K.: A novel electrophotographic system. Appl Opt. **8**, 192 (1969). https://doi.org/10.1364/AO.8.S1.000192
66. Kim, M., Jung, I.D., Kim, Y., Yun, J., Gao, C., Lee, H.-W., Lee, S.W.: An electrochromic alarm system for smart contact lenses. Sens. Actuators B Chem. **322**, 128601 (2020). https://doi.org/10.1016/j.snb.2020.128601
67. Hu, L., Chen, L., Du, N., Takamatsu, T., Xiao, T., Miyake, T.: Electrochromic soft contact lenses with built-in non-interfering, high-efficient dual-band wireless power transfer system. Sens. Actuators A Phys. **344**, 113766 (2022). https://doi.org/10.1016/j.sna.2022.113766
68. Deb, S.K.: Optical and photoelectric properties and colour centres in thin films of tungsten oxide. Philos. Mag. **27**, 801–822 (1973). https://doi.org/10.1080/14786437308227562
69. Chiou, J.-C., Huang, Y.-C., Yeh, G.-T.: A capacitor-based sensor and a contact lens sensing system for intraocular pressure monitoring. J. Micromech. Microeng. **26**, 015001 (2016). https://doi.org/10.1088/0960-1317/26/1/015001
70. Chen, G.-Z., Chan, I.-S., Lam, D.C.C.: Capacitive contact lens sensor for continuous non-invasive intraocular pressure monitoring. Sens. Actuators Phys. **203**, 112–118 (2013). https://doi.org/10.1016/j.sna.2013.08.029
71. Kim, J., Kim, M., Lee, M.-S., Kim, K., Ji, S., Kim, Y.-T., Park, J., Na, K., Bae, K.-H., Kyun Kim, H., Bien, F., Young Lee, C., Park, J.-U.: Wearable smart sensor systems integrated on soft contact lenses for wireless ocular diagnostics. Nat. Commun. **8**, 14997 (2017). https://doi.org/10.1038/ncomms14997
72. Keum, D.H., Kim, S.-K., Koo, J., Lee, G.-H., Jeon, C., Mok, J.W., Mun, B.H., Lee, K.J., Kamrani, E., Joo, C.-K., Shin, S., Sim, J.-Y., Myung, D., Yun, S.H., Bao, Z., Hahn, S.K.: Wireless smart contact lens for diabetic diagnosis and therapy. Sci. Adv. **6**, eaba3252 (2020). https://doi.org/10.1126/sciadv.aba3252
73. Kim, J., Cha, E., Park, J.: Recent advances in smart contact lenses. Adv. Mater. Technol. **5**, 1900728 (2020). https://doi.org/10.1002/admt.201900728
74. Mirzajani, H., Mirlou, F., Istif, E., Singh, R., Beker, L.: Powering smart contact lenses for continuous health monitoring: recent advancements and future challenges. Biosens. Bioelectron. **197**, 113761 (2022). https://doi.org/10.1016/j.bios.2021.113761

75. Sebbag, L., Mochel, J.P.: An eye on the dog as the scientist's best friend for translational research in ophthalmology: focus on the ocular surface. Med. Res. Rev. **40**, 2566–2604 (2020). https://doi.org/10.1002/med.21716
76. Chiou, J.-C., Hsu, S.-H., Huang, Y.-C., Yeh, G.-T., Liou, W.-T., Kuei, C.-K.: A wirelessly powered smart contact lens with reconfigurable wide range and tunable sensitivity sensor readout circuitry. Sensors **17**, 108 (2017). https://doi.org/10.3390/s17010108
77. Song, H., Shin, H., Seo, H., Park, W., Joo, B.J., Kim, J., Kim, J., Kim, H.K., Kim, J., Park, J.: Wireless non-invasive monitoring of cholesterol using a smart contact lens. Adv. Sci. **9**, 2203597 (2022). https://doi.org/10.1002/advs.202203597
78. Takamatsu, T., Sijie, Y., Shujie, F., Xiaohan, L., Miyake, T.: Multifunctional high-power sources for smart contact lenses. Adv. Funct. Mater. **30**, 1906225 (2020). https://doi.org/10.1002/adfm.201906225
79. Takamatsu, T., Sijie, Y., Miyake, T.: Wearable, implantable, parity-time symmetric bioresonators for extremely small biological signal monitoring. Adv. Mater. Technol. 2201704 (2023). https://doi.org/10.1002/admt.202201704
80. Leonardi, M., Pitchon, E.M., Bertsch, A., Renaud, P., Mermoud, A.: Wireless contact lens sensor for intraocular pressure monitoring: assessment on enucleated pig eyes. Acta Ophthalmol. **87**, 433–437 (2009). https://doi.org/10.1111/j.1755-3768.2008.01404.x
81. Høiby, N.: A short history of microbial biofilms and biofilm infections. APMIS **125**, 272–275 (2017). https://doi.org/10.1111/APM.12686
82. Butlin, K.R., Adams, M.E., Thomas, M.: Sulphate-reducing bacteria and internal corrosion of ferrous pipes conveying water. Nature **163**(4131), 26–27 (1949). https://doi.org/10.1038/163026a0
83. Rabaey, K., Verstraete, W.: Microbial fuel cells: novel biotechnology for energy generation. Trends Biotechnol. **23**, 291–298 (2005). https://doi.org/10.1016/J.TIBTECH.2005.04.008
84. Atkinson, J.T., Su, L., Zhang, X., Bennett, G.N., Silberg, J.J., Ajo-Franklin, C.M.: Real-time bioelectronic sensing of environmental contaminants. Nature **611**(7936), 548–553 (2022). https://doi.org/10.1038/s41586-022-05356-y
85. Kuss, S., Amin, H.M.A., Compton, R.G.: Electrochemical detection of pathogenic bacteria—recent strategies. Adv. Challenges. Chem Asian J. **13**, 2758–2769 (2018). https://doi.org/10.1002/ASIA.201800798
86. de Vos, W.M., Tilg, H., Van Hul, M., Cani, P.D.: Gut microbiome and health: mechanistic insights. Gut **71**, 1020–1032 (2022). https://doi.org/10.1136/gutjnl-2021-326789
87. Kelly, J.R., Borre, Y., O' Brien, C., Patterson, E., El Aidy, S., Deane, J., Kennedy, P.J., Beers, S., Scott, K., Moloney, G., Hoban, A.E., Scott, L., Fitzgerald, P., Ross, P., Stanton, C., Clarke, G., Cryan, J.F., Dinan, T.G.: Transferring the blues: depression-associated gut microbiota induces neurobehavioural changes in the rat. J. Psychiatr. Res. **82**, 109–118 (2016). https://doi.org/10.1016/j.jpsychires.2016.07.019
88. Rutsch, A., Kantsjö, J.B., Ronchi, F.: The gut-brain axis: how microbiota and host inflammasome influence brain physiology and pathology. Front. Immunol. **11**, 604179 (2020). https://doi.org/10.3389/fimmu.2020.604179
89. Miran, W., Naradasu, D., Okamoto, A.: Pathogens electrogenicity as a tool for in-situ metabolic activity monitoring and drug assessment in biofilms. iScience **24**, 102068 (2021). https://doi.org/10.1016/j.isci.2021.102068
90. Chen, M., Zhou, X., Liu, X., Zeng, R.J., Zhang, F., Ye, J., Zhou, S.: Facilitated extracellular electron transfer of Geobacter sulfurreducens biofilm with in situ formed gold nanoparticles. Biosens. Bioelectron. **108**, 20–26 (2018). https://doi.org/10.1016/j.bios.2018.02.030
91. Song, R.-B., Wu, Y., Lin, Z.-Q., Xie, J., Tan, C.H., Loo, J.S.C., Cao, B., Zhang, J.-R., Zhu, J.-J., Zhang, Q.: Living and conducting: coating individual bacterial cells with in situ formed polypyrrole. Angew. Chem. Int. Ed. **56**, 10516–10520 (2017). https://doi.org/10.1002/anie.201704729
92. Zajdel, T.J., Baruch, M., Méhes, G., Stavrinidou, E., Berggren, M., Maharbiz, M.M., Simon, D.T., Ajo-Franklin, C.M.: PEDOT:PSS-based multilayer bacterial-composite films for bioelectronics. Sci. Rep. **8**, 15293 (2018). https://doi.org/10.1038/s41598-018-33521-9

93. Freyman, M.C., Kou, T., Wang, S., Li, Y.: 3D printing of living bacteria electrode. Nano Res. **13**, 1318–1323 (2020). https://doi.org/10.1007/s12274-019-2534-1
94. Sawa, M., Fantuzzi, A., Bombelli, P., Howe, C.J., Hellgardt, K., Nixon, P.J.: Electricity generation from digitally printed cyanobacteria. Nat. Commun. **8**, 1327 (2017). https://doi.org/10.1038/s41467-017-01084-4
95. Méhes, G., Roy, A., Strakosas, X., Berggren, M., Stavrinidou, E., Simon, D.T.: Organic microbial electrochemical transistor monitoring extracellular electron transfer. Adv. Sci. **7**, 2000641 (2020). https://doi.org/10.1002/advs.202000641
96. Gross, B.J., El-Naggar, M.Y.: A combined electrochemical and optical trapping platform for measuring single cell respiration rates at electrode interfaces. Rev. Sci. Instrum. **86**, 064301 (2015). https://doi.org/10.1063/1.4922853
97. Spyropoulos, G.D., Savarin, J., Gomez, E.F., Simon, D.T., Berggren, M., Gelinas, J.N., Stavrinidou, E., Khodagholy, D.: Transcranial electrical stimulation and recording of brain activity using freestanding plant-based conducting polymer hydrogel composites. Adv. Mater. Technol. **5**, 1900652 (2020). https://doi.org/10.1002/admt.201900652
98. Diacci, C., Lee, J.W., Janson, P., Dufil, G., Méhes, G., Berggren, M., Simon, D.T., Stavrinidou, E.: Real-time monitoring of glucose export from isolated chloroplasts using an organic electrochemical transistor. Adv. Mater. Technol. **5**, 1900262 (2020). https://doi.org/10.1002/admt.201900262
99. Gao, Y., Mohammadifar, M., Choi, S.: From microbial fuel cells to biobatteries: moving toward on-demand micropower generation for small-scale single-use applications. Adv. Mater. Technol. **4**, 1900079 (2019). https://doi.org/10.1002/admt.201900079
100. Mohammadifar, M., Tahernia, M., Yang, J.H., Koh, A., Choi, S.: Biopower-on-Skin: Electricity generation from sweat-eating bacteria for self-powered E-Skins. Nano Energy **75**, 104994 (2020). https://doi.org/10.1016/j.nanoen.2020.104994
101. Mimee, M., Nadeau, P., Hayward, A., Carim, S., Flanagan, S., Jerger, L., Collins, J., McDonnell, S., Swartwout, R., Citorik, R.J., Bulović, V., Langer, R., Traverso, G., Chandrakasan, A.P., Lu, T.K.: An ingestible bacterial-electronic system to monitor gastrointestinal health. Science **1979**(360), 915–918 (2018). https://doi.org/10.1126/science.aas9315
102. Signore, M.A., De Pascali, C., Giampetruzzi, L., Siciliano, P.A., Francioso, L.: Gut-on-chip microphysiological systems: latest advances in the integration of sensing strategies and adoption of mature detection mechanisms. Sens Biosensing Res. **33**, 100443 (2021). https://doi.org/10.1016/j.sbsr.2021.100443
103. Shah, P., Fritz, J.V., Glaab, E., Desai, M.S., Greenhalgh, K., Frachet, A., Niegowska, M., Estes, M., Jäger, C., Seguin-Devaux, C., Zenhausern, F., Wilmes, P.: A microfluidics-based in vitro model of the gastrointestinal human–microbe interface. Nat. Commun. **7**, 11535 (2016). https://doi.org/10.1038/ncomms11535
104. Mohammadifar, M., Choi, S.: A papertronic, on-demand and disposable biobattery: saliva-activated electricity generation from lyophilized exoelectrogens preinoculated on paper. Adv. Mater. Technol. **2**, 1700127 (2017). https://doi.org/10.1002/admt.201700127
105. Mannoor, M.S., Tao, H., Clayton, J.D., Sengupta, A., Kaplan, D.L., Naik, R.R., Verma, N., Omenetto, F.G., McAlpine, M.C.: Graphene-based wireless bacteria detection on tooth enamel. Nat. Commun. **3**, 763 (2012). https://doi.org/10.1038/ncomms1767
106. Ling, W., Wang, Y., Lu, B., Shang, X., Wu, Z., Chen, Z., Li, X., Zou, C., Yan, J., Zhou, Y., Liu, J., Li, H., Que, K., Huang, X.: Continuously quantifying oral chemicals based on flexible hybrid electronics for clinical diagnosis and pathogenetic study. Research (2022). https://doi.org/10.34133/2022/9810129
107. Lee, Y., Howe, C., Mishra, S., Lee, D.S., Mahmood, M., Piper, M., Kim, Y., Tieu, K., Byun, H.-S., Coffey, J.P., Shayan, M., Chun, Y., Costanzo, R.M., Yeo, W.-H.: Wireless, intraoral hybrid electronics for real-time quantification of sodium intake toward hypertension management. Proc. Natl. Acad. Sci. **115**, 5377–5382 (2018). https://doi.org/10.1073/pnas.1719573115
108. Shi, Z., Lu, Y., Shen, S., Xu, Y., Shu, C., Wu, Y., Lv, J., Li, X., Yan, Z., An, Z., Dai, C., Su, L., Zhang, F., Liu, Q.: Wearable battery-free theranostic dental patch for wireless intraoral sensing and drug delivery. npj Flex. Electron. **6**, 49 (2022). https://doi.org/10.1038/s41528-022-00185-5

109. Naradasu, D., Miran, W., Sakamoto, M., Okamoto, A.: Isolation and characterization of human gut bacteria capable of extracellular electron transport by electrochemical techniques. Front Microbiol. **9** (2019). https://doi.org/10.3389/fmicb.2018.03267
110. Light, S.H., Su, L., Rivera-Lugo, R., Cornejo, J.A., Louie, A., Iavarone, A.T., Ajo-Franklin, C.M., Portnoy, D.A.: A flavin-based extracellular electron transfer mechanism in diverse Gram-positive bacteria. Nature **562**, 140–144 (2018). https://doi.org/10.1038/s41586-018-0498-z
111. Tahernia, M., Plotkin-Kaye, E., Mohammadifar, M., Gao, Y., Oefelein, M.R., Cook, L.C., Choi, S.: Characterization of Electrogenic Gut Bacteria. ACS Omega **5**, 29439–29446 (2020). https://doi.org/10.1021/acsomega.0c04362

Wearable Device for Daily Continuous Blood Pressure Estimation Based on Pulse Rate Measurement

Toshiyuki Hayase

1 Introduction

Blood pressure is a most basic vital sign, and recently, the importance of the daily continuous blood pressure (DCBP) has been pointed out [1–3]. Studies have been made on the clinical usefulness of ambulatory blood pressure monitoring (ABPM), in which daily blood pressure is automatically measured with a cuff-based portable sphygmomanometer [4–7]. There is also a measurement method called arterial line monitoring in which a catheter is clinically inserted into a blood vessel to measure the blood pressure continuously [8]. This method, however, is not suitable for DCBP measurement because it is highly invasive.

The author proposed a new DCBP estimation method and showed its validity in a previous report [9]. As shown in Fig. 1, the novelty of this method is that a continuous blood pressure estimation is obtained as a solution of the forward problem of a simple model representing the circulatory dynamics and the circulatory control with the input of pulse rate. On the other hand, cuff-type blood pressure measurement [10–12] which is an existing representative noninvasive blood pressure measurement method, and cuffless blood pressure measurement based on pulse waveform [13–18] or pulse propagation time [19–25], obtain the blood pressure as the inverse problem with the input of cuff pressure or pulse waveform. It is difficult to use these existing methods for DCBP measurement devices because of discomfort due to cuff pressure and degradation of measurement accuracy by body movement, respectively. In the previous study, we examined the validity of the proposed method by comparing the analysis results with the measurement results of the automatic sphygmomanometer at 30 min or 1 h intervals for one subject in four days and four subjects in one daytime

T. Hayase (✉)
President-Appointed Extraordinary Professor, Director, Frontier Research Institute for Interdisciplinary Sciences, Tohoku University, 6-3 Aramaki Aza-Aoba, Aoba-Ku, Sendai 980-8578, Japan
e-mail: hayase@fris.tohoku.ac.jp

K. Mitsubayashi (ed.), *Wearable Biosensing in Medicine and Healthcare*,
https://doi.org/10.1007/978-981-99-8122-9_9

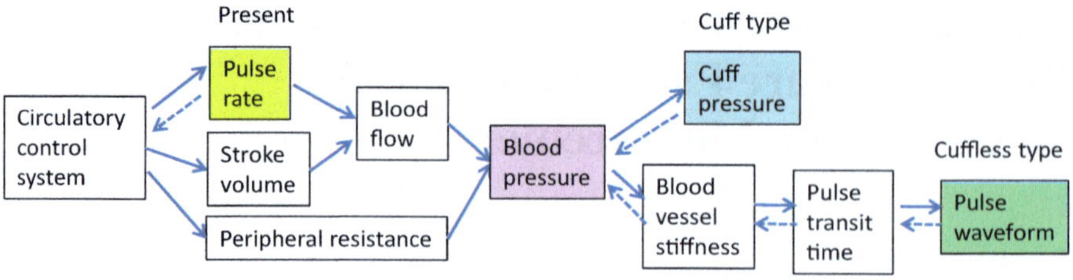

Fig. 1 Principles of blood pressure measurement for the proposed method, conventional cuff type, and cuffless one

[9]. As a result, it was shown that this estimation method gives a reasonable blood pressure estimate.

However, some unresolved problems remained in the previous study. Two parameters of the inverse model of the circulatory control system were determined through a one-parameter optimization problem by applying an empirical constraint for the parameters, the validity of which is not verified yet. Further, the statistical properties of the parameters are unknown since the number of subjects was small. The other two parameters of the circulatory dynamic system model were determined using half of the daily or daytime measurements of the automatic sphygmomanometer, but the effect of the number of measurement data on the parameters is also unknown. In order to apply this estimation method to practical blood pressure estimation devices, it is necessary to solve these problems to clarify the effect of constraints among model parameters and the number of the data of the standard measurement in parameters determination on the accuracy of estimation, realizing good accuracy, small computational load, and small parameter determination measurement number in the estimation method.

In the subsequent article [26], therefore, we focused on the experimental verification and optimization of the continuous blood pressure estimating method. In the article we simultaneously performed a 25-h continuous pulse rate measurement using a commercially available wearable device and blood pressure measurement within 30 min interval using an ABPM device for 29 subjects. Under four conditions for the constraint of model parameters, including the case where two parameters of the inverse model of the circulatory control system are independently changed, blood pressure estimations were performed to determine the optimal parameters and evaluated the estimation error for each condition by comparing the estimation results with the ABPM measurements. From these results, the validity of each condition for the constraint and the statistical properties of the parameters were clarified. Among these conditions, optimum parameter determination method was determined from the viewpoints of accuracy and computational costs. For the optimum parameter determination method of the circulatory control system, the effect of the number of the ABPM data on the accuracy of pressure estimation was investigated and the number of measurement data necessary for appropriate parameter determination was

(a) (b)

Fig. 2 Wearable device for **a** pulse rate measurement (http://www.epson.jp/products/wgps/sf710s/) and **b** DCBP estimation

obtained. From these results we clarified the applicability of the present estimation method to practical blood pressure estimation devices.

In this article we present this new DCBP estimation method and its experimental validation and optimization based on our former works [9, 26]. First the novel circulatory system model for the present blood pressure estimation with the input of pulse rate is presented in Sect. 2. In the following Sect. 3, we show a computational result for 24 h blood flow dynamics in which values of blood pressure, blood flow, and blood volume in left/right atrium/ventricle and pulmonary/systemic arteries/veins are obtained from pulse rate measurement data with a wearable device (Fig. 2a). Section 4 shows the experimental verification and optimization of the blood pressure estimation method. The blood pressure estimation program was installed in a smart watch (Fig. 2b). The summary of former sections and future expectations are given in Sect. 5.

2 Circulatory System Model

2.1 Introduction

In this section we show the circulatory system model for DCBP estimation. The following model is that of the second article [26], which is practically the same as that of the first report [9]. The circulatory system model consists of the circulatory dynamics model and the circulatory control inverse model. The real circulatory dynamics is a very complex fluid–structure coupled system [27], but the present study models it as a simple lumped parameter dynamical system consisting of eight elastic containers representing a left atrial (1), a left ventricle (2), systemic arteries (3), systemic veins (4), a right atrial (5), a right ventricle (6), pulmonary arteries (7), and pulmonary veins (8); and eight liner resistors connecting these containers (right side of Fig. 3a). The model can be described as basic equations of elastic mechanics

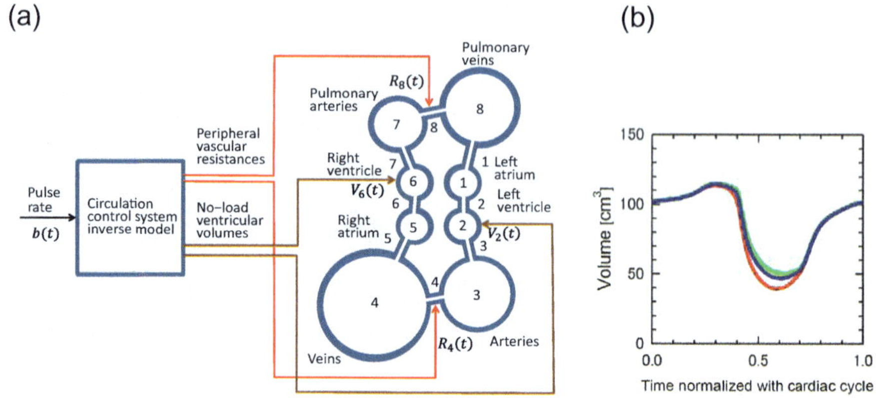

Fig. 3 **a** Circulatory system model consisting of circulatory dynamics model and circulatory control inverse model with the input of pulse rate and **b** variation of ventricular volume by the reference [29] (green), and those at zero pressure, or no-load ventricular volumes, derived from the reference values for left ventricle (red) and right ventricle (blue)

and fluid dynamics, Windkessel model [28], with variables of pressures and volumes of the containers $P_i(t)$ and $V_i(t)$, and flow rates between the containers $Q_i(t)$. Blood flow circulation is generated by introducing the variation of the ventricular volume at zero pressure, or *no-load ventricular volume*, based on the reference [29] to the left and right ventricles (Fig. 3b). Parameters affecting the systemic arterial blood pressure $P_3(t)$ in this circulatory dynamics model is the standard pulse rate b_0, the elasticity E_{30} and the peripheral resistance coefficient R_{40} of the lumped systemic arteries at the standard pulse rate.

The real circulatory control system is a very complex one including the short-term regulation by the autonomic nervous system and the long-term hormonal regulation, etc. [30], but the present study models it as a simple dynamical model with the input of pulse rate $b(t)$ and the outputs of peripheral vascular resistance of the systemic and pulmonary arteries, $R_4(t)$ and $R_8(t)$, and the no-load ventricular volumes, $V_2(t)$ and $V_6(t)$ (left side of Fig. 3a). The present model represents an inverse system of the real circulatory control system in which the pulse rate is also an output. The circulatory control inverse model is constructed based on the following characteristics: (1) the circulatory control system maintains blood pressure constant, (2) baroreceptors have differential characteristics to effectively respond to short-term changes of blood pressure [31]. (3) the ventricular stroke volume increases with increase of the pulse rate [32]. Parameters of the circulatory control inverse model are the rate s_a of variation of ventricular stroke volumes against the change of the pulse rate (cf. characteristic 3), the rate s_r of variation of peripheral vascular resistances against the change of the low-frequency component of the pulse rate (cf. characteristic 1), and the time constant T_c of the low pass filter characteristics of the control system with the input of the pulse rate (cf. characteristic 2). A total of six parameters of the present circulatory dynamics and circulatory control inverse models are determined by comparing the measured and calculated blood pressure values.

2.2 Circulatory Dynamics Model

The circulatory dynamics model consists of eight elastic containers representing a left atrial (1), a left ventricle (2), systemic arteries (3), systemic veins including organs (4), a right atrial (5), a right ventricle (6), pulmonary arteries (7), and pulmonary veins (8); and eight liner resistors connecting these containers (Fig. 3a). The numbers assigned to resistors are identical to those of downstream side containers.

Dynamics of the pressures in the containers are represented by the following equations.

$$\frac{dP_i}{dt} = E_i\left(Q_i - Q_{i+1} - \frac{dV_i}{dt}\right) \quad (i = 1, \cdots, 8) \tag{1}$$

$V_i(t)(i = 2, 6)$ represent the variations of the volume at zero pressure, or *no-load volume*, of the left and right ventricles, respectively, given by

$$V_i(t) = f_i(0) + a(t)\left\{f_i\left(\frac{b(t)\tau(t)}{b_0\tau_0}\right) - f_i(0)\right\} \quad (i = 2, 6) \tag{2}$$

where $b(t)$ is the pulse rate of the pulse including the time point t, $\tau(t)$ is the elapsed time from the beginning of this pulse, b_0 and τ_0 are the standard pulse rate and the corresponding cardiac cycle, respectively, $f_i(\tau(t)/\tau_0)$ are the variations of the no-load volume of the left and right ventricles for the standard pulse rate derived by reference to the literature [29] (Fig. 3b), $a(t)$, we call it as *no-load stroke volume ratio*, is the ratio of the ventricular volume change to that for the standard pulse rate, determination of the value of which by the circulatory control inverse model will be explained later. No-load volumes of the other containers are constant.

In Eq. (1), $E_i(i = 1 \cdots 8)$ are the elasticities of the containers. The elasticity of the left ventricle E_2 is assumed to take a relatively lower value for the internal pressure lower than a threshold due to buckling. The elasticity of lumped systemic arteries E_3 is defined as a function of the arterial pressure P_3 considering the characteristics of blood vessels [33]. Refer to the reference for details [9]. Elasticities of the other containers were set constant values for simplicity.

The flow rates Q_i through the resistors are given as

$$Q_i = \frac{C_i(P_{i-1} - P_i)}{R_i} \quad (i = 1 \cdots 8) \tag{3}$$

where C_i represent check valve characteristics to prevent reverse flow.

$$C_i = \begin{cases} 0 & P_{i-1} \leq P_i \\ 1 & P_{i-1} > P_i \end{cases} \tag{4}$$

Indices $i = 2, 3, 6, 7$ correspond to the mitral, aortic, tricuspid, and pulmonary valves, respectively. In the other resistors there is no possibility of reverse flow.

$R_i (i = 1 \cdots 8)$ are the resistance coefficients. Those of the peripheral systemic (4) and pulmonary (8) arteries, respectively, are modeled by the following expression.

$$R_i = R_{i0} r(t)/a(t) \quad (i = 4, 8) \tag{5}$$

where $r(t)$, we call it as *peripheral vascular resistance ratio,* is the ratio of the peripheral vascular resistance coefficient to that for the standard pulse rate multiplied with the no-load stroke volume ratio, determination of the value of which by the circulatory control inverse model will be explained later. The other resistance coefficients are set to constant values.

2.3 Circulatory Control Inverse Model

We explain the models for the no-load stroke volume ratio $a(t)$ and the peripheral vascular resistance ratio $r(t)$ in the followings. Taking into account of the characteristics of the circulatory control system that the ventricular stroke volume increases with increase of the pulse rate [32], the no-load stroke volume ratio $a(t)$ (see Eq. (2)) is modeled by the interpolation of the linear function $a(t) = b(t)/b_0$ of the pulse rate $b(t)$ and the constant $a(t) = 1$ with a weighting factor s_a, we call it as *stroke volume change rate.*

$$a(t) = s_a \frac{b(t)}{b_0} + (1 - s_a) \tag{6}$$

Since the circulatory control system maintains blood pressure constant [30], and baroreceptors have differential characteristics to effectively respond to short-term changes of blood pressure [31], the peripheral vascular resistance ratio $r(t)$ (see Eq. (5)) is modeled as the multiplication of the effects of the lower and higher frequency components of the pulse rate variation.

$$r(t) = r_{LF}(t) r_{HF}(t) \tag{7}$$

where the effect of the lower frequency component of the pulse rate variation $r_{LF}(t)$ is modeled by an inverse function of an interpolation of the linear function $b_{LF}(t)/b_0$ of the lower frequency component of the pulse rate variation $b_{LF}(t)$ and the constant 1 with a weighting factor s_r, we call it as *peripheral resistance change rate* whereas the effect of the higher frequency component of the pulse rate variation $r_{HF}(t)$ is modeled by the inversely proportional function of the pulse rate.

$$r_{LF}(t) = 1/(s_r b_{LF}(t)/b_0 + (1 - s_r)) \tag{8}$$

$$r_{HF}(t) = b_{LF}(t)/b(t) \tag{9}$$

It is noted that in the first paper [9] $r_{LF}(t)$ in Eq. (8) was defined as an interpolation of $b_0/b_{LF}(t)$, inverse function of the low frequency component of the pulse rate change $b_{LF}(t)$, and a constant value of 1. While the previous paper assumed the linearity between the pulse rate change and the peripheral resistance coefficient change, the second article [26] assumes the linearity between the pulse rate change and the peripheral vessel diameter change.

The lower frequency component of the pulse rate variation $b_{LF}(t)$ is expediently modeled by the following second-order low-pass filter with the cut-off frequency of ω_c and the time constant $T_c = 1/\omega_c$, we call it the time constant of slow pulse rate variation.

$$\begin{gathered} \frac{d^2 b_{LF}(t)}{dt^2} + 2\omega_c \frac{db_{LF}(t)}{dt} + \omega_c^2 b_{LF}(t) = \omega_c^2 b(t) \\ b_{LF}(0) = b(0) \end{gathered} \tag{10}$$

3 Summary

In this section the circulatory system model for DCBP estimation was presented. The circulatory system model consists of the circulatory dynamics model and the circulatory control inverse model. The real circulatory dynamics is a very complex fluid–structure coupled system, but the present study models it as a simple lumped parameter dynamical system. The real circulatory control system also is a very complex one including the short-term regulation by the autonomic nervous system and the long-term hormonal regulation, etc. [30], but the present study models it as a simple dynamical model with the input of pulse rate and the outputs of peripheral vascular resistance of the systemic and pulmonary arteries, and the no-load ventricular volumes.

4 Analysis Result of Circulatory System

4.1 Introduction

In this section an analysis result of the circulatory system model for DCBP estimation for one subject in one day is presented based on the first paper [9]. The input of the model or the continuous pulse rate variation was measured by a wearable device. Systolic and diastolic pressures and pulse rate were also measured with the sphygmomanometer with the interval of 30 or 60 min. Differential equations of the model were integrated using the measured pulse rate data. Values of six model parameters were determined for the data by comparing the result of the sphygmomanometer measurement and those of the computations for various combinations of these parameters.

4.2 Methods

Computation

Differential equations for the circulatory dynamics model and the circulatory control inverse model were numerically integrated with the 4-th order Runge–Kutta method. In order to prevent the accumulation of numerical errors, computational results are modified to maintain the total blood volume constant with the interval of one minute of the model time.

Subjects

The subject was a healthy male volunteer of 60 s (subject 1 of the former study [9]). Informed consent was obtained from the subject. The study was approved by the Ethics Committee of Graduate School of Engineering, Tohoku University (15A-9). All research methods were performed in accordance with relevant guidelines and regulations.

Verification Experiments

As the input of the present model, the pulse rate was measured for the subjects by a commercially available wearable device (Wristable GPS, SF-810, EPSON, Japan, Fig. 2a) with the measurement interval of one second in one day. As the purpose of comparison, systolic and diastolic pressures and pulse rate were also measured in sitting position with an automatic sphygmomanometer (HEM-1025, OMRON, Japan) in one day with the interval of 30 min (wake up hours) or 60 min (sleeping hours). Differential equations of the present model were integrated using the measurement data of the pulse rate by the wearable device. The computational time step was fixed to $\Delta t = 0.0002$ s according to preliminary calculations. Calculation was performed by a server (HPCT W215s, Intel Xeon Gold 6132, 2.6 GHz 14 Core $\times$ 2, 192 GB memory, HPC Tec, Japan) with a typical computational time of 370 s for a 24 h calculation. Values of the model parameters were determined for the day by comparing half of the sphygmomanometer measurements and the corresponding computations obtained with various combinations of these parameters. The validity of the present DCBP estimation method was then examined by comparing the other half of measurements and those of DCBP computations with the determined parameters.

Determination of the Optimum Model Parameters

Parameters affecting the systemic arterial blood pressure $P_3(t)$ in this circulatory dynamics model is the standard pulse rate b_0, the elasticity E_{30} and the peripheral resistance coefficient R_{40} of the lumped systemic arteries at the standard pulse rate. Parameters of the circulatory control inverse model are the stroke volume change rate s_a, the peripheral resistance change rate s_r, and time constant of the slow pulse rate variation T_c. Optimum values of these parameters were determined by comparing the result of the sphygmomanometer measurement and those of the computations for various combinations of these parameters. For detailed explanation refer to Sect. 4.2 or the paper [9].

4.3 Results

We show the results of the subject. Variation of the pulse rate in 24 h in one day is shown in Fig. 4a. Approximate times of activities in the day are, sleeping in 0:00–7:30, wake up in 7:30–24:00, meals at 8:00, 12:00, and 18:30, driving car at 9:00 and 18:00, office work in 10:00–17:30. Measurements of the wearable device (line) and those of the sphygmomanometer (circles) agree well. Differential equations of the present model were integrated using the measurement data of the pulse rate by the wearable device. Values of the model parameters, b_0, E_{30}, R_{40}, s_a, s_r, and T_c, were determined by comparing the result of the sphygmomanometer measurements and those of the computations performed with various combinations of these parameters. Variations of the no-load ventricular volume ratio $a(t)$ and the peripheral vascular resistance ratio $r(t)$ are shown in Fig. 4b. Computational results for the variations of pressures $P_i(t)$ and volumes $V_i(t)$ of eight elastic containers in the model are shown in Fig. 4c and 4d, respectively. Lines in Fig. 4e show computational results of the daily continuous blood pressure (DCBP) estimation for variation of the systolic (blue), average (red), diastolic (green), and pulse (brown) pressures at all pulses in 24 h obtained from the result of the arterial pressure $P_3(t)$ (Fig. 4c). Circles in the figure are corresponding measurement data by the sphygmomanometer.

In order to obtain the blood pressure estimation at a fixed time point, computation is necessary using pulse rate data in a certain period. According to preliminary calculations, it was confirmed that the computational result in the last 60 s of 120 s calculation starting from the time point 60 s ahead of the target time point using the initial value of the low frequency component of the pulse rate variation evaluated by the data in the period of 1,600 s, which is eight times of the time constant T_c, agreed well with the corresponding results of the 24 h calculation. The average values in the last 60 s were used in the comparison with the measurement data of sphygmomanometer with one minute temporal resolution.

Errors of the computations from measurements of the sphygmomanometer at 40 time points are shown in Fig. 4f for the systolic and diastolic pressures. Colors of the plots correspond to those in Fig. 4e, and closed and cross circles show the data used

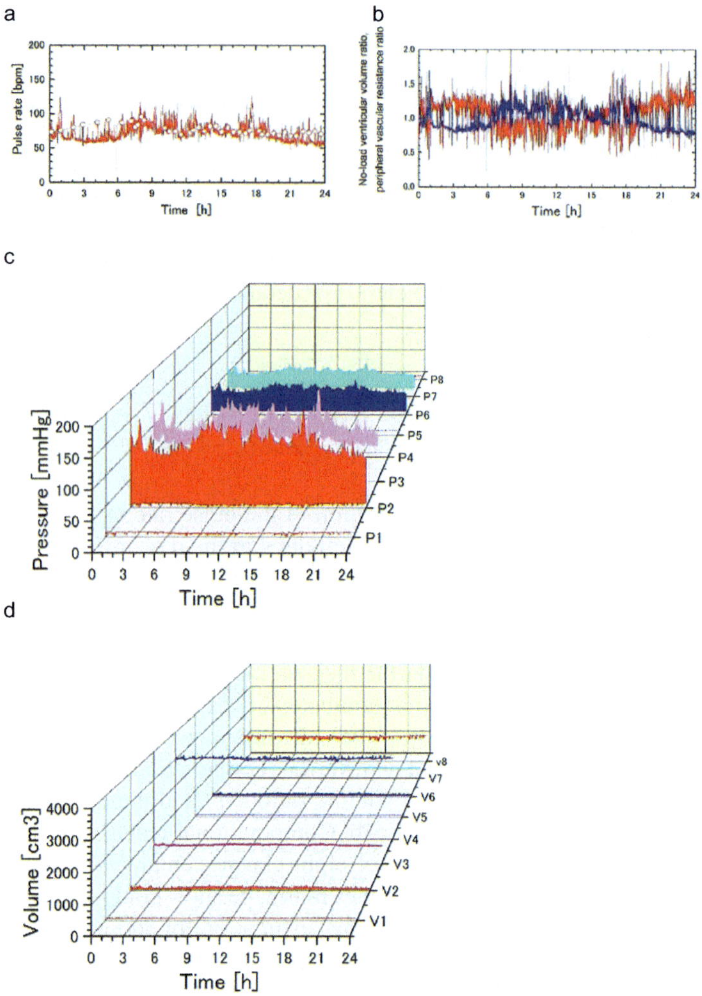

Fig. 4 Daily continuous blood pressure (DCBP) estimation for the subject 1 in the former study [9]. **a** Pulse rate measurements of wearable device (line) and those of sphygmomanometer (circles). **b** Computational results for no-load ventricular volume ratio (blue) and peripheral vascular resistance ratio (red). Pressures (**c**) and volumes (**d**) of eight elastic containers. **e** 24 h computations (lines) and measurements by sphygmomanometer (circles) for systolic (blue), average (red), diastolic (green), and pulse (brown) pressures. Same colors are used in **f**–**j**. **f** Variation of estimation errors of computations for systolic and diastolic pressures at time points for parameter determination (closed circles) and those for validation (cross circles). **g** Bland–Altman plot showing the relation between errors and averages of estimations and measurements for systolic and diastolic pressures with mean values (middle lines) and mean values $\pm 2 \times$ standard deviation (upper and lower lines) with data for parameter determination data (closed circles, solid lines) and validation data (cross circles, broken lines). **h** Correlation between measurements and calculations for systolic, mean, diastolic, and pulse pressures using the same symbols as those in **g**. Measurements (open circles), corresponding computations (closed circles), and all 24 h computations (light color circles) plotted with pulse rate (**i**) and with low-frequency component of pulse rate (**j**)

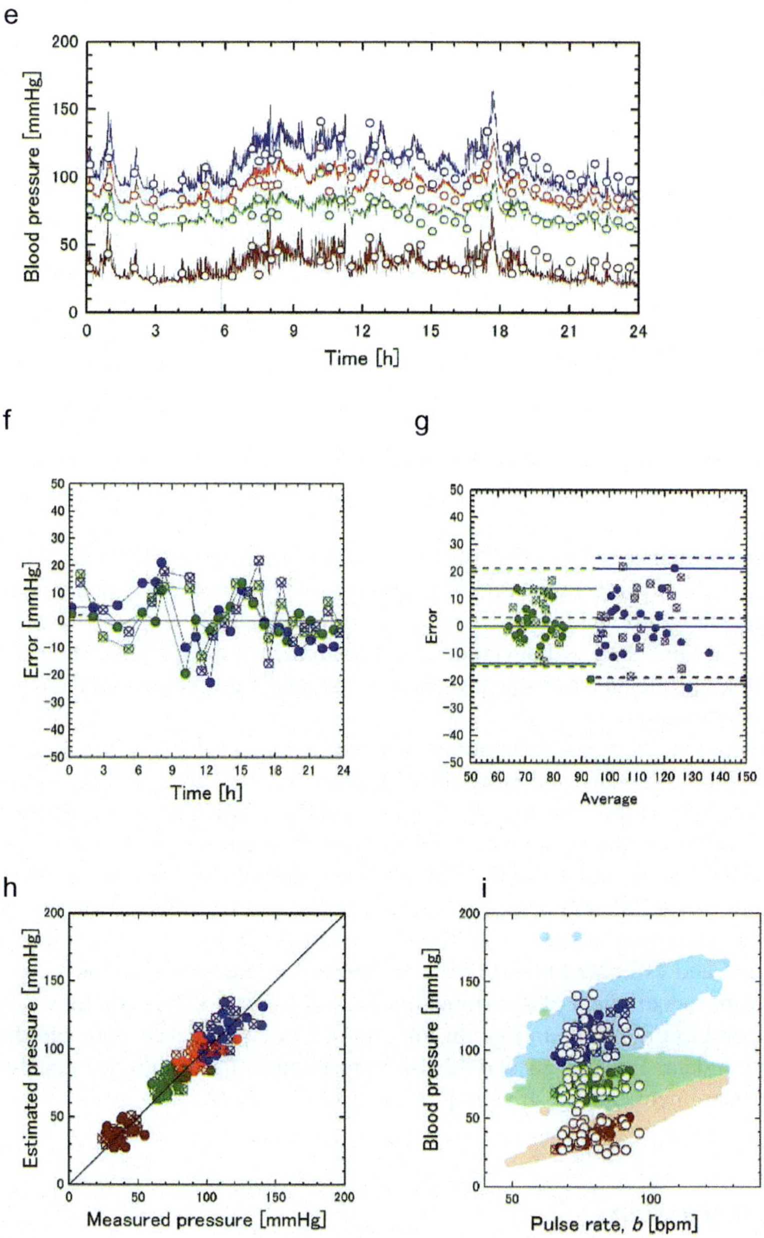

Fig. 4 (continued)

Fig. 4 (continued)

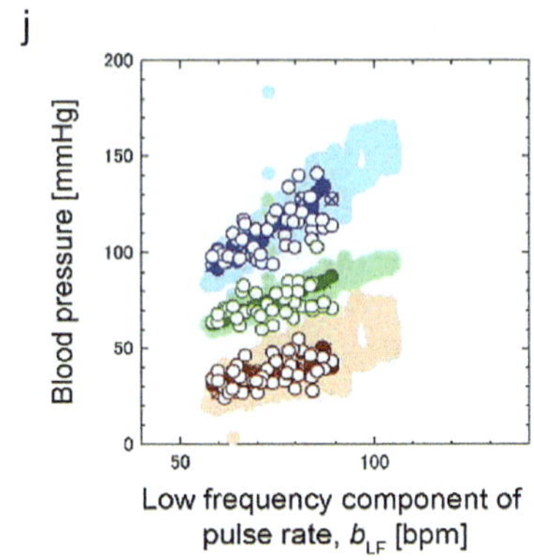

for parameter determination (20 time points) and those for validation (20 time points), respectively. Figure 4g (Bland–Altman plot) shows the relation between errors and averages for estimation and measurement for systolic (blue) and diastolic (green) pressures with mean values (middle lines) and mean values $\pm 2 \times$ standard deviation (upper and lower lines) with data for parameter determination (closed circles, solid lines) and those for validation (cross circles, broken lines). Rows 1–4 in Table 1 show mean values and standard deviations of measurements, estimations, and estimation errors for systolic and diastolic pressures in parameter determination data and those for validation ones.

Correlations between estimations and measurements for systolic (blue), mean (red), diastolic (green), and pulse (brown) pressures are shown in Fig. 4h using the same symbols as those in Fig. 4g. Figure 4i and 4j show the measurements (open circles), corresponding calculations (closed and cross circles), and all the 24 h calculations (light color circles) for the systolic (blue), diastolic (green), and pulse (brown) pressures plotted with the pulse rate and with the low-frequency component of the pulse rate variation, respectively. Rows 5–14 in Table 1 show correlations among measured and estimated blood pressures and pulse rate with Pearson's correlation coefficient r, coefficient of determination R^2, and slope of regression line. Rows 5–8 correspond to the results in Fig. 4h for parameter determination data and those for validation data. Rows 9–11 and 12–14 correspond to the results for computations and measurements in Fig. 4i and 4j, respectively.

4.4 Discussion

Computational results for systolic, average, diastolic, and pulse pressures for the subject agree with those of measurements with a sphygmomanometer as shown in Fig. 4e and 4h. The computational results for daily variation of systolic pressure show reduced blood pressure in the sleeping hours [4] and fluctuation of the blood pressure in the wake up hours [34] (Fig. 4e). Computational results for the variations

Table 1 DCBP estimation result for the subject in the former study [9]. Rows 1–4 show mean values and standard deviations of measurements, estimations, and estimation errors for systolic and diastolic pressures for parameter determination data and those for validation ones. Following rows show correlations among measured and estimated blood pressures and pulse rate with Pearson's correlation coefficient *r*, coefficient of determination R^2, and slope of regression line. Rows 5–8 correspond to the results in Fig. 4h for parameter determination data and those for validation ones. Rows 9–11 and 12–14 correspond to the results for computations and measurements in Fig. 4i and 4j, respectively

		Param. determ. Data			Validation data		
		Meas	Estim	Error	Meas	Estim	Error
Systolic	MEAN	112.0	112.6	0.0	109.4	112.0	3.2
	SD	13.3	12.6	10.6	10.9	11.8	11.0
Diastolic	MEAN	74.9	75.1	0.0	71.9	74.9	3.3
	SD	9.7	7.5	6.9	6.5	6.7	9.0
Independent variables	Dependent variables	Param. determ. data			Validation data		
		r	R^2	slope	r	R^2	slope
Meas. data	P_{sys}	0.646	0.417	0.576	0.570	0.325	0.660
	P_{dia}	0.702	0.492	0.485	0.177	0.031	0.205
	P_{ave}	0.713	0.509	0.588	0.440	0.194	0.551
	P_{pulse}	0.114	0.013	0.097	0.705	0.496	0.505
Independent variables	Dependent variables	Estimations			Measurements		
		r	R^2	slope	r	R^2	slope
Pulse rate	P_{sys}	0.635	0.403	1.173	0.083	0.007	0.145
	P_{dia}	0.361	0.130	0.388	0.069	0.005	0.083
	P_{pulse}	0.878	0.771	0.785	0.057	0.003	0.062
Low-frequency component of pulse rate	P_{sys}	0.969	0.940	1.307	0.606	0.368	0.817
	P_{dia}	0.997	0.994	0.782	0.454	0.206	0.420
	P_{pulse}	0.805	0.648	0.525	0.471	0.222	0.397

of pressures $P_i(t)$ and volumes $V_i(t)$ of eight elastic containers in the model are qualitatively in good agreement with those of the literature [35, 36] (Fig. 4c and 4d). Variation of the no-load ventricular volume ratio $a(t)$ and that of the peripheral vascular resistance ratio $r(t)$ show increase of the no-load ventricular volume and decrease in the peripheral vascular resistance in the systemic and pulmonary arteries with increasing pulse rate through the circulatory control inverse model, respectively (Fig. 4b).

As shown in Table 1, the standard deviation of measurements for the systolic pressure with parameter determination data and that with validation data were 13.3 mmHg and 10.9 mmHg, respectively, and those for the diastolic pressure were 9.7 mmHg and 6.5 mm Hg, respectively. Corresponding values for the errors of computations from measurements were 10.6 mmHg and 11.0 mmHg for the systolic pressure, and those for diastolic pressure were 6.9 mmHg and 9.0 mmHg, respectively, which are comparable with those of measurements. These standard deviations of the errors are less than 11.3 mmHg, which is evaluated assuming the standard deviation of a common sphygmomanometer (8 mmHg) for those measurements and computations, and independency between them.

$$\sigma_{cal-meas} = \sqrt{\sigma_{cal}^2 + \sigma_{meas}^2} = \sqrt{8^2 + 8^2} = 11.3(\text{mmHg}) \tag{11}$$

Mean value for the errors evaluated with validation data was 3.2 mmHg for the systolic pressure, and that for diastolic pressure was 3.3 mmHg, respectively, which are less than the tolerance of common sphygmomanometers (5 mmHg).

The correlation coefficients and the coefficients of determination have large values to show the effectiveness of the present estimation method as shown in Fig. 4h and rows 5–8 of Table 1. The correlation between the pulse rate and the blood pressure has been considered to be low [37]. As to the measurements and corresponding calculations, correlations with the pulse rate are low as shown in Fig. 4i and rows 9–11 of Table 1, being consistent with former studies. On the other hand, they have significant correlations with the low-frequency component of the pulse rate variation in Fig. 4j and rows 12–14 of Table 1, in accordance with the circadian cycle of the pulse rate [38] and that of the blood pressure [39]. As to all the 24 h calculations, the systolic and diastolic pressures have more significant correlations with the low-frequency component of the pulse rate whereas the pulse pressure has the one with the pulse rate, reflecting the present circulatory control inverse model.

5 Summary

Our results suggest that a fundamental part of DCBP can be represented by continuous pulse rate data and the simple circulatory dynamics and circulatory control inverse model with six model parameters. It is obviously easier to perform DCBP estimation by this method than by the other methods.

Although the present verification is very limited, the mean absolute error was comparable with that of the standard for wearable, cuffless blood pressure measuring devices [40]. Our results demonstrate how DCBP is appropriately estimated by the simple circulatory system model and the pulse rate measurement. We anticipate our methodology to be a starting point of new diagnosis based on DCBP [4–7]. Studies to clarify the relation between DCBP and diseases are important in many clinical departments. Furthermore, present six model parameters can be used as reliable personal vital signs relating the blood pressure, measurement of which often experiences large fluctuations [11].

6 Experimental Validation and Optimization

6.1 Introduction

In this section we show the experimental validation and optimization of the DCBP estimation method based on the second article [26]. We simultaneously performed a 25-h continuous pulse rate measurement using a commercially available wearable device and blood pressure measurement with 30 min interval using an ABPM device for 29 subjects. Under four conditions for the constraint of model parameters, including the case where two parameters of the inverse model of the circulatory control system are independently changed, blood pressure estimations are performed to determine the optimal parameters and evaluate the estimation error for each condition by comparing the estimation results with the ABPM measurements. From these results, the validity of each condition for the constraint and the statistical properties of the parameters are clarified. Among these conditions, optimum parameter determination method is determined from the viewpoints of accuracy and computational costs. For the optimum parameter determination method of the circulatory control system, the effect of the number of the ABPM data on the accuracy of pressure estimation is investigated and the number of measurement data necessary for appropriate parameter determination is obtained. From these results we clarify the applicability of the present estimation method to practical blood pressure estimation devices.

6.2 Method

In this study, we use a circulatory system model of the second article [26] explained in Sect. 2 which is slightly modified from that in the first paper [9] consisting of a circulatory dynamics model and a circulatory control inverse model with the input of pulse rate.

Computations

Differential equations for the circulatory dynamics model and the circulatory control inverse model were numerically integrated with the 4-th order Runge–Kutta method. In order to prevent accumulation of numerical errors, computational results are modified to maintain the total blood volume constant with the interval of one minute of the model time.

Subjects

The subjects were 29 volunteers with an average age of 44 years and a standard deviation of 13 years (Table 2). They are classified to classes 1–3 and none of them to class 4 in the classification of IEEE criteria for cuffless blood pressure measurement devices [40]. Informed consent was obtained from the subjects. The study was approved by the Ethics Committee of Graduate School of Engineering, Tohoku University (15A-9). All research methods were performed in accordance with relevant guidelines and regulations.

Verification Experiment

We developed a wearable DCBP estimation device (Fig. 2b). A DCBP estimation program based on the algorithm explained in Sect. 2 was installed in a commercially available wearable pulse rate measurement device (M600, POLAR, Sweden), and by specifying model parameters it is possible to perform real-time analysis to display estimated blood pressure. In this study, however, off-line analysis was performed in order to examine the validity of the DCBP estimation method including model parameters determination. As the input of the model, the pulse rate was measured for the subjects by the wearable device with the measurement interval of the pulse period for 25 h from the wake-up time. As the purpose of comparison, systolic and diastolic pressures and pulse rate were also measured in daily life condition with an ambulatory blood pressure (ABP) monitor (TM2433, A&D, Japan) in the above-mentioned time with the interval of 30 min. After the measurement by the wearable device and the ABP monitor, differential equations of the model were integrated using the measurement data of the pulse rate by the wearable device. The computational time step was fixed to $\Delta t = 0.0002$ s according to the former study [9]. Calculation was performed by a server (HPCT W215s, Intel Xeon Gold 6132, 2.6 GHz 14 Core $\times$ 2, 192 GB memory, HPC Tec, Japan) with a typical computational time of 385 s for a 25-h calculation. For each subject, values of the model parameters were determined by comparing half of the ABP monitor measurements and the corresponding computations obtained with various combinations of these parameters. The validity of the present DCBP estimation method was then examined by comparing the other half of the measurements and those of DCBP computations with the determined parameters.

Determination of the Model Parameters

In this study, the model parameters were determined for conditions denoted by Case 1 to Case 4 in Table 3, and the accuracy of blood pressure estimation was investigated

Table 2 Subjects data [26]

Total number		29
Gender	Male	26
	Female	3
Age	MEAN, SD	43.9, 12.6
	Max, Min	63, 23
Mean of ABPM measurement for systolic pressure	1 $\leq$ 119	10
	2 120–139	12
	3 140–160	7
	4 161 $\leq$	0

for each case. Parameters affecting the systemic arterial blood pressure $P_3(t)$ in this circulatory dynamics model is the standard pulse rate b_0, the elasticity E_{30} and the peripheral resistance coefficient R_{40} of the lumped systemic arteries at the standard pulse rate. Those of the circulatory control inverse model are the stroke volume change rate s_a, the peripheral resistance change rate s_r, and the time constant of the slow pulse rate variation T_c.

In Case 1, the results of odd number measurements of the ABP monitor in 25 h period (*parameter determination data*) for the systolic and diastolic pressures and the pulse rate, $P_{sysm}(t_n)$, $P_{diam}(t_n)$, $b_m(t_n)$ and quantities derived from these measurement results for the average pressure and the pulse pressure, $P_{avem}(t_n) = (P_{sysm}(t_n) + P_{diam}(t_n))/2$, $P_{pulsem}(t_n) = P_{sysm}(t_n) - P_{diam}(t_n)$, were used to determine above-mentioned model parameters by comparing them with the corresponding results of calculation, $P_{sysc}(t_n)$, $P_{diac}(t_n)$, $b_c(t_n)$, $P_{avec}(t_n)$, and $P_{pulsec}(t_n)$ in the following conditions whereas the results of the even number measurements (*validation data*) were used to verify the validity of the estimation results.

(1) b_0 is expediently determined as the average value of the pulse rate data in the parameter determination data of the ABP monitor whereas it was determined as the average value of measurement data of the wearable device in 24 h in former study [9].
(2) E_{30} and R_{40} are determined in the same way as the former study so that the average values of the computation for $P_{pulsec}(t_n)$ and $P_{avec}(t_n)$ are the same as the corresponding results of the measurement. A fixed point iterative method was used for the parameter determination.
(3) s_a and s_r are determined to minimize the following cost function by multiple conditions that the standard deviation of $P_{pulsec}(t_n)$ and that of $P_{avec}(t_n)$ are the same as the corresponding results of the measurement and that the mean square error of $P_{pulsec}(t_n)$ and thazt of $P_{avec}(t_n)$ are the minimum. The cost function J_{sum} is defined as the weighted sum of two functions corresponding to the above-mentioned conditions. The weighting factor $\alpha = 0.25[\text{Pa}^{-1}]$ was set to the same value as the former study.

Table 3 Definition of cases in verification experiment [26]

	Case 1	Case 2	Case 3	Case 3 (n)	Case 4
s_a, s_r	2D optimization	1D optimization	Mean of parameters in Case 2	Mean of parameters in Case 2	Mean of parameters in Case 2
E_{30}, R_{40}	2D optimization	2D optimization	2D optimization	2D optimization using first n data	Mean of parameters in Case 2
b_0	Mean of ABPM measurement for each subject	Mean of ABPM measurement for each subject	Mean of parameters in Case 2	Mean of parameters in Case 2	Mean of parameters in Case 2

$$J_{sum} = J_{sd} + \alpha J_{rms} \tag{12}$$

where

$$J_{sd} = \left(\sigma_{pulsec}/\sigma_{pulsem} - 1\right)^2 + \left(\sigma_{avec}/\sigma_{avem} - 1\right)^2$$

$$J_{rms} = \left[\frac{1}{2N}\sum\left\{\left(P_{pulsem}(t_n) - P_{pulsec}(t_n)\right)^2 + \left(P_{avem}(t_n) - P_{avec}(t_n)\right)^2\right\}\right]^{1/2} \tag{13}$$

The former function evaluates the degree of agreement for the standard deviations σ_{pulsec} and σ_{avec} of $P_{pulsec}(t_n)$ and $P_{avec}(t_n)$ with those of measurement σ_{pulsem} and σ_{avem}, respectively, and the latter function evaluates the mean square errors of $P_{pulsec}(t_n)$ and $P_{avec}(t_n)$. It is noted that a power of 1/2 was missing in the righthand side of the second formulation of Eq. (13), and the unit of $\alpha[\text{Pa}^{-1}]$ in Eq. (12) was incorrectly defined as $[\text{Pa}^{-2}]$ in the former study [9]. The parameter determination was performed as a two-dimensional optimization problem; a round-robin method was used to determine the values of s_a and s_r to minimize the cost function in Eq. (12) by changing these parameters between 0 and 1 with the increment of 0.1, respectively.

(4) T_c was set to 200 s which was determined in the former study [9] so that the mean absolute error of $P_{sysc}(t_n)$ and $P_{diac}(t_n)$ is small and the variations of the computed 25 h blood pressure properly represent the characteristics of the measurement.

Other cases are explained in the followings. Case 2, corresponding to the condition of the former study [9], assumes a simpler condition than Case 1 in which the parameter determination was performed as a one-dimensional optimization problem under the empirical constraint of Eq. (14) between two parameters and a round-robin method was used to determine the value of s_r to minimize the cost function in Eq. (12) by changing the parameter between 0 and 1 with the increment of 0.1.

$$s_a = 1 - s_r \tag{14}$$

Case 3 assumes a simpler condition than Case 2 in which values of the parameters s_a, s_r and b_0 are given as the mean values of those of all subjects in Case 2, respectively. Case 3(n) is basically the same as Case 3 but the first n data of the parameter determination data was used for determination of E_{30} and R_{40}. Case 4 assumes the simplest condition in this study in which all the model parameters are constant. In addition to s_a, s_r and b_0, parameters E_{30} and R_{40} are defined as the mean values of those of all subjects in Case 2, respectively.

From Eqs. (1) and (3), we obtain the approximate expressions of E_{30} and R_{40} by measurement values, P_{sysm}, P_{diam}, and b_m; and the stroke volume ΔV.

$$\begin{aligned} E_{30} &= (P_{sysm} - P_{diam})/\Delta V \\ R_{40} &= 60(P_{sysm} + P_{diam})/(2\Delta V b_m) \end{aligned} \tag{15}$$

Values of the other model parameters were determined in reference to the literature [35] since their effect on the systemic arterial pressure is relatively low.

6.3 Results

First, the results of Case 1 are shown in the followings. Figure 5a–e and f–i show the model parameters and the mean errors of estimated blood pressure with respect to the validation data of ABPM with the daily average value of systolic pressures measured by ABPM for all 29 subjects, respectively. Figure 5a shows the standard pulse rate b_0 determined as the average value of the pulse rate data in the parameter determination data of ABPM, Fig. 5b, c parameters of the circulatory control inverse model, or the stroke volume change rate s_a and the peripheral resistance change rate s_r, and Fig. 5d, e parameters of the circulatory dynamics model, or the elasticity E_{30} and the peripheral resistance coefficient R_{40} of the lumped systemic arteries at the standard pulse rate b_0, respectively. Figure 5f, g and h, i show the mean absolute errors for systolic and diastolic pressures, and the mean errors, respectively. Numbers at the plots show the subject numbers, and solid and broken lines show means and means $\pm$ standard deviations, respectively.

DCBP analysis results and measurements with the wearable device and the ABP monitor are compared in Fig. 6 for five subjects corresponding to representative results of Case 1 in Fig. 5. Left figures show measurements of pulse rate by the wearable device (lines) and the ABP monitor (symbols). In the followings, triangles and circles show the parameter determination data and the validation data of ABPM, respectively. Second column figures show daily variations of DCBP for systolic (blue), average (red), diastolic (green), and pulse (brown) pressures by computation (line) and ABP monitor measurement (symbols). Same colors are used in the following results. Third column figures show errors of computed systolic and diastolic pressures with respect to measured ones with ABPM. Fourth figures show correlations between the computations and the measurements for systolic, average, diastolic, and pulse pressures.

Figure 7 shows the correlation between the stroke volume change rate s_a and the peripheral resistance change rate s_r in Case 1 for all 29 subjects. Numbers at symbols show subject numbers, horizontal and vertical solid lines the mean values of s_a and s_r, respectively, and the inclined line is the direction of the first principal component vector of the covariance matrix.

Next, we compare the results of Cases 2–4 with those of Case 1 for all 29 subjects. Figure 8a–e show the relations between the model parameters of Case 1 and those of Cases 2–4 for (a) the standard pulse rate b_0, (b) the stroke volume change rate s_a, (c) the peripheral resistance change rate s_r, (d) the elasticity E_{30} and (e) the peripheral resistance coefficient R_{40} of the lumped systemic arteries at the standard pulse rate, respectively. Figures 8f-i show the relations between the errors of estimated blood pressure of Case 1 and those of Cases 2–4 for (f, g) the mean absolute errors and (h, i) the mean errors of estimated systolic and diastolic pressures, respectively, with

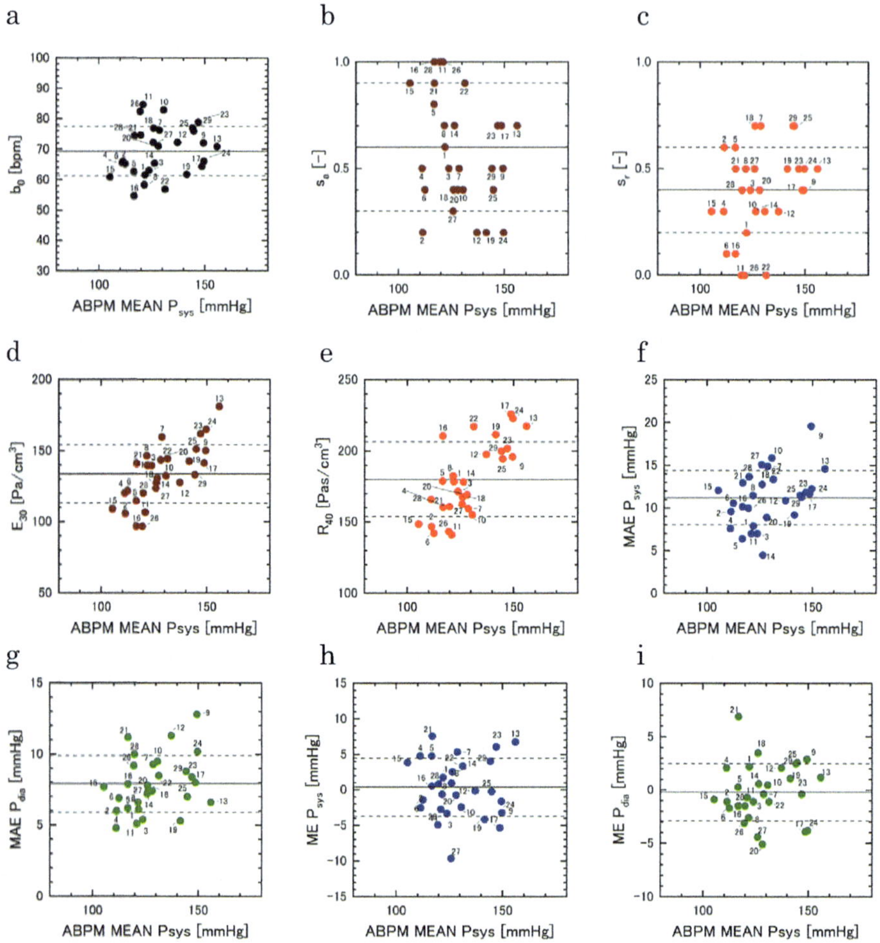

Fig. 5 Analysis results of Case 1. Model parameters and mean errors of estimated blood pressure with respect to the validation data of ABPM are shown with the daily average value of systolic pressure measured by ABPM for all 29 subjects for **a** the standard pulse rate b_0 determined as the average value of the pulse rate data in the parameter determination data of ABPM, **b**, **c** parameters of the circulatory control inverse model, the stroke volume change rate s_a and the peripheral resistance change rate s_r, **d**, **e** parameters of the circulatory dynamics model, the elasticity E_{30} and the peripheral resistance coefficient R_{40} of the lumped systemic arteries at the standard pulse rate, **f**, **g** the mean absolute errors for systolic and diastolic pressures, **h**, **i** the mean errors. Numbers at the plots show the subject numbers, and solid and broken lines show means and means $\pm$ standard deviations, respectively

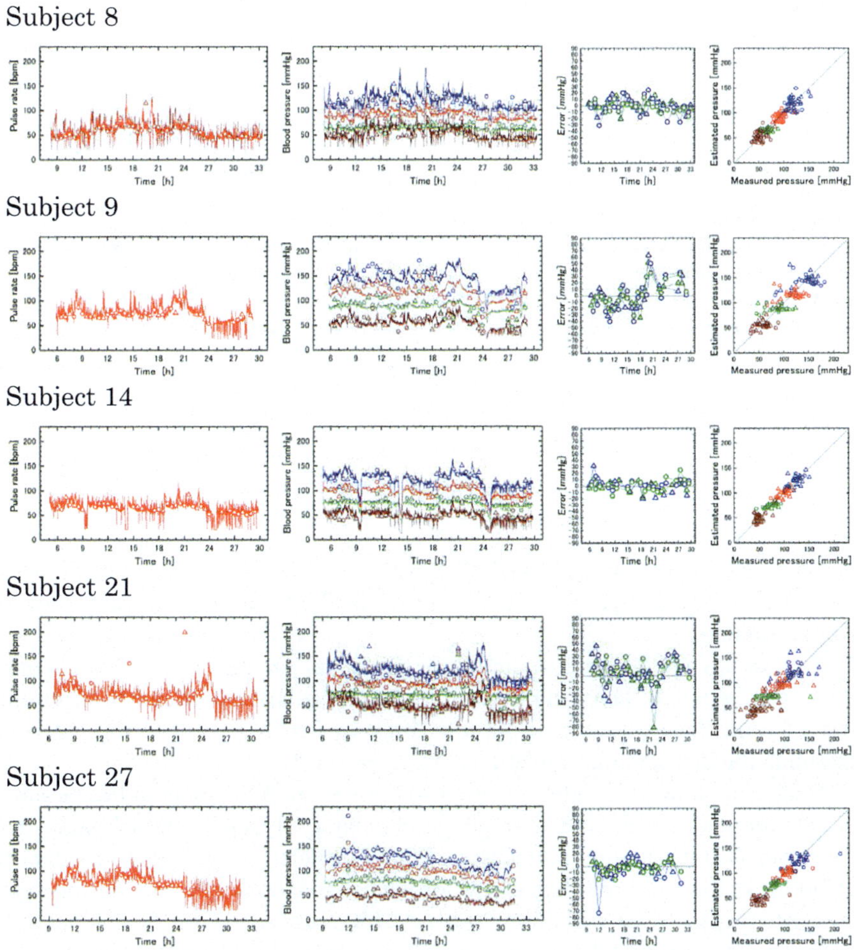

Fig. 6 Comparison among DCBP analysis results and measurements with the wearable device and the ABP monitor for five subjects corresponding to representative results of Case 1 in Fig. 5. First column figures: measurements of pulse rate by the wearable device (lines) and the ABP monitor (symbols). Triangles and circles show the parameter determination data and the validation data of ABPM, respectively. Second column figures: daily variations of DCBP for systolic (blue), average (red), diastolic (green), and pulse (brown) pressures by computation (line) and ABP monitor measurement (symbols). Same colors are used in the following results. Third column figures: errors of computed systolic and diastolic pressures with respect to measured ones with ABPM. Fourth column figures: correlations between computations and measurements for systolic, average, diastolic, and pulse pressures

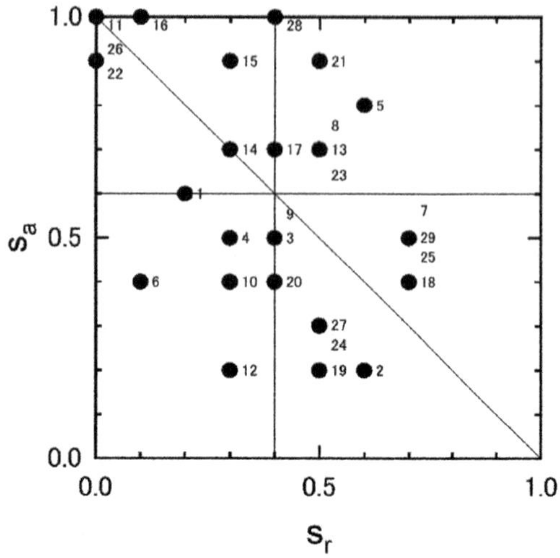

Fig. 7 Correlation between the stroke volume change rate s_a and the peripheral resistance change rate s_r in Case 1 for all 29 subjects. Numbers at symbols show subject numbers, horizontal and vertical lines the mean values of s_a and s_r, respectively, and the inclined line is the direction of the first principal component vector of the covariance matrix

respect to measurements with ABPM. Results of Cases 2, 3, and 4 are shown by closed, light-colored, and open symbols, respectively. Numbers at the plots show the subject numbers.

Mean values and standard deviations of the model parameters are summarized in Table 4 for Cases 1–4. Those of the mean error, standard deviation, and mean absolute error of estimated systolic and diastolic pressures with respect to the parameter determination data and the validation data, respectively, are in Table 5.

Finally, we show the results of Case 3(n) in which n, the number of the parameter determination data used to determine the circulatory dynamics model parameters, was changed in the condition of Case 3. For all subjects, mean values and standard deviations of the mean absolute errors for the estimated systolic and diastolic pressures with respect to the validation data of ABPM are plotted with the number of the parameter determination data in Fig. 9a and b, respectively. Corresponding results for the mean errors are shown in Fig. 9c and d, respectively. For the purpose of reference in these figures, the results at $n = 0$ show those of Case 4, in which the parameters were determined by averaging those of Case 3 for all the subjects, and the results at $n = 20$ show those of Case 3, in which all parameter determination data were used.

6.4 Discussion

In this study we showed that the continuous blood pressure estimating method based on the simple circulatory system model with the input of pulse rate proposed in the former study [9] can be applied to practical DCBP estimation devices. As the evidence for that, a 25-h continuous pulse rate measurement using a commercially available wearable device and blood pressure measurement with 30 min interval

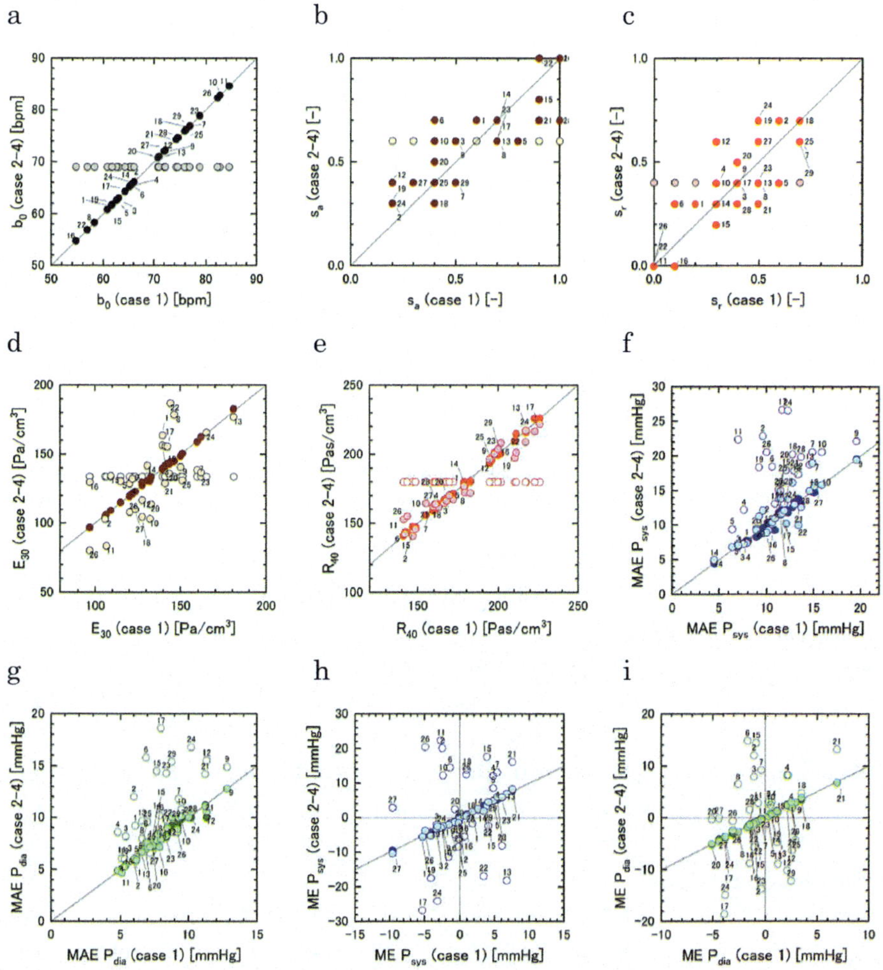

Fig. 8 Comparison of the results of Cases 2–4 with those of Case 1 for all 29 subjects for **a** the standard pulse rate b_0, **b** the stroke volume change rate s_a, **c** the peripheral resistance change rate s_r, **d** the elasticity E_{30} and **e** the peripheral resistance coefficient R_{40} of the lumped systemic arteries at the standard pulse rate, **f**, **g** the mean absolute errors and **h**, **i** the mean errors of estimated systolic and diastolic pressures, respectively, with respect to measurements with ABPM. Results of Cases 2, 3, 4 are shown by closed, light-colored, and open symbols, respectively. Numbers at the plots show the subject numbers

using an ABPM device were simultaneously performed for 29 subjects. Blood pressure estimations were performed to determine optimal parameters and evaluate the estimation error for each of four conditions for the constraint of model parameters by comparing the estimation results with the ABPM measurements. From these results, the validity of the various conditions of the constraint and the statistical properties of

Table 4 Mean values and standard deviations of the model parameters for Cases 1–4

	Case 1		Case 2		Case 3		Case 4	
	MEAN	SD	MEAN	SD	MEAN	SD	MEAN	SD
b_0 [bpm]	69.4	8.1	69.4	8.1	69	0	69	0
E_{30} [Pa/cm^3]	133.7	20.5	133.5	20.7	133.4	27.0	133.8	0
R_{40} [Pa·s/cm^3]	180.0	26.3	180.0	26.5	179.1	24.1	180.1	0
s_a [−]	0.6	0.3	0.6	0.2	0.6	0	0.6	0
s_r [−]	0.4	0.2	0.4	0.2	0.4	0	0.4	0
T_c [s]	200	0	200	0	200	0	200	0
P_{sys0} [mmHg]	124.5	14.3	124.6	14.4	124.2	13.6	125	0
P_{dia0} [mmHg]	77.6	11.4	77.8	11.4	77.4	10.1	78	0

Table 5 Mean values and standard deviations of the mean error, standard deviation, and mean absolute error of estimated systolic and diastolic pressures with respect to the parameter determination data and the validation data for Cases 1–4

		Case 1		Case 2		Case 3		Case 4	
		MEAN	SD	MEAN	SD	MEAN	SD	MEAN	SD
Estimation errors [mmHg]									
Systolic	MEAN	0.0	0.1	0.0	0.1	0.0	0.1	1.2	14.2
	SD	13.5	4.4	13.6	4.5	14.6	4.8	14.8	5.0
	MA	10.4	3.2	10.4	3.2	11.2	3.4	16.6	5.3
Diastolic	MEAN	0.0	0.1	0.0	0.1	0.0	0.1	0.2	9.2
	SD	10.0	4.2	10.1	4.2	10.2	4.3	10.1	4.3
	MA	7.7	2.7	7.8	2.8	7.8	2.9	11.1	3.6
Validation errors [mmHg]									
Systolic	MEAN	0.4	4.1	0.4	4.2	0.3	4.3	1.4	14.1
	SD	14.0	4.3	13.9	4.3	14.5	4.7	14.6	4.8
	MA	11.2	3.2	11.2	3.3	11.6	3.6	16.4	5.8
Diastolic	MEAN	-0.2	2.7	-0.1	2.7	-0.1	2.7	0.1	9.3
	SD	10.2	3.2	10.2	3.0	10.3	3.2	10.2	3.1
	MA	7.9	2.0	7.9	1.9	8.0	2.0	10.8	3.6

the parameters were clarified. Among these conditions, optimum parameter determination method was determined from the viewpoints of accuracy and computational costs. For the optimum parameter determination method of the circulatory control system, the effect of the number of the ABPM data on the accuracy of pressure estimation was clarified and the number of measurement data necessary for appropriate parameter determination was clarified. The relations between model parameters and the daily average of systolic pressures by ABPM were clarified for all subjects in Case 1, in which four model parameters were changed independently (Fig. 5a–e).

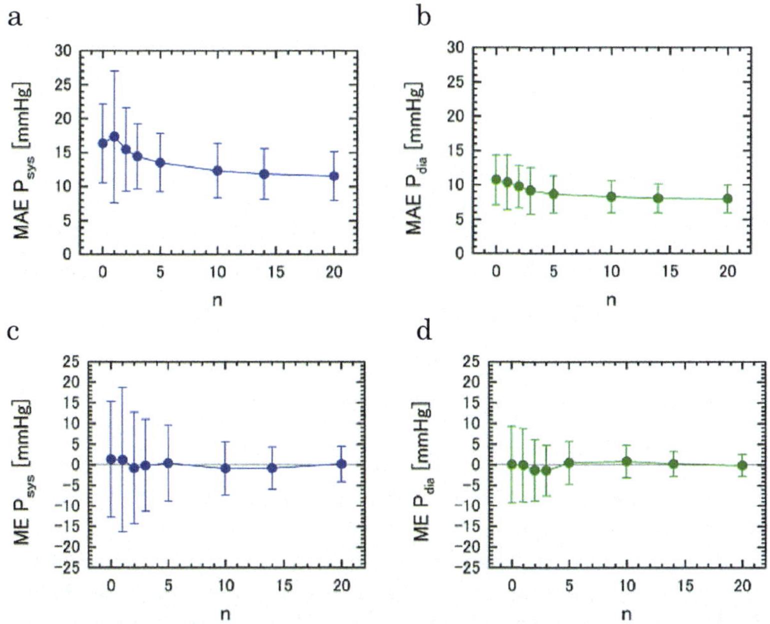

Fig. 9 Results of Case 3(n) for mean values (symbols) and standard deviations (error bars) for all the subjects of **a**, **b** the mean absolute errors, and **c**, **d** the mean errors for the estimated systolic and diastolic pressures with respect to the validation data of ABPM, respectively, plotted with the number of the parameter determination data. For the purpose of reference, the results at $n = 0$ show those of Case 4, and the results at $n = 20$ show those of Case 3, respectively

For the stroke volume change rate s_a and the peripheral resistance change rate s_r, no correlation was found to the blood pressure (Fig. 5b, c). In Fig. 7, which shows the correlation between the stroke volume change rate s_a and the peripheral resistance change rate s_r, mean values of both the parameters satisfy Eq. (14), the constraint proposed in the former study, and the direction of the first principal component vector of the covariance matrix agrees with that of Eq. (14), verifying the effectiveness of the constraint. On the other hand, we obtained a physiologically reasonable result that the parameters of the circulatory dynamics model, the elasticity E_{30} and the peripheral resistance coefficient R_{40} of the lumped systemic arteries at the standard pulse rate, respectively, have positive correlations with the blood pressure (Fig. 5d, e). No correlation was found between the standard pulse rate b_0 determined as the average value of the daily pulse rate data in the parameter determination data of ABPM and the blood pressure (Fig. 5a).

Next, we discuss the error of analysis results for all the subjects in Case 1 with respect to the verification data of ABPM. Mean absolute errors and mean errors of systolic and diastolic pressures have no correlation with the blood pressure (Fig. 5 f–i). As to statistical data of estimation errors with respect to validation data of ABPM, mean errors of systolic and diastolic pressures, 0.4 ± 4.1/-0.2 ± 2.7 mmHg, respectively, were within a standard tolerance of those for common sphygmomanometers, ±5 mmHg. The standard deviations 14.0 ± 4.3/10.2 ± 3.2 mmHg and mean absolute errors 11.2 ± 3.2/7.9 ± 2.0 mmHg were larger than those for the sphygmomanometers, 8 mmHg and 7 mmHg, respectively. A reason for the larger standard deviations and the mean absolute errors than those of common sphygmomanometer is possibly the errors in the standard validation data. If the errors in the validation data and those in the computation are uncorrelated with zero means and standard deviations of 8 mmHg corresponding to that of common sphygmomanometer, the standard deviation is analytically obtained as Eq. (11) in the former section and the mean absolute error as.

$$E = \sqrt{2(\sigma_{cal}^2 + \sigma_{meas}^2)/\pi} = 9.0(\text{mmHg}) \tag{16}$$

Mean values of standard deviations and mean absolute errors in systolic and diastolic pressures in this study are comparable to those in above expressions.

Next, we discuss five typical analysis results in Case 1 shown in Fig. 6. The result in the top row in the figure for subject 8 is the one the value of the mean absolute error in systolic pressure of which is almost average of those of all subjects (see Fig. 5f). Good agreement is shown in the figures in the top row for the daily variation (second figure), the error (third figure), and the correlation (fourth figure), respectively, between the estimated blood pressures and measured ones with ABPM. On the other hand, the result in the second row for subject 9 is the one the values of mean absolute errors in systolic and diastolic pressures of which are largest, respectively, among those of all subjects. Large error appears around 20 h in the third figure. The result in the third row for subject 14 is the one the value of the mean absolute error in systolic pressure of which is minimum among those of all subjects, and good agreement is shown in the figures. The result in the fourth row for subject 21 and that in the fifth row for subject 27 corresponds to that with the largest mean error in diastolic pressure (Fig. 5i) and that with the negative smallest mean error in systolic pressure (Fig. 5h), respectively. These degradations in accuracy are mostly attributed to the outlier of ABPM data around 22 h and that around 12 h, respectively.

Cases 2–4 are defined by adding several constraints to the parameter determination method in Case 1, resulting in the reduction of the degree of freedom in the method and of the computational load. We discuss the results of these cases by using the plots of the model parameters and the estimation errors of all subjects for each of Cases 2–4 with those of Case 1 (Fig. 8), the statistical data of the model parameters for Cases 1–4 (Table 4), and those of the estimation errors (Table 5). As to the model parameters, values of the stroke volume change rate s_a and those of the peripheral resistance change rate s_r of Case 2 distribute around those of Case 1, and their mean values $s_a = 0.6$ and $s_r = 0.4$ are identical (Fig. 8b, c, Table 4). Values of the elasticity E_{30} of the lumped systemic arteries at the standard pulse rate of Case 2 are close to those of Case 1 although those of Case 3 are far different from those of Case 1 (Fig. 8d, Table 4). Values of the peripheral resistance coefficient R_{40} of the lumped systemic arteries at the standard pulse rate of Cases 2 and 3 are close to those of Case 1 (Fig. 8e, Table 4). As to the errors with respect to the validation data of ABPM (Fig. 8f–i), Table 5), the results of Case 2 are almost identical to those of Case 1 for all of mean absolute errors and mean errors in systolic and diastolic pressures. In Case 3, the mean absolute errors in systolic pressure in some data deviate from those of Case 1 although the other results are almost identical to those of Case 1. In Case 4, the errors are larger than those of case 1 for most of the results (Fig. 8f–i) with the standard deviations of the mean errors in systolic/diastolic pressures of 14.1/9.3 mmHg (cf. 4.1/2.7 mmHg in Case 1) and the mean values of the mean absolute errors of 16.4/10.8 mmHg (cf. 11.2/7.9 mmHg in Case 1). According to the above discussion, it was revealed that the condition of Case 3, in which the elasticity E_{30} and the peripheral resistance coefficient R_{40} are optimized while the stroke volume change rate s_a and the peripheral resistance change rate s_r are fixed to 0.6 and 0.4, respectively, realizes a blood pressure estimation with an accuracy comparable with that of Case 1 in substantially reduced computational load.

Next, we discuss about the number of ABPM data necessary to determine appropriate values of the elasticity E_{30} and the peripheral resistance coefficient R_{40} in Case 3. Variations of the statistical values of the estimation errors with respect to the number of parameter determination data of ABPM used to determine the model parameters were clarified in Fig. 9 for Case 3(n), in which the first n data were used in parameter determination in Case 3. As to the mean absolute error of the estimated systolic pressure, the mean value and the standard deviation first increase at n = 1 from those at n = 0, which corresponds to Case 4 with fixed E_{30} and R_{40}, and then monotonically decrease with increasing n to converge to those of Case 3 at n = 20 (Fig. 9a). For the mean absolute error of the estimated diastolic pressure, the mean value and the standard deviation almost monotonically decrease with increasing n from those of Case 4 at n = 0 to converge to those of Case 3 at n = 20 (Fig. 9b). As to the mean errors of the estimated systolic and diastolic pressures, the mean values are close to zero at all values of n in both results, while the standard deviations monotonically decreases to converge to that of Case 3 (n = 20) except for the first increase at n = 1 for the systolic pressure (Fig. 9c, d). It is confirmed that the accuracy of the estimated blood pressure in Case 3(n) is comparable to that of Case 3 if the number of the data n is set to 10 or larger.

Limitations of this study are discussed in the followings. As to subject groups, this study was performed for volunteer subjects, who were classified into either of groups 1–3 of IEEE standard [40] based on a range of systolic pressure, but not into group 4 which is a hypertension group with a systolic pressure $\geq$161 mmHg. In order to apply the present blood pressure estimation scheme to practical DCBP estimation devices, verification experiments covering all groups 1–4 should be performed in future.

As to variation of the model parameters among days, the former study [9] investigated the variation in five days for one subject. The present study clarified the statistical characteristics of the parameters among subjects but not those among days. Experiment should be performed for sufficient number of subjects in sufficient number of days to clarify the statistics of the parameters among subjects and days.

As to the cost function to determine the model parameters, this study adopted the one in Eq. (12) which is the same as the former study [9], not dealing with the effect of the cost function on the results of blood pressure estimation. Especially, the effect of the weighting factor α in Eq. (12) should be clarified in future.

Concerning improvement of the circulatory system model, it is difficult to evaluate the dynamical characteristics of the model in detail since the validation in this study was done in comparison with ABPM measurement data of 30 min interval, and continuous measurement data are not available. Validation experiment using a continuous blood pressure data obtained with a device such as an arterial line monitoring should be done in future.

In this study the range of parameter determination is limited between 0 and 1 for the stroke volume change rate s_a and the peripheral resistance change rate s_r. As shown in Fig. 5, optimum parameter values are obtained at the boundary of the range as $s_a = 1$ or $s_r = 0$ for five cases in 29 subjects, implying that real optimum values may exist outside the range of $s_a > 1$ or $s_r < 0$. Parameter values in these regions, however, correspond to the computational results with large standard deviations for the pulse pressure or the average pressure, respectively, and are possibly ascribed to outliers in ABPM data. Therefore, it seems reasonable to put above-mentioned limitation in the parameter range.

7 Summary

In this section it was shown that the continuous blood pressure estimating method based on the simple circulatory system model with the input of pulse rate can be applied to practical DCBP estimation devices. A 25-h continuous pulse rate measurement using a wearable device and blood pressure measurement with 30 min interval using an ABPM device were simultaneously performed for 29 subjects. Blood pressure estimations were performed for four conditions modified by adding constraints to model parameters to reduce computational load ranging from Case 1 in which all parameters are changed independently to Case 4 in which all parameters are fixed. Determination of the optimal parameters and evaluation of estimation errors were performed for each of the four conditions by comparing the estimation results with the parameter determination data and the verification data of the ABPM measurement, respectively. Comparison of these results confirmed that the condition of Case 3, in which circulatory dynamics parameters are optimized while the circulatory control system parameters are fixed to the average values of those for all subjects, is suitable for DCBP estimating devices due to comparable accuracy and reduced computational load with respect to those of Case 1. In the results of Case 3, mean values $\pm$ standard deviations of the mean errors for systolic and diastolic pressures were 0.3 ± 4.3/-0.1 ± 2.7 mmHg, and those of the mean absolute errors were 11.6 ± 3.6/8.0 ± 2.0 mmHg, respectively, which were reasonable values taking the errors in the standard validation data of ABPM in consideration. For this condition, circulatory dynamics model parameters can be determined appropriately if about 10 ABPM data is used for parameter determination. In conclusion, the present estimation method can be applied to practical blood pressure estimation devices with good accuracy, small computational load, and small parameter determination measurement data number.

8 Conclusions

In this article we presented the new DCBP estimation method and its experimental validation and optimization based on our former works [9, 26]. First the circulatory system model with the input of pulse rate for the present blood pressure estimation was presented. The present circulatory system model consists of the circulatory dynamics model and the circulatory control inverse model. The circulatory dynamics model was given as a simple lumped parameter dynamical system. The circulatory control inverse model was defined as a simple dynamical model with the input of pulse rate and the outputs of peripheral vascular resistance of the systemic and pulmonary arteries, and the no-load ventricular volumes.

Next we presented computational result for 24 h blood flow dynamics in which values of blood pressure, blood flow, and blood volume in left/right atrium/ventricle and pulmonary/systemic arteries/veins are obtained from pulse rate measurement data with a wearable device. The results suggest that a fundamental part of DCBP can be represented by continuous pulse rate data and the simple circulatory dynamics and circulatory control inverse model with six model parameters.

Then we discussed the experimental validation and optimization of the DCBP estimation method. A 25-h continuous pulse rate measurement using a wearable device and blood pressure measurement with 30 min interval using an ABPM device were simultaneously performed for 29 subjects. Blood pressure estimations were performed for four conditions modified by adding constraints to model parameters to reduce computational load ranging from Case 1 in which all parameters are changed independently to Case 4 in which all parameters are fixed. Determination of the optimal parameters and evaluation of estimation errors were performed for each of the four. Comparison of these results confirmed that the condition of Case 3, in which circulatory dynamics parameters are optimized while the circulatory control system parameters are fixed to the average values of those for all subjects, is suitable for DCBP estimating devices due to comparable accuracy and reduced computational load. In the results of Case 3, mean values $\pm$ standard deviations of the mean errors for systolic and diastolic pressures were $0.3 \pm 4.3/\text{-}0.1 \pm 2.7$ mmHg, and those of the mean absolute errors were $11.6 \pm 3.6/8.0 \pm 2.0$ mmHg, respectively, which were reasonable values taking the errors in the standard validation data of ABPM in consideration. For this condition, circulatory dynamics model parameters can be determined appropriately if about 10 ABPM data are used for parameter determination.

In conclusion, the present estimation method can be applied to practical blood pressure estimation devices with good accuracy, small computational load, and small parameter determination measurement data number.

Acknowledgements This work was supported in part by JST COI Grant Number JPMJCE1303. The author acknowledges Prof. R. Nagatomi of Graduate School of Biomedical Engineering, Tohoku University and Prof. D. Ito of Sendai Seiyo Gakuin University for valuable discussions, Mr. H. Kiso, Mr. A. Sato, Mr. T. Ogasawara for their technical assistance, Mr. O. Iwamoto of Elecom Corp. and Mr. S. Pak for their technical support.

References

1. Andreu-Perez, J., et al.: From wearable sensors to smart implants-toward pervasive and personalized healthcare. IEEE Trans. Biomed. Eng. **62**(12), 2750–2762 (2015)
2. Mukkamala, R., et al.: Toward ubiquitous blood pressure monitoring via pulse transit time: theory and practice. IEEE Trans. Biomed. Eng. **62**(8), 1879–1901 (2015)
3. Hosanee, M., et al.: Cuffless single-site photoplethysmography for blood pressure monitoring. J. Clin. Med. **9**(3) (2020)
4. Staessen, J.A., et al.: Predicting cardiovascular risk using conventional vs ambulatory blood pressure in older patients with systolic hypertension. Jama-J. Am. Med. Assoc. **282**(6), 539–546 (1999)
5. Verdecchia, P., et al.: Ambulatory blood-pressure—an independent predictor of prognosis in essential-hypertension. Hypertension **24**(6), 793–801 (1994)
6. Anstey, D.E., et al.: Diagnosing masked hypertension using ambulatory blood pressure monitoring, home blood pressure monitoring, or both? Hypertension **72**(5), 1200–1207 (2018)
7. Parati, G., et al.: Relationship of 24-hour blood-pressure mean and variability to severity of target-organ demage in hypertension. J. Hypertens. **5**(1), 93–98 (1987)
8. Saugel, B., et al.: How to measure blood pressure using an arterial catheter: a systematic 5-step approach. Critical Care 24(1) (2020)
9. Hayase, T.: Blood pressure estimation based on pulse rate variation in a certain period. Sci. Rep. **10**(1) (2020)
10. Forouzanfar, M., et al.: Coefficient-free blood pressure estimation based on pulse transit time-cuff pressure dependence. IEEE Trans. Biomed. Eng. **60**(7), 1814–1824 (2013)
11. Pickering, T.G., et al.: Recommendations for blood pressure measurement in humans and experimental animals—part 1: blood pressure measurement in humans—a statement for professionals from the subcommittee of professional and public education of the American heart association council on high blood pressure research. Hypertension **45**(1), 142–161 (2005)
12. Verberk, W.J., et al.: Home blood pressure measurement—a systematic review. J. Am. Coll. Cardiol. **46**(5), 743–751 (2005)
13. Chandrasekhar, A., et al.: An iPhone application for blood pressure monitoring via the oscillometric finger pressing method. Sci. Rep. **8** (2018)
14. Chandrasekhar, A., et al.: Formulas to explain popular oscillometric blood pressure estimation algorithms. Front. Physiol. **10** (2019)
15. Peng, R.C., et al.: Cuffless and continuous blood pressure estimation from the heart sound signals. Sensors **15**(9), 23653–23666 (2015)
16. Suzuki, A.: Inverse-model-based cuffless blood pressure estimation using a single photoplethysmography sensor. Proc. Inst. Mech. Eng. Part H-J. Eng. Med. **229**(7), 499–505 (2015)
17. Kachuee, M., et al.: Cuffless blood pressure estimation algorithms for continuous health-care monitoring. IEEE Trans. Biomed. Eng. **64**(4), 859–869 (2017)
18. Chandrasekhar, A., et al.: PPG sensor contact pressure should be taken into account for cuff-less blood pressure measurement. IEEE Trans. Biomed. Eng. **67**(11), 3134–3140 (2020)
19. Ding, X.R., et al.: Continuous cuffless blood pressure estimation using pulse transit time and photoplethysmogram intensity ratio. IEEE Trans. Biomed. Eng. **63**(5), 964–972 (2016)

20. Huynh, T.H., Jafari, R., Chung, W.Y.: Noninvasive cuffless blood pressure estimation using pulse transit time and impedance plethysmography. IEEE Trans. Biomed. Eng. **66**(4), 967–976 (2019)
21. Ma, Y.J., et al.: Relation between blood pressure and pulse wave velocity for human arteries. Proc. Natl. Acad. Sci. U.S.A. **115**(44), 11144–11149 (2018)
22. Mase, M., et al.: Feasibility of cuff-free measurement of systolic and diastolic arterial blood pressure. J. Electrocardiol. **44**(2), 201–207 (2011)
23. Sharma, M., et al.: Cuff-less and continuous blood pressure monitoring: a methodological review. Technologies **5**(2) (2017)
24. Wong, M.Y.M., Poon, C.C.Y., Zhang, Y.T.: An evaluation of the cuffless blood pressure estimation based on pulse transit time technique: a half year study on normotensive subjects. Cardiovasc. Eng. **9**(1), 32–38 (2009)
25. Yavarimanesh, M., et al.: Commentary: relation between blood pressure and pulse wave velocity for human arteries. Front. Physiol. **10** (2019)
26. Hayase, T.: Yamada, I. (ed.) Development of Blood Flow Dynamics Sensing Device, in Smart Healthcare, pp. 217–229. NTS publishing, Tokyo (2023) (in Japanese)
27. Lu, K., et al.: Cerebral autoregulation and gas exchange studied using a human cardiopulmonary model. Am. J. Physiol.-Hear. Circ. Physiol. **286**(2), H584–H601 (2004)
28. Shi, Y.B., Lawford, P., Hose, R.: Review of zero-D and 1-D models of blood flow in the cardiovascular system. Biomed. Eng. Online **10** (2011)
29. Klabunde, R.E.: Cardiovascular Physiology Concepts, 2nd edn., xi, 243 p. Lippincott Williams & Wilkins/Wolters Kluwer, Philadelphia, PA (2012)
30. Sun, M.K.: Central neural organization and control of sympathetic nervous system in mammals. Prog. Neurobiol. **47**(3), 157–233 (1995)
31. Malpas, S.C., Ninomiya, I.: THE AMPLITUDE AND PERIODICITY OF SYNCHRONIZED RENAL SYMPATHETIC-NERVE DISCHARGES IN ANESTHETIZED CATS - DIFFERENTIAL EFFECT OF BARORECEPTOR ACTIVITY. J. Auton. Nerv. Syst. **40**(3), 189–198 (1992)
32. Weissler, A.M., Peeler, R.G., Roehll, W.H.: Relationships between left ventricular ejection time, stroke volume, and heart rate in normal individuals and patients with cardiovascular disease. Am. Heart J. **62**(3), 367–370 (1961)
33. Hayashi, K., et al.: Clinical assessment of arterial stiffness with cardio-ankle vascular index: theory and applications. J. Hypertens. **33**(9), 1742–1757 (2015)
34. Hayashi, T., et al.: Seasonal influence on blood pressure in elderly normotensive subjects. Hypertens. Res. **31**(3), 569–574 (2008)
35. Cooney, D.O.: Biomedical engineering principles : an introduction to fluid, heat, and mass transport processes. In: Biomedical Engineering and Instrumentation Series, vol. 2, xvi, 458 p. M. Dekker, New York (1976)
36. Ugander, M., Jense, E., Arheden, H.: Pulmonary intravascular blood volume changes through the cardiac cycle in healthy volunteers studied by cardiovascular magnetic resonance measurements of arterial and venous flow. J. Cardiovasc. Magn. Reson. **11** (2009)
37. Cohen, Z., Haxha, S.: Optical-based sensor prototype for continuous monitoring of the blood pressure. IEEE Sens. J. **17**(13), 4258–4268 (2017)
38. Weibel, L., et al.: Comparative effect of night and daytime sleep on the 24-hour cortisol secretory profile. Sleep **18**(7), 549–556 (1995)
39. Shea, S.A., et al.: Existence of an endogenous circadian blood pressure rhythm in humans that peaks in the evening. Circ. Res. **108**(8), 980-U207 (2011)
40. IEEE Standard for Wearable, Cuffless Blood Pressure Measuring Devices. IEEE Std., pp. 1708–2014. New York, USA (2014)

Wet Interface Technologies for Wearable Sweat Sensors

Kuniaki Nagamine and Shizuo Tokito

1 Introduction

The recent key issue is the development of chemical sensors to support the realization of personalized healthcare that can track the daily physiological conditions of individuals and provide personalized advice to encourage behavioral changes. Portable biosensors that can easily detect chemical substances with a signal converter equipped with biological receptors have been developed for point-of-care-testing applications, especially for self-management of blood glucose in diabetic patients [1]. Recent dramatic improvements in the sensitivity of biosensors have enabled the detection of ultra-trace amounts of chemical components in tears, saliva, sweat, breath, and skin gas. Besides, comprehensive omics analysis of the chemical compounds has uncovered useful biomarkers in these biological samples [2, 3]. The more important point is that these fluids and gases can be noninvasively and easily collected from the human body, enabling daily self-health monitoring based on the possible biomarkers in these bodily fluids and gases avoiding an invasive blood sampling process. Among these bodily fluids and gases, sweat is one of the most favorable fluids because it is easily accessible from the skin. Sweat is made up of 99% water, with the remaining 1% containing chemicals that contain useful biomarkers [4, 5]. Therefore, biosensors capable of multi-sensing possible biomarkers are essential for highly reliable sweat-based healthcare. In this chapter, we describe the basic physiology of perspiration

K. Nagamine (✉) · S. Tokito
Graduate School of Organic Materials Science, Yamagata University, 4-3-16 Jonan, Yonezawa 992-8510, Yamagata, Japan
e-mail: nagamine@yz.yamagata-u.ac.jp

S. Tokito
e-mail: tokito@yz.yamagata-u.ac.jp

S. Tokito
Research Center of Organic Electronics (ROEL), Yamagata University, 4-3-16 Jonan, Yonezawa 992-8510, Yamagata, Japan

K. Mitsubayashi (ed.), *Wearable Biosensing in Medicine and Healthcare*,
https://doi.org/10.1007/978-981-99-8122-9_10

and its components and then summarize the current methods of sweat collection. In particular, the promising non-invasive sweat collection method, wet-interfacing method, and sweat sensor that implements this method are summarized to introduce and discuss recent advances and issues of this type of sensor for non-invasive at-rest sweat-based healthcare.

2 Sweat Components

Sweat, which has a similar composition to serum, contains various biomarkers. These biomarkers are related to some diseases listed in Table 1. For example, sweat glucose is a typical biomarker that many researchers have found to be correlated with blood glucose levels [13]. On the other hand, the origin of sweat components is still under research, and the correlation between blood components has not been sufficiently proven [4]. In order to realize highly reliable healthcare using sweat components as biomarkers, the study of the correlation between blood components should be continued.

3 Sweat Collection Methods

3.1 Passive Collection

Sweat is excreted from sweat glands under some circumstances including exercise, bathing, hot weather, or the skin under the occluded environment. Visible sweat can be easily sampled by absorbing it with filter paper, cotton, gauze, or towels (Fig. 1a). PharmChek, a semi-occlusive dressing consisting of a cellulose-based collection pad, is a commercially available passive sweat collection system that has been utilized for sweat analysis [33]. The passive processes for sweat collection are straightforward and useful for disposable use. However, it takes a relatively long term to collect enough amount of sweat for biosensing purposes. In addition, the method of analyzing accumulated sweat makes it difficult to track continuous changes in sweat composition.

The on-skin microfluidic system is a promising device for continuous sampling and monitoring of sweat components (Figs. 1a, 2a) [34]. Poly(dimethylsiloxane) (PDMS)-based microfluidic channel is generally flexible and conformable to human skin. Therefore, its inlet can be tightly faced to the skin surface containing sweat glands in order to directly introduce the secreted sweat into the microchannel. This device configuration suppresses the volatilization of the collected subtle sweat (typical perspiration rate is a few nL min^{-1} $gland^{-1}$) [35]. The sweat collected into the microchannel is continuously discharged from the outlet of the microchannel to keep the sweat in the channel fresh. However, owing to the slow perspiration rate, the sweat

Table 1 A list of sweat biomarkers

Diseases	Sweat biomarkers	References
Active tuberculosis	Proteins	[6]
Atopic dermatitis	Dermcidin, Amino acids	[7, 8]
Cardiovascular diseases	Na^+, Ascorbic acid, Neuropeptide Y	[9]
Chronic anxiety disorders, Major depressive disorder	Neuropeptide Y	[10]
Chronic hepatitis C	Hepatitis C virus	[11]
Cystic fibrosis	Cl^-	[12]
Diabetes	Glucose	[13]
Fatigue	Lactate	[14]
Fatigue	NH_4^+	[15]
Gout	Uric acid, Tyrosine	[16]
Hand surface infection barrier	Lactic acid	[17]
Heart failure	Lactate	[18]
Hepatic encephalopathy Diagnosis	NH_4^+	[19]
Heat stroke	Na^+	[20]
Hypoxia	Lactate	[21]
Inflammatory Bowel Disease	IL-1β, CRP	[22]
Inflammatory/infectious diseases	IL-6, IL-8, IL-10, TNFα	[23]
Kidney disorder	Creatinine, Urea	[24]
Lung cancer	Monogluceride, Muconic acid, Suberic acid, Tetrahexose, Nonanedioic acid, Urocanic acid	[25]
Major depressive disorder	Cytokines, Neuropeptide Y, Substance P, Calcitonin-gene-related peptide, Vasoactive intestinal peptide	[26]
Mental stress	Cortisol	[27]
Nutritional imbalance	Vitamin C	[28]
Ocular Behcet's disease	l-Citrulline, l-Pyroglutamic acid, Urocanic acid, 2-Oxoadipic acid, Cholesterol 3-sulfate, Pentadecanoic acid	[29]
Psoriasis	Choline, Glutamic acid, Phenylalanine, Lactic acid, Urocanic acid, Citrulline	[30]
Schizophrenia	Proteins	[31]
Vogt-Koyanagi-Harada diseases	Amino acids	[32]

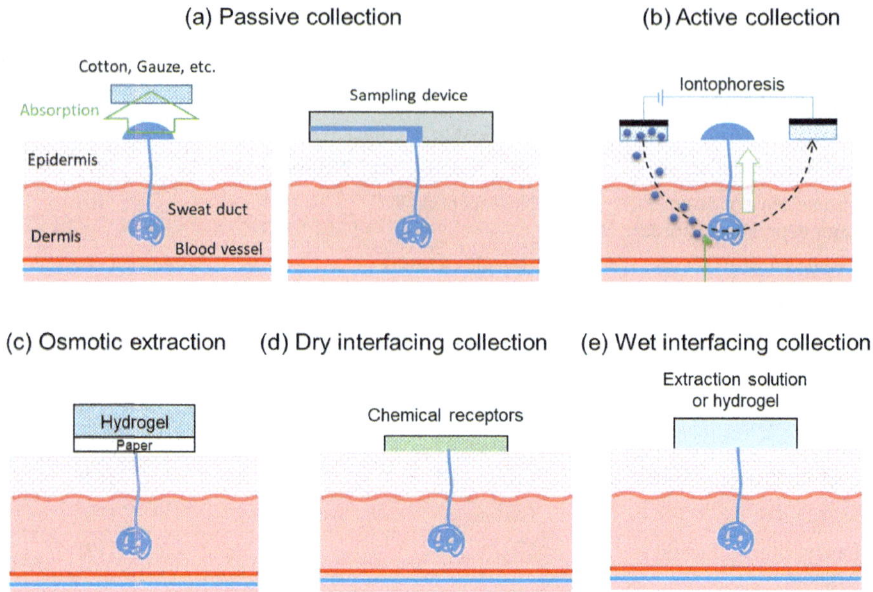

Fig. 1 Summary of the sweat collection methods. **a** Passive collection, **b** active collection, **c** osmotic collection, **d** dry-interfacing collection, and **e** wet-interfacing collection

collected earlier is more likely to be mixed with the sweat collected later, resulting in the concentration of the sweat components averaging. To solve this issue, the active transport of passively collected sweat was achieved using digital microfluidic technology. Electrowetting on dielectrics is a typical digital microfluidic technology that employs the electric field generated by a high voltage to transport droplets of sweat collected at different times in real time [36]. The other strategy developed by Kim et al. was an open, truncated cone-shaped vertical sweat microchannel (height: 1 mm, bottom diameter: 1.5 mm, top diameter: 0.5 mm) and a sweat-clearing structure composed of a hydrophilic carbon nanotube-PDMS sponge at the top of the channel (Fig. 2b) [37]. When the sweat excreted from the sweat gland fills this vertical microchannel, the sponge-based top layer quickly wicks the sweat to clear the channel. This process was repeated for the continuous collection of fresh sweat into the channel.

3.2 Active Collection

For the daily use of the sweat sensors, it is necessary to develop a mechanism that can collect sufficient sweat at any time. To solve this issue, an iontophoretic device was developed to actively stimulate perspiration at any time (Figs. 1b, 2c) [38]. This device can enhance the transdermal administration of pilocarpine, a cholinergic

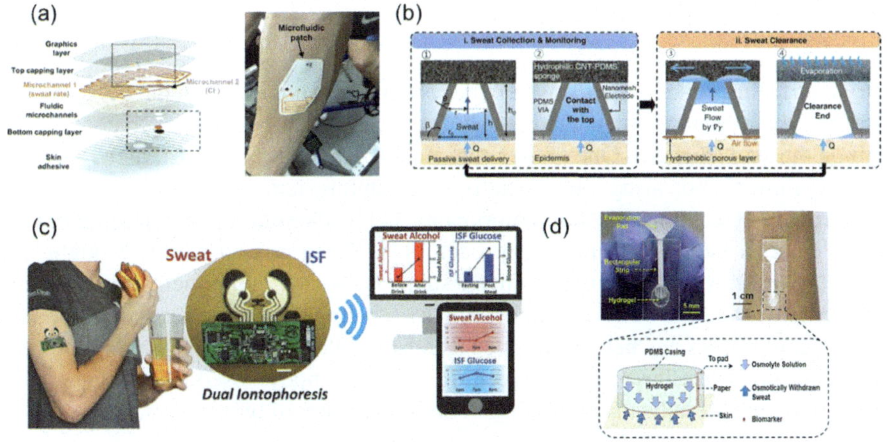

Fig. 2 Research examples of the sweat collection methods. **a** Passive collection using the on-skin microfluidic system (reproduced with permission from Ref. 34, http://creativecommons.org/licenses/by/4.0). **b** Passive collection using the open, truncated cone-shaped vertical sweat microchannel (reproduced with permission from Ref. 37, http://creativecommons.org/licenses/by/4.0). **c** Active collection using the iontophoretic technique (reproduced with permission from Ref. 38, http://creativecommons.org/licenses/by/4.0). **d** Osmotic pressure-driven paper-based microfluidics (reprinted with permission from Ref. 39. Copyright (2021) American Chemical Society)

agonist to activate eccrine sweat glands, through transcutaneous delivery of ionic current. However, some people may experience discomfort due to the application of ionic current. Besides, daily drug administration may also be unacceptable to the users.

3.3 *Osmotic Extraction*

Saha et al. developed osmotic pressure-driven paper-based microfluidics to achieve continuous collection and detection of the sweat biomarkers, even at rest (Figs. 1c, 2d) [39, 40]. The device is composed of a PDMS-based microfluidic channel filled with Whatman 542 paper, and its inlet contains a hydrogel disk with a high concentration of glucose or glycerin. The hydrogel disk contacts the human skin via paper. The difference in osmotic pressure between sweat and the hydrogel disk enables the effective collection of sweat into the paper. Continuous collection and detection of sweat lactate was achieved by utilizing a passive osmotic pressure-driven pump and the capillary action of the paper-based channel even under a resting state [40]. In this principle, the collected sweat is mixed with solutes in the hydrogel; therefore, it is necessary to adjust the composition of the hydrogel according to the target analytes.

3.4 Dry Interfacing Collection

Dry interfacing collection means that the receptors immobilized on the dry substrate are directly in contact with the human skin surface to capture the secreted compounds left on the skin (Fig. 1d). One of the representative devices is the ELIPatch (enzyme-linked immunospot array on a patch). The capture antibodies are arrayed on a substrate to detect multiple skin components [41]. A similar dry-interfacing collection device is the fingerprinting approach [42]. When human skin is brought into contact with the antibody-immobilized substrate, sweat components are captured by the antibody, followed by visualization by labeling detectable second antibodies. This assay process has been developed as a lateral flow device and is commercially available for sweat-based drug screening [43]. Antibody-based sweat sensors are potentially useful in healthcare using sweat protein-based biomarkers.

3.5 Wet Interfacing Collection

Here, we introduce a simple sweat collection method using 1% aqueous ethanol as an extraction solution [44]. Simply by contacting the skin with 1% aqueous ethanol, sweat components can be diffused from the sweat glands into the solution and collected (Fig. 1e). This method has also been utilized for the glucose analysis of sweat [45, 46]. Considering biological safety, the extraction solution has recently been replaced with phosphate-buffered saline (PBS) [18] or hydrogel patches in PBS [47–49]. A similar technique using PBS as an extraction solution has also been applied to the immunoassay of dermal biomarkers [50–52]. This immunoassay method evolved into a transdermal analysis patch, which is an antibody-immobilized flexible membrane in contact with the human skin surface for in-situ capture of sweat biomarkers [53]. Heavy metal ions dissolved in human sweat were detected using a similar technique called the "finger immersion method" [54]. The fingers of the subjects were immersed in high-purity water containing 0.1% nitric acid to extract nickel from the skin, followed by quantification using induced coupled plasma-optical emission spectrometry. The drawback of this method is that extremely small amounts of sweat components are diluted with the extraction solution. Therefore, a highly sensitive analysis with liquid chromatography and mass spectrometry, for example, is essential for quantifying the extracted components. Such expensive and large size of systems are unsuitable for daily use. The combination of at-rest sweat collection technology using the wet interface and portable biosensors is expected to create advanced sweat sensors for daily use.

4 Hydrogel-Based Wearable Touchpad Biosensors for Extracting and Detecting At-Rest Human Sweat Components

As described above, the most important issue in sweat biosensors is the development of an easy way of collecting at-rest sweat that can be performed anytime and anywhere. We proposed that one of the critical solutions to this issue is the use of a novel wet-interfacing collection-based biosensor, named the hydrogel touchpad-based sweat sensor for the first time (Fig. 3a) [55]. Our first model of this type of biosensor was a lactate oxidase (LOx)–based sweat lactate sensor. LOx and Prussian blue (PB)-modified working electrode and an Ag/AgCl reference electrode are covered by an agarose hydrogel in PBS as a sweat extraction pad. When the human skin is in contact with the hydrogel-based sweat extraction pad, the sweat lactate diffuses from the sweat glands into the gel and is extracted. The sweat lactate in the gel was electrochemically detected using the LOx-PB-modified electrode. Our subsequent sensor was equipped with a liquid-junction reference electrode that exhibits a stable reference potential to detect Cl^- ion in at-rest sweat [56].

Hydrogel touchpad-based sweat sensors have continued to evolve after our invention to this day as listed in Table 2. Lin et al. developed an agarose hydrogel touchpad embedding an electrochemical lactate sensor [57]. Their sensor had a sensitivity of 1.88 ± 0.24 μA/(mM cm^2) in the lactate concentration range of 0 to 4 mM. The

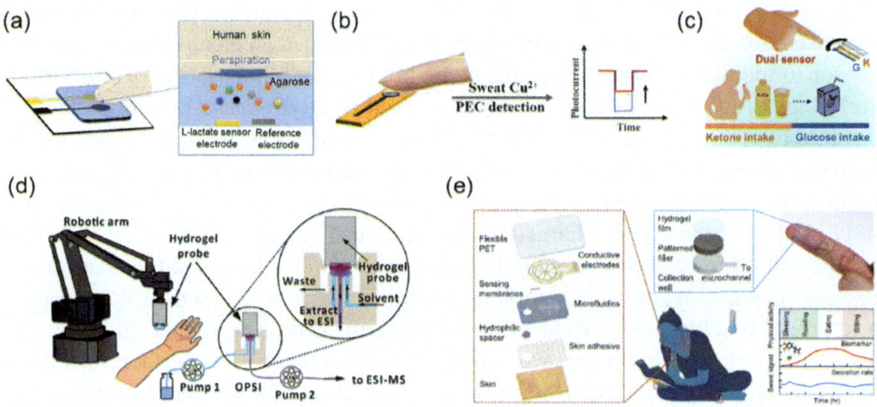

Fig. 3 Representative examples of the hydrogel-based touchpad sweat biosensors. **a** Electrochemical sweat lactate sensor using an agarose hydrogel-based touchpad (reproduced with permission from Ref. 55, http://creativecommons.org/licenses/by/4.0/). **b** Photoelectrochemical sweat Cu^{2+} sensor (reproduced with permission from Ref. 64, http://creativecommons.org/licenses/by/4.0/). **c** Sweat β-hydroxybutyrate (HB) sensor (reprinted with permission from Ref. 65. Copyright (2022) American Chemical Society). **d** Agarose hydrogel touchpad equipped with an electrospray ion source (reprinted with permission from Ref. 68. Copyright (2023) American Chemical Society). **e** Combination of a hydrogel touchpad inlet with a microfluidic channel (reproduced from Ref. 70, http://creativecommons.org/licenses/by/4.0/)

data collected by this sensor was wirelessly transferred to a cloud system for further analysis. Yin et al. developed a hydrogel touchpad-based self-powered biosensor for sweat lactate detection [58]. They used a porous polyvinyl alcohol (PVA) hydrogel as a touchpad due to its softness, mechanical strength, and the diffusibility of the sweat components. The concentration of sweat lactate was visually understood as a change in color of poly(3,4-ethylenedioxy)thiophene:poly(styrene sulfonate) (PEDOT:PSS) used as an electrochromic display, which was driven by the self-powered sensor depending on the sweat lactate concentration. Sempionatto et al. developed a thin porous PVA hydrogel touchpad-based electrochemical biosensor for sweat glucose [59]. Sweat glucose is expected to be an important biomarker that can reflect blood glucose levels. They determined a parameter that takes into account individual differences in the correlation between sweat glucose concentration and blood glucose level and demonstrated the possibility of indirect blood glucose level prediction using their sensor. Lin et al. developed a sweat glucose sensor using a PB-doped PEDOT nanocomposite as a mediator between glucose oxidase and the electrode [60]. The quantitative range of this sensor was 6.25 μM to 0.8 mM, and the limit of detection was 4 μM, which is high sensitivity enough to detect low concentrations of sweat glucose extracted in the agarose hydrogel. They demonstrated that there is a correlation between diurnal fluctuations in sweat glucose and of blood sugar levels. Tang et al. developed an electrochemical sensor using a molecularly imprinted polymer (MIP)/PB composite membrane for the detection of sweat cortisol [61]. The quantitative range of this sensor was 10 nM to 1 μM with a sensitivity of 60.31 nA/log [nM] in artificial sweat, which is enough sensitivity to detect sweat cortisol extracted in the hydrogel. They successfully monitored diurnal fluctuations in sweat cortisol using this sensor. Moon et al. developed an electrochemical L-dopa sensor to monitor its sweat concentrations after oral administration of the L-dopa/carbidopa drug [62]. The sensor was composed of a tyrosinase-modified electrode covered by a thin porous PVA hydrogel. The concentration range of linear amperometric response to L-dopa was 1–30 μM with a limit of detection of 300 nM. They demonstrated intermittent monitoring of sweat L-dopa while measuring blood L-dopa concentrations. They claimed that continuous monitoring data of sweat L-dopa can be helpful in establishing guidelines for individualized treatment of Parkinson's disease patients. Wang et al. fabricated a colorimetric sweat sensor composed of a hydrogel touchpad composed of arrayed chemical probes for pH, glucose, Cl^-, and Ca^{2+} [63]. The hydrogel was adhesive to the human skin surface in order to stably monitor the color change attributed to the change in the concentration of the extracted sweat components. Zhang et al. developed a highly sensitive photoelectrochemical sensor for the detection of natural sweat Cu^{2+}, which is a possible marker for diseases such as Wilson's disease, Menkes disease, hematological abnormalities, and kidney and cardiovascular diseases (Fig. 3b) [64]. Laser-induced graphene and In-doped CdS (LIG-In-CdS) were fabricated by laser engraving of a chitosan/Cd^{2+}/In^{3+}/ cysteine composite membrane coated on a polyimide film. The active area was covered with a PVA hydrogel. The photocurrent of the LIG-In-CdS electrode changed depending on the concentration of sweat Cu^{2+} in the range of 1.28 ng/mL to 5.12 g/mL. Moon et al. developed a natural sweat β-hydroxybutyrate (HB) sensor

(Fig. 3c) [65]. β-hydroxybutyrate is a promising biomarker for diabetic ketoacidosis diagnosis. A dual HB/glucose sensor was composed of an Au nanoparticle-modified screen-printed carbon electrode. The surface of the electrode was modified by a chitosan polymer containing toluidine blue O (as a mediator), β-hydroxybutyrate dehydrogenase, and nicotinamide adenine dinucleotide (as a cofactor), followed by entirely covering with a PVA hydrogel. The sensitivity of the sensor was 3.15 nA mM^{-1} cm^{-2} in the HB concentration range from 0.1 to 2.0 mM. The limit of detection was 14.43 μM. The glucose sensor also showed a current response depending on the glucose concentration in the concentration range from 0.1 to 2 mM. They demonstrated non-invasive dual monitoring of sweat HB and glucose during two kinds of drinks, the ketone beverage and the sweetened fruit beverage, by comparing blood ketone and glucose levels. Hu et al. developed a physical/chemical multi-sensing device composed of a reduced graphene oxide-modified hydrogel-based strain sensor and a hydrogel touchpad-based sweat sensor [66]. A sweat sensor based on a closed bipolar electrode was composed of a reduced graphene oxide-modified hydrogel as the cathode and a luminol-modified hydrogel as the anode. The self-healing ability of these hydrogels allows them to adhere to each other to construct a closed bipolar cell. Chemical reactions between sweat components and the reaction reagents at the cathode were coupled with the electrochemiluminescence (ECL) reaction at the anode to give a highly sensitive ECL signal. In vivo measurements were conducted by attaching the reduced graphene oxide-modified hydrogel containing reaction reagents to the subject's skin during exercise to collect sweat components into the gel. This was followed by the peeling off of the skin and wrapping one end of the luminol-modified hydrogel to measure the ECL signal correlated with the analyte concentration. Lin et al. fabricated a hydrogel touchpad-based sensor combined with photoplethysmography (to detect the subject's heart rate and oxygen saturation level) and a finger scanner to translate the touch-based input into encrypted bioinformation [67]. Two types of biosensors, an alcohol oxidase-based ethanol sensor, and a hydrogen-terminated boron-doped diamond-based acetaminophen sensor, were developed. They demonstrated the integration of a sensing system on a car handle to check the ethanol level in the driver's sweat. Another application was a pill case integrated with an acetaminophen sensor to monitor medication intake and record personal health. This multimodal sensing system will open a new era in bio-human–machine interfacing engineering. Yu et al. developed an agarose hydrogel touchpad equipped with a time-of-flight mass spectrometer (Fig. 3d) [68]. The hydrogel touchpad was held using a robotic arm for automatic sampling and highly sensitive analysis of arginine, citrulline, and histidine extracted in the agarose hydrogel. Various metabolites extracted from human skin were quantified using this system.

It is also possible to integrate the wet interface into commercially available chemical sensors (Fig. 4) [69]. For example, a filter paper impregnated with PBS as a touchpad was put on the sensing part of the commercially available sodium ion sensor (LAQUAtwin Na-11, Horiba) to establish the touchpad-based sweat sodium sensor. We hope that the practical use of the touchpad-based sweat sensor will be accelerated by actively using commercially available chemical sensors.

Table 2 A summary of representative hydrogel touchpad-based sweat biosensors

Biomarkers	Types of sensors	References
Lactate	Electrochemical sensor	[55]
Lactate	Electrochemical sensor	[57]
Lactate	Self-powered electrochemical sensor	[58]
D-glucose	Electrochemical sensor	[59]
D-glucose	Electrochemical sensor	[60]
Cl^-	Electrochemical sensor	[56]
Cortisol	Electrochemical sensor	[61]
L-Dopa	Electrochemical sensor	[62]
pH, D-glucose, Cl^-, Ca^{2+}	Colorimetric sensor	[63]
Cu^{2+}	Photoelectrochemical sensor	[64]
β-hydroxybutyrate D-glucose	Electrochemical sensor	[65]
Urea, Lactic acid, Cl^-	Electrochemiluminescence sensor	[66]
Ethanol, Acetaminophen	Electrochemical sensor	[67]
Arginine, Citrulline, Histidine, Caffeine	quadrupole-time-of-flight mass spectrometer	[68]
Sweat pH, Cl^-, D-glucose, L-Dopa, perspiration rate	Electrochemical sensor	[70]
Cortisol, Mg^{2+}, sweat pH	Electrochemical sensor	[71]

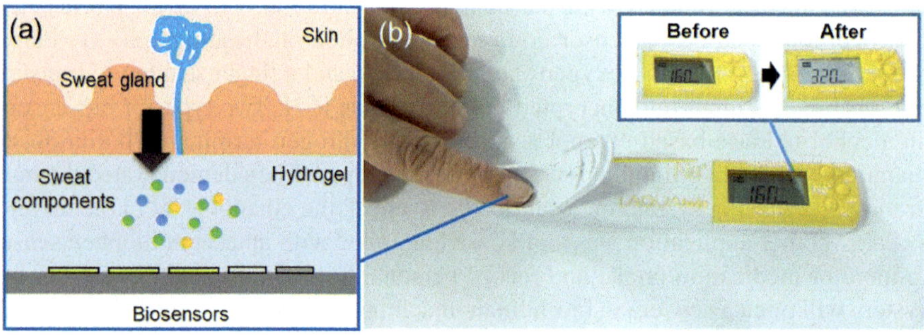

Fig. 4 **a** Illustration of the interface between human skin and a touchpad-based sweat biosensor. **b** Photograph of a touchpad-based sweat Na^+ sensor fabricated by using the commercially available sodium ion meter (reproduced with permission from Ref. 69, http://creativecommons.org/licenses/by/4.0/)

5 Limitations of the Hydrogel Touchpad-Based Sweat Sensors

Quantifying the amount of sweat extracted into hydrogels is still a major challenge. Therefore, it is difficult to distinguish whether the change in the concentration of sweat components detected by the hydrogel touchpad-based sensor is due to the amount of sweat extracted or the change in the concentration of sweat components. One straightforward strategy to solve this problem is to ignore the contribution of the perspiration volume over a relatively short sampling time compared to the volume of the hydrogel [67]. The other quantitative solution is the combination of a hydrogel touchpad inlet with a microfluidic channel (Fig. 3e) [70, 71]. In this configuration, the at-rest sweat extracted into the hydrogel touchpad was directly introduced into the microchannel to calculate the extracted sweat volume from the microchannel dimensions. Using this device, Nyein et al. calculated the rate of sweating in human fingers in contact with the hydrogel to be 0.1–1 $\mu L\ min^{-1}\ cm^{-2}$. However, because the rate of sweating at rest is so slow, the sweat previously introduced into the channel can be mixed with new sweat that has just been introduced, averaging the concentration detected by the sensor. Challenges remain in real time and simultaneous quantification of both the at-rest sweat volume and the concentration of at-rest sweat components in the hydrogel touchpad.

6 Conclusions and Future Perspectives

This chapter summarized the advances in sweat sensors and the relevant sweat collection methods. We consider that hydrogel touchpad-based sweat sensors are now among the best solutions for the quantitative collection and detection of at-rest sweat components because of their safety, and simple construction and use. On the other hand, detailed research is still required on the relationship between changes in the concentration of sweat components and diseases. By simultaneously analyzing sweat components and developing sensors for the required biomarkers, we hope to establish sweat component sensors that can contribute to the realization of individualized healthcare in the future.

Acknowledgements We would like to acknowledge each of our laboratory members and colleagues for their productive discussions and contributions to these experiments. This chapter presents the results of several projects financially supported by JSPS KAKENHI Grant Number JP21422580 and the foundation of the YU-COE(C) program at Yamagata University.

References

1. Lee, H., Hong, Y.J., Baik, S., Hyeon, T., Kim, D.H.: Enzyme-based glucose sensor: from invasive to wearable device. Adv. Healthc. Mater. **7**, 1701150 (2018). https://doi.org/10.1002/adhm.201701150
2. Sempionatto, J.R., Lasalde-Ramírez, J.A., Mahato, K., Wang, J., Gao, W.: Wearable chemical sensors for biomarker discovery in the omics era. Nat. Rev. Chem. **6**, 899 (2022). https://doi.org/10.1038/s41570-022-00439-w
3. Heikenfeld, J., Jajack, A., Feldman, B., Granger, S.W., Gaitonde, S., Begtrup, G., Katchman, B.A.: Accessing analytes in biofluids for peripheral biochemical monitoring. Nat. Biotechnol. **37**, 407 (2019). https://doi.org/10.1038/s41587-019-0040-3
4. Baker, L.B.: Physiology of sweat gland function: The roles of sweating and sweat composition in human health. Temperature (Austin) **6**, 211 (2019). https://doi.org/10.1080/23328940.2019.1632145
5. Delgado-Povedano, M.M., Castillo-Peinado, L.S., Calderón-Santiago, M., Luque de Castro, M.D., Priego-Capote, F.: Dry sweat as sample for metabolomics analysis. Talanta **208**, 120428 (2020). https://doi.org/10.1016/j.talanta.2019.120428
6. Adewole, O.O., Erhabor, G.E., Adewole, T.O., Ojo, A.O., Oshokoya, H., Wolfe, L.M., Prenni, J.E.: Proteomic profiling of eccrine sweat reveals its potential as a diagnostic biofluid for active tuberculosis. Proteomics Clin. Appl. **10**, 547–553 (2016). https://doi.org/10.1002/prca.201500071
7. Mark, H., Harding, C.R.: Amino acid composition, including key derivatives of eccrine sweat: potential biomarkers of certain atopic skin conditions. Int. J. Cosmet. Sci. **35**, 163–168 (2013). https://doi.org/10.1111/ics.12019
8. Rieg, S., Steffen, H., Seeber, S., Humeny, A., Kalbacher, H., Dietz, K., Garbe, C., Schittek, B.: Deficiency of dermcidin-derived antimicrobial peptides in sweat of patients with atopic dermatitis correlates with an impaired innate defense of human skin in vivo. J. Immunol. **174**, 8003–8010 (2005). https://doi.org/10.4049/jimmunol.174.12.8003
9. Wei, J., Zhang, X., Mugo, S.M., Zhang, Q.: A portable sweat sensor based on carbon quantum dots for multiplex detection of cardiovascular health biomarkers. Anal. Chem. **94**(37), 12772–12780 (2022). https://doi.org/10.1021/acs.analchem.2c02587
10. Churcher, N.K.M., Upasham, S., Rice, P., Bhadsavle, S., Prasad, S.: Development of a flexible, sweat-based neuropeptide Y detection platform. RSC Adv. **10**, 23173–23186 (2020). https://doi.org/10.1039/D0RA03729J
11. Ortiz-Movilla, N., Lázaro, P., Rodríguez-Iñigo, E., Bartolomé, J., Longo, I., Lecona, M., Pardo, M., Carreño, V.: Hepatitis C virus replicates in sweat glands and is released into sweat in patients with chronic hepatitis C. J. Med. Virol. **68**, 529–536 (2002). https://doi.org/10.1002/jmv.10238
12. Emaminejad, S., Gao, W., Wu, E., Davies, Z.A., Nyein, H.Y.Y., Challa, S., Ryan, S.P., Fahad, H.M., Chen, K., Shahpar, Z., Talebi, S., Milla, C., Javey, A., Davis, R.W.: Autonomous sweat extraction and analysis applied to cystic fibrosis and glucose monitoring using a fully integrated wearable platform. Proc. Natl. Acad. Soc. **114**, 4625–4630 (2017). https://doi.org/10.1073/pnas.1701740114
13. Bae, C.W., Toi, P.T., Kim, B.Y., Lee, W.I., Lee, H.B., Hanif, A., Lee, E.H., Lee, N.E.: Fully stretchable capillary microfluidics-integrated nanoporous gold electrochemical sensor for wearable continuous glucose monitoring. ACS Appl. Mater. Interfaces **11**, 14567–14575 (2019). https://doi.org/10.1021/acsami.9b00848
14. Seki, Y., Nakashima, D., Shiraishi, Y., Ryuzaki, T., Ikura, H., Miura, K., Suzuki, M., Watanabe, T., Nagura, T., Matsumato, M., Nakamura, M., Sato, K., Fukuda, K., Katsumata, Y.: A novel device for detecting anaerobic threshold using sweat lactate during exercise. Sci. Rep. **11**, 4929 (2021). https://doi.org/10.1038/s41598-021-84381-9
15. Guinovart, T., Bandodkar, A.J., Windmiller, J.R., Andrade, F.J., Wang, J.: A potentiometric tattoo sensor for monitoring ammonium in sweat. Analyst **138**, 7031–7038 (2013). https://doi.org/10.1039/C3AN01672B

16. Yang, Y., Song, Y., Bo, X., Min, J., Pak, O.S., Zhu, L., Wang, M., Tu, J., Kogan, A., Zhang, H., Hsiai, T.K., Li, Z., Gao, W.: A laser-engraved wearable sensor for sensitive detection of uric acid and tyrosine in sweat. Nat. Biotechnol. **38**, 217–224 (2020). https://doi.org/10.1038/s41587-019-0321-x
17. Hayashi, K., Mori, I., Takeda, K., Okada, Y., Hayase, A., Mori, T., Nishioka, Y., Manabe, K.: Analysis of hand environment factors contributing to the hand surface infection barrier imparted by lactic acid. Skin Res. Technol. **27**(6), 1135–1144 (2021). https://doi.org/10.1111/srt.13078
18. Stoffers, K.M., Cronkright, A.A., Huggins, G.S., Baleja, J.D.: Noninvasive epidermal metabolite profiling. Anal. Chem. **92**, 12467–12472 (2020). https://doi.org/10.1021/acs.analchem.0c02274
19. Kim, S.B., Koo, J., Yoon, J., Hourlier-Fargette, A., Lee, B., Chen, S., Jo, S., Choi, J., Oh, Y.S., Lee, G., Won, S.M., Aranyosi, A.J., Lee, S.P., Model, J.B., Braun, P.V., Ghaffari, R., Park, C., Rogers, J.A.: Soft, skin-interfaced microfluidic systems with integrated enzymatic assays for measuring the concentration of ammonia and ethanol in sweat. Lab Chip **20**, 84–92 (2020). https://doi.org/10.1039/C9LC01045A
20. Cazalé, A., Sant, W., Ginot, F., Launay, J.C., Savourey, G., Revol-Cavalier, F., Lagarde, J.M., Heinry, D., Launay, J., Temple-Boyer, P.: Physiological stress monitoring using sodium ion potentiometric microsensors for sweat analysis. Sens. Actuators B **225**, 1–9 (2016). https://doi.org/10.1016/j.snb.2015.10.114
21. Daboss, E.V., Tikhonov, D.V., Shcherbacheva, E.V., Karyakin, A.A.: Ultrastable lactate biosensor linearly responding in whole sweat for noninvasive monitoring of hypoxia. Anal. Chem. **94**(25), 9201–9207 (2022). https://doi.org/10.1021/acs.analchem.2c02208
22. Jagannath, B., Lin, K.C., Pali, M., Sankhala, D., Muthukumar, S., Prasad, S.: A sweat-based wearable enabling technology for real-time monitoring of IL-1β and CRP as potential markers for inflammatory bowel disease. Inflamm. Bowel Dis. **26**(10), 1533–1542 (2020). https://doi.org/10.1093/ibd/izaa191
23. Jagannath, B., Lin, K.C., Pali, M., Sankhala, D., Muthukumar, S., Prasad, S.: Temporal profiling of cytokines in passively expressed sweat for detecting of infection using wearable device. Bioeng. Trans. Med. **6**(3), e10220 (2021). https://doi.org/10.1002/btm2.10220
24. Zhang, Y., Guo, H., Kim, S.B., Wu, Y., Ostojich, D., Park, S.H., Wang, X., Weng, Z., Li, R., Bandodkar, A.J., Sekine, Y., Choi, J., Xu, S., Quaggin, S., Ghaffari, R., Rogers, J.A.: Passive sweat collection and colorimetric analysis of biomarkers relevant to kidney disorders using a soft microfluidic system. Lab Chip **19**, 1545–1555 (2019). https://doi.org/10.1039/C9LC00103D
25. Delgado-Povedano, M.M., Calderón-Santiago, M., Priego-Capote, F., Jurado-Gámez, B., Luque de Castro, M.D.: Recent advances in human sweat metabolomics for lung cancer screening. Metabolomics **12**, 166 (2016). https://doi.org/10.1007/s11306-016-1116-4
26. Cizza, G., Marques, A.H., Eskandari, F., Christie, I.C., Torvik, S., Silverman, M.N., Phillips, T.M., Sternberg, E.M.: Elevated neuroimmune biomarkers in sweat patches and plasma of premenopausal women with major depressive disorder in remission: the POWER study. Biol. Psychiatry **64**, 907–911 (2008). https://doi.org/10.1016/j.biopsych.2008.05.035
27. Torrente-Rodríguez, R.M., Tu, J., Yang, Y., Min, J., Wang, M., Song, Y., Yu, Y., Xu, C., Ye, C., IsHak, W.W., Gao, W.: Investigation of cortisol dynamics in human sweat using a graphene-based wireless mHealth system. Matter **2**(4), 921–937 (2020). https://doi.org/10.1016/j.matt.2020.01.021
28. Zhao, J., Nyein, H.Y.Y., Hou, L., Lin, Y., Bariya, M., Ahn, C.H., Ji, W., Fan, Z., Javey, A.: A wearable nutrition tracker. Adv. Mater. **33**(1), 2006444 (2021). https://doi.org/10.1002/adma.202006444
29. Cui, X., Zhang, L., Su, G., Kijlstra, A., Yang, P.: Specific sweat metabolite profile in ocular Behcet's disease. Int. Immunopharmacol. **97**, 107812 (2021). https://doi.org/10.1016/j.intimp.2021.107812
30. Dutkiewicz, E.P., Hsieh, K.T., Wang, Y.S., Chiu, H.Y., Urban, P.L.: Hydrogel micropatch and mass spectrometry-assisted screening for psoriasis-related skin metabolites. Clin. Chem. **62**, 1120–1128 (2016). https://doi.org/10.1373/clinchem.2016.256396

31. Raiszadeh, M.M., Ross, M.M., Russo, P.S., Schaepper, M.A., Zhou, W., Deng, J., Ng, D., Dickson, A., Dickson, C., Strom, M., Osorio, C., Soeprono, T., Wulfkuhle, J.D., Petricoin, E.F., Liotta, L.A., Kirsch, W.M.: Proteomic analysis of eccrine sweat: implications for the discovery of schizophrenia biomarker proteins. J. Proteome Res. **11**(4), 2127–2139 (2012). https://doi.org/10.1021/pr2007957
32. Cui, X., Su, G., Zhang, L., Yi, S., Cao, Q., Zhou, C., Kijlstra, A., Yang, P.: Integrated omics analysis of sweat reveals an aberrant amino acid metabolism pathway in Vogt–Koyanagi–Harada disease. Clin. Exp. Immunol. **200**(3), 250–259 (2020). https://doi.org/10.1111/cei.13435
33. Katchman, B.A., Zhu, M., Christen, J.B., Anderson, K.S.: Eccrine sweat as a biofluid for profiling immune biomarkers. Proteomics Clin. Appl. **12**, 1800010 (2018). https://doi.org/10.1002/prca.201800010
34. Baker, L.B., Model, J.B., Barnes, K.A., Anderson, M.L., Lee, S.P., Lee, K.A., Brown, S.D., Reimel, A.J., Roberts, T.J., Nuccio, R.P., Bonsignore, J.L., Ungaro, C.T., Carter, J.M., Li, W., Seib, M.S., Reeder, J.T., Aranyosi, A.J., Rogers, J.A., Ghaffari, R.: Skin-interfaced microfluidic system with personalized sweating rate and sweat chloride analytics for sports science applications. Sci. Adv. **6**, eabe3929 (2020). https://doi.org/10.1126/sciadv.abe3929
35. Sonner, Z., Wilder, E., Heikenfeld, J., Kasting, G., Beyette, F., Swaile, D., Sherman, F., Joyce, J., Hagen, J., Kelley-Loughnane, N., Naik, R.: The microfluidics of the eccrine sweat gland, including biomarker partitioning, transport, and biosensing implications. Biomicrofluidics **9**, 031301 (2015). https://doi.org/10.1063/1.4921039
36. Shen, H., Lei, H., Gu, M., Miao, S., Gao, Z., Sun, X., Sun, L., Chen, G., Huang, H., Chen, L., Wen, Z.: A wearable electrowetting on dielectrics sensor for real-time human sweat monitor by triboelectric field regulation. Adv. Funct. Mater. **32**(34), 2204525 (2022). https://doi.org/10.1002/adfm.202204525
37. Kim, S., Park, S., Choi, J., Hwang, W., Kim, S., Choi, I.S., Yi, H., Kwak, R.: An epifluidic electronic patch with spiking sweat clearance for event-driven perspiration. Nat. Commun. **13**, 6705 (2022). https://doi.org/10.1038/s41467-022-34442-y
38. Kim, J., Sempionatto, J.R., Imani, S., Hartel, M.C., Barfidokht, A., Tang, G., Campbell, A.S., Mercier, P.P., Wang, J.: Simultaneous monitoring of sweat and interstitial fluid using a single wearable biosensor platform. Adv. Sci. (Weinh) **5**, 1800880 (2018). https://doi.org/10.1002/advs.201800880
39. Saha, T., Fang, J., Mukherjee, S., Dickey, M.D., Velev, O.D.: Wearable osmotic-capillary patch for prolonged sweat harvesting and sensing. ACS Appl. Mater. Interfaces **13**, 8071–8081 (2021). https://doi.org/10.1021/acsami.0c22730
40. Saha, T., Songkakul, T., Knisely, C.T., Yokus, M.A., Daniele, M.A., Dickey, M.D., Bozkurt, A., Velev, O.D.: Wireless wearable electrochemical sensing platform with zero-power osmotic sweat extraction for continuous lactate monitoring. ACS Sens. **7**, 2037–2048 (2022). https://doi.org/10.1021/acssensors.2c00830
41. Oh, D.Y., Na, H., Song, S.W., Kim, J., In, H., Lee, A.C., Jeong, Y., Lee, D., Jang, J., Kwon, S.: ELIPatch, a thumbnail-size patch with Immunospot array for multiplexed protein detection from human skin surface. Biomicrofluidics **12**, 031101 (2018). https://doi.org/10.1063/1.5032170
42. Leggett, R., Lee-Smith, E.E., Jickells, S.M., Russell, D.A.: "Intelligent" fingerprinting: simultaneous identification of drug metabolites and individuals by using antibody-functionalized nanoparticles. Angew. Chem. Int. Ed. Engl. **46**, 4100–4103 (2007). https://doi.org/10.1002/anie.200700217
43. Hudson, M., Stuchinskaya, T., Ramma, S., Patel, J., Sievers, C., Goetz, S., Hines, S., Menzies, E., Russell, D.A.: Drug screening using the sweat of a fingerprint: lateral flow detection of Δ9-tetrahydrocannabinol, cocaine, opiates and amphetamine. J. Anal. Toxicol. **43**, 88–95 (2019). https://doi.org/10.1093/jat/bky068
44. Tsuda, T., Noda, S., Kitagawa, S., Morishita, T.: Proposal of sampling process for collecting human sweat and determination of caffeine concentration in it by using GC/MS. Biomed. Chromatogr. **14**, 505–510 (2000). https://doi.org/10.1002/1099-0801(200012)14:8%3c505::AID-BMC17%3e3.0.CO;2-7

45. Ge, X., Rao, G., Kostov, Y., Kanjananimmanont, S., Viscardi, R.M., Woo, H., Tolosa, L.: Detection of trace glucose on the surface of a semipermeable membrane using a fluorescently labeled glucose-binding protein: a promising approach to noninvasive glucose monitoring. J. Diabetes Sci. Technol. **7**, 4–12 (2013). https://doi.org/10.1177/193229681300700102
46. Kanjananimmanont, S., Ge, X., Mupparapu, K., Rao, G., Potts, R., Tolosa, L.: Passive diffusion of transdermal glucose: non-invasive glucose sensing using a fluorescent glucose binding protein. J. Diabetes Sci. Technol. **8**, 291–298 (2014). https://doi.org/10.1177/1932296813519994
47. Dutkiewicz, E.P., Lin, J.D., Tseng, T.W., Wang, Y.S., Urban, P.L.: Hydrogel micropatches for sampling and profiling skin metabolites. Anal. Chem. **86**, 2337–2344 (2014). https://doi.org/10.1021/ac4039338
48. Dutkiewicz, E.P., Chiu, H.Y., Urban, P.L.: Micropatch-arrayed pads for non-invasive spatial and temporal profiling of topical drugs on skin surface. J. Mass Spectrom. **50**, 1321–1325 (2015). https://doi.org/10.1002/jms.3702
49. Dutkiewicz, E.P., Chiu, H.Y., Urban, P.L.: Probing skin for metabolites and topical drugs with hydrogel micropatches. Anal. Chem. **89**, 2664–2670 (2017). https://doi.org/10.1021/acs.analchem.6b04276
50. Portugal-Cohen, M., Oron, M., Ma'or, Z., Boaz, M., Shtendik, L., Biro, A., Cernes, R., Barnea, Z., Kazir, Z., Kohen, R.: Noninvasive skin measurements to monitor chronic renal failure pathogenesis. Biomed. Pharmacother. **65**, 280–285 (2011). https://doi.org/10.1016/j.biopha.2011.02.001
51. Portugal-Cohen, M., Horev, L., Ruffer, C., Schlippe, G., Voss, W., Ma'or, Z., Oron, M., Soroka, Y., Frušić-Zlotkin, M., Milner, Y., Kohen, R.: Non-invasive skin biomarkers quantification of psoriasis and atopic dermatitis: cytokines, antioxidants and psoriatic skin auto-fluorescence. Biomed. Pharmacother. **66**, 293–299 (2012). https://doi.org/10.1016/j.biopha.2011.12.009
52. Portugal-Cohen, M., Kohen, R.: Non-invasive evaluation of skin cytokines secretion: an innovative complementary method for monitoring skin disorders. Methods **61**, 63–68 (2013). https://doi.org/10.1016/j.ymeth.2012.10.002
53. Orro, K., Smirnova, O., Arshavskaja, J., Salk, K., Meikas, A., Pihelgas, S., Rumvolt, R., Kingo, K., Kazarjan, A., Neuman, T., Spee, P.: Development of TAP, a non-invasive test for qualitative and quantitative measurements of biomarkers from the skin surface. Biomark. Res. **2**, 20 (2014). https://doi.org/10.1186/2050-7771-2-20
54. Staton, I., Ma, R., Evans, N., Hutchinson, R.W., McLeod, C.W., Gawkrodger, D.J.: Dermal nickel exposure associated with coin handling and in various occupational settings: assessment using a newly developed finger immersion method. British J. Dermatol. **154**, 658–664 (2006). https://doi.org/10.1111/j.1365-2133.2006.07128.x
55. Nagamine, K., Mano, T., Nomura, A., Ichimura, Y., Izawa, R., Furusawa, H., Matsui, H., Kumaki, D., Tokito, S.: Noninvasive sweat-lactate biosensor emplsoying a hydrogel-based touch pad. Sci. Rep. **9**, 10102 (2019). https://doi.org/10.1038/s41598-019-46611-z
56. Ichimura, Y., Kuritsubo, T., Nagamine, K., Nomura, A., Shitanda, I., Tokito, S.: A fully screen-printed potentiometric chloride ion sensor employing a hydrogel-based touchpad for simple and non-invasive daily electrolyte analysis. Anal. Bioanal. Chem. **413**, 1883-1891 (2021). https://doi.org/10.1007/s00216-021-03156-3
57. Lin, S., Wang, B., Zhao, Y., Shih, R., Cheng, X., Yu, W., Hojaiji, H., Lin, H., Hoffman, C., Ly, D., Tan, J., Chen, Y., Carlo, D.D., Milla, C., Emaminejad, S.: Natural perspiration sampling and in situ electrochemical analysis with hydrogel micropatches for user-identifiable and wireless chemo/biosensing. ACS Sens. **5**, 93–102 (2020). https://doi.org/10.1021/acssensors.9b01727
58. Yin, L., Moon, J.M., Sempionatto, J.R., Lin, M., Cao, M., Trifonov, A., Zhang, F., Lou, Z., Jeong, J.M., Lee, S.J., Xu, S., Wang, J.: A passive perspiration biofuel cell: High energy return on investment. Joule **5**(7), 1888–1904 (2021). https://doi.org/10.1016/j.joule.2021.06.004
59. Sempionatto, J.R., Moon, J.M., Wang, J.: Touch-based fingertip blood-free reliable glucose monitoring: personalized data processing for predicting blood glucose concentrations. ACS Sens. **6**(5), 1875–1883 (2021). https://doi.org/10.1021/acssensors.1c00139

60. Lin, P.H., Sheu, S.C., Chen, C.W., Huang, S.C., Li, B.R.: Wearable hydrogel patch with noninvasive, electrochemical glucose sensor for natural sweat detection. Talanta **241**, 123187 (2022). https://doi.org/10.1016/j.talanta.2021.123187
61. Tang, W., Yin, L., Sempionatto, J.R., Moon, J.M., Teymourian, H., Wang, J.: Touch-based stressless cortisol sensing. Adv. Mater. **33**(18), 2008465 (2021). https://doi.org/10.1002/adma.202008465
62. Moon, J.M., Teymourian, H., De la Paz, E., Sempionatto, J.R., Mahato, K., Sonsa-Ard, T., Huang, N., Longardner, K., Litvan, I., Wang, J.: Non-invasive sweat-based tracking of L-dopa pharmacokinetics profiles following an oral tablet administration. Angew. Chem. Int. Ed. Engl. **60**(35), 19074–19078 (2021). https://doi.org/10.1002/anie.202106674
63. Wang, L., Xu, T., He, X., Zhang, X.: Flexible, self-healable, adhesive and wearable hydrogel patch for colorimetric sweat detection. J. Mater. Chem. C **9**, 14938–14945 (2021). https://doi.org/10.1039/D1TC03905A
64. Zhang, S., Liu, Y., Wang, J., Liu, Z.: A laser-induced photoelectrochemical sensor for natural sweat Cu^{2+} detection. Chemosensors **10**(5), 169 (2022). https://doi.org/10.3390/chemosensors10050169
65. Moon, J.M., Caño, R.D., Moonla, C., Sakdaphetsiri, K., Saha, T., Mendes, L.F., Yin, L., Chang, A.Y., Seker, S., Wang, J.: Self-testing of ketone bodies, along with glucose, using touch-based sweat analysis. ACS Sens. **7**(12), 3973–3981 (2022). https://doi.org/10.1021/acssensors.2c02369
66. Hu, Y., Li, J., Liu, J., Yu, X., Yang, J., Li, Y.: A simple yet multifunctional sensing platform inspired by healing-assembly hydrogels serving motion and sweat monitoring. Sens. Actuators B. Chem. **378**, 133173 (2023). https://doi.org/10.1016/j.snb.2022.133173
67. Lin, S., Zhu, J., Yu, W., Wang, B., Sabet, K.A., Zhao, Y., Cheng, X., Hojaiji, H., Lin, H., Tan, J., Milla, C., Davis, R.W., Emaminejad, S.: A touch-based multimodal and cryptographic bio-human-machine interface. Proc. Natl. Acad. Sci. U.S.A. **119**(15), e2201937119 (2022). https://doi.org/10.1073/pnas.220193711
68. Yu, K.C., Hsu, C.Y., Prabhu, G.R.D., Chiu, H.Y., Urban, P.L.: Vending-machine-style skin excretion sensing. ACS Sens. **8**(1), 326-334 (2023). https://doi.org/10.1021/acssensors.2c02325
69. Nagamine, K.: Wet interface engineering for at-rest sweat analysis. Sens. Mater. **34**(8), 3147–3154 (2022). https://doi.org/10.18494/SAM3899
70. Nyein, H.Y.Y., Bariya, M., Tran, B., Ahn, C.H., Brown, B.J., Ji, W., Davis, N., Javey, A.: A wearable patch for continuous analysis of thermoregulatory sweat at rest. Nat. Commun. **12**, 1823 (2021). https://doi.org/10.1038/s41467-021-22109-z
71. Zhao, H., Zhang, X., Qin, Y., Xia, Y., Xu, X., Sun, X., Yu, D., Mugo, S.M., Wang, D., Zhang, Q.: An integrated wearable sweat sensing patch for passive continuous analysis of stress biomarkers at rest. Adv. Funct. Mater. **33**(9), 2212083 (2023). https://doi.org/10.1002/adfm.202212083

Wearable Core Body Temperature Sensor and Its Application

Yujiro Tanaka

1 Introduction

The core body temperature (CBT) is the temperature deep inside the body (e.g., the rectum, brain, heart, and liver). The CBT is essential for our daily health management. The CBT changes depending on the health condition, age, individual, body's internal clock, and behavior. The heatstroke or catching a cold can lead to a high temperature, while a low temperature can lead to hypothermia. In terms of the CBT's time dependence, it has a 24-h rhythm that is strongly related to the body's internal clock [1]. The CBT reaches its minimum (nadir) in the night or few times before waking up and its maximum is between 2 and 7 pm, as shown schematically in Fig. 1. The time of the nadir is used as a parameter for evaluating the body's internal clock, which is strongly related to the blood melatonin level [2]. Disorders of the internal clock, which are related to problems such as diabetes, cancer, and sleep disorders, induce metabolic aberrations and depression. Such disorders also decrease exercise and work performance [1–9]. Therefore, continuous CBT measurement could be a valuable tool for managing the daily physical condition and detecting diseases in their early stages.

2 Core Body Temperature

The temperature in the pulmonary artery is the gold standard for the core temperature, but the rectal temperature (described below) is used in clinical practices. In recent years, the esophageal, tympanic, and rectal temperatures have been used as the main

Y. Tanaka (✉)
NTT Basic Research Laboratories, 3-1, Morinosato Wkamiya, Atsugi-Shi, Kanagawa 243-0198, Japan
e-mail: yujiro.tanaka@ntt.com

K. Mitsubayashi (ed.), *Wearable Biosensing in Medicine and Healthcare*,
https://doi.org/10.1007/978-981-99-8122-9_11

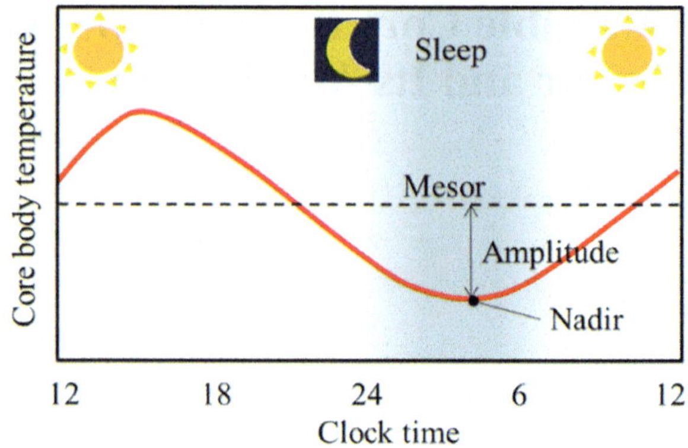

Fig. 1 CBT behavior over the day

CBT indicators. In addition, the sublingual mouth and axilla have been used as CBT measurement sites. Below, we summarize the characteristics of each site [10–13].

2.1 *Rectal Temperature*

As stated before, the temperature of the rectum is used as a gold standard in clinical practices and research on the CBT and circadian rhythm. The rectal temperature is measured by inserting a sensor probe a few centimeters into the rectum. It indicates the temperature of internal organs located deep in the body, with a heat source derived from the iliac artery and iliac vein. The rectal temperature is higher than the temperatures of other sites because of the rectum's isolation from ambient conditions and low heat dissipation. It also changes more slowly than the other temperatures. Although it provides an accurate CBT, the measurement is stressful and potentially risky because of rectal damage and infection; moreover, it is affected by the presence of stool, inflammation around the rectum, and heat-producing activities by microorganisms.

2.2 *Oral Temperature*

The oral temperature is measured at the posterior sublingual pockets. The heat source is a branch of the external carotid artery. This temperature can be affected by the sensor position, vascular activity in the sublingual area, salivation, food and fluid intake, and breathing. Stabilization of the temperature requires a few minutes.

2.3 *Axillary Temperature*

The axillary temperature is relatively lower than the temperatures at the other sites. While it can be measured easily, it is strongly affected by the ambient conditions, blood flow, sweat, and sensor position. Also, temperature differences of up to 1.4 °C have been reported between the right and left axillas. Moreover, the axillary temperature takes approximately 5 min to stabilize.

2.4 *Tympanic Temperature*

The tympanic temperature is the temperature of the eardrum, whose heat source is the blood supply from the internal and external carotid arteries. Compared with the other sites, it has less of a relationship to metabolic activity. An indirect monitoring technique was developed by using infrared radiation from the eardrum. Because of its usability and accuracy, it has been used for monitoring CBT changes during physical exercise. However, the ambient temperature and convection affect the tympanic temperature.

3 Noninvasive Core Body Temperature Monitoring Technique

3.1 *Noninvasive Techniques*

As mentioned above, the rectal temperature is reliable and stable against ambient conditions. However, a natural body cavity such as the rectum is not suitable for direct, long-term measurement of the core body temperature (CBT) by insertion of a sensor probe. Such placement of a sensor probe is stressful for a person, especially while awake, and it has the accompanying risks of infection and damage to the rectal cavity. To address this issue, alternative noninvasive techniques have been explored, including in-ear, ingestible, and skin-attachable sensors [14–26]. In-ear sensors cause discomfort, and the measurements are sometimes influenced by environmental conditions and movements. Ingestible sensors indicate the CBT similarly to the rectal temperature; however, they are limited by the duration of monitoring and the difficulty of collecting the sensor after use. On the other hand, skin-attachable sensors, which indirectly estimate the CBT by using the skin temperature and heat flux, avoid these disadvantages.

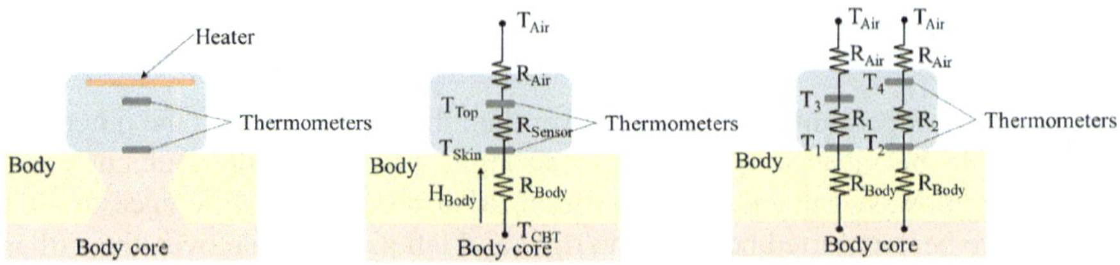

(a) Zero heat flux method (b) Single heat flux method (c) Dual heat flux method

Fig. 2 Comparison of skin-attachable CBT sensors

3.2 *Skin-Attachable Sensing Techniques*

The temperature of the skin does not satisfy the state of thermal equilibrium with the body core. Therefore, it forms a partial temperature gradient from body core to skin [14]. Therefore, it is necessary to devise a way to compensate for this temperature gradient.

3.2.1 Zero-Heat-Flux Method

A representative technique for noninvasive CBT measurement, called the zero-heat-flux method, was proposed by Togawa et al. [18] and by Fox et al. [19]. This technique makes use of a heater to heat the skin and compensates the dissipation of heat with respect to the ambient conditions, as shown in Fig. 2a. It provides an isothermal region from the skin to the body core, which makes the skin temperature close to the CBT. This technique is used for monitoring the CBT during surgery and in intensive care units [20–22]. Sensors based on this technology have been commercialized by TERUMO Co. (CoreTemp), and 3 M (Bair Hugger). However, it requires a heater, which consumes considerable electrical power and usually requires an AC power supply. Thus, it is difficult to monitor the CBT all day long. Moreover, it has been pointed out that this technique might measure the temperature of the hottest part of the measurement site.

3.2.2 Heat-Flux Method

The technique with sensors containing two or four thermometers was proposed [23–26]. The heat-flux method uses the skin temperature and the heat flux obtained from a sensor attached to the skin, as shown in Fig. 2b, c. The CBT is obtained from the following equation based on the body's unknown thermal resistance:

$$T_{CBT} = T_{\text{Skin}} + R_{\text{Body}} H_{\text{Body}}, \tag{1}$$

where R_{Body} is a constant coefficient related to the thermal properties of the body and the sensor, T_{Skin} is a temperature measured at the skin surface, and the H_{Body} is the heat flux measured at the skin surface. The heat flux H_{Body} is sometimes assumed to be proportional to the temperature difference between two points T_{Skin} and T_{Top}. On the other hand, it is necessary to calibrate R_{Body} at the beginning of monitoring by using a reference temperature (e.g., the tympanic temperature) in the single heat flux method (Fig. 2b). To obtain the R_{Body}, a sensor with four thermometers solves this issue by using two different thermal resistances, as shown in Fig. 2c, which is called the dual-heat-flux method. As reported by Kitamura et al., the CBT is estimated by assuming a 1-dimensional thermal equivalent circuit with two thermally isolated heat paths [23], as shown in (Fig. 2c). From the skin temperatures and heat fluxes along the two heat paths, Eq. (1) can be obtained for each path. Then, the body's unknown thermal resistance can be obtained by combining the two equations via the following equation:

$$T_{CBT} = T_1 + \frac{(T_1 - T_2)(T_1 - T_3)}{K(T_2 - T_4) - (T_1 - T_3)}, \; K = \frac{R_1}{R_2}. \tag{2}$$

This technique provides fully passive CBT measurements and is suitable for daily continuous measurement, although it assumes that the sensors are used in a stable environment or a hospital. Also, the heat is assumed to flow vertically from the point of skin contact, although lateral heat flow is also commonly caused by traveling, air conditioning, and other factors.

3.3 Problems of Skin-Attachable Sensor

According to Pennes' models of heat transfer in humans [27], the CBT estimation procedure is expressed as follows:

$$\frac{\partial T}{\partial t} = \frac{k_{\text{tissue}}}{C_{\text{tissue}}\rho_{\text{tissue}}}\Delta T + \rho_{\text{Blood}} C_{\text{Blood}} w\left(T_{\text{Array}} - T\right) + Q, \tag{3}$$

where T, t, k, ρ, C, w, and Q denote the temperature, time, thermal conductivity, density, specific heat, circulation, and internal heating, respectively, and Δ is the Laplace operator. The left term on the right side of Eq. (3) can be ignored if the body temperature changes slowly. Moreover, by choosing a site where the circulation and internal heating are small, the second and third terms on the right can also be ignored. Finally, by assuming that the tissue has uniform thermal properties and the heat travels vertically, the following simplified heat transfer equation can be obtained:

$$\frac{d^2T}{dz^2} = 0, \tag{4}$$

where z is the depth coordinate. This result indicates that the temperature distribution is linear from the body core to the skin. Thus, the CBT can be obtained from the skin temperature and the heat flux at the skin, as mentioned above. However, while the heat-flux method enables fully passive CBT estimation without any external heater, it is difficult to satisfy the above assumptions for a person being monitored in daily life. Ambient convection induces a lateral heat flow, which immediately increases the estimation error. Moreover, the sensor's presence also causes heat to travel laterally.

3.4 Design of Skin-Attachable Sensor [33]

To leverage the advantages of the heat-flux method while addressing the issue of lateral heat flow, we designed a novel structure for a CBT sensor. Ambient convection causes the heat flux to be overestimated or underestimated as shown in Eq. (5) and Fig. 3. Such convection results in heat dissipation, where heat is lost passively because of ambient air flow due to external sources such as air conditioning and activities such as walking or running. For continuous CBT monitoring in daily life, it is thus necessary to reduce the heat loss. Hence, we first focus on the problem of mitigating lateral heat transfer by designing the sensor's structure accordingly. Specifically, the design was developed through topology optimization, which provides a systematic design procedure and has become widely used in various fields (e.g., thermo-fluid dynamics) [28–31].

$$T_{CBT} = T_{\mathrm{Skin}} + R_{\mathrm{Body}}\left(H_{\mathrm{Body}} - H_{\mathrm{Loss}}\right), \tag{5}$$

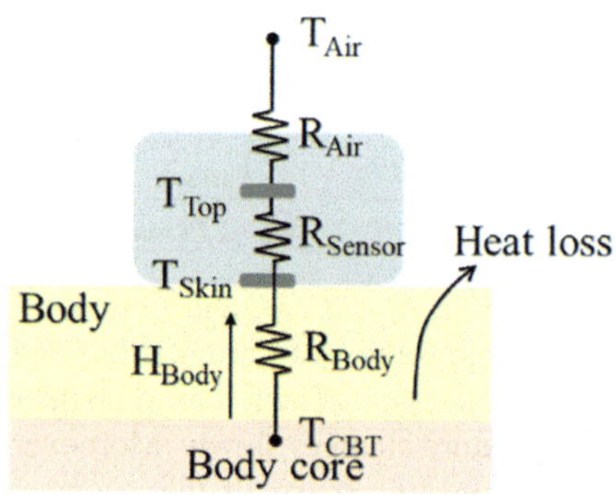

Fig. 3 Schematic images of the heat loss

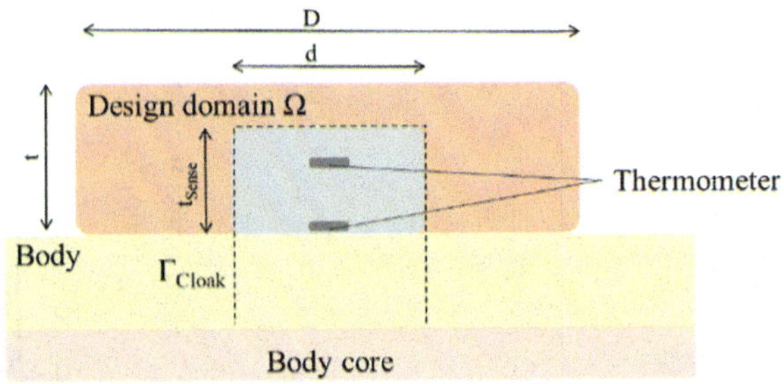

Fig. 4 Schematic cross section of the CBT sensor

3.4.1 Numerical Computation

Model

Figure 4 shows a schematic cross section of a model for numerical computation of the CBT sensor. The sensor has a cylindrical chassis with diameter D = 30 mm and height t = 5 mm, which were determined as optimal values for mobility and usability. The chassis contains a heat-resistant substrate which has two thermometers. In the figure, the aluminum structure is allocated in the design domain Ω for reducing the heat loss. The boundary conditions of the following model were considered for estimating the heat transfer coefficient H at the boundary. The CBT sensor was assumed to be exposed to air flow due to air conditioning, walking, or running. Specifically, the wind speed V_{In} was assumed to be 5 m/s (corresponding to a running speed of 18 km/h). The distance from the skin to the body core, t_{Body}, was assumed to be 10 mm. To simplify the optimization for the fluid and heat transfer interaction system, we used the heat transfer coefficient H at the boundary with the ambient air to incorporate convectional heat transfer. The relationship between the wind speed and H was estimated from the Nusselt number Nu by its definition, Prandtl number Pr, and Reynolds number Re, as follows:

$$\mathrm{Pr} = \frac{Cp_{\mathrm{Air}}\mu_{\mathrm{Air}}}{\lambda_{\mathrm{Air}}}, \tag{6}$$

$$\mathrm{Re} = \frac{\rho_{\mathrm{Air}} V_{\mathrm{In}} t_{\mathrm{Body}}}{\mu_{\mathrm{Air}}}, \tag{7}$$

where V_{In}, L, ρ_{Air}, μ_{Air}, and λ_{Air} denote the wind speed, characteristic length, air density, air viscosity, and thermal conductivity, respectively. For laminar flow, the Nusselt number Nu is given by [32]

$$\mathrm{Nu} = 0.664 Re^{1/2} Pr^{1/3}. \tag{8}$$

Finally, by combining Eqs. (6), (7), and (8), we obtain the heat transfer coefficient as a function of the wind speed V_{In}:

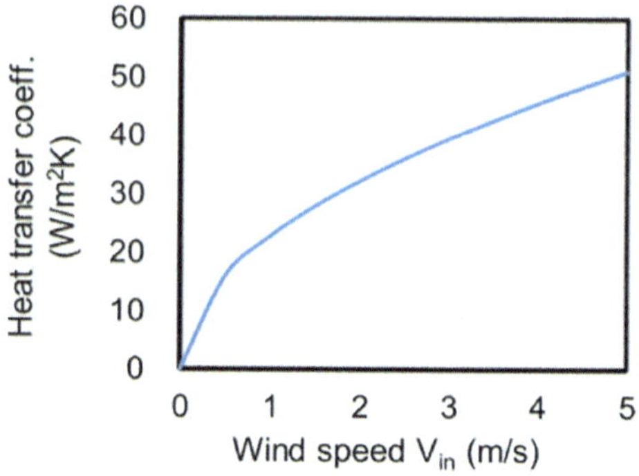

Fig. 5 Relationship between the heat transfer coefficient and wind speed

$$\mathrm{H} = 0.664 Pr^{1/3}\sqrt{\frac{\rho_{\mathrm{Air}}}{\mu_{\mathrm{Air}}}}\sqrt{\frac{V_{\mathrm{in}}}{L}}. \tag{9}$$

Figure 5 shows the relationship between the heat transfer coefficient and the wind speed, with the diameter of the CBT sensor used as the characteristic length L. As seen in the figure, the heat transfer coefficient ranges from 0 to 50 $\mathrm{W/m^2K}$ when $\mathrm{V_{In}}$ ranges from 0 to 5 m/s.

Problem Setting for the Topology Optimization

The structure's design is explained in this section. To reduce the heat loss, an aluminum structure was designed by using topology optimization to reduce the estimation error. In this optimization, the objective was to minimize the heat flux at the boundary:

$$J = \int_{\Gamma_{\mathrm{Cloak}}} H d\Gamma. \tag{10}$$

This objective aims to directly reduce the heat loss. The sensitivity of the objective to changes in the design variables was required for the optimization and was obtained by adjoint sensitivity analysis. The optimization procedure was the method of moving asymptotes (MMA). The numerical computation was performed using COMSOL Multiphysics 5.5.

The steady-state heat conduction equation is used as follows.

$$\nabla\left(k_\gamma \nabla T\right) = 0. \tag{11}$$

where k is the thermal conductivity (with a subscript indicating the material, e.g., Al for aluminum), γ is the design variable, and q is a parameter for the nonlinearity of interpolation, such that the nonlinearity increases as q decreases. The material's thermal conductivity distribution was optimized via a spatially defined density design

Table 1 Parameters in the computation

Variable	Variable	Value
Sensor height	t	5 mm
Sensor diameter	D	30 mm
Substrate height	t_{Sub}	4 mm
Substrate diameter	d	8 mm
Body depth	t_{Body}	10 mm
Ambient temp	T_{in}	20 °C
CBT	T_{Body}	37 °C
Wind speed	V_{In}	5 m/s
Calculation diameter	L	80 mm

Table 2 Thermo and fluid physical parameters

Part	Thermal conductivity k (W/mK)
Air	0.02
Body	0.37
Aluminum	204

variable γ, which ranged from 0 to 1. The thermal conductivity k_γ was interpolated as follows:

$$k_\gamma = k_{Al} + (k_{\mathrm{Air}} - k_{Al})\gamma \frac{1+q}{\gamma+q}, \tag{12}$$

Table 1 lists the parameters used in the optimization. Table 2 lists the thermophysical parameters used in the numerical computation. A typical problem that arises in topology optimization is a gray scale for the material distribution, which means a porous structure and a lower fabrication tolerance. To avoid this issue, a Helmholtz PDE filter,

$$\gamma_f = \gamma + R_{\min}^2 \Delta \gamma_f, \tag{13}$$

and a smoothed Heaviside projection are used [28],

$$\gamma = \frac{\tanh(\beta(\gamma_f - \theta_\beta)) + \tanh(\beta\theta_\beta)}{\tanh(\beta(1-\theta_\beta)) + \tanh(\beta\theta_\beta)}, \tag{14}$$

where β is a parameter determining the projection's steepness, and θ_β is a threshold parameter, which was fixed at 0.5. In this work, we set β as 15, and the initial design variable γ was 0.5.

3.4.2 Optimized Structure and Accuracy of Topology Optimization

The initial structure and the resulting optimized aluminum structure are shown in Fig. 6a, b. The red region represents the optimized structure, which resembles a truncated cone with a hole at the top. Unfortunately, the structure's complex curves and inconsistent thickness made fabrication difficult and decreased the accuracy of numerical computation. To improve the accuracy and fabrication tolerance, we simplified the structure to have a uniform thickness of 0.5 mm and a smooth surface. The diameter of the hole in the truncated cone was 2 mm. Figure 6c shows the estimation errors for various diameters of the cone's base and hole. The estimation error gradually decreased with a smaller base diameter D or hole diameter d, and it bottomed out at approximately 2 mm. For evaluation, as shown in Fig. 7, the (a) heat loss and (b) estimation error were calculated numerically as a function of the wind speed V_{In}. The maximum wind speed was $V_{\mathrm{In}} = 5$ m/s. The coefficient for estimating the CBT, R_{Sensor}, was obtained from T_{skin} and H_{Body} at $V_{\mathrm{In}} = 1$ m/s. The blue and orange points respectively indicate the results with and without the optimized aluminum structure. In either case, the heat loss H_{Loss} and the estimation error increased with the wind speed, but they were effectively reduced when the optimized structure was used. The maximum estimation errors with and without the structure were 0.1 and 0.4 °C, respectively, at $V_{\mathrm{In}} = 5$ m/s.

3.5 *In-Vitro Experimental Verification [34]*

The proposed structure's effectiveness was preliminarily verified through an in-vitro experiment using a phantom to simulate the body's thermal conductivity.

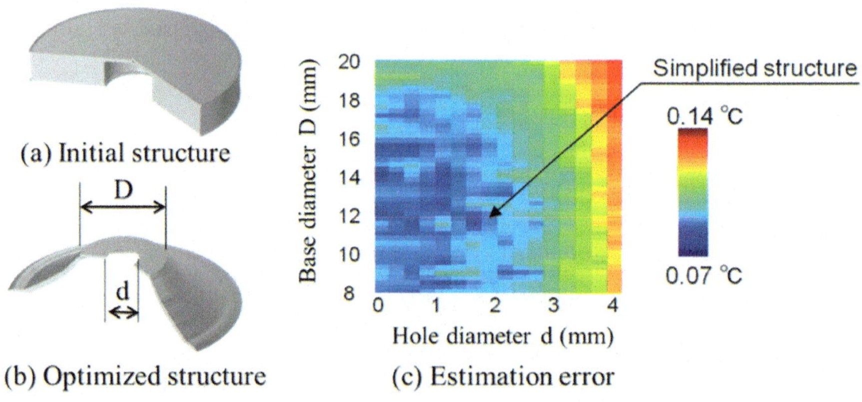

Fig. 6 Optimized aluminum structure. **a** and **b** are the initial structure and the resulting optimized aluminum structure. **c** Estimation errors for various values of the hole diameter d and the base diameter D

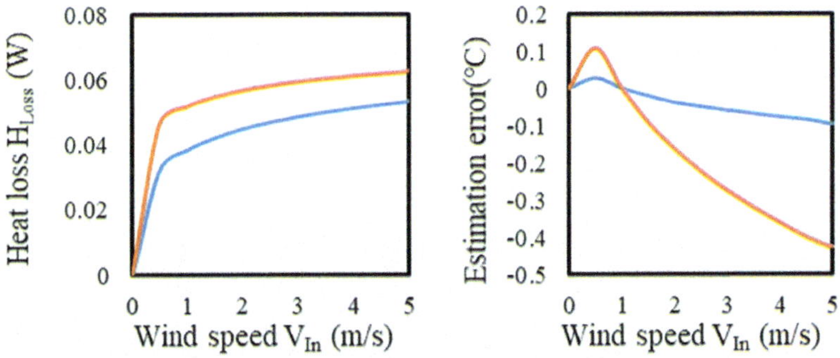

Fig. 7 Comparison of the **a** heat loss and **b** estimation error with (blue) and without (orange) the optimized aluminum structure, as a function of the wind speed V_{In}

3.5.1 Experimental Setup

The proposed CBT sensor with the aluminum structure to reduce the heat loss was fabricated as shown in Fig. 8. The sensor chassis' diameter and height were 31 and 5.1 mm, respectively. Its top and bottom were made of 0.1-mm-thick polyethylene terephthalate (PET), and its sides were made of 0.5-mm-thick polylactic acid (PLA). The thickness of the aluminum structure was 0.5 mm across the entire structure. The height of the heat resistance substrate was 5 mm, and the diameter of the hole in the truncated cone was 2 mm. The heat-resistant substrate was made of PLA and had two platinum resistance thermometers integrated into it. The gap d between the two thermometers was 2 mm, and each thermometer's diameter and length were 0.4 and 1 mm, respectively. The temperature of a hotplate, $T_{Hotplate}$, was measured every 1 s by a thermometer and read out from the thermometer by a data logger. The coefficient R_{Sensor} was estimated from the temperatures T_{skin} and T_{top} measured by the two thermometers. The heat flux H_{Body} was assumed proportional to the temperature gap T_{skin}-T_{top}:

$$H_{Body} \propto T_{Skin} - T_{top}. \tag{15}$$

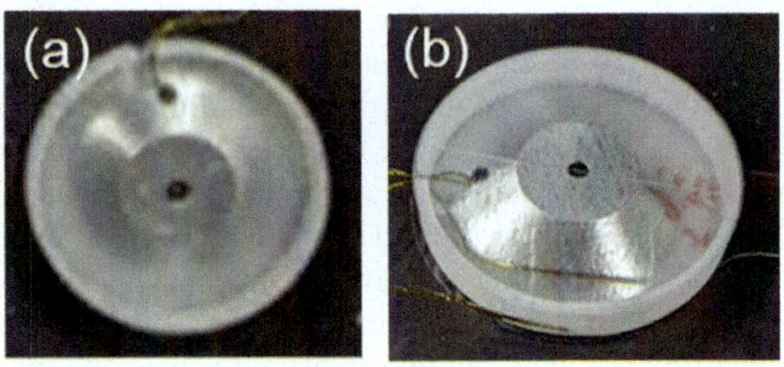

Fig. 8 Illustrations of the fabricated CBT sensor: **a** top view image, **b** bird's eye view image

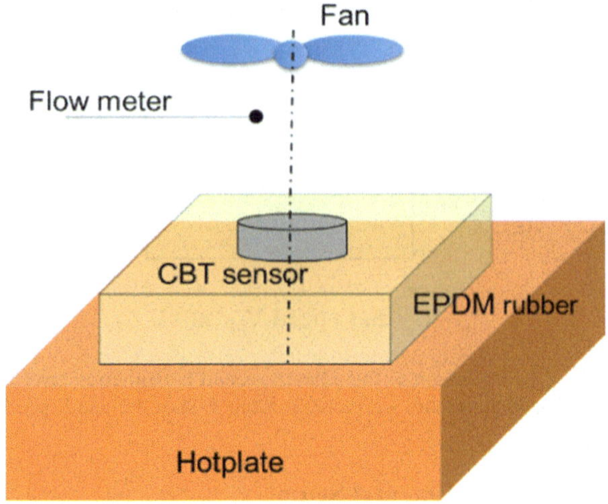

Fig. 9 In-vitro experimental configuration with a hotplate to simulate the CBT

By substituting Eq. (15) into Eq. (1), we obtained the coefficient R_{Sensor} from the hotplate temperature at a specific time:

$$R_{\text{Sensor}} = \frac{T_{\text{Hotplate}} - T_{\text{Skin}}}{T_{\text{Skin}} - T_{\text{Top}}}. \quad (16)$$

As for the phantom representing the body, shown in Fig. 9, we used EPDM rubber with a thickness of 10 mm because its thermal conductivity of 0.36 W/m/K is similar to that of the human body (0.37 W/m/K). The phantom was placed on the hotplate, and the CBT sensor was placed on the phantom. The ambient temperature was room temperature (20 °C). Convection was induced by a fan placed above the sensor, and the wind speed was measured with a hot-wire anemometer.

3.5.2 In-Vitro Experimental Results and Discussion

The hotplate temperature T_{Hotplate} was varied from 32 to 42 °C, and the wind speed V_{In} was varied from 0 to 5 m/s. The coefficient R_{Sensor} was 2.45 in the experiment, whereas the computed value was 2.5. This discrepancy may have been due to the contact heat resistance between the sensor and phantom and the size of the thermometers. The relationship between the estimated and hotplate temperatures is shown in Fig. 10a, while the root-mean-square error (RMSE) with respect to the wind speed is shown in Fig. 10b. The amplitude of the CBT was about 1 °C [34], which shows that the proposed sensor can be used to monitor the CBT over the course of a day. In addition, Fig. 10c shows a Bland–Altman plot, which is a tool to assess whether measurements are contaminated with systematic errors. Specifically, it comprises a scatter plot of the difference between two measured values on the y-axis and the mean of the two values on the x-axis. The scatter plot enables visual evaluation of the presence and degree of systematic errors. If there is no systematic error, i.e., only random error, the plot will show a distribution that varies both positively and negatively from the x-axis. If there is a fixed systematic error, the plot will show a distribution that is biased either positively or negatively from the x-axis, as seen in the figure here. Lastly, a proportional error will yield a fan-shaped distribution where the difference between the two measurements increases as the value on the x-axis increases. As seen in Fig. 10c, the systematic error increased with the hotplate temperature. To summarize these results, although systematic errors could arise depending on the hotplate temperature, the proposed structure successfully decreased the wind's effect.

Lastly, we discuss the heat loss reduction with the optimized aluminum structure comprising a truncated cone with a hole. Figure 11 shows a schematic image of the heat flow around the CBT sensor on temperature field. The heat flux H_{Body} comes from the body core, and estimation error is induced by the heat loss H_{Loss}. However, the aluminum structure collects heat H_{Plus} from the sensor's surrounding region. The collected heat is then isolated from the heat flux because the foot of the truncated cone is far enough from the heat-resistant substrate. Thus, the collected heat H_{Plus} increases the temperature of the region surrounding the substrate, and it mitigates the heat loss H_{Loss}. Additionally, the hole in the truncated cone enables the heat flux H_{Body} to be easily released into the ambient air.

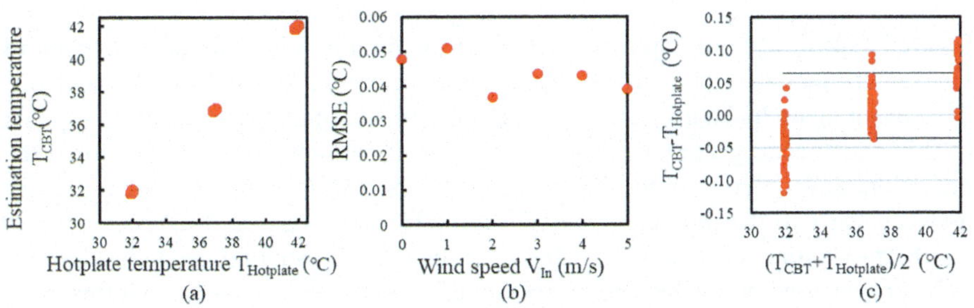

Fig. 10 In-vitro experimental evaluation of the CBT sensor: **a** response curve of the CBT sensor with respect to the hotplate temperature, **b** root-mean-square error (RMSE) versus wind speed, and **c** Bland–Altman plot

Fig. 11 Schematic image of the heat flow around the CBT sensor on temperature field

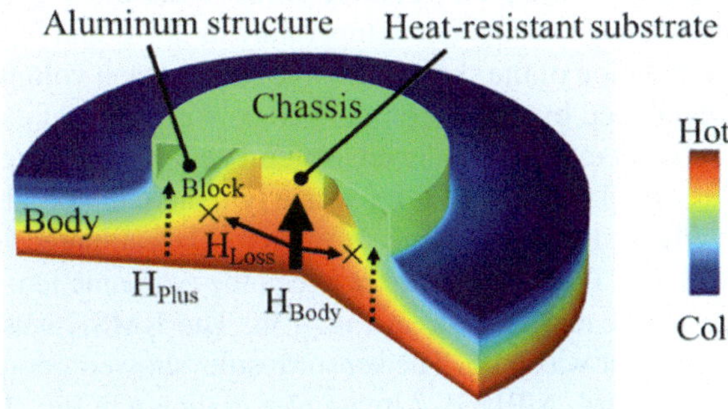

3.6 *In-Vivo Experimental Verification*

Through the in-vitro experiment described above, we demonstrated that a truncated cone structure made of a material with high thermal conductivity can reduce the lateral heat flow. Next, to address the assumptions of steady approximation and that circulation and internal heating can be ignored, we conducted an in-vivo experiment.

3.6.1 Examination of Measurement Site

It is pointed out that the method of estimating CBT from the skin surface might measure the temperature of the hottest part of the measurement site [35]. If there are many muscles at the position where the sensor is attached, the estimated temperature may be affected by heat production due to muscle activity. As the forehead does not have large muscles and is close to the cerebrum, where the thermoregulatory center exists, it is thought to reflect the brain temperature, thus making it a suitable sensor attachment site.

3.6.2 CBT Monitoring During Fanning

Figure 12 shows a photograph and a schematic of the skin-attachable sensor. In this in-vivo experiment, the sensor's diameter and thickness were 30 and 5 mm, respectively. The sensor again had a truncated cone structure made of aluminum with a thickness of 0.5 mm to reduce the lateral heat flow. Two thermometers were integrated into the cylindrical polymer, which had a diameter of 8 mm, and the vertical distance between the two thermometers was 2.5 mm. The temperature difference between the thermometers was taken to be H_{Body} in the CBT estimation, as it was assumed to be proportional to the heat flux. Six healthy volunteers participated in this study. The skin-attachable sensor was placed on the forehead with double-sided tape. The tympanic temperature was monitored by an infrared thermometer to provide a reference temperature, because it is easily measured and highly correlated with the CBT [36, 37]. Because the tympanic temperature is easily affected by the ambient environment [34, 38, 39], the volunteers wore earmuffs to prevent this effect. The study was performed at an ambient temperature of 28 °C and 50% humidity. After 1 h delay for stabilization of the skin-attachable sensor, each volunteer was exposed to wind (1 m/s) produced by an electric fan and performed light exercise on a bicycle ergometer. The wind exposure induced a lateral heat flow. At the start of the experiment, $\mathrm{R}_{\mathrm{Body}}$ was calculated from the reference tympanic temperature.

Figure 13a shows typical results for the estimated CBT and tympanic temperature behavior. The relationship between the tympanic temperature and CBT estimated at the forehead is shown in Fig. 13b. The RMSE was 0.14 °C, and the correlation coefficient was 0.96. The sensor results showed good agreement with the tympanic temperature. A Bland–Altman plot is shown in Fig. 13c.

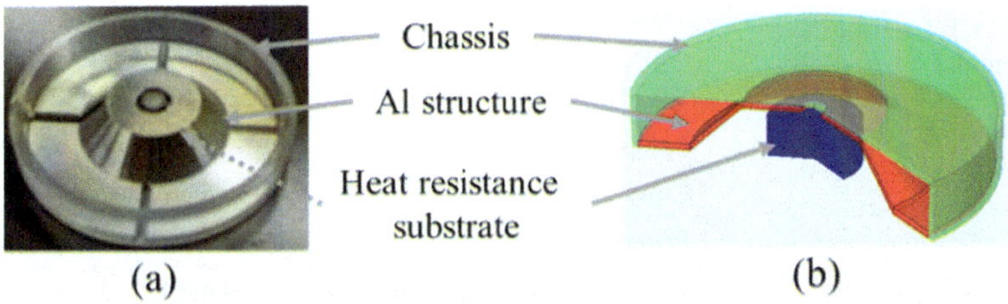

Fig. 12 **a** Photograph and **b** schematic of the skin-attachable sensor designed via topology optimization

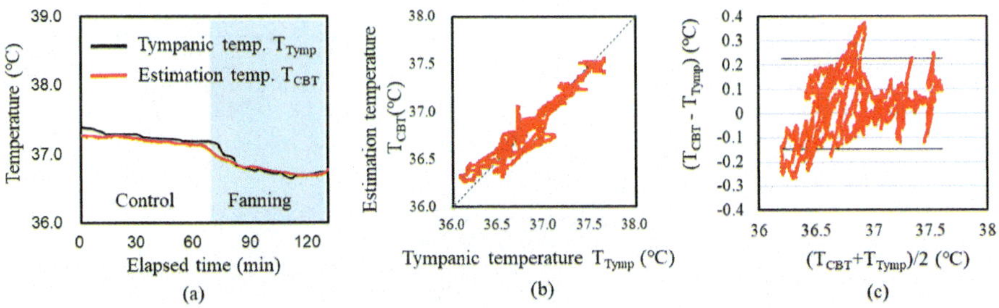

Fig. 13 In-vivo evaluation of the CBT sensor: **a** Typical CBT behavior during fanning, **b** response curve of the CBT sensor with respect to the tympanic temperature, and **c** Bland–Altman plot

3.7 Applications

In this section, we describe an application of system integration with the proposed skin-attachable CBT sensor and demonstrations of its use for monitoring sleep, jet lag induced by a long flight, and fever after COVID-19 vaccination.

3.7.1 System Integration

To monitor the CBT in daily life, we developed a sensor device that integrates a data storage memory, communication module, and battery. Table 3 lists the device specifications. As shown in Fig. 14, the circuit board and battery are stacked on the proposed sensor structure. The sensor has a diameter of 33 mm, a height of 8 mm, and a weight of 10 g. It can communicate with a computer or a smartphone, as seen in Fig. 14, via Bluetooth Low Energy (BLE). The coin-type battery (RC2032) enables continuous monitoring of the core body temperature for up to 60 days. In addition, the device is waterproof to allow its use in everyday life.

Table 3 CBT sensor device specifications

Size	33 mm Φ × 8 mm thickness
Weight	10 g (including battery)
Sampling rate	1–60 s
Communication	Bluetooth low energy
Power supply	Coin-type battery (CR2032)
Continuous measurement time	4–60 days

Fig. 14 Photograph of the CBT sensor device, and a screenshot of the smartphone application

3.7.2 Sleep Monitoring

As mentioned above, the CBT drops during sleep, and the time when it reaches its minimum is an index for evaluating the circadian rhythm. Accordingly, we attached a sensor to the forehead during sleep to measure changes in the CBT, along with a tympanic thermometer for comparison. In the results shown in Fig. 15a, the red line indicates the eardrum temperature, and the black line indicates the temperature measured by the proposed sensor. We could successfully obtain the time when the CBT reached its lowest point, which was at about 5 o'clock, after which it gradually increased toward morning. The mean square error for the tympanic temperature was 0.05 °C, and the correlation coefficient was 0.93. As seen from the Bland–Altman plot in Fig. 15b, the proposed sensor had a small systematic error but a high correlation with the tympanic temperature during sleep.

3.7.3 Jet Lag Monitoring

Next, Fig. 16 shows the CBT changes in a person who returned to Japan after staying in the UK for about a week. The red line represents the normal daily CBT variation for the person in Japan, while the blue line represents the CBT variation immediately after returning from the UK. The time when the CBT reached its minimum shifted from 4:00 to 9:00, a difference of about 5 h, whereas the time difference between

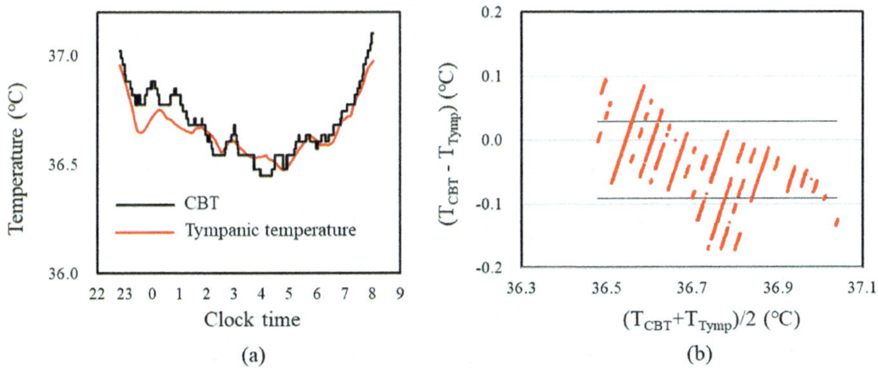

Fig. 15 Sleep demonstration: **a** behavior of the CBT and the reference tympanic temperature, and **b** Bland–Altman plot

the UK and Japan was 7 h. Shifts in circadian rhythms can be visualized from CBT changes, as illustrated in Fig. 17. It has been reported that phase shifts in the circadian rhythm are modulated by external stimuli such as light and food [40]. For example, a phase response curve provides an index for evaluating how much the circadian rhythm shifts depending on the time of light exposure [41]. In particular, Minors et al. reported the shift time for the circadian rhythm with respect to the difference between the time when the CBT reaches its minimum and the time of light exposure. According to this notion, if the body is exposed to light before the time when the CBT is at its lowest, the phase is delayed, whereas if the body is exposed to light after that time, the phase is advanced. Here, the time of the lowest CBT in England was about 9 o'clock, which suggests that the phase was delayed when the person was exposed to the morning sun in Japan. In other words, the jet lag, which should be set back only 5 h, would be expected to shift back 19 h. As this example shows, visualization of the time of the lowest CBT can provide an efficient method for correcting jet lag. This application has a high potential to help improve the quality of life for people who lead irregular lives, such as workers with varying shifts.

3.7.4 Fever Monitoring

Since 2019, vaccination has been a focus in fighting the COVID-19 pandemic, but many people have developed fever after being vaccinated. Thus, we also demonstrated CBT measurement after COVID-19 vaccination. Figure 18a, b show the CBT changes after vaccination and a Bland–Altman plot, respectively. The tympanic temperature was also measured and compared in the same manner as in the previous demonstration. The measurements were started after vaccination, and an antipyretic was taken around 16:00 when the CBT was at its maximum. The results confirmed that the CBT rose for several hours after vaccination and then dropped after the antipyretic was taken.

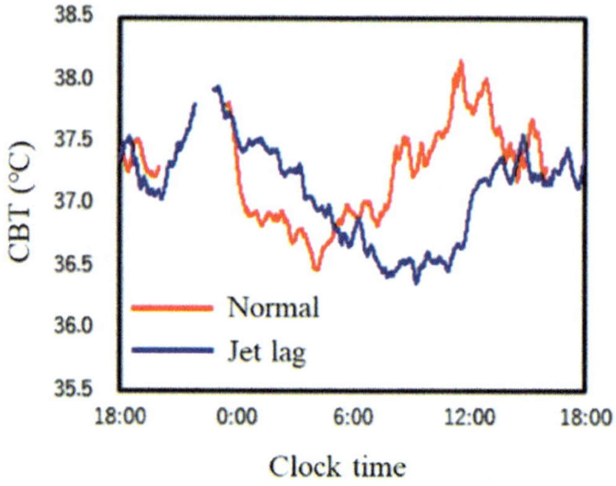

Fig. 16 Comparison of CBT behavior under normal and jet-lag conditions

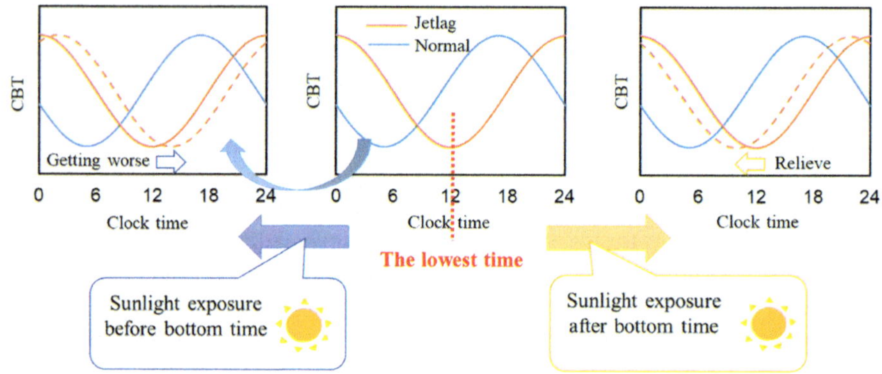

Fig. 17 Illustration of the CBT's circadian phase shift with sunlight exposure

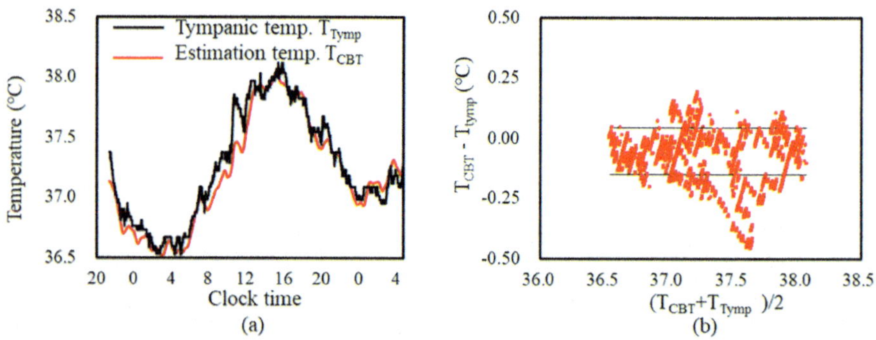

Fig. 18 Fever demonstration: **a** behavior of the CBT and the reference tympanic temperature, and **b** Bland–Altman plot

3.8 Conclusion

We have introduced a technology for estimating the core body temperature (CBT) by using heat flux, with the goal of developing a noninvasive CBT thermometer that can be attached to the body surface. This CBT measurement technique for daily life would be expected to find use in health management systems and new medical applications such as chronopharmacology. Furthermore, if we could visualize the circadian rhythm from CBT fluctuations and understand its deviations from daily rhythms, it would be possible to develop applications that appropriately maintain circadian rhythms through environmental control linked to smart homes.

References

1. Costa, G.: The impact of shift and night work on health. Appl. Ergon. **27**(1), 9–16 (1996)
2. Komarzynski, S., Bolborea, M., Huang, Q., Finkenstädt, B., Lévi, F.: Predictability of individual circadian phase during daily routine for medical applications of circadian clocks. JCI Insight **4**(18) (2019)
3. Scales, W.E., Vander, A.J., Brown, M.B., Kluger, M.J.: Human circadian rhythms in temperature, trace metals, and blood variables. J. Appl. Physiol. **65**(4), 1840–1846 (1988)
4. Tobaldini, E., Fiorelli, E.M., Solbiati, M., Costantino, G., Nobili, L., Montano, N.: Short sleep duration and cardiometabolic risk: from pathophysiology to clinical evidence. Nat. Rev. Cardiol. **16**(4), 213–224 (2019)
5. Hashimoto, S., Nakamura, K., Honma, S., Tokura, H., Honma, K.: Melatonin rhythm is not shifted by lights that suppress nocturnal melatonin in humans under entrainment. Am. J. Physiol.-Regul., Integr. Comp. Physiol. **270**(5), R1073–R1077 (1996)
6. Lévi, F., et al.: Circadian timing in cancer treatments. Annual Rev. Pharmacol. Toxicol. **50**, 377–421 (2010)
7. Higuchi, S., Motohashi, Y., Liu, Y., Ahara, M., Kaneko, Y.: Effects of VDT tasks with a bright display at night on melatonin, core temperature, heart rate, and sleepiness. J. Appl. Physiol. **94**(5), 1773–1776 (2003)
8. Takasu, N.N., Hashimoto, S., Yamanaka, Y., Tanahashi, Y., Yamazaki, A., Honma, S., Honma, K.I.: Repeated exposures to daytime bright light increase nocturnal melatonin rise and maintain circadian phase in young subjects under fixed sleep schedule. Am. J. Physiol.-Regul., Integr. Comp. Physiol. **291**(6), R1799–R1807 (2006)
9. Jean-Louis, G., Kripke, D.F., Cole, R.J., Elliott, J.A.: No melatonin suppression by illumination of popliteal fossae or eyelids. J. Biol. Rhythms **15**(3), 265–269 (2000)
10. Talo, H., Macknin, M.L., VanderBrug Medendorp, S.: Tympanic membrane temperatures compared to rectal and oral temperatures. Clin. Pediatrics **30**(4_suppl), 30–33 (1991)
11. Sund-Levander, M., Forsberg, C., Wahren, L.K.: Normal oral, rectal, tympanic and axillary body temperature in adult men and women: a systematic literature review. Scand. J. Caring Sci. **16**(2), 122–128 (2002)
12. Sund-Levander, M., Grodzinsky, E.: Time for a change to assess and evaluate body temperature in clinical practice. Int. J. Nurs. Pract. **15**(4), 241–249 (2009)
13. Mazerolle, S.M., Ganio, M.S., Casa, D.J., Vingren, J., Klau, J.: Is oral temperature an accurate measurement of deep body temperature? A systematic review. J. Athl. Train. **46**(5), 566–73 (2011)
14. Taylor, N.A., Tipton, M.J., Kenny, G.P.: Considerations for the measurement of core, skin and mean body temperatures. J. Therm. Biol. **46**, 72–101 (2014)

15. Yoshida, S., Miyaguchi, H., Nakamura, T.: Development of tablet-shaped ingestible core-body thermometer powered by gastric acid battery. IEEE Sens. J. **18**(23), 9755–9762 (2018)
16. Kalantar-Zadeh, K., Ha, N., Ou, J.Z., Berean, K.J.: Ingestible sensors. ACS Sens. **2**(4), 468–483 (2017)
17. Byrne, C., Lim, C.L.: The ingestible telemetric body core temperature sensor: a review of validity and exercise applications. Br. J. Sports Med. **41**(3), 126–133 (2007)
18. Nemoto, T., Togawa, T.: Improved probe for a deep body thermometer. Med. Biol. Eng. Comput. **26**, 456–459 (1988)
19. Fox, R.H., Solman, A.J., Isaacs, R., Fry, A.J., MacDonald, I.C.: A new method for monitoring deep body temperature from the skin surface. Clin. Sci. **44**(1), 81–86 (1973)
20. Yamakage, M., Namiki, A.: Deep temperature monitoring using a zero-heat-flow method. J. Anesth. **17**, 108–115 (2003)
21. Mäkinen, M. T., Pesonen, A., Jousela, I., Päivärinta, J., Poikajärvi, S., Albäck, A., ... Pesonen, E.: Novel zero-heat-flux deep body temperature measurement in lower extremity vascular and cardiac surgery. J. Cardiothorac. Vasc. Anesth. **30**(4), 973–978 (2016)
22. Eshraghi, Y., Nasr, V., Parra-Sanchez, I., Van Duren, A., Botham, M., Santoscoy, T., Sessler, D.I.: An evaluation of a zero-heat-flux cutaneous thermometer in cardiac surgical patients. Anesth. Analg. **119**(3), 543–549 (2014)
23. Kitamura, K.I., Zhu, X., Chen, W., Nemoto, T.: Development of a new method for the noninvasive measurement of deep body temperature without a heater. Med. Eng. Phys. **32**(1), 1–6 (2010)
24. Zhang, Y., Chad Webb, R., Luo, H., Xue, Y., Kurniawan, J., Cho, N.H., Rogers, J.A.: Theoretical and experimental studies of epidermal heat flux sensors for measurements of core body temperature. Adv. Healthc. Mater. **5**(1), 119–127 (2016)
25. Huang, M., Tamura, T., Tang, Z., Chen, W., Kanaya, S.: A wearable thermometry for core body temperature measurement and its experimental verification. IEEE J. Biomed. Health Inf. **21**(3), 708–714 (2016)
26. Tamura, T., Huang, M., Yoshimura, T., Umezu, S., Ogata, T.: An Advanced internet of things system for heatstroke prevention with a noninvasive dual-heat-flux thermometer. Sensors **22**(24), 9985 (2022)
27. Pennes, H.H.: Analysis of tissue and arterial blood temperatures in the resting human forearm. J. Appl. Physiol. **1**(2), 93–122 (1948)
28. Yoon, G.H.: Topological design of heat dissipating structure with forced convective heat transfer. J. Mech. Sci. Technol. **24**(6), 1225–1233 (2010)
29. Lazarov, B.S., Schevenels, M., Sigmund, O.: Topology optimization considering material and geometric uncertainties using stochastic collocation methods. Struct. Multidiscip. Optim. **46**, 597–612 (2012)
30. Wang, F., Lazarov, B.S., Sigmund, O.: On projection methods, convergence and robust formulations in topology optimization. Struct. Multidiscip. Optim. **43**, 767–784 (2011)
31. Sigmund, O., Maute, K.: Topology optimization approaches: a comparative review. Struct. Multidiscip. Optim. **48**(6), 1031–1055 (2013)
32. Incropera's Principles of Heat and Mass Transfer. Western John Wiley & Sons Inc., pp. 405–407 (2017)
33. Tanaka, Y., Matsunaga, D., Tajima, T., Seyama, M.: Robust skin attachable sensor for core body temperature monitoring. IEEE Sens. J. **21**(14), 16118–16123 (2021)
34. Moran, J.L., Peter, J.V., Solomon, P.J., Grealy, B., Smith, T., Ashforth, W., Wake, M., Peake, S.L., Peisach, A.R.: Tympanic temperature measurements: are they reliable in the critically ill? A clinical study of measures of agreement. Crit. Care Med. **35**(1), 155–164 (2007)
35. Brajkovic, D., Ducharme, M.B.: Confounding factors in the use of the zero-heat-flow method for non-invasive muscle temperature measurement. Eur. J. Appl. Physiol. **94**, 386–391 (2005)
36. Mariak, Z., Lewko, J., Luczaj, J., Polocki, B., White, M.D.: The relationship between directly measured human cerebral and tympanic temperatures during changes in brain temperatures. Eur. J. Appl. Physiol. **69**, 545–549 (1994)

37. Fulbrook, P.: Core body temperature measurement: a comparison of axilla, tympanic membrane and pulmonary artery blood temperature. Intensive Crit. Care Nurs. **13**(5), 266–272 (1997)
38. Casa, D.J., Becker, S.M., Ganio, M.S., Brown, C.M., Yeargin, S.W., Roti, M.W., ... Maresh, C.M.: Validity of devices that assess body temperature during outdoor exercise in the heat. J. Athl. Train. **42**(3), 333 (2007)
39. Vongsavan, N., Matthews, B.: Some aspects of the use of laser Doppler flow meters for recording tissue blood flow. Exp. Physiol.: Transl. Integrat. **78**(1), 1–14 (1993)
40. Tahara, Y., Shibata, S.: Circadian rhythms of liver physiology and disease: experimental and clinical evidence. Nat. Rev. Gastroenterol. Hepatol. **13**(4), 217–226 (2016)
41. Minors, D.S., Waterhouse, J.M., Wirz-Justice, A.: A human phase-response curve to light. Neurosci. Lett. 133(1), 36–40 (2016) (1991)

Wearable Artificial Pancreas Device Technology

Akira Matsumoto

1 Introduction

Diabetes is a major global health threat that poses a devastating impact on society [1, 2]. Insulin therapy remains to be a primary option for both palliative and preventive purposes in clinical practice. Long-term intensive insulin therapy has been shown to suppress the progression of diabetic microvascular complications [3–5]. Recent studies have identified blood glucose fluctuations, especially postprandial hyperglycemia, as a major cause of diabetic complications [6]. Indeed, it has been observed that patients with similar HbA1c (hemoglobin A1c) levels often exhibit distinct blood glucose variation patterns [7]. These findings came to develop a consensus that care must be taken to control the daily blood glucose fluctuations, so-called "glucose spikes", as much as the commonly focused fasted blood glucose levels and HbA1c, in order to strengthen the long-term therapeutic outcomes. At present, the predominant approach to this treatment involves patients administering insulin themselves, known as "open-loop" insulin delivery. Nevertheless, this method lacks precision in dose regulation, necessitating careful avoidance of overdosing to prevent the occurrence of severe and potentially fatal hypoglycemia.

This chapter will detail recent attempts in creating a "closed-loop" insulin delivery system, which is self-regulated and continuously administers precise insulin amounts based on fluctuations in blood glucose levels. These systems, often termed "artificial pancreas" systems (APSs), combine implanted glucose sensors with automated algorithm-driven insulin infusion units. These units can communicate either through

A. Matsumoto (✉)
Institute of Biomaterials and Bioengineering, Tokyo Medical and Dental University, 2-3-10 Kanda-Surugadai, Chiyoda-Ku, Tokyo 101-0062, Japan
e-mail: matsumoto.bsr@tmd.ac.jp

Department of Materials Engineering, Graduate School of Engineering, The University of Tokyo, Tokyo, Japan

Kanagawa Institute of Industrial Science and Technology, Ebina, Japan

K. Mitsubayashi (ed.), *Wearable Biosensing in Medicine and Healthcare*,
https://doi.org/10.1007/978-981-99-8122-9_12

electrical or wireless means. First, such an electronics-based APS concept will be outlined by summarizing the up-to-date development and validation status. Then, an ever-growing research effort to develop electronics-free, chemically-controlled APSs will be mentioned, with a special focus on the recent attempts of making it "wearable" format.

2 Electronics-Based APS

2.1 Basic Concept of the Artificial Pancreas Systems (APSs)

The first insulin pump was invented by Arnold Kadish in the early 1960s, which used an auto analyzer for continuous blood glucose monitoring via an intravenous catheter and two intravenous syringe pumps containing insulin and glucose or glucagon (Fig. 1a) [8]. The first APSs were developed in the early 1970s which was computer programmed to respond to continuous venous glucose monitoring to control the intravenous delivery of insulin and/or dextrose [9–12]. The first "wearable" APS was developed in the early 1980s, which used a subcutaneous glucose sensor paired with intravenous pumps for insulin and glucagon infusions [13, 14].

In 2005, the Juvenile Diabetes Research Foundation (JDRF) established the Artificial Pancreas Project to promote research, regulatory approval, and eventual adoption of closed-loop technologies [15]. Since then, progress toward a fully closed-loop system has been accelerating, with the development and commercialization of numerous glucose-sensing and insulin-delivery systems of increasing sophistication

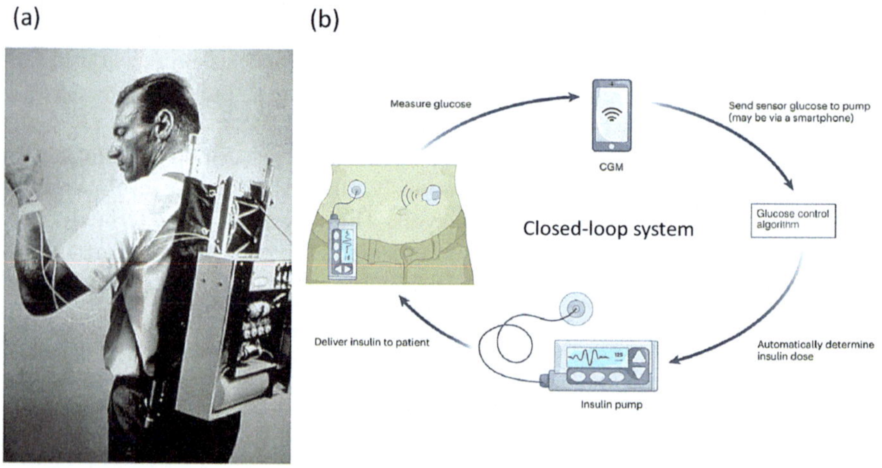

Fig. 1 The first insulin pump; Kadish's device. Reproduced with permission from Ref. [12]

[16, 17]. A workable closed-loop system for diabetes management requires a continuous glucose monitoring system (CGM), an insulin pump, and autonomous control algorithms (Fig. 1b) [16, 17]. The development of closed-loop systems has been ongoing since the early 2000s, with early versions being PC-based and later versions being portable and smartphone-based [18, 19]. Sensor-augmented pumps (SAPs) were developed in the late 2000s and were followed by SAPs with low glucose suspension (LGS) [20–22], an ability to suspend insulin when sensor glucose values are below the prespecified threshold, and predictive low glucose suspension (PLGS) functions [23, 24]. The first closed-loop system was developed in the early 2000s and has since been developed into a portable form [25, 26]. There are also dual-hormone closed-loop systems that dispense glucagon as well as insulin, and advanced single-hormone closed-loop systems with additional auto-correction bolus features [27]. Other peptide hormones such as amylin analogues [27, 28] and glucagon-like peptide 1 (GLP-1) receptor (GLP-1R) agonists have been investigated [29], and intraperitoneal insulin delivery systems have also been in development [30].

2.2 Control Algorithms

The main control algorithms used for closed loops in APS [31] are PID [32] (Proportional-Integral–Differential Controller), MPC (Model Predictive Control), and fuzzy logic algorithms (Fig. 2) [33]. PID detects the present glucose level, the past deviation, and predicts the future direction. MPC predicts changes by adjusting independent variables such as body mass index (BMI), carbohydrate intake, and insulin-on-board (insulin that is still active in the body from previous bolus doses). Fuzzy logic produces correctness as a matter of degree instead of as a yes or no result and operates through supervised learning based on expert opinions to establish a specific decision. In addition to these, some modified algorithms have also been used. These include PD (proportional and derivative) algorithm, which is a PID without the integral feature, and fading memory PD [34]. Medtronic, CamDiab, Diabeloop, Insulet, and TypeZero groups used PID or MPC algorithms for insulin-only systems, while DreaMed group adopted a fuzzy logic algorithm. Inreda used PID for both insulin and glucagon, Oregon used a fading memory PD algorithm, and Beta Bionics and IRCM groups used an MPC algorithm for insulin, but for glucagon [34].

2.3 Commercial Devices and Validation Status

The Medtronic 670G Hybrid Closed-loop System is a commercially available closed-loop system that comprises an insulin pump with an embedded PID algorithm and the Guardian sensor 3 (Fig. 3a). It requires a minimum of 4–6 finger prick calibrations per day and accurate basal rates, insulin-to-carbohydrate ratios, insulin sensitivity

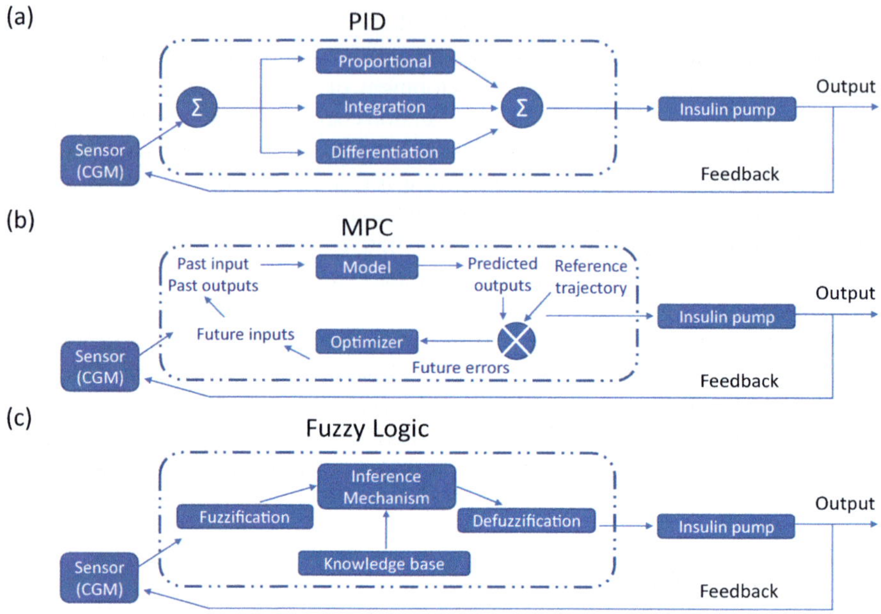

Fig. 2 "Artificial pancreas" systems (APSs) consisting of an insulin pump, a continuous glucose monitor (CGM) for interstitial fluid glucose monitoring, and a glucose control algorithm. The CGM communicates the glucose concentration data to the insulin pump either directly or by a smartphone application, and an algorithm on the phone or pump directs the insulin delivery from the pump. Reproduced with permission from Ref. [70]

setting, and active insulin time for the setup. The system automatically determines basal insulin delivery and adjusts it every 5 min according to the sensor glucose values. A fixed target of 6.7 mmol/L (glucose concentration) is used, but there is an optional activity mode that raises the target to 8.3 mmol/L. Studies have shown that this closed-loop system significantly improves time in target range (the amount of time the patient spent in the target blood glucose range) and reduces hypoglycemia compared to usual care, with no significant difference in severe hypoglycemia events [35]. The Medtronic 780G Advanced Hybrid Closed-Loop (AHCL) is the next-generation system that has an enhanced PID algorithm with an added fuzzy logic component and requires fewer finger prick calibrations per day (Fig. 3b). This system has customizable glucose level targets and provides automated correction boluses. Studies have shown that the AHCL system performs better than the 670G system in terms of glycemic outcomes and closed-loop usage [36].

Tandem's Control-IQ system is a hybrid closed-loop system for diabetes management, consisting of the t:slim X2 insulin pump with an embedded MPC algorithm and a Dexcom G6 CGM (Fig. 3c). It is licensed for use in patients aged 6 years and above and requires accurate basal rates, insulin-to-carbohydrate ratios, and insulin sensitivity information for the setup. The system delivers automated correction boluses and has an optional activity mode and sleep mode. Clinical trials have shown that

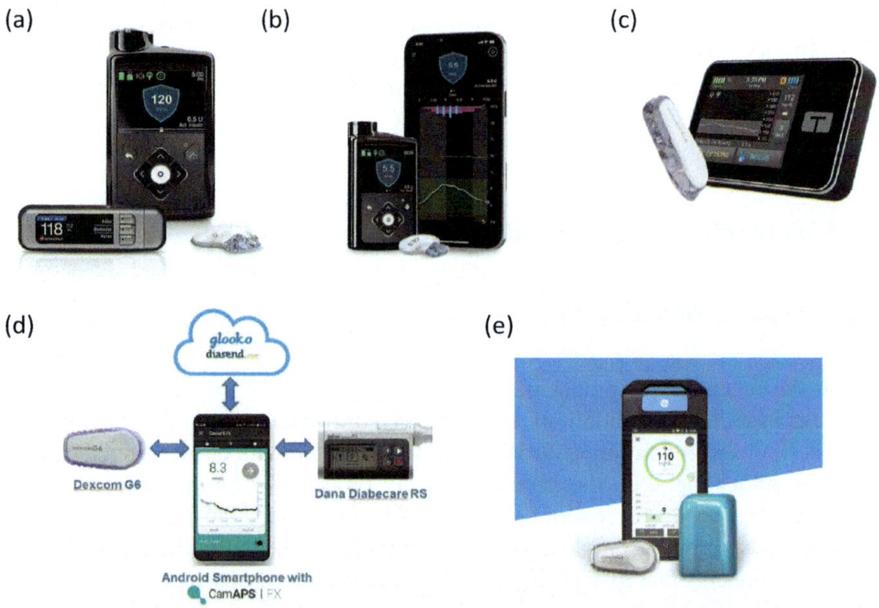

Fig. 3 APSs on market. **a** The Medtronic MiniMed 670G with Guardian Link 3 sensor/transmitter. **b** The Medtronic MiniMed 780G paired with Guardian Link 3 sensor/smartphone. **c** Tandem t:slim X2 pump paired with Dexcom G6 sensor. **d** CamAPS FX algorithm hosted on Android smartphone. **e** Diabeloop DLBG1 algorithm with Kaleido patch-pump and Dexcom G6 sensor. This Figure includes works that are distributed in the Apache License 2.0. and the CC licensing

the Control-IQ system significantly improves time in target glucose range, reduces time in hypoglycemia and HbA1c levels [37]. Safety has been demonstrated in pilot studies and larger randomized controlled trials for older children and adults [38].

The CamAPS FX system, developed by the University of Cambridge, is a regulatory-approved hybrid closed-loop app that communicates wirelessly with certain insulin pumps and CGM devices (Fig. 3d). The algorithm is highly adaptive, which takes into account a total daily insulin, postprandial, and diurnal insulin requirements. Studies have shown that the system is beneficial in improving time in target range, reducing hypoglycemia and HbA1c levels across a wide range of populations [39]. The Cambridge algorithm is the only closed-loop algorithm to have been evaluated in a longer randomized trial in very young children aged 1–7 years [40].

The Diabeloop system, which uses an MPC algorithm called DBLG1, is an interoperable closed-loop system for adults in Europe (Fig. 3e). It can be used with the Accu-Check Insight insulin pump and Dexcom G6 CGM, and has been used with the Kaleido insulin patch-pump in clinical trials. The algorithm is adaptive and has five user-adjustable settings. In a pivotal trial, DBLG1 hybrid closed-loop was compared to SAP therapy in 68 adults, resulting in an increase in time in target range and a reduction in time in hypoglycemia [41]. A longer, single-arm, non-randomized study showed the DLBG1 system was safe in 25 adults over 6 months, with no reported severe hypoglycemia or diabetic ketoacidosis adverse (DKA) events [42]. Recently, a 6-week randomized crossover study in 17 children aged 6–12 years showed improvements in time in range and a reduction in time in hypoglycaemia [43].

3 Electronics-Free, Chemically Controlled APSs

3.1 Protein-Based APSs

While the recent development of electronics-based APSs has seen a remarkable progress with clinical validations, it poses some drawbacks, including the high cost, the risk of electronic failures, the discomfort associated with needle-tipped catheter insertion under the skin, the time-demanding process for establishing a learning curve, and the need of frequent calibrations [16, 34]. Since the 1990s, a great deal of efforts has sprung up to develop electronics-free, chemically driven APSs. Two major components that historically prevailed are glucose oxidase and Concanavalin A, a sugar-binding protein or lectin [44–47]. Typically, the former is immobilized in a pH-responsive polymer gel (or membrane), and the insulin permeability of the immobilized matrix is changed by the enzymatic reaction. For the latter approach, a representative strategy is the use of a sugar-modified insulin, which can be recognized by the lectins, in the semipermeable membrane and is released when endogenous insulin comes inside the membrane. Although these approaches have been investigated extensively, they have not yet been put to practical use due to the limitations inherent to these biologically derived materials, that is, intolerance of long-term use and storage associated with the protein denaturation and immunotoxicity. In contrast, an emerging approach using boronic acids chemistry that the author's group has pioneered is a fully synthetic system offering the potential to solve all the above problems [48, 49].

3.2 Boronic Acid-Based Bioengineering Related to Diabetes Care and Treatment

Boronic acids are able to reversibly interact with the diol groups, a commonly found motif in biomolecules including sugars, ribose, and catechols. For their carbohydrate-binding properties, they can be regarded as a synthetic mimic of lectins, often termed "borono-lectins" [50]. There are a wide variety of bioengineering attempts that utilize glucose recognition capability by boronic acids. A recent noteworthy example is a miniaturized blood glucose monitoring device based on a diboronic acid-type fluorescent sensor that can be used continuously for up to six months. This "wearable" glucose sensor is subcutaneously implantable with a small LED, which is now available in the United States and Europe (Fig. 4) [51, 52].

Efforts persist to create insulin variants with tunable kinetics, especially long-acting ones in order for less frequent administration. However, these lack the ability of acute (spike-like) response. Anderson et al. developed a glucose-responsive, "smart" insulin based on a creative phenylboronic boronic acid (PBA)-based engineering [53]. Aliphatic molecules were conjugated to insulin, mimicking a structure of a clinically used long-acting Insulin detemir, but with the end PBA function for the glucose-sensitivity (Fig. 5). The best-performing derivative was able to rapidly ameliorate blood glucose in a diabetic mouse model following glucose challenge with reduced

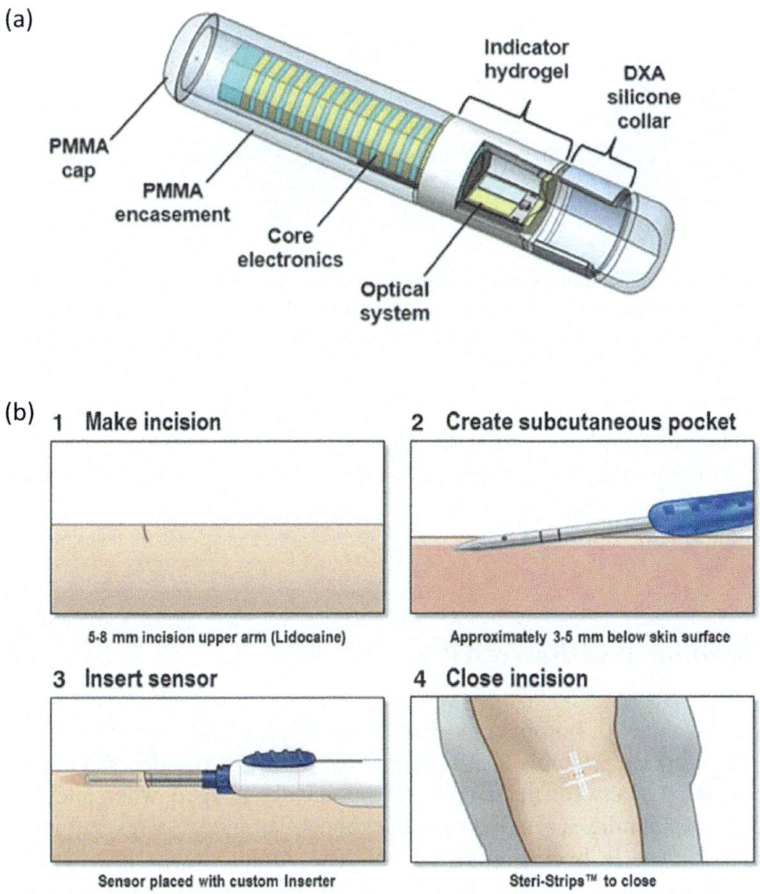

Fig. 4 Eversense sensor. **a** The sensor structure; DXA, dexamethasone acetate; PMMA, polymethyl methacrylate. **b** The sensor insertion procedure. Reproduced with permission from Ref. [52]

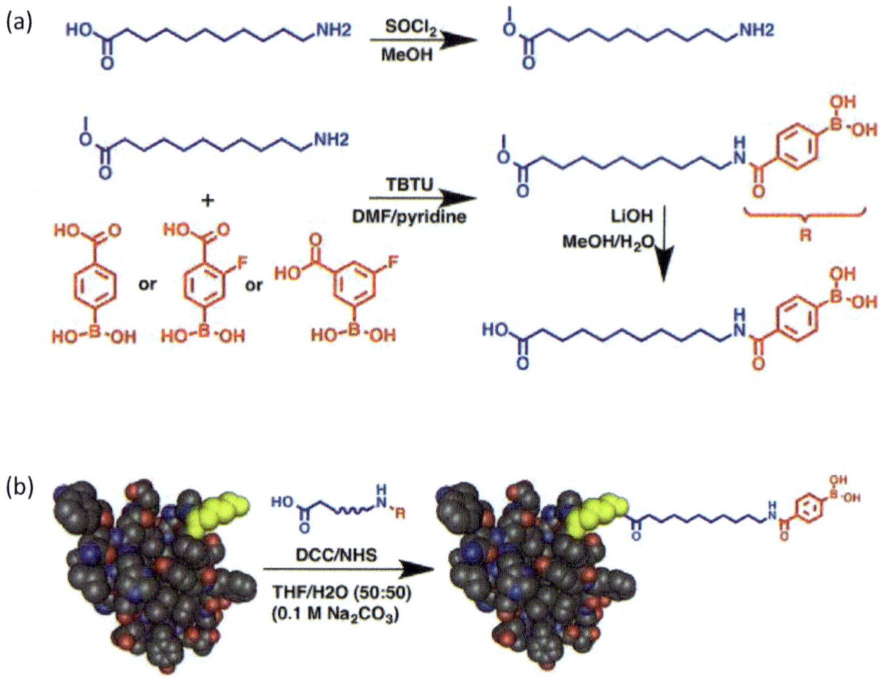

Fig. 5 Schematic illustration of the strategy for insulin modification. **a** Generalized scheme to prepare PBA-modified insulin. **b** Final carboxylic acid-containing small molecule is conjugated to the native human insulin protein through the ε-amine of the lysine at the B29 position (yellow). Reproduced with permission from Ref. [53]

events of hypoglycemia, providing the glycemic control that is superior even to native insulin. Although the glucose-responsive mechanism and safety of such modified insulin need further clarification, the approach may offer an attractive adjunct to the current therapeutic options.

3.3 *Boronic Acid-Based APSs*

We have developed a boronate gel-based, totally synthetic and "electronics-free" APSs. The working principle is as follows (Fig. 6). A glucose-dependent shift in the equilibria of PBA between uncharged and anionically charged, when integrated with optimally amphiphilic acrylamide gel backbone, can induce a reversible, glucose-dependent change in hydration of the gel. The abrupt change in hydration leads to the formation of a gel surface–emerging, microscopically dehydrated layer, so-called "skin layer", which provides a mode for the diffusion control of the gel-loaded insulin [54, 55]. The chemical structure of the gel was further refined to ensure the occurrence of the above-mentioned performance under physiologically

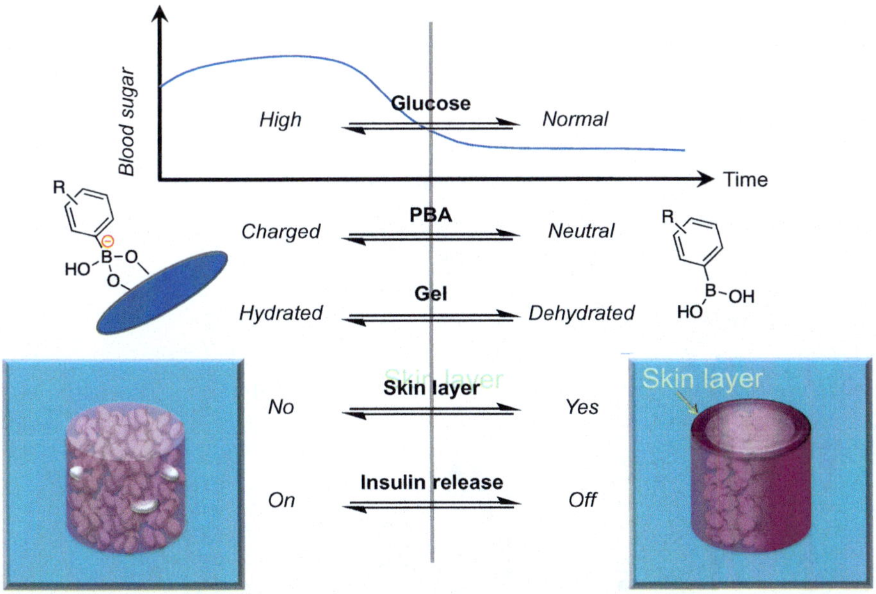

Fig. 6 Schematic illustration of a boronate gel-based "electronics-free" and "protein-fee" APS. Reproduced with permission from Ref. [60]

relevant conditions, while also establishing a remarkably "gated" manner response to the level of normoglycemia. The new structure also achieved a performance that remains stable regardless of temperature fluctuations [56–59].

The first in vivo validation was made by using a catheter-combined device that was suitably scaled for mouse model experiments (Fig. 7) [60]. Subcutaneous implantation of the device in healthy and diabetic mice established a closed-loop system composed of "continuous glucose sensing" and the "skin layer"–regulated insulin release (Fig. 7a). Glucose metabolism was controllable in response to interstitial glucose fluctuation under both insulin-deficient and insulin-resistant conditions with at least 3-week durability. We then combined the gel with hemodialysis hollow fibers and scaled it suitable for rats (Fig. 8) [61]. This device not only normalized average glucose level of rats, but also markedly ameliorated the fluctuations over timescale of a day without inducing hypoglycemia, the first successful in vivo validation with no electronics (Fig. 8a) [61]. This finding is worth emphasizing because of the fact that hypoglycemia and blood glucose fluctuations are major causes for diabetic complications possibly through oxidative stress and proinflammatory cytokines. With inherent stability, diffusion-dependent scalability, and week-long and acute glucose-responsiveness, such gel-based technology could offer a low-cost alternative to the current electronics-based APSs.

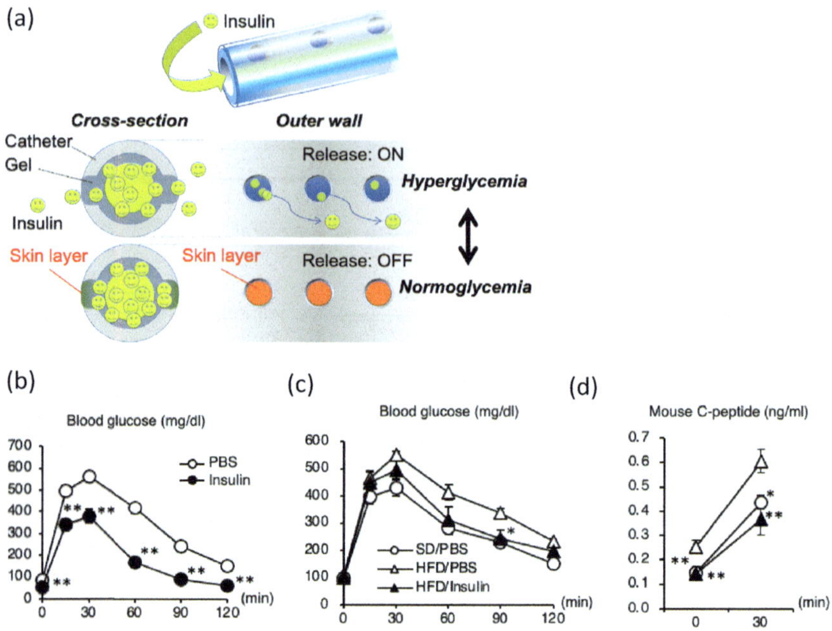

Fig. 7 Catheter-combined device. **a** Schematic of glucose-responsive insulin regulated by skin layer. **b** Glucose tolerance test (GTT) of healthy mice implanted with catheter devices containing PBS and human insulin. **c** Blood glucose concentration and **d** mouse C-peptide concentration of type II diabetic mice in the GTT. Reproduced with permission from Ref. [60]

3.4 Microneedle Engineering as a Promising Route for "Wearable" APSs

Traditional insulin injection involves pain, inconvenience, and risks of microbial contamination and nerve damage [62, 63]. However, the groundbreaking microneedle (MN) technology offers a promising solution to overcome these limitations. The MNs boast sharper needles compared to their hypodermic counterparts, with a length range of several hundred micrometers [64]. These characteristics enable MNs to effectively disrupt the stratum corneum, the main barrier to successful transdermal insulin delivery. Since MNs only form micro-scale channels on the skin, and their needle length is short enough not to reach the nerve endings in the dermis, an MNs patch is minimally invasive, eliminating the injection pain [65]. The incorporation of those glucose-sensitive elements described above into the MNs could render them capable of achieving on-demand, painless insulin delivery without human intervention in a "wearable" fashion.

Based on this concept, we have reported a PBA gel-based MN device capable of achieving both at once weekly sustained and a remarkably acute response in a matter of tens of seconds (Fig. 9) [66–69]. In order to ensure sufficient mechanical strength for the skin insertion, biocompatible and mechanically robust silk fibroin,

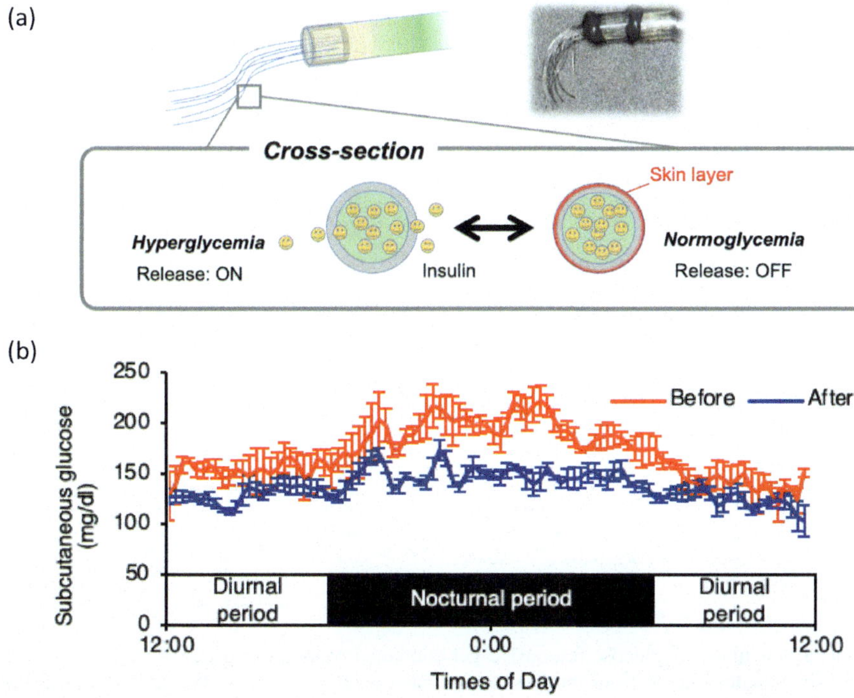

Fig. 8 Hollow fiber-combined device. **a** Schematic of the closed-loop functionality. **b** Daily glucose fluctuations in mildly diabetic rats before and after device implantation. Reproduced with permission from Ref. [61]

derived from the silkworm Bombyx mori, was incorporated into the hydrogel to form a semi-interpenetrating network. Fibroin is a biomaterial with remarkable mechanical properties, biocompatibility, and chemically controllable biodegradability, and is FDA-approved as a surgical thread and hard tissue defect replacement (implantation) material. By controlling the microphase-separation structure during the gelation reaction, the material could show both high mechanical properties and insulin permeability [64].

Recently, Gu et al. have also reported a MN patch device made of the bortonate gel [70]. In vitro experiments showed the ability to switch the insulin release on and off in response to glucose. Further, a patch of about 5 cm^2 size enabled effective regulation of the plasma glucose levels of 25 kg minipigs for over 20 h, during which minipigs were subject to glucose tolerance tests.

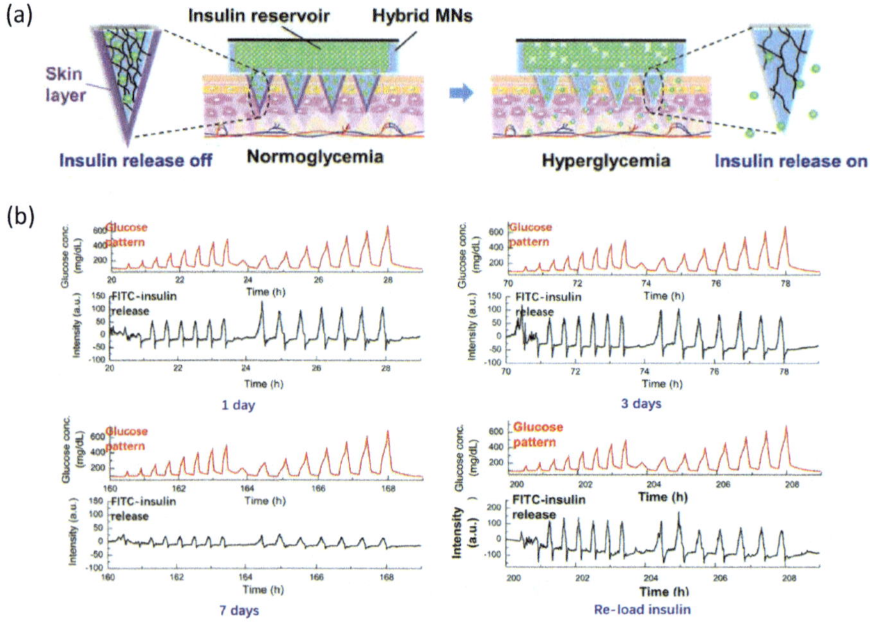

Fig. 9 **a** Schematic of glucose-responsive gel combined MNs array patch. **b** In vitro glucose-responsive insulin release from the MNs array patch on day 1, day 3, day 7, and after insulin re-loading. Reproduced with permission from Ref. [66]

4 Summary and Future Perspectives

Since the 2000s, numerous APSs have been developed and clinically validated for Type 1 Diabetes patients, but a fully closed-loop (FCL) system that enables glycemic control in every situation without meal or exercise announcements has not yet been established. Future research should consider Type 2 Diabetes patients as a new target, which outnumbers Type 1 Diabetes and involves a large number of elderly patients and those pregnant or gestational diabetes patients [71]. In order to make APSs widely acceptable by those patients, efforts must continue of downsizing, reducing the cost, and improving the comfort. Toward the improvement of user comfort, the development has been directed toward an all-in-one "wearable" APS system that combines CGM, a pump, and the control algorithm into one device. Indeed, a new APS named Omnipod 5 has recently been authorized by FDA as the first tubeless and wearable system for individuals 6 and older. For inherent challenges of APSs, time lags in insulin infusion and the difficulty of controlling glucose without manual bolus control in situations where blood glucose changes rapidly (such as glucose spikes) are recognized. To solve these problems, possibilities such as the use of faster-acting insulin, an artificially intelligent camera for automatic carbohydrate counting, and a dual-hormone closed-loop system that can actively improve hypoglycemia have been debated [72].

Currently, in developing countries where 80% of people with diabetes will soon live, not enough funds are spent to provide even the least expensive life-saving treatments, largely due to limited access to an adequate health care infrastructure. In this regard, the electronics-free approach described in the latter part of this chapter, especially microneedle-based device engineering, should offer a potential to provide more user-friendly technology that would be acceptable even by whom otherwise may not be strongly motivated, such as the elderly, infants, patients in need of nursing care, patients averse to electrical and mechanical medical devices.

Noteworthy is that a recent clinical trial investigated the use of microneedles to administer insulin intradermally (ID) for a continuous period of 72 h. It was found that ID bolus infusion of insulin aspart was associated with a shorter Tmax (time to maximum blood concentration) and less intra-subject variability compared to subcutaneous (SC) infusion [73]. Although the mechanism is not clear at this point, these results suggest that a microneedle-based, painless and "wearable" ID delivery of insulin is a viable delivery route alternative providing reduced time for insulin absorption with less intra-subject variability and lower glycemic response, an encouraging finding for the community pursuing the electronics-free and "wearable" approaches.

References

1. Cefalu, W.T., Buse, J.B., Tuomilehto, J., Fleming, G.A., Ferrannini, E., Gerstein, H.C., Bennett, P.H., Ramachandran, A., Raz, I., Rosenstock, J., Kahn, S.E.: Update and next steps for real-world translation of interventions for type 2 diabetes prevention: reflections from a diabetes care editors' expert forum. Diabetes Care **39**(7), 1186–1201 (2016). https://doi.org/10.2337/dc16-0873
2. Ginter, E., Simko, V.: Type 2 diabetes mellitus, pandemic in 21st century. Adv. Exp. Med. Biol. **771**, 42–50 (2012). https://doi.org/10.1007/978-1-4614-5441-0_6
3. The Diabetes, C., Complications Trial Research, G.: The effect of intensive diabetes therapy on measures of autonomic nervous system function in the Diabetes Control and Complications Trial (DCCT). Diabetologia **41**(4), 416–423 (1998). https://doi.org/10.1007/s001250050924
4. Diabetes, C., Complications Trial Research, G., Nathan, D.M., Genuth, S., Lachin, J., Cleary, P., Crofford, O., Davis, M., Rand, L., Siebert, C.: The effect of intensive treatment of diabetes on the development and progression of long-term complications in insulin-dependent diabetes mellitus. N Engl. J. Med. **329**(14), 977–986 (1993). https://doi.org/10.1056/NEJM199309303291401
5. Ohkubo, Y., Kishikawa, H., Araki, E., Miyata, T., Isami, S., Motoyoshi, S., Kojima, Y., Furuyoshi, N., Shichiri, M.: Intensive insulin therapy prevents the progression of diabetic microvascular complications in Japanese patients with non-insulin-dependent diabetes mellitus: a randomized prospective 6-year study. Diabetes Res. Clin. Pract. **28**(2), 103–117 (1995). https://doi.org/10.1016/0168-8227(95)01064-k
6. Decode Study Group, t.E.D.E.G.: Glucose tolerance and cardiovascular mortality: comparison of fasting and 2-hour diagnostic criteria. Arch. Intern. Med. **161**(3), 397–405 (2001). https://doi.org/10.1001/archinte.161.3.397
7. Fleischer, J., Cichosz, S.L., Hansen, T.K.: Comment on Lachin et al. Association of glycemic variability in type 1 diabetes with progression of microvascular outcomes in the diabetes control and complications trial. Diabetes Care **40**:777–783, e164 (2017). https://doi.org/10.2337/dc17-1339

8. Kadish, A.H.: Automation control of blood sugar a servomechanism for glucose monitoring and control. T Am. Soc. Art. Int. Org. **9**, 363 (1963)
9. Albisser, A.M., Leibel, B.S., Ewart, T.G., Davidovac, Z., Botz, C.K., Zingg, W.: An artificial endocrine pancreas. Diabetes **23**(5), 389–396 (1974). https://doi.org/10.2337/diab.23.5.389
10. Albisser, A.M., Leibel, B.S., Ewart, T.G., Davidovac, Z., Botz, C.K., Zingg, W., Schipper, H., Gander, R.: Clinical control of diabetes by the artificial pancreas. Diabetes **23**(5), 397–404 (1974). https://doi.org/10.2337/diab.23.5.397
11. Pfeiffer, E.F.: On the way to the automated (blood) glucose regulation in diabetes: the dark past, the grey present and the rosy future. In: XII Congress of the International Diabetes Federation, Madrid, 22–28 September 1985. Diabetologia 30(2), 51–65 (1987). https://doi.org/10.1007/BF00274572
12. Alsaleh, F.M., Smith, F.J., Keady, S., Taylor, K.M.: Insulin pumps: from inception to the present and toward the future. J. Clin. Pharm. Ther. **35**(2), 127–138 (2010). https://doi.org/10.1111/j.1365-2710.2009.01048.x
13. Shichiri, M., Kawamori, R., Yamasaki, Y., Hakui, N., Abe, H.: Wearable artificial endocrine pancrease with needle-type glucose sensor. Lancet **2**(8308), 1129–1131 (1982). https://doi.org/10.1016/s0140-6736(82)92788-x
14. Shichiri, M., Kawamori, R., Hakui, N., Yamasaki, Y., Abe, H.: Closed-loop glycemic control with a wearable artificial endocrine pancreas. Variations in daily insulin requirements to glycemic response. Diabetes **33**(12), 1200–1202 (1984). https://doi.org/10.2337/diab.33.12.1200
15. Hovorka, R.: Closed-loop insulin delivery: from bench to clinical practice. Nat. Rev. Endocrinol. **7**(7), 385–395 (2011). https://doi.org/10.1038/nrendo.2011.32
16. Templer, S.: Closed-loop insulin delivery systems: past, present, and future directions. Front. Endocrinol. (Lausanne) **13**, 919942 (2022). https://doi.org/10.3389/fendo.2022.919942
17. Ware, J., Hovorka, R.: Recent advances in closed-loop insulin delivery. Metabolism **127**, 154953 (2022). https://doi.org/10.1016/j.metabol.2021.154953
18. Kovatchev, B.P., Renard, E., Cobelli, C., Zisser, H.C., Keith-Hynes, P., Anderson, S.M., Brown, S.A., Chernavvsky, D.R., Breton, M.D., Farret, A., Pelletier, M.J., Place, J., Bruttomesso, D., Del Favero, S., Visentin, R., Filippi, A., Scotton, R., Avogaro, A., Doyle, F.J., 3rd.: Feasibility of outpatient fully integrated closed-loop control: first studies of wearable artificial pancreas. Diabetes Care **36**(7), 1851–1858 (2013). https://doi.org/10.2337/dc12-1965
19. Bergenstal, R.M., Garg, S., Weinzimer, S.A., Buckingham, B.A., Bode, B.W., Tamborlane, W.V., Kaufman, F.R.: Safety of a hybrid closed-loop insulin delivery system in patients with type 1 diabetes. JAMA **316**(13), 1407–1408 (2016). https://doi.org/10.1001/jama.2016.11708
20. Agrawal, P., Welsh, J.B., Kannard, B., Askari, S., Yang, Q., Kaufman, F.R.: Usage and effectiveness of the low glucose suspend feature of the medtronic paradigm Veo insulin pump. J. Diabetes Sci. Technol. **5**(5), 1137–1141 (2011). https://doi.org/10.1177/193229681100500514
21. Bergenstal, R.M., Klonoff, D.C., Garg, S.K., Bode, B.W., Meredith, M., Slover, R.H., Ahmann, A.J., Welsh, J.B., Lee, S.W., Kaufman, F.R., Group, A.I.-H.S.: Threshold-based insulin-pump interruption for reduction of hypoglycemia. N Engl. J. Med. **369**(3), 224-232 (2013).https://doi.org/10.1056/NEJMoa1303576
22. Ly, T.T., Nicholas, J.A., Retterath, A., Lim, E.M., Davis, E.A., Jones, T.W.: Effect of sensor-augmented insulin pump therapy and automated insulin suspension vs standard insulin pump therapy on hypoglycemia in patients with type 1 diabetes: a randomized clinical trial. JAMA **310**(12), 1240–1247 (2013). https://doi.org/10.1001/jama.2013.277818
23. Forlenza, G.P., Li, Z., Buckingham, B.A., Pinsker, J.E., Cengiz, E., Wadwa, R.P., Ekhlaspour, L., Church, M.M., Weinzimer, S.A., Jost, E., Marcal, T., Andre, C., Carria, L., Swanson, V., Lum, J.W., Kollman, C., Woodall, W., Beck, R.W.: Predictive low-glucose suspend reduces hypoglycemia in adults, adolescents, and children with type 1 diabetes in an at-home randomized crossover study: results of the PROLOG trial. Diabetes Care **41**(10), 2155–2161 (2018). https://doi.org/10.2337/dc18-0771
24. Chen, E., King, F., Kohn, M.A., Spanakis, E.K., Breton, M., Klonoff, D.C.: A review of predictive low glucose suspend and its effectiveness in preventing nocturnal Hypoglycemia. Diabetes Technol. Ther. **21**(10), 602–609 (2019). https://doi.org/10.1089/dia.2019.0119

25. Hovorka, R., Chassin, L.J., Wilinska, M.E., Canonico, V., Akwi, J.A., Federici, M.O., Massi-Benedetti, M., Hutzli, I., Zaugg, C., Kaufmann, H., Both, M., Vering, T., Schaller, H.C., Schaupp, L., Bodenlenz, M., Pieber, T.R.: Closing the loop: the adicol experience. Diabetes Technol. Ther. **6**(3), 307–318 (2004). https://doi.org/10.1089/152091504774197990
26. Weinzimer, S.A., Steil, G.M., Swan, K.L., Dziura, J., Kurtz, N., Tamborlane, W.V.: Fully automated closed-loop insulin delivery versus semiautomated hybrid control in pediatric patients with type 1 diabetes using an artificial pancreas. Diabetes Care **31**(5), 934–939 (2008). https://doi.org/10.2337/dc07-1967
27. Peters, T.M., Haidar, A.: Dual-hormone artificial pancreas: benefits and limitations compared with single-hormone systems. Diabet. Med. **35**(4), 450–459 (2018). https://doi.org/10.1111/dme.13581
28. Haidar, A., Tsoukas, M.A., Bernier-Twardy, S., Yale, J.F., Rutkowski, J., Bossy, A., Pytka, E., El Fathi, A., Strauss, N., Legault, L.: A Novel dual-hormone insulin-and-Pramlintide artificial pancreas for type 1 diabetes: a randomized controlled crossover trial. Diabetes Care **43**(3), 597–606 (2020). https://doi.org/10.2337/dc19-1922
29. Ilkowitz, J.T., Katikaneni, R., Cantwell, M., Ramchandani, N., Heptulla, R.A.: Adjuvant Liraglutide and insulin versus insulin monotherapy in the closed-loop system in type 1 diabetes: a randomized open-labeled crossover design trial. J. Diabetes Sci. Technol. **10**(5), 1108–1114 (2016). https://doi.org/10.1177/1932296816647976
30. Dassau, E., Renard, E., Place, J., Farret, A., Pelletier, M.J., Lee, J., Huyett, L.M., Chakrabarty, A., Doyle, F.J., 3rd., Zisser, H.C.: Intraperitoneal insulin delivery provides superior glycaemic regulation to subcutaneous insulin delivery in model predictive control-based fully-automated artificial pancreas in patients with type 1 diabetes: a pilot study. Diabetes Obes. Metab. **19**(12), 1698–1705 (2017). https://doi.org/10.1111/dom.12999
31. Lal, R.A., Ekhlaspour, L., Hood, K., Buckingham, B.: Realizing a closed-loop (Artificial Pancreas) system for the treatment of type 1 diabetes. Endocr. Rev. **40**(6), 1521–1546 (2019). https://doi.org/10.1210/er.2018-00174
32. Wang, Y.Q., Fang, M.Q., Jiang, X., Bequette, B.W., Xie, H.Z.: Intensive insulin therapy for critically ill subjects based on direct data-driven model predictive control. J. Process Contr. **24**(5), 493–503 (2014). https://doi.org/10.1016/j.jprocont.2013.12.012
33. Mauseth, R., Hirsch, I.B., Bollyky, J., Kircher, R., Matheson, D., Sanda, S., Greenbaum, C.: Use of a "Fuzzy Logic" controller in a closed-loop artificial pancreas. Diabetes Technol. Ther. **15**(8), 628–633 (2013). https://doi.org/10.1089/dia.2013.0036
34. Moon, S.J., Jung, I., Park, C.Y.: Current advances of artificial pancreas systems: a comprehensive review of the clinical evidence. Diabetes Metab. J. **45**(6), 813–839 (2021). https://doi.org/10.4093/dmj.2021.0177
35. Abraham, M.B., de Bock, M., Smith, G.J., Dart, J., Fairchild, J.M., King, B.R., Ambler, G.R., Cameron, F.J., McAuley, S.A., Keech, A.C., Jenkins, A., Davis, E.A., O'Neal, D.N., Jones, T.W., Australian Juvenile Diabetes Research Fund Closed-Loop Research, g.: Effect of a hybrid closed-loop system on glycemic and psychosocial outcomes in children and adolescents with type 1 diabetes: a randomized clinical trial. JAMA Pediatr. **175**(12), 1227–1235 (2021). https://doi.org/10.1001/jamapediatrics.2021.3965
36. Collyns, O.J., Meier, R.A., Betts, Z.L., Chan, D.S.H., Frampton, C., Frewen, C.M., Hewapathirana, N.M., Jones, S.D., Roy, A., Grosman, B., Kurtz, N., Shin, J., Vigersky, R.A., Wheeler, B.J., de Bock, M.I.: Improved glycemic outcomes with medtronic MiniMed advanced hybrid closed-loop delivery: results from a randomized crossover trial comparing automated insulin delivery with predictive low glucose suspend in people with type 1 diabetes. Diabetes Care **44**(4), 969–975 (2021). https://doi.org/10.2337/dc20-2250
37. Breton, M.D., Kanapka, L.G., Beck, R.W., Ekhlaspour, L., Forlenza, G.P., Cengiz, E., Schoelwer, M., Ruedy, K.J., Jost, E., Carria, L., Emory, E., Hsu, L.J., Oliveri, M., Kollman, C.C., Dokken, B.B., Weinzimer, S.A., DeBoer, M.D., Buckingham, B.A., Chernavvsky, D., Wadwa, R.P., i, D.C.L.T.R.G.: A randomized trial of closed-loop control in children with type 1 diabetes. N Engl. J. Med. **383**(9), 836–845 (2020). https://doi.org/10.1056/NEJMoa2004736

38. Ekhlaspour, L., Schoelwer, M.J., Forlenza, G.P., Deboer, M.D., Norlander, L., Hsu, L.A., Kingman, R., Boranian, E., Berget, C., Emory, E., Buckingham, B.A., Breton, M.D., Wadwa, R.P.: Safety and performance of the Tandem t:slim X2 with Control-IQ automated insulin delivery system in toddlers and preschoolers. Diabetes Technol. Ther. **23**(5), 384–391 (2021). https://doi.org/10.1089/dia.2020.0507
39. Tauschmann, M., Thabit, H., Bally, L., Allen, J.M., Hartnell, S., Wilinska, M.E., Ruan, Y., Sibayan, J., Kollman, C., Cheng, P., Beck, R.W., Acerini, C.L., Evans, M.L., Dunger, D.B., Elleri, D., Campbell, F., Bergenstal, R.M., Criego, A., Shah, V.N., Leelarathna, L., Hovorka, R., Consortium, A.P.: Closed-loop insulin delivery in suboptimally controlled type 1 diabetes: a multicentre, 12-week randomised trial. Lancet **392**(10155), 1321–1329 (2018).https://doi.org/10.1016/S0140-6736(18)31947-0
40. Tauschmann, M., Allen, J.M., Nagl, K., Fritsch, M., Yong, J., Metcalfe, E., Schaeffer, D., Fichelle, M., Schierloh, U., Thiele, A.G., Abt, D., Kojzar, H., Mader, J.K., Slegtenhorst, S., Barber, N., Wilinska, M.E., Boughton, C., Musolino, G., Sibayan, J., Cohen, N., Kollman, C., Hofer, S.E., Frohlich-Reiterer, E., Kapellen, T.M., Acerini, C.L., de Beaufort, C., Campbell, F., Rami-Merhar, B., Hovorka, R., Kids, A.P.C.: Home use of day-and-night hybrid closed-loop insulin delivery in very young children: a multicenter, 3-Week. Random. Trial. Diabetes Care **42**(4), 594–600 (2019). https://doi.org/10.2337/dc18-1881
41. Benhamou, P.Y., Franc, S., Reznik, Y., Thivolet, C., Schaepelynck, P., Renard, E., Guerci, B., Chaillous, L., Lukas-Croisier, C., Jeandidier, N., Hanaire, H., Borot, S., Doron, M., Jallon, P., Xhaard, I., Melki, V., Meyer, L., Delemer, B., Guillouche, M., Schoumacker-Ley, L., Farret, A., Raccah, D., Lablanche, S., Joubert, M., Penfornis, A., Charpentier, G., Investigators, D.W.T.: Closed-loop insulin delivery in adults with type 1 diabetes in real-life conditions: a 12-week multicentre, open-label randomised controlled crossover trial. Lancet Digit. Health **1**(1), e17–e25 (2019). https://doi.org/10.1016/S2589-7500(19)30003-2
42. Amadou, C., Franc, S., Benhamou, P.Y., Lablanche, S., Huneker, E., Charpentier, G., Penfornis, A., Diabeloop, C.: Diabeloop DBLG1 closed-loop system enables patients with type 1 diabetes to significantly improve their glycemic control in real-life situations without serious adverse events: 6-month follow-up. Diabetes Care **44**(3), 844–846 (2021). https://doi.org/10.2337/dc20-1809
43. Kariyawasam, D., Morin, C., Casteels, K., Le Tallec, C., Godot, C., Sfez, A., Garrec, N., Polak, M., Charpentier, G., Franc, S., Beltrand, J.: Diabeloop DBL4K hybrid closed loop system improves time-in-range without increasing time-in Hypoglycemia in children aged 6–12 years. Diabetes **70** (2021). https://doi.org/10.2337/db21-98-LB
44. Miyata, T., Uragami, T., Nakamae, K.: Biomolecule-sensitive hydrogels. Adv. Drug Deliver. Rev. **54**(1), 79–98 (2002). https://doi.org/10.1016/S0169-409x(01)00241-1
45. Qiu, Y., Park, K.: Environment-sensitive hydrogels for drug delivery. Adv. Drug Deliver. Rev. **53**(3), 321–339 (2001). https://doi.org/10.1016/S0169-409x(01)00203-4
46. Veiseh, O., Tang, B.C., Whitehead, K.A., Anderson, D.G., Langer, R.: Managing diabetes with nanomedicine: challenges and opportunities. Nat. Rev. Drug Discov. **14**(1), 45–57 (2015). https://doi.org/10.1038/nrd4477
47. Wang, J.Q., Wang, Z.J., Yu, J.C., Kahkoska, A.R., Buse, J.B., Gu, Z.: Glucose-responsive insulin and delivery systems: innovation and translation. Adv. Mater. **32**(13) (2020). ARTN 1902004. https://doi.org/10.1002/adma.201902004
48. Matsumoto, A., Chen, S.Y.: A boronate gel-based synthetic platform for closed-loop insulin delivery systems. Polym. J. **53**(12), 1305–1314 (2021). https://doi.org/10.1038/s41428-021-00525-8
49. Banach, L., Williams, G.T., Fossey, J.S.: Insulin delivery using dynamic covalent boronic acid/ester-controlled release. Adv. Ther-Germany **4**(11) (2021). ARTN 2100118. https://doi.org/10.1002/adtp.202100118
50. Matsumoto, A., Miyahara, Y.: Borono-lectin' based engineering as a versatile platform for biomedical applications. Sci. Technol. Adv. Mater. **19**(1), 18–30 (2018). https://doi.org/10.1080/14686996.2017.1411143

51. Kropff, J., Choudhary, P., Neupane, S., Barnard, K., Bain, S.C., Kapitza, C., Forst, T., Link, M., Dehennis, A., DeVries, J.H.: Accuracy and longevity of an implantable continuous glucose sensor in the PRECISE study: a 180-day, prospective, multicenter. Pivotal Trial. Diabetes Care **40**(1), 63–68 (2017). https://doi.org/10.2337/dc16-1525
52. Christiansen, M.P., Klaff, L.J., Brazg, R., Chang, A.R., Levy, C.J., Lam, D., Denham, D.S., Atiee, G., Bode, B.W., Walters, S.J., Kelley, L., Bailey, T.S.: A Prospective multicenter evaluation of the accuracy of a novel implanted continuous glucose sensor: PRECISE II. Diabetes Technol. Ther. **20**(3), 197–206 (2018). https://doi.org/10.1089/dia.2017.0142
53. Chou, D.H., Webber, M.J., Tang, B.C., Lin, A.B., Thapa, L.S., Deng, D., Truong, J.V., Cortinas, A.B., Langer, R., Anderson, D.G.: Glucose-responsive insulin activity by covalent modification with aliphatic phenylboronic acid conjugates. Proc. Natl. Acad. Sci. USA **112**(8), 2401–2406 (2015). https://doi.org/10.1073/pnas.1424684112
54. Kataoka, K., Miyazaki, H., Bunya, M., Okano, T., Sakurai, Y.: Totally synthetic polymer gels responding to external glucose concentration: their preparation and application to on-off regulation of insulin release. J. Am. Chem. Soc. **120**(48), 12694–12695 (1998)
55. Matsumoto, A., Kurata, T., Shiino, D., Kataoka, K.: Swelling and shrinking kinetics of totally synthetic, glucose-responsive polymer gel bearing phenylborate derivative as a glucose-sensing moiety. Macromolecules **37**(4), 1502–1510 (2004)
56. Matsumoto, A., Ikeda, S., Harada, A., Kataoka, K.: Glucose-responsive polymer bearing a novel phenylborate derivative as a glucose-sensing moiety operating at physiological pH conditions. Biomacromolecules **4**(5), 1410–1416 (2003)
57. Matsumoto, A., Yoshida, R., Kataoka, K.: Glucose-responsive polymer gel bearing phenylborate derivative as a glucose-sensing moiety operating at the physiological pH. Biomacromol **5**(3), 1038–1045 (2004)
58. Matsumoto, A., Yamamoto, K., Yoshida, R., Kataoka, K., Aoyagi, T., Miyahara, Y.: A totally synthetic glucose responsive gel operating in physiological aqueous conditions. Chem. Commun. **46**(13), 2203–2205 (2010)
59. Matsumoto, A., Ishii, T., Nishida, J., Matsumoto, H., Kataoka, K., Miyahara, Y.: A synthetic approach toward a self-regulated insulin delivery system. Angew. Chem. Int. Edit. **51**(9), 2124–2128 (2012)
60. Matsumoto, A., Tanaka, M., Matsumoto, H., Ochi, K., Moro-Oka, Y., Kuwata, H., Yamada, H., Shirakawa, I., Miyazawa, T., Ishii, H., Kataoka, K., Ogawa, Y., Miyahara, Y., Suganami, T.: Synthetic "smart gel" provides glucose-responsive insulin delivery in diabetic mice. Sci. Adv. **3**(11), eaaq0723 (2017). https://doi.org/10.1126/sciadv.aaq0723
61. Matsumoto, A., Kuwata, H., Kimura, S., Matsumoto, H., Ochi, K., Moro-Oka, Y., Watanabe, A., Yamada, H., Ishii, H., Miyazawa, T., Chen, S., Baba, T., Yoshida, H., Nakamura, T., Inoue, H., Ogawa, Y., Tanaka, M., Miyahara, Y., Suganami, T.: Hollow fiber-combined glucose-responsive gel technology as an in vivo electronics-free insulin delivery system. Commun. Biol. **3**(1), 313 (2020). https://doi.org/10.1038/s42003-020-1026-x
62. Zhang, Y., Yu, J., Kahkoska, A.R., Wang, J., Buse, J.B., Gu, Z.: Advances in transdermal insulin delivery. Adv. Drug Deliv. Rev. **139**, 51–70 (2019). https://doi.org/10.1016/j.addr.2018.12.006
63. Pillai, O., Panchagnula, R.: Insulin therapies—past, present and future. Drug Discov. Today **6**(20), 1056–1061 (2001). https://doi.org/10.1016/s1359-6446(01)01962-6
64. Halder, J., Gupta, S., Kumari, R., Gupta, G.D., Rai, V.K.: Microneedle array: applications, recent advances, and clinical pertinence in transdermal drug delivery. J. Pharm. Innov. **16**(3), 558–565 (2021). https://doi.org/10.1007/s12247-020-09460-2
65. Singh, P., Carrier, A., Chen, Y., Lin, S., Wang, J., Cui, S., Zhang, X.: Polymeric microneedles for controlled transdermal drug delivery. J. Control. Release **315**, 97–113 (2019). https://doi.org/10.1016/j.jconrel.2019.10.022
66. Chen, S.Y., Matsumoto, H., Moro-oka, Y., Tanaka, M., Miyahara, Y., Suganami, T., Matsumoto, A.: Microneedle-array patch fabricated with enzyme-free polymeric components capable of on-demand insulin delivery. Adv. Funct. Mater. **29**(7) (2019). ARTN 1807369. https://doi.org/10.1002/adfm.201807369

67. Chen, S.Y., Matsumoto, H., Moro-oka, Y., Tanaka, M., Miyahara, Y., Suganami, T., Matsumoto, A.: Smart microneedle fabricated with silk fibroin combined semi interpenetrating network hydrogel for glucose-responsive insulin delivery. Acs Biomater. Sci. Eng. **5**(11), 5781–5789 (2019). https://doi.org/10.1021/acsbiomaterials.9b00532
68. Chen, S.Y., Miyazaki, T., Itoh, M., Matsumoto, H., Moro-oka, Y., Tanaka, M., Miyahara, Y., Suganami, T., Matsumoto, A.: Temperature-stable boronate gel-based microneedle technology for self-regulated insulin delivery. Acs Appl. Polym. Mater. **2**(7), 2781–2790 (2020). https://doi.org/10.1021/acsapm.0c00341
69. Chen, S.Y., Miyazaki, T., Itoh, M., Matsumoto, H., Moro-Oka, Y., Tanaka, M., Miyahara, Y., Suganami, T., Matsumoto, A.: A porous reservoir-backed boronate gel microneedle for efficient skin penetration and sustained glucose-responsive insulin delivery. Gels-Basel **8**(2) (2022). ARTN 74. https://doi.org/10.3390/gels8020074
70. Yu, J.C., Wang, J.Q., Zhang, Y.Q., Chen, G.J., Mao, W.W., Ye, Y.Q., Kahkoska, A.R., Buse, J.B., Langer, R., Gu, Z.: Glucose-responsive insulin patch for the regulation of blood glucose in mice and minipigs. Nat. Biomed. Eng. **4**(5), 499–506 (2020). https://doi.org/10.1038/s41551-019-0508-y
71. Beck, R.W.: Closing in on closed-loop systems for type 2 diabetes. Nat. Med. **29**(1), 33–34 (2023). https://doi.org/10.1038/s41591-022-02127-0
72. Alfonsi, J.E., Choi, E.E.Y., Arshad, T., Sammott, S.S., Pais, V., Nguyen, C., Maguire, B.R., Stinson, J.N., Palmert, M.R.: Carbohydrate counting app using image recognition for youth with type 1 diabetes: pilot randomized control trial. JMIR Mhealth Uhealth **8**(10), e22074 (2020). https://doi.org/10.2196/22074
73. Rini, C.J., McVey, E., Sutter, D., Keith, S., Kurth, H.J., Nosek, L., Kapitza, C., Rebrin, K., Hirsch, L., Pettis, R.J.: Intradermal insulin infusion achieves faster insulin action than subcutaneous infusion for 3-day wear. Drug Deliv. Transl. Res. **5**(4), 332–345 (2015). https://doi.org/10.1007/s13346-015-0239-x

Wearable Biosensors on Sutures and Threads

Atul Sharma, Cihan Asci, Jean Louis Marty, and Sameer Sonkusale

1 Introduction

Efforts in academia and industry have increased to explore and create smart, flexible systems for wearable applications. Several laboratory prototypes, testbeds, and industrial products have already been manufactured [1]. These adaptable/wearable devices aim to monitor the health of, and consequently improve the quality of life for those with impaired physical and cognitive abilities or suffering from various diseases. Examples range from wearable heart rate monitors for those with cardiac conditions, to fall detectors for the elderly. As a result, these systems promote independent living for the elderly, postoperative rehabilitation for patients to speed recovery, and the evaluation or enhancement of individual sporting or technical abilities [1–4]. The ideal smart flexible system could include a variety of wearable components, such as sensors, actuators, power sources, wireless communication networks, processing units, multimedia devices, user interfaces, software, and algorithms for data collection, processing, and decision support [5–7]. These systems may keep an eye on vital signals like heart rate [6], breathing rate, blood, or wound oxygenation [8], arterial blood pressure [9], body mobility and strain sensing [10], and brain activity [11]. A wireless sensor network can transmit the measurements to a central connection node, such as a personal digital assistant (PDA), or a hospital. Advances in microfabrication have enabled several biosensors for POC testing and wearable real-time

A. Sharma (✉) · C. Asci · S. Sonkusale (✉)
Nano Lab, Advanced Technology Laboratory, Tufts University, Medford, USA
e-mail: atul.sharma@tufts.edu

S. Sonkusale
e-mail: sameer@ece.tufts.edu

Department of Electrical and Computer Engineering, Tufts University, Medford, USA

J. L. Marty
Université de Perpignan Via Domitia, Perpignan, France

K. Mitsubayashi (ed.), *Wearable Biosensing in Medicine and Healthcare*,
https://doi.org/10.1007/978-981-99-8122-9_13

monitoring [12–15]. Despite the significant advancements in wearable sensors for physiological monitoring, non-invasive chemical sensing for assessing analytes and biomarkers in bodily fluids has received very little attention.

Compared to paper as a substrate for sensing, textile is a more resilient material system that is more durable, widely accessible, and reasonably priced. It also allows capillary-based sample conveyance similar to capillary transport in paper. Additionally, textile-based electronics and sensors allow for simple integration with wearable clothing and materials, which may be utilized to create wearable sensor systems [5, 13]. The incorporation of sensors into textiles to track physiological variables like electrolytes and small molecules [16], heart rate [17–19], temperature [20], and breathing [21] has been the focus of recent research in wearable sensing. In the last decade, several research groups reported the development of electrochemical sensors on threads or smart fabrics for health monitoring. For example, Wang's group has developed electrochemical sensors on fabrics for monitoring the environment [22], wounds [23], and health [24]. A wearable electrochemical sensor for sweat analysis based on textiles has also been shown by our group (Sonkusale's group) [16]. We have reported a thread-based multiplexed sensing patch for continuous and simultaneous on-skin sweat monitoring. The sweat patch is capable and measures crucial physiological indicators such as metabolites (lactate), electrolytes (sodium and ammonium ions), and acidity (pH) in sweat. In this work, sensor fabrication was realized on flexible threads coated with conductive inks as sensing electrodes. Diamond's group has also demonstrated wearable, textile-based electrochemical sensors for sweat analysis [25]. The use of screen-printed sensors, which are frequently structurally brittle and difficult to connect with textile-based electrical components, is what allows these devices to undertake delicate analytical measurements. Recently, it was claimed that conductive thread [7] or silk yarn [26] was woven into the fabric to fabricate electrochemical sensors that were based on textiles. This approach offers improved robustness compared with screen-printed sensors but is limited to woven electrode geometries and substrates.

This chapter explores the design, fabrication, and validation of wearable electrochemical sensors on sutures/ threads for health monitoring. This chapter aims to review the necessary background and knowledge for designing and fabricating flexible sensors on sutures/threads with several key applications in sweat sensing, smart bandage, and smart sutures while emphasizing the key qualities that make them the best possible candidates for flexible bioelectronics. Last but not least, we assess the electrochemical readout circuitry's performance, which displays the sensor's capacity to provide reliable measurements under such circumstances.

2 Threads and Sutures

2.1 *Types of Threads and Sutures*

Fibers may be broadly categorized into two groups: natural, which come from biological sources including plants and animals, and synthetic, which are created by chemical processes. Both natural and synthetic fibers' molecular structures may be used to predict how their surfaces will respond to certain treatments, like dyeing, or to evaluate other factors, such as mechanical tunability and biofriendliness.

Natural fibers are frequently derived from plants and animals. Cotton, linen (flax), and jute are all cellulose-based fibers, with 94%, 80%, and 74% cellulose content, respectively [27]. Bioproteins make up animal fibers such as silk and wool. Silk is made up of the fibroin protein, which is made up of recurring sequences of glycine, serine, and alanine amino acids [28]. Wool is also formed of keratin-type protein, but its outer surface (the cuticle) defines many of its features. The epicuticle, the cuticle's topmost layer, includes fatty acids covalently bound to the wool fiber via a thioester bond, providing a hydrophobic barrier [29]. Its complex surface structure makes it less dyeable and wettable than other natural fibers [30]. Natural fibers are an excellent choice for biological sensors that contact with tissue because of their high specific strength, biodegradability, and inexpensive production costs [7, 31]. Another benefit of using natural fiber sources is their capacity to be recycled, which is becoming more and more crucial in light of the present environmental issues. Given the current state of climate change and the demand for sustainable resources, the benefit of renewable natural fiber sources is growing over time.

Polymers used to create conventional synthetic fibers often come from petroleum [32]. These include acrylic [33], polyurethane, polyester fibers, and nylon [34]. Most frequently, they are made by a "melt-spinning" procedure in which fibers are extracted from strands that have been heated to their melting point [31]. Natural fibers tend to be less durable than synthetic fibers, which also quickly take the dye. Due to their strong elastic properties, synthetic fibers like polyurethane become excellent candidates for uses like strain sensing [7, 35].

2.2 *Processing Methods*

Traditional twisting techniques have been used to create both natural and synthetic fibers. Natural fibers like cotton or wool are drawn out while being twisted to form threads and yarns, while traditional synthetic fibers like nylon are drawn out during a thermal process to form continuous threads. Extruding precursor solutions into fibers is a technique used in more recent methods including electrospinning, wet-spinning, and direct writing [36, 37]. Due to their fine control over manufactured fiber form, sub-micrometer to nanoscale regime widths, and compatibility with several inks, these methods are being employed more often in biomedical, tissue engineering, and better versions of conventional sensing applications.

Electrospinning is a method of drawing polymer fibers from an injecting needle onto a plate while exposed to an electric field. Electrospinning produces fibers with diameters ranging from 100 nm to a few micrometers, resulting in high fiber density in a simple procedure [38, 39]. Liu et al. used electrospinning to construct a stretchy strain sensor based on a flexible polymer nanowire, whereas earlier devices based on inorganic materials revealed lesser strain sensitivity [40]. Researchers have also shown that fibers with increased modulus and tensile strength may be electrospun into fiber composites using carbon nanotubes (CNTs).

In wet spinning, a polymer precursor solution is pumped into an insoluble coagulation bath, resulting in the continuous polymerization of long fiber [38]. Wetspinning generates fibers with lengths ranging from 30 to 600 m and may make tissue-friendly fibers by using appropriate inks such as collagen and polycaprolactone. Wetspun fibers' mechanical strength may be improved by orders of magnitude by adding CNTs or graphene oxide, just as electrospun fibers. Because the fibers are greater in size here than in electrospinning, the scaffolds produced by this process have higher porosity and larger pore sizes [38].

Insects produce silk fibers through a process known as biospinning. Silk is biodegradable and has a high tensile strength ranging from 460–972 MPa [38]. These fibers consist of a fibroin core surrounded by sericin, a glue protein. After chemical processing, silk remains non-cytotoxic and non-inflammatory. The diameter of silk fibers can vary from 25 to 70 m, depending on whether they are extracted from the insect or from the cocoon. Silk fiber finds extensive use in biomedical applications including implants, tissue scaffolds, and drug delivery systems [41–43]. These applications are particularly relevant for load-bearing tissues, where a mechanical strength of around 150 MPa is required. For instance, Karageorgiou and Kaplan's work on 3D scaffolds contributed to osteogenesis [41].

2.3 Surface Modification Techniques

Surface modification techniques play a crucial role in improving the performance of threads for sensing applications. Biorecognition components (enzymes, aptamers, antibodies, etc.) are attached to the immobilization surface using various techniques. When selecting an immobilization approach, several aspects such as the immobilization surface, sample matrix, protein characteristics, buffer components, and assay performance metrics are considered. In an ideal world, reaction partners should be able to reach active regions for antibody binding or enzymatic conversion (i.e., the binding site should face away from the immobilization surface to reduce steric hindrance and should not be sterically hindered by nearby immobilized proteins). For a high-performance, repeatable experiment, receptor conformation must remain unaltered upon immobilization to maintain receptor binding properties, permit diffusion of substrates and products, and allow excellent electron transfer. These techniques can enhance the sensitivity, selectivity, and stability of the threads, making them more suitable for specific sensing tasks. Here are some commonly used surface modification techniques for modifying threads for sensing applications.

a. **Physical Adsorption**

The physisorption method is the simplest method for immobilizing a bio-recognition component on a surface (i.e., physical adsorption). In physical adsorption, the baroreceptors are immobilized by van der Waals forces, hydrophobic contacts, or hydrogen bonds, whereas enzymes are joined by salt linkages in electrostatic interaction (Fig. 1a, b). The process of non-covalent immobilization is caused by the nature of the forces involved, and it can be stopped by altering the factors that determine how strong the contact is (e.g., pH, salt concentration, solvent polarity, temperature, etc. [44]. The functioning of the biomolecule is often maintained during immobilization by adsorption, which is a simple procedure. Weak attachment and unpredictable orientation, which result in desorption and poor reproducibility, are the adsorption mechanism's drawbacks [45].

b. **Covalent Immobilization**

One of the most prevalent techniques for immobilizing receptors involves the utilization of covalent bonding to anchor them in place (Fig. 1c–e). In the grafting process, stable bonds are formed between the biomolecule and the substrate, effectively preventing any leakage of the bioreceptor into the surrounding solution. To ensure the optimal preservation of protein activity, careful attention must be paid to the covalent attachment chemistry to the substrate, to safeguard the biomolecule's active

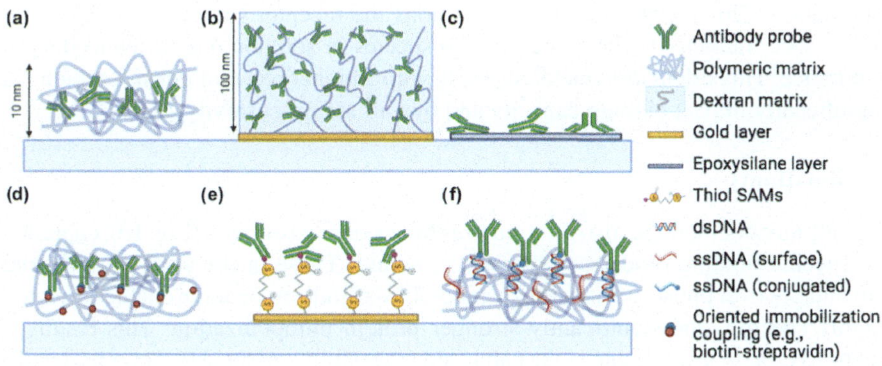

Fig. 1 Common immobilization methods for microarray applications. **a** Copoly(DMA-NAS-MAPS), **b** Carboxymethyl dextran, **c** Epoxysilane, **d** Oriented immobilization (biotin-streptavidin, click chemistry), **e** Thiol-gold coupling, **f** DNA-directed immobilization. Figure reproduced with permission from [51]

region. A critical consideration is to prevent any interference or hindrance of vital amino acid residues that govern enzyme activity or antibody recognition. However, achieving this delicate balance could prove challenging under certain circumstances [46]. Tailored to the functional groups inherent to the target molecule, several distinct reaction pathways have been developed. Despite the intricate structural complexity of biomolecules, only a limited selection of functional groups can be effectively leveraged as discriminating targets for efficient immobilization. For instance, thiol groups (-SH), integral components of cysteine residues' side chains, have demonstrated noteworthy efficacy. In the context of sugars, the Fc region represents a primary site for glycosylation. While the oxidation of diol moieties to aldehydes entails additional steps, this approach often yields antibody conjugates with heightened activity due to the specific positioning of the carbohydrate moiety. Alternatively, the reductive amination process provides a direct means to react sugars with primary amine or hydrazine-based groups. Illustrative instances of covalent immobilization are depicted in Fig. 1a–e, showcasing the versatility and application of this methodology.

Utilizing a blocking solution, such as bovine serum albumin (BSA) in PBS, can deactivate functional groups that are not bound to the receptors [47]. The inherent randomness in receptor orientation, a common outcome of covalent immobilization, ensures the irreversible attachment of receptors to surfaces. However, this randomness might influence their biological activity and their capacity to effectively bind analytes from the solution. Covalent immobilization can also be time-consuming, particularly when dealing with low-concentration receptor solutions during the attachment phase, often due to their cost considerations. In situations where protein immobilization is necessary, the combination of non-covalent and covalent immobilization can be achieved by introducing a spacer between the surface and the receptors [48]. An alternative approach termed "random immobilization" involves placing receptors on a surface without specific consideration for their orientation. In contrast, "direct immobilization" refers to attaching receptors with controlled orientation. This method frequently positions the receptor's active site away from the surface, making it more accessible to species in solution due to reduced steric hindrance. This orientation yields surfaces with a higher receptor density, which can significantly impact binding capacity and overall assay sensitivity.

c. **Bioaffinity**

Bio-affinity interactions, also referred to as biospecific adsorption (Fig. 1d), capitalize on specific binding processes observed in nature. These interactions offer distinct advantages over physisorption. For instance, bio-affinity association enables directed, highly specific, and significantly stronger protein immobilization. This results in reduced protein leakage due to the enhanced accessibility of fixed receptor molecules to their binding partners, unlike random orientation strategies. Furthermore, bio-affinity immobilization can be reversed through chemical means, pH adjustment, or thermal treatment. The concept of natural affinity between molecules, such as lectin-sugar, antibody-antigen, and biotin-avidin pairs, has been employed for biomolecule immobilization. However, often this requires covalent attachment of costly affinity ligands like antibodies or lectins to the support. Notably, interactions like strepta-vidin–biotin and the utilization of Protein A or G for antibody immobilization stand out as the most widely adopted approaches [44].

d. **Sulphur-Atom Linkage or Disulfide Bond**

In the case of surfaces like gold threads or textiles doped with gold nanoparticles, although a sturdy covalent connection is established between the support and the bioreceptor, it can be cleaved through a gentle reaction with a suitable chemical, such as dithiothreitol (DTT) (Fig. 1e). Furthermore, when utilizing a thiol-reactive adsorbent with heightened specificity, the activity yield of procedures involving the formation of disulfide bonds is often significantly elevated. This enhanced yield can be attributed to the precisely controllable reactivity of thiol groups, which can be finely regulated by adjusting the pH [49].

e. **Metal-Link Binding/Chelation**

The metal salt or hydroxide, primarily titanium and zirconium salts, undergo precipitation and coordination with nucleophilic groups on the surface, which may include carriers based on cellulose, chitin, alginic acid, and silica [50]. Some of these metal binding sites remain unoccupied due to steric hindrances, creating opportunities for interactions with biomolecular groups. While the procedure itself is straightforward, challenges arise due to the non-uniform distribution of adsorption sites, resulting in limited repeatability, alongside notable leakage of metal ions.

f. **2D Versus 3D Surface Immobilization**

The surface geometry, which can be either planar (2D) or three-dimensional (3D), has a considerable impact on immobilization techniques [52]. 2D surface immobilization on threads involves the attachment of bioreceptors onto a flat, planar substrate. This approach, although relatively straightforward, comes with limitations. Microfluidic channels within the thread's structure can be constructed from various substrates, including those coated with functional materials like gold, platinum, or graphene. This method is efficient for certain applications and allows for controlled and uniform immobilization of receptors. However, the available surface area for receptor attachment is inherently constrained due to the planar nature of the substrate. Consequently, the density and quantity of immobilized receptors are limited, potentially affecting the sensitivity and performance of the biosensor. On the other hand, 3D surface immobilization introduces a three-dimensional architecture to the thread's surface. This configuration provides several distinct advantages. Threads with embedded microstructures, such as microposts or porous membranes, offer an increased surface area for receptor attachment [53]. This enables higher receptor density, thereby enhancing the biosensor's sensitivity and allowing for the detection of lower analyte concentrations [52]. Additionally, the reduced diffusion length between reactants on a 3D surface can expedite the binding kinetics, leading to faster response times [47]. Such enhancements are particularly valuable in rapid and real-time detection scenarios. The selection between 2 and 3D surface immobilization on thread-based biosensors depends on the specific application requirements. If uniform and controlled immobilization is sufficient and high sensitivity is not a critical factor, 2D surface immobilization may be suitable. On the other hand, if enhanced sensitivity, faster response times, and the ability to detect lower analyte concentrations are essential, 3D surface immobilization provides a more advantageous solution.

In conclusion, the decision between 2 and 3D surface immobilization on thread-based biosensors involves a trade-off between simplicity and enhanced performance. While 2D immobilization is straightforward, 3D immobilization offers superior sensitivity and binding kinetics, making it a valuable choice for applications demanding rapid and highly sensitive detection.

3 Flexible-Sensing Platform

This section provides an overview of recent advancements in utilizing thread-based fabrics as electrochemical sensors, with a specific emphasis on biochemical sensing [8, 35]. Electrochemical sensors are devices that combine a recognition component with an electrochemical transducer to generate electronic signals. In electrochemical sensing, the signal readout is directly linked to analyte binding, enabling rapid detection. Due to these attributes, electrochemical sensors find extensive applications in daily life, ranging from household carbon monoxide detectors to research laboratory pH meters. In the realm of biochemical sensing, particularly in point-of-care settings, the demand for smaller and more compact sensors is imperative to minimize invasiveness. Conversely, thread-based fabrics have garnered significant attention for creating wearable electrochemical sensors, owing to their distinctive physical and chemical attributes, including flexibility, high porosity, and mechanical stability [5, 7, 16, 20, 54–56]. The diameter of a single fiber, constituting the smallest unit of fabric, is approximately a few micrometers, promising ultra-small sensor design. Simultaneously, individual threads can be incorporated into clothing using industrial textile techniques like weaving, knitting, and embroidering, offering unparalleled mass production capabilities.

These sensors can be classified into two types based on fabrication approaches:

- **Using Fabrics as Substrates for Screen-Printed Electrodes**: This approach involves utilizing fabrics as substrates for screen-printed electrodes. This method leverages the inherent properties of the fabric as a base and integrates screen-printed electrodes for sensing purposes.
- **Using Individual Functionalized Threads as Sensing Electrodes**: Alternatively, individual functionalized threads are employed as sensing electrodes and assembled into wearable sensors. This approach enables precise customization and assembly of sensors, tailoring them to specific applications.

In summary, the integration of thread-based fabrics into electrochemical sensors presents exciting possibilities, particularly in the realm of biochemical sensing. The inherent characteristics of threads and fabrics, combined with innovative fabrication approaches, are paving the way for smaller, more versatile, and wearable sensors with applications ranging from healthcare to environmental monitoring.

3.1 Principle of Sensing

Electrochemical detection hinges on the conversion of biochemical signals into electronic ones. In the realm of electrochemical sensing, a recognition layer selectively captures the analyte, while a transducer layer translates this interaction into an electronic readout. Electrochemical methods encompass potentiometric, coulometric, and voltammetric techniques. In potentiometric measurements, the potential difference between the working and counter electrodes is gauged without the passage of current. In coulometric measurements, the cumulative current during an electrolysis process is quantified. Voltammetric methods involve measuring the current elicited by either a constant or varying voltage application. Thread-based fabrics are typically rendered electrically conductive through coatings of materials such as carbon, carbon nanotubes, or graphene oxide. Within the context of thread-based electrochemical sensors, ionic sensors designed to detect pH, sodium, potassium, calcium, ammonium, chloride, and lithium primarily rely on potentiometric approaches. Meanwhile, enzymatic sensors targeting analytes like glucose, lactate, uric acid, and ascorbic acid predominantly employ voltammetric methods. Additionally, gas sensors that measure resistance changes have garnered significant attention and will be elaborated upon. In conclusion, the integration of thread-based fabrics into electrochemical sensing holds significant promise. The diverse array of methods available, coupled with the materials used for sensor enhancement, is driving advancements in various sensing applications, ranging from ions to enzymes and gases.

3.2 Biomarker Monitoring

3.2.1 Screen-Printed Electrodes on Fabric

The application of industrial ink printing technology to fabric materials has facilitated the creation of fabric-based wearable electrochemical sensors. Screen printing stands out among these technologies, especially for disposable paper-based electrodes. Following its success in producing disposable glucose test strips [57, 58], screen-printed electrodes (SPE) have been directly applied to fabrics for a variety of biochemical analytes. Wang's group pioneered screen printing onto textiles and garments for biosensing purposes [22, 24, 59]. In their initial paper, they developed an amperometric sensor by screen printing a carbon working electrode onto the elastic waistband of underwear. They used a platinum wire and Ag/AgCl as the counter and reference electrodes, respectively [24]. These printed sensors demonstrated high tolerance to mechanical stress caused by cloth deformation. In a subsequent study, all three electrodes were screen-printed onto textiles to detect nitroaromatic explosives [22]. The authors revealed that differences in the hydrophobicity of the textile substrate had a significant impact on the electrochemical behavior of the SPE. A waterproof fabric, GORE-TEX, was used due to its hydrophobic nature, which enabled minimal sample absorption, ensured good printing quality, and maintained the reproducible electrochemical activity of the SPE. In a subsequent investigation, the group introduced a wearable screen-printed sensor on synthetic rubber neoprene to detect pollutants and security threats in marine environments [59]. Once again, the superhydrophobic nature of the selected fabric contributed to excellent electrochemical performance. Importantly, this study also presented the first enzyme-based screen-printed sensor on fabric. Overall, these initial works demonstrated the feasibility of applying SPE to fabrics for biochemical sensing and examined how fabric properties influenced SPE performance.

In another study by Wang's group, they developed a smart bandage for sensing uric acid in a wound environment. They accomplished this by screen printing a Prussian, blue-mediated carbon electrode onto a commercial bandage, followed by chemical modification of the working electrode with uricase (Fig. 2a–c) [23]. Beyond hydrophobic, water-repellent fabrics, more common cotton- and nylon-based textiles were used as substrates for SPE. For instance, Corcoles and coworkers reported an electrochemical lactate sensor by printing electrodes onto cotton fabric [60]. They employed the wax-patterning technique to create a hydrophobic barrier serving as the sample placement/reaction zone on the hydrophilic cotton substrate. This procedure is similar to the approach widely adopted in electrochemical paper-based assays [61]. Lactate oxidase (LOx) is immobilized on the working electrode to enable amperometric measurement of lactate. The authors also explored the potential of incorporating sample collection functionality using a hydrophilic cotton thread. While most of the aforementioned studies predominantly used amperometric methods for sensing, Prasad and coworkers developed an affinity-based sensor by screen printing silver conductive electrodes and a graphene oxide transduction film onto utility textiles. The electrodes were then chemically modified with an antibody to sense influenza protein (Fig. 2d–f) [62]. The authors validated the reproducibility and stability of the fabrication procedure and the affinity-based assay using electrochemical impedance spectroscopy (EIS). Notably, EIS offers the advantage of miniaturizing the signal readout component, as the output of EIS has negligible memory consumption. However, the study did not demonstrate the sensor's performance in biological samples with possible interference from non-specific binding.

Nomura and coworkers reported a blood leakage sensor using a screen-offset printing technique on cotton textiles (Fig. 2g) [63]. Applying conventional screen printing on textile surfaces would result in ink blurring due to the highly porous textile structure unless the substrate was hydrophobic [22, 59]. To address this challenge, the authors employed screen-offset printing, adapted from screen printing. In this method, the ink was initially printed onto a silicon blanket that absorbed the

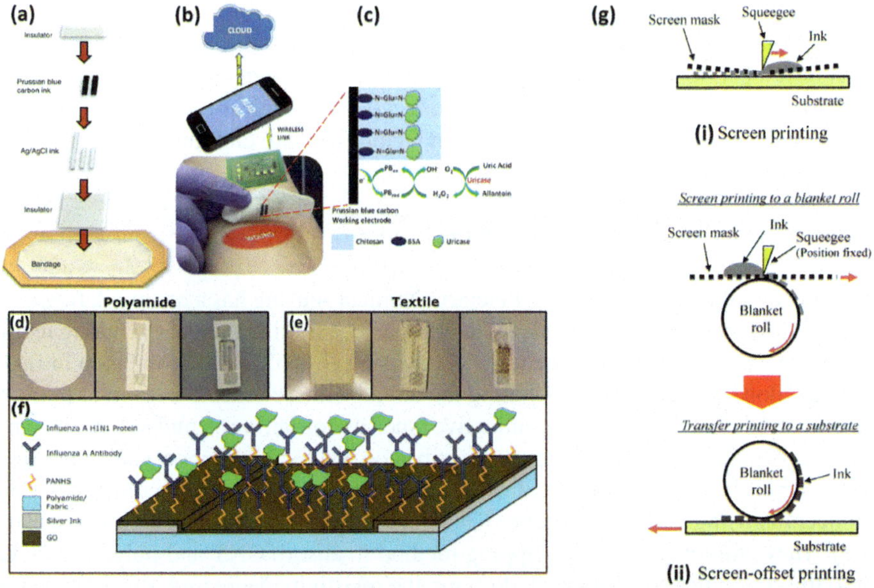

Fig. 2 **a** Screen printing the smart bandage. **b** Wearable potentiostat determines UA concentration and wirelessly communicates with a computer or Smartphone. **c** Schematic showing amperometric detection of UA with uricase immobilized on PB working electrode [23]. **d** Fabrication process flow for the flexible screen-printed electrodes on polyamide. From left to right, raw polyamide sheets are screen-printed with a ~100 μm thick pattern of conductive silver ink to form two electrodes. After, a layer of graphene oxide is screen-printed in a complimentary manner completing the fabrication. Shown are individual sensors, while fabrication occurs in sheets of 14 electrodes. **e** Fabrication process flow for the flexible screen-printed electrodes on textile. From left to right, textile sheets are screen-printed with a ~100 μm thick pattern of conductive silver ink to form two electrodes. After, a layer of graphene oxide is screen-printed in a complimentary manner completing the fabrication. **f** Cross-section schematic visually detailing the affinity assay for influenza where PANHS crosslinker binds to the graphene oxide allowing for subsequent binding of influenza protein antibody and influenza protein for detection [62]. **g** Schematic images of (i) screen printing and (ii) screen-offset printing [63]. Figure reproduced with permissions

organic solvent in the ink, making it more viscous. The pattern was then transferred onto textile substrates with improved resolution. The authors detected blood leakage volumes as low as 15 μL by measuring the frequency dependence of blood and other liquid parameters.

3.2.2 Functionalized Thread as Individual Electrodes

As the fundamental unit of fabrics, threads can be pre-functionalized as sensing electrodes and incorporated into textiles through weaving, knitting, and embroidering methods. This "bottom-up" approach facilitates multiplex sensing with various thread electrode types, offering greater stability compared to SPE. This stability arises from the natural embedding of sensing elements within the fabric. The mass production of such sensors is cost-effective and has high throughput, thanks to the well-developed textile industry. In this section, we will review recent advancements in crafting single thread-based electrochemical sensors, with an emphasis on different functionalization methods and integration techniques into clothing.

Methods for Individual Thread Functionalization

The functionalization of thread for electrochemical sensing has a history dating back to the 1990s when Reynolds et al. and Collins et al. experimented with coating textile threads with conductive polymers for gas sensing [64, 65]. The initial step in functionalization typically involves coating the thread with a conductive layer, as conventional thread materials are non-conductive (such as cotton, wool, and nylon). In some studies, conductive threads like carbon fiber or conductive carbon yarn have been used due to their strong electrical conductivity [66, 67]. Shim and colleagues were pioneers in using carbon nanotube (CNT)-coated cotton thread for immunosensing [68]. They employed a simple dipping method to absorb CNT and polyelectrolytes onto the cotton thread. With each dipping, the resistivity decreased, reaching as low as 20 Ω/cm. The absorbed CNT exhibited stability against washing and heating, likely due to interactions between polyelectrolytes and cotton-based materials. Anti-albumin antibodies were further coated onto the CNT layer to detect albumin by measuring changes in thread resistance. Similarly, Andrade and coworkers coated cotton yarns with CNT and ion-selective membranes for pH, K^+, and NH_4^+ measurements in body fluids [69]. Whitesides' group demonstrated an ion sensor for Cl^-, K^+, Na^+, and Ca^{2+} using cotton thread [56]. They developed a surfactant-free ink consisting of carbon black as the conductive material and an organic polymer as the ink binder. This modification addressed the poor stability of the previous CNT-coated thread. Carbon black's high specific capacitance established a stable interfacial potential between the transducer and the ion-selective layer, minimizing signal drift. A detailed sensor fabrication process is illustrated in Fig. 3a–c, where the thread tip is functionalized through dip-coating and sealed with heat-shrink tubing or nail polish.

Another approach was developed by Sonkusale's group, who created a continuous coating device to sequentially deposit various inks onto a single thread fiber with drying capabilities [7]. This method enabled the fabrication of meters of functionalized thread, collected on rotating spools (Fig. 3d) [7]. Besides CNT and carbon black ink, functionalized graphene oxide (FGO) was also explored as a conductive and recognition layer for gas sensing [70]. Graphene oxide was modified with organic molecules to distinguish specific gas species. The functionalized FGO was dip-coated onto a single cotton yarn as the sensing electrode. FGO, with available sites for gas molecule absorption, showed exceptional mechanical and chemical stability. Tang et al. [71] reported the conversion of non-conductive silk fibers into conductive substrates through carbonization under 950 °C. They further enhanced the fabric electrode's electrochemical performance by dip-coating CNT and modifying the fabric

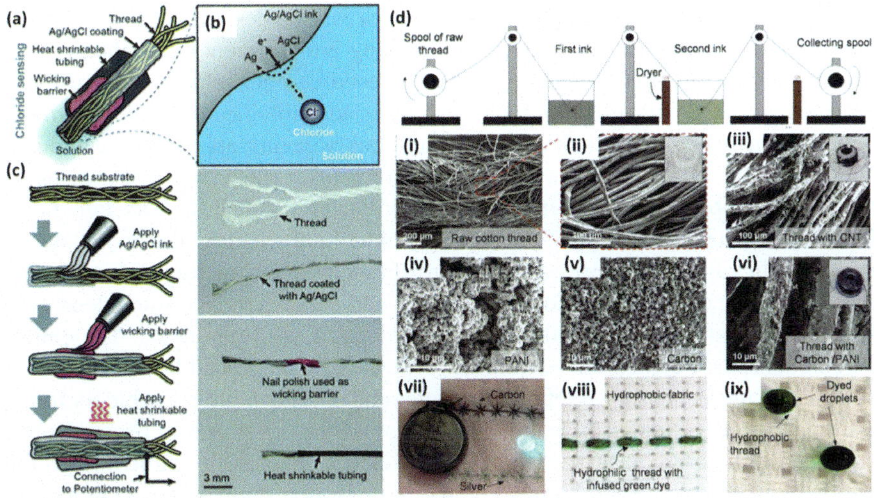

Fig. 3 Design (**a**), working principle (**b**), and fabrication process (**c**) for the thread-based Cl^- ISE [56]. **d** Fabrication of functional threads. (i) Schematic representation of the coating system for preparation of conductive threads. (ii) SEM image of the cotton raw thread. (iii) SEM image of a nano-infused thread coated with CNTs. (iv–vi) SEM images of the surface of PANI-, carbon-, and carbon/PANI-coated threads. (vii) Pattern conductive threads interconnect on a woolen fabric to illuminate an LED. (viii) Embroidered hydrophilic threads on a hydrophobic fabric after the green dye was infused. (ix) Hydrophobic threads repelling water. Green food dye was used to improve visualization. LED, light-emitting diode most [7]. Figure reproduced with permission

surface with platinum microspheres and glucose oxidase (GOx) for glucose sensing [72]. This combination offered good electrical conductivity and flexibility, with platinum microspheres significantly enhancing sensitivity. Sekar et al., used conductive carbon yarn (CCY) to fabricate an immunosensor for sweat cortisol detection [66]. They directly grew Fe_2O_3 nanoparticles on CCY using a hydrothermal method. These "binder-free" electrodes eliminated the need for binder materials added to conductive ink, enhancing substrate adhesion [73]. The directly grown Fe_2O_3 nanoparticles, shaped like ellipsoids, shortened the ion and electron transport path due to their high aspect ratio. These nanoparticles, decorated with anti-cortisol antibodies, enabled the detection of fg-level cortisol using high-sensitivity differential pulse voltammetry.

Methods for Integrating Individual Thread into Fabric and Clothing

Threads, as the fundamental building blocks of textiles, can be utilized and processed using various techniques to create the final clothing product. Established textile technologies are readily available for integrating individual thread electrodes into wearable sensors. Dendukuri's group explored textile weaving to mass-produce thread-based electrochemical sensors [26, 73]. Using a handloom, enzyme-functionalized threads were woven into fabric patches, allowing for the creation of a sensing zone with defined hydrophilic and hydrophobic silk yarns. This design confined the sample flow to hydrophilic yarns only, avoiding wicking into adjacent hydrophobic yarns. The one-layer structure offered advantages in resource-poor settings and minimized ink wastage. In contrast to top-down patterning technologies like screen printing, which often result in multi-layer structures, this weaving approach maintained a constant electrode surface area during measurements. Lillehoj's group investigated the use of embroidery to craft thread-based electrochemical sensors into fabrics like cotton gauze and t-shirts [74, 75]. Embroidery allowed for custom geometries, providing flexibility for electrochemical instrumentation. Computerized embroidery machines were used with predefined patterns, and parameters like stitch density and length were optimized for uniformity and reduced thread resistance. The embroidered dual sensor demonstrated a linear response to glucose and lactate due to precoated enzymes, enduring 100 bending cycles and proving highly durable for in vitro testing. Similarly, Sonkusale's group [7] reported the development of a pH sensor array and a glucose sensor by embroidering functionalized hydrophilic threads onto highly hydrophobic woven fabric (Fig. 4). In this work, a thread-based toolkit was developed to measure chemical (pH and glucose) markers and physical (strain and temperature) in an integrated thread-based diagnostic device (TDD) platform. The hydrophobic woven fabric was adorned with hydrophilic threads, ingeniously fashioned into microfluidic channels designed to precisely transport bodily fluids to specific sensing zones. To measure key physiological parameters such as glucose levels, pH, temperature, and strain, conductive threads enhanced with nanomaterials like carbon nanotubes (CNTs), carbon nanopowders, polyaniline (PANI), and various combinations thereof, were employed as the foundation for thread-based electrodes, both in vitro and in vivo. These sensor outputs were seamlessly integrated with a separate layer housing signal processing electronics and wireless communication capabilities, facilitated by conductive threads serving as interconnections, thus enabling seamless data transmission to a smartphone or computer.

3.3 Physical Sensors

Biophysical sensors, which monitor parameters like temperature, strain, and pressure, find another application in thread sensors. Utilizing the dip-and-dry coating technique depicted in Fig. 5, ordinary threads crafted from materials like cotton, polyester, or polyurethane can be endowed with functional conductive and dielectric layers. This conductive layer imparts strain-sensitive conductance to the thread, while the dielectric layer ensures insulation from other conductors and safeguards against physical harm. We will discuss them in this section.

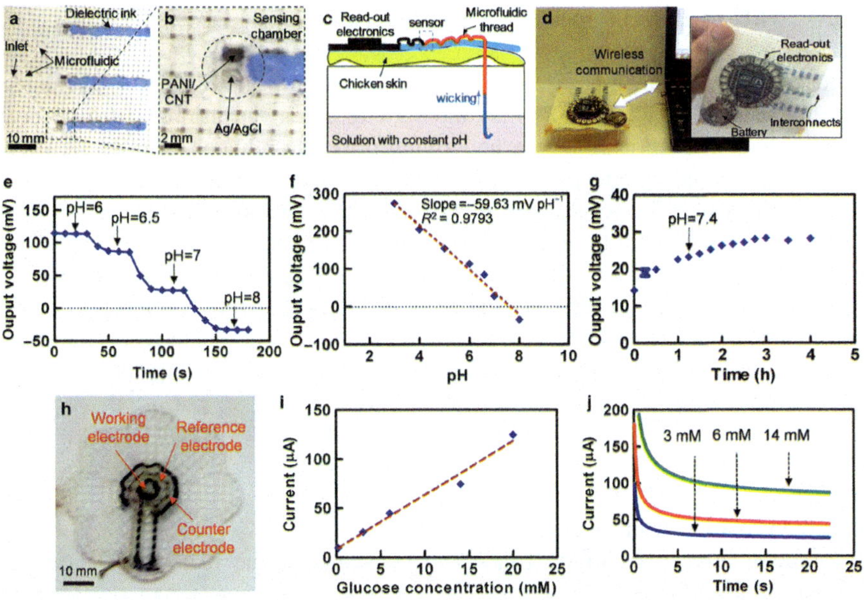

Fig. 4 Characterization of chemical sensors. **a** and **b** Optical image of a multiplexed microfluidic pH sensors assay. **c** Schematic illustration of measuring pH in an in vitro skin model. **d** Sensing system communicating with an external computer via a wireless system. **e** Transient response of the pH sensor to different pH values. **f** Calibration plot of the pH sensor. **g** Continuous pH measurement for four hours. **h** Optical image of the glucose sensor. **i** Calibration plot of the glucose sensor. **j** Transient response of the glucose sensor to different glucose concentrations in the PBS solution. PBS, phosphate-buffered saline [7]. Figure reproduced with permission

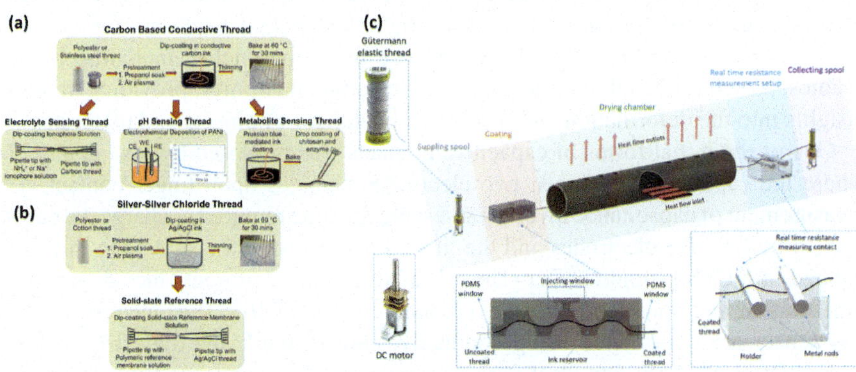

Fig. 5 Steps in preparation of **a** carbon-based conductive threads and sensing threads, **b** PVB-coated Ag/AgCl/PE thread [7], **c** Reel-to-reel automated fabrication of smart threads showing the coating process involving a specialized cartridge with ink that also enables pre-stretching, drying, real-time impedance monitoring and spool collection [76]. Figure reproduced with permission

3.3.1 Thread-Based Capacitive Strain/Pressure Sensors

Thread-based pressure and strain sensors commonly utilize a capacitive mechanism for their design. These sensors may incorporate either intrinsic or extrinsic coatings of diverse materials, yet the underlying principle remains uniform across all variations. Thread-based capacitive strain sensors operate based on the principle of changes in capacitance resulting from mechanical strain applied to the sensor. These sensors are constructed by integrating conductive threads into a flexible substrate, with a dielectric layer between them. When the sensor is subjected to mechanical deformation or strain, the distance between the conductive threads changes, leading to a change in capacitance. The fundamental equation governing the capacitance (C) of a capacitive strain sensor is similar to the general capacitance equation.

$$C = \varepsilon_0 * \varepsilon_r * A/d$$

where:
ε_0 is the vacuum permittivity
ε_r is the relative permittivity (dielectric constant) of the material between the conductive threads.
A is the area of overlap between the conductive threads.
d is the distance between the conductive threads.

As strain is applied to the sensor, the separation distance (d) between the conductive threads alters, causing a corresponding change in capacitance. This change in capacitance is directly proportional to the magnitude of the applied strain. By measuring the resulting capacitance change, the strain or deformation experienced by the sensor can be accurately determined. Thread-based capacitive strain sensors offer advantages such as flexibility, lightweight design, and suitability for wearable and conformable applications. They find use in various fields, including biomechanics, structural health monitoring, and wearable electronics, providing valuable insights into the deformation and mechanical behavior of objects and materials.

One of the initial forms of capacitive contacts was proposed by Post et al. [77], where the capacitance between two electrodes changes upon finger contact. The measurement of capacitance involves assessing the time required to charge the capacitor (formed by the electrodes and the finger) from zero to the switching threshold voltage of a CMOS logic buffer. This capacitive contact presents the advantage of time-domain measurement, offering a dynamic range of 10^4. The electrodes within this sensor are constructed using conductive thread, while the dielectric material is composed of cloth. For ideal yarn or textile in fabric circuits, it would possess fully adjustable electrical properties, maintaining these characteristics during sewing, bending, and wear. In their study, the authors employed stainless steel yarns as conductive electrodes. Stainless steel yarn holds certain advantages, notably its inert nature, rendering it washable and impervious to sweat-related effects. These stainless-steel yarns exhibit varying compositions, ranging from 100% continuous conductive steel fibers to feltings or composites intertwined with polyester and short steel fibers. Altering the ratio between these constituent fibers yields variations in resistivity.

Sergio et al. [78] introduced an alternative capacitive platform for pressure sensing that utilizes conductive fabric. This approach involves mapping the distribution of applied pressure across the fabric surface by detecting changes in capacitance between rows and columns of conductive fibers arranged on opposing sides of an elastic synthetic foam layer. The application of pressure to the smart fabric leads to a reduction in the dielectric layer's thickness between corresponding rows and columns, thereby increasing the coupling capacitance. To process the data, a dedicated chip is positioned at the corner of the smart fabric. The change in induced charge is transformed into a voltage value through readout circuitry integrated into the chip. The central region of the pressure sensor occupies an area of 32 mm^2. In a similar vein, Meyer et al. [79] employed a comparable approach in their pressure sensor design. They employed conductive textiles on both sides of a compressible spacer. Single electrodes were arrayed on one side of the spacer, while the other side featured a single common electrode. This configuration forms capacitors between each individual electrode and the common electrode. The single electrodes are skillfully embroidered with conductive yarn, while the common electrode is composed of a woven textile coated with silver. Hoffman et al. [80] devised a capacitive textile force sensor tailored for respiration monitoring. Their force sensor comprises a compressible 3D textile situated at its core, dictating its sensitivity to applied forces. Functioning as a dielectric layer, the 3D textile is sandwiched between two layers of conductive fabrics, which serve as electrodes. Notably, they opted for a fabric containing 75% silver. In a similar vein, Holleczek et al. [81] engineered a pressure insole with an extended range of pressure detection. For its electrodes, they utilized silver-coated textiles, while a compressible spacer crafted from CrosliteTM (known as Proprietary Closed Cell Resin or PCCR) served as the dielectric layer. However, the sensor's larger size rendered it inconvenient for sock integration. The employed silicone sealant exhibited quick peeling, and the connecting wires were susceptible to breakage. A significant concern was the lack of washability for the socks bearing the assembled sensors.

3.3.2 Fiber and Yarn Strain-Based Sensors

Textile sensors emerge as potent tools for monitoring bio-signals. Catrysse et al. [82] introduced a textile sensor for monitoring electrocardiogram (ECG) and respiration rate. They employed woven and knitted stainless steel electrodes named "Textrodes" for ECG signal measurement (Fig. 6a, b). The Textrodes were advantageous for their integration into garments and skin-friendly nature. However, they exhibited higher skin–electrode impedance (1–5 MW cm^2) compared to commercially available gel electrodes (10 kW cm^2). Similarly, Cho et al. [83] proposed various conductive fabrics as electrodes using three approaches: (1) Sputtering on Polyurethane (PU) laminated and PU dry-coated nylon fabrics (2) electroless Cu/Ni plated fabrics (3) Embroidering with stainless steel filament yarns over woven or knitted fabrics.

Three types of electrodes were developed using different materials by Cho's group: metal yarn embroidered on plain-woven fabric, blended knitted fabric of metal yarns and cotton, and metal yarn embroidered on blended knitted fabric. Knitted fabric (100% stainless steel yarns) and braided fabric (spandex core covered with polyester filament yarns and stainless-steel multifilament yarns) were employed as motion sensors. Zhang et al. [84] utilized carbon yarn and steel fibers to fabricate strain sensors. Carbon fiber, due to its smooth surface and intrinsic properties, outperformed stainless steel for gauge fabrication. However, steel fiber's strength made it suitable for frequency sensor application. Pacelli et al. [85] introduced textile piezoresistive sensors for post-stroke rehabilitation and cardiovascular disease monitoring. They developed knitted piezoresistive fabric (KPF) sensors using the intarsia technique and Printed Piezoresistive Fabric (PPF) sensors through a coating printing process of Conductive Elastomer (CE) material. Huang et al. [86] innovatively fashioned yarns using piezoresistive, elastic, and regular polyester fibers, which were then employed to create cloth, dresses, and sensing textiles. Various wrapping methods were employed, with double wrapping proving more linear. Different fibers, including piezoresistive, elastic, and polyester, were incorporated into the yarn to create the composite core yarn, followed by the application of the carbon-coated fiber using single and double wrapping techniques.

Recently, Sonkusale's group demonstrated the preliminary finding and fabrication of a sensing pillow to monitor head movement [87]. In this paper, we incorporate strain-sensing threads comprised of 64% polyester, 36% polyurethane, and a polydimethylsiloxane (PDMS) layer. A memory foam pillow is then created using them. The threads' covering gives them resilience, enabling the smart threads to withstand exposure to moisture from washing or perspiration. The gauge factor of the smart threads is approximately 2.5 for 50% strain. With a thread resistance range of 0.5–20.0 M, it displays linearity for applied weights of 50.0–5000.0 g. The resistance readout circuitry and wireless module are linked to the smart threads to wirelessly record and communicate strain data to a computer. Initial findings show that the smart pillow is functional and can measure a person's head motion in real time.

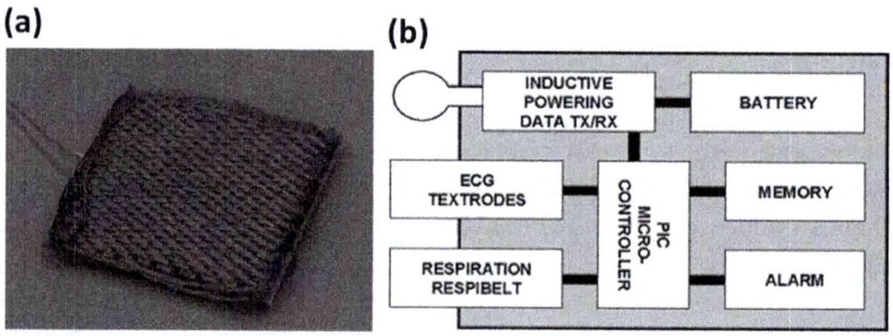

Fig. 6 **a** Schematic overview of the wearable electronics. **b** Photograph of a knitted 3 × 3 cm Textrode. Figure reproduced with permission [82]

4 Thread-Based Transistors, Antennas, and Circuits for Thread-Based Sensors

4.1 E-Textile and Thread-Based Antennas

Over the past couple of decades, there has been a growing interest in wireless technologies which enable many electronic devices to connect via WiFi, WLAN, or Bluetooth. As wireless devices and body-centric communication systems gained importance owing to the advancements in semiconductor research and nanotechnology, the future of electronics is moving toward wearable devices. Even though the design and verification of such systems are challenging due to cost, integration, and space requirements, it is now possible to shrink the size of the most sophisticated electronic components. Thus, the same idea applies to high-frequency circuits as well. One of the most crucial components in wearable devices is the antenna. This section is thereby devoted to the overview of textile-based antennas, and the design, fabrication, and discussion of thread-based antennas.

Being seen virtually everywhere, the antennas have lots of applications in automobiles, spacecraft, held mobile devices, the military, radar, and so on. The design of antennas for wearable devices has necessitated a change toward flexible materials that comprise textiles, fabrics, and threads instead of utilizing rigid and flexible antenna elements and structures. It must be noted that wearable technology and devices are not the only application area of textile- and thread-based antennas despite being the most used ones. As presented in the literature, wearable devices can be used in many applications including imaging, sensing, and healthcare [88–93], and wireless communications [94–97]. and radio-frequency identification (RFID) [98, 99]. The antennas designed for wearable applications in which wireless connectivity is required are mostly textile-based. For body-centric communications, antennas are designed to be wearable such that they can be integrated into clothes or woven fabrics. These materials should be suitable for sustaining an optimal level of RF antenna performance. Wearable and textile-based antennas occasionally comprise conductive textiles (e-textiles) in which a polymer is coated with a conductive material [100–102]. Antennas fabricated with electro-textiles offer good RF performance, flexibility, compactness, and mechanical durability [103–105].

E-textile materials and e-textile-based antennas have been developed owing to the improvements in the Internet of Things (IoT) and the demand for wearable devices in the field of sensing and electronics. Nowadays, it is possible to find many off-the-shelf e-textile materials that have various conductivity, flexibility, and mechanical durability. Being mainly used as EM shielding materials [106], e-textiles have found an important place in wearable antennas [103, 106–109]. In general, textile materials and fabrics are inherently nonconductive which necessitates techniques for obtaining conductive areas or surfaces to realize antennas. Briefly, there are two methods to make such materials conductive [110]. The first method is coating the nonconductive surface after the weaving process to form a cloth or fabric. The other technique is using ready-made threads or conductive fibers and integrating them into the cloth or fabric by interweaving or embroidery process. Textile structures and fabrics can be made electrically conductive via knitting, weaving, or embroidery [110, 111]. To be able to fabricate antennas with optimal electromagnetic performance, surface resistivity must be kept below 1 Ω/sq [112].

It has been investigated the relationship between woven densities on a fabric and antenna performance [110]. It has been shown that the e-textile antennas with woven patterns whose grid length is less than 2 mm have antenna performance in terms of radiation efficiency and pattern. Once the distance between two conductive fibers exceeds 2 mm, an offset in antenna resonant frequency is observed. Thus, if the grid length is small enough, it is shown that the conductive e-textile antenna can be regarded as a homogeneous metal plate. In addition to the woven density, the type of the e-textile material is also effective on antenna performance. Four e-textiles are utilized in testing various types of materials that have different resistivities within 0.03 Ω/m^2. Having fabricated e-textiles on FR-4 substrate, antennas made of different materials show that some of the conductive e-textiles are only suitable for decent antenna performance.

Textile-based antennas that comprise electro-textiles and e-threads are primarily based on macro-level functionalization of the radiating structures using fabrication techniques such as embroidery and screen- or inkjet-printing. In general, off-the-shelf threads, yarns, or fibers are utilized in the production of antennas as described in the previous section. However, the important factor that separates the thread-based approach from the textile-based antennas is the functionalization of each fiber in the thread structure. There have been numerous studies that focus on the device-level fabrication and analysis of thread-based antennas, especially using carbon nanotubes (CNTs) even though the number of published works in this field is still relatively low. The key advantage of using CNT fibers is the possibility of attaining higher fiber conductivity which can result in a huge antenna radiation performance [113].

Current research is primarily focused on the utilization of CNT bundles and CNT threads for the implementation of antennas. Owing to their mechanical, electrical, and thermal properties, CNTs have become potential materials for achieving mechanically stable and lightweight antennas for wireless communications. Aside from the thread-based antennas, a few studies have investigated microstrip antennas in which CNTs are utilized as basic RF signal radiators [114–116]. The performance of nanotubes and nanowires has been intensively investigated in terms of input impedance and antenna efficiency in which mathematical models are developed for near- and far-field radiation patterns [117]. The analysis showed that individual CNTs possess weak radiation performance which can be attributed to the value of quantum capacitance and poor resistive part of the input impedance. Since the analyzed antenna is of a dipole type, self-impedance (0.02 Ω) and mutual impedance (1.2 Ω) have been calculated to determine the radiation region in which the antenna has the highest efficiency. In other words, it is observed that there is a remarkable amount of coupling between the wires which concludes that antennas are better radiators in the near field. The physical measurements that the high-frequency impedance of carbon nanotube bundles improve by the scaling of inductance with the number of each CNTs [118].

There are two basic techniques to fabricate carbon nanotube threads: dry spinning and wet spinning. The first step of the dry spinning method is the synthesis of vertically aligned CNT arrays (spinnable arrays) that comprise double or multiwall nanotubes. The density and length of CNTs inside the array are two important factors as there should be a sufficient number of contact points to ensure thread assembly by van der Waals interaction. In other words, multiwall Nanotubes are gathered and twisted into a thread structure in which every single nanotube adheres via van der Waals forces [119]. Following the synthesis of a vertically aligned CNT array, a spinnable array can be obtained using chemical vapor deposition (CVD) in which CNTs can grow under the mixture of Ar, C_2H_4, H_2, and H_2O [119, 120]. The width of the spinnable array is effective on the diameter of the final version of the thread which is also known as 1-ply (single-ply) thread. The dry spinning method and the usage of a spinnable array provide controllability to the fabrication process such that the desired density, length, and CNT thread diameter can be attained. Further weaving processes can also be used to obtain 3-ply CNT threads and ropes. Additionally, it was also proposed and validated in a postprocessing method called densification to increase the contact area between the neighboring nanotubes using dimethyl sulfoxide (DMSO) to improve electrical conductivity and mechanical strength [120]. The results of Keller's investigation of the thread spacing on the E-plane radiation pattern are related to the spacing and geometry.

The other CNT thread fabrication method, wet-spinning, can also be used to fabricate highly efficient thread-based monopole antennas [113]. It is reported that the dry-spun CNT fibers possess low radiation efficiency which is highly attributed to the density of the CNT thread. However, threads that possess very high density and conductivity can be utilized in obtaining antennas with better radiation performance. In contrast to dry spinning, much denser and highly ordered CNT fibers can be obtained via the wet-spinning method. CNT fibers from chlorosulfonic acid using single-walled carbon nanotubes (SWCNTs) are fabricated before the purification of the fibers that can create the antenna structure [113]. The CNT fibers are purified with H_2O_2 and HCl to obtain high conductivity and stabilize the overall fiber structure. Various l/4-long monopole antennas have been fabricated to investigate the effect of different design parameters such as the number of CNT fibers and the twist of the thread. The major outcomes of this work are worth mentioning as the overall study is very important in terms of pioneering the thread-based approach to wireless connectivity. The twist of the thread which can be classified as low (~1 turn/mm) and high twist (~2 turns/mm) has no major influence on the DC resistance or antenna radiation performance. DC resistance of the CNT-thread-based antenna is inversely proportional to the cross-sectional area. The conductivity of the single fiber is almost an order of magnitude smaller than multifilament high-conductivity fibers. Concordantly, thread-based antennas made of CNT fibers have an improved S_{11} performance over 30 AWG copper wire at least an order of magnitude. The radiation efficiency of the antenna is validated through 2-port VNA measurements inside a reverberation chamber with a standard horn antenna. The radiation efficiency of the CNT thread antenna significantly improves with increased cross-sectional area and lower operational frequencies as the antennas are tested at 1 GHz and 2.4 GHz.

It has also been reported that the frequency dependence of the radiation efficiency can be attributed to an intrinsic RF property of the CNT thread antenna because there is no impedance matching between the VNA and the antenna. Therefore, a modified version of radiation efficiency is proposed to quantitatively compare the threaded antenna with the copper wire antenna.

4.2 Readout Circuits for Thread-Based Sensors

There has been a growing interest in the utilization of thread-based sensors in various platforms such as health monitoring and wearable devices in the form of minimally invasive or non-invasive fashion. It is often desired to obtain information related to biophysical markers such as respiration rate, electrocardiogram (ECG), electrooculogram (EOG), core body temperature, or metabolism of an individual from biological fluids, sweat, and saliva which might be utilized to provide an assessment of individual's physical performance [16, 54, 121, 122]. In addition to health monitoring, flexible sensor systems are broadly used in scientific and industrial applications in which monitoring pressure, acceleration, force, gas concentration, and humidity is of primary importance [87, 122–127]. The block diagram of a flexible sensing system is shown in Fig. 7. A physical quantity (or measurand) that is being measured is sensed by the flexible sensor before feeding the signal into the signal conditioning circuit. Usually, sensors convert the measurand into a signal that could be in the form of voltage, current, capacitance, or resistance which could further be processed by the signal conditioning circuit. Different sensors could generate different outputs in which the signal amplitude could be either too weak or moderately weak which necessitates amplification. One major function of the signal conditioning circuit is to be able to amplify the signal at the output of the sensor. Furthermore, the signal generated by the sensor could possess unwanted harmonics and noise which could be suppressed and filtered out by the signal conditioning circuits [128]. Hence, the sensor front-end system can preprocess the sensed signal which can now be accepted by the digital signal processor for further digitization, visualization, and eventually data analysis.

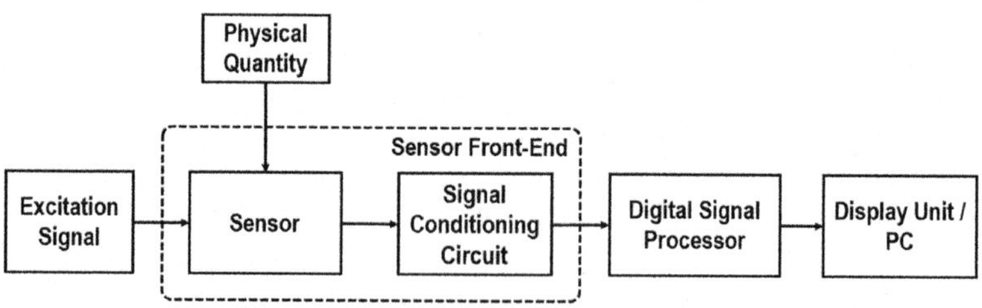

Fig. 7 Block diagram of a sensor readout system

5 Conclusions and Perspectives

The next generation of personal healthcare devices and medical therapies requires seamless integration of biosensors and power generation/storage technologies into flexible and wearable formats. In this context, threads and sutures emerge as highly promising solutions. Threads offer a plethora of advantages, including their universal availability, cost-effectiveness, material versatility, and straightforward textile-based processing. Moreover, they inherently lend themselves to wearability and seamless integration with the human body. Additionally, the thread can function as a smart flexible substrate for sensing, computing, communication, and drug delivery. Leveraging the process of surface modification by physical coating or reel-to-reel fabrication techniques, involving sequential dip-coating, and drying, enables the production of functional smart threads tailored for various sensing, therapeutic, and electronic applications. For example, smart threads have been specifically engineered to sense chemical and physical biomarkers to monitor health and well-being. Furthermore, these smart threads can be implanted in tissue as sutures or incorporated into wearable kits, such as band-like patches, to monitor multiple electrolytes and metabolites in sweat. The above-mentioned applications marked the unique ability of threads to facilitate three-dimensional, tissue-embedded bioelectronic interfaces, a feat unattainable with other flexible bioelectronic platforms.

Nevertheless, there is substantial room for further exploration and innovation in the development of smart threads capable of sensing physio-chemical activity. The flexible nature of threads or sutures provides a distinctive substrate that can be effortlessly sewn or affixed onto any item of wearable clothing or serve as a standalone device. In the near future, we can envision clothing in which threads interconnect to create an intelligent system capable of continuously monitoring an individual's health and real-time environmental changes. In the realm of fiber or thread substrates, novel fabrication, and modification techniques for creating hybrid or nanocomposite thread materials hold significant potential, providing control over material composition and functionality.

For the future, smart threads hold promise in applications such as neural implants, drug delivery, and the development of cardiac implants or patches. Ongoing research and recent findings suggest that threads can be harnessed for on-demand drug delivery and serve as scaffold substrates for cell growth, thus opening up exciting avenues for treatment. Looking ahead, it may become feasible to seamlessly integrate sensing and therapeutic functions within the same wearable platform, where threads play a dual role as a sensor and as a drug delivery device to close the loop. Lastly, there is still a need for improvement in terms of developing tighter integration and more sophisticated packaging for thread-based platforms with flexible electronics for remote health and environmental monitoring.

References

1. Chan, M., Estève, D., Fourniols, J.Y., Escriba, C., Campo, E.: Smart wearable systems: current status and future challenges. Artif. Intell. Med. **56**, 137–156 (2012). https://doi.org/10.1016/J.ARTMED.2012.09.003
2. Becerik-Gerber, B., Lucas, G., Aryal, A., Awada, M., Bergés, M., Billington, S., Boric-Lubecke, O., Ghahramani, A., Heydarian, A., Höelscher, C., Jazizadeh, F., Khan, A., Langevin, J., Liu, R., Marks, F., Mauriello, M.L., Murnane, E., Noh, H., Pritoni, M., Roll, S., Schaumann, D., Seyedrezaei, M., Taylor, J.E., Zhao, J., Zhu, R.: The field of human building interaction for convergent research and innovation for intelligent built environments. Sci. Reports **12**(1), 1–19 (2022). https://doi.org/10.1038/s41598-022-25047-y
3. Stanberry, B.: Legal ethical and risk issues in telemedicine. Comput. Methods Programs Biomed. **64**, 225–233 (2001). https://doi.org/10.1016/S0169-2607(00)00142-5
4. Gatzoulis, L., Iakovidis, I.: Wearable and portable eHealth systems. IEEE Eng. Med. Biol. Mag. **26**, 51–56 (2007). https://doi.org/10.1109/EMB.2007.901787
5. Sonkusale, S.: Smart threads for tissue-embedded bioelectronics. In: Proceedings of the Custom Integrated Circuits Conference. 2022-April (2022). https://doi.org/10.1109/CICC53496.2022.9772846
6. Servati, A., Zou, L., Jane Wang, Z., Ko, F., Servati, P.: Novel flexible wearable sensor materials and signal processing for vital sign and human activity monitoring. Sensors **17**, 1622 (2017). https://doi.org/10.3390/S17071622
7. Mostafalu, P., Akbari, M., Alberti, K.A., Xu, Q., Khademhosseini, A., Sonkusale, S.R.: A toolkit of thread-based microfluidics, sensors, and electronics for 3D tissue embedding for medical diagnostics. Microsyst. Nanoeng. **2**(1), 1–10 (2016). https://doi.org/10.1038/micronano.2016.39
8. Mostafalu, P., Lenk, W., Dokmeci, M.R., Ziaie, B., Khademhosseini, A., Sonkusale, S.R.: Wireless flexible smart bandage for continuous monitoring of wound oxygenation. IEEE Trans. Biomed. Circuits Syst. **9**, 670–677 (2015). https://doi.org/10.1109/TBCAS.2015.2488582
9. Chen, S., Qi, J., Fan, S., Qiao, Z., Yeo, J.C., Lim, C.T.: Flexible wearable sensors for cardiovascular health monitoring. Adv. Healthc. Mater. **10**, 2100116 (2021). https://doi.org/10.1002/ADHM.202100116
10. Qin, H., Owyeung, R.E., Sonkusale, S.R., Panzer, M.J.: Highly stretchable and nonvolatile gelatin-supported deep eutectic solvent gel electrolyte-based ionic skins for strain and pressure sensing. J. Mater. Chem. C Mater. **7**, 601–608 (2019). https://doi.org/10.1039/C8TC05918G
11. Sun, W., Guo, Z., Yang, Z., Wu, Y., Lan, W., Liao, Y., Wu, X., Liu, Y.: A review of recent advances in vital signals monitoring of sports and health via flexible wearable sensors. Sensors **22**, 7784 (2022). https://doi.org/10.3390/S22207784
12. Punjiya, M., Rezaei, H., Zeeshan, M.A., Sonkusale, S.: A flexible pH sensing smart bandage with wireless CMOS readout for chronic wound monitoring. In: TRANSDUCERS 2017—19th International Conference on Solid-State Sensors, Actuators and Microsystems, pp. 1700–1702 (2017). https://doi.org/10.1109/TRANSDUCERS.2017.7994393
13. Sonkusale, S.: Sutures for the wireless sensing of deep wounds. Nat. Biomed. Eng. **5**(10), 1113–1114 (2021). https://doi.org/10.1038/s41551-021-00806-w
14. Gao, W., Emaminejad, S., Nyein, H.Y.Y., Challa, S., Chen, K., Peck, A., Fahad, H.M., Ota, H., Shiraki, H., Kiriya, D., Lien, D.H., Brooks, G.A., Davis, R.W., Javey, A.: Fully integrated wearable sensor arrays for multiplexed in situ perspiration analysis. Nature **529**:7587, 509–514 (2016). https://doi.org/10.1038/nature16521
15. Mondal, S., Zehra, N., Choudhury, A., Iyer, P.K.: Wearable sensing devices for point of care diagnostics. ACS Appl. Bio Mater. **4**, 47–70 (2021). https://doi.org/10.1021/ACSABM.0C00798/ASSET/IMAGES/LARGE/MT0C00798_0015.JPEG
16. Terse-Thakoor, T., Punjiya, M., Matharu, Z., Lyu, B., Ahmad, M., Giles, G.E., Owyeung, R., Alaimo, F., Shojaei Baghini, M., Brunyé, T.T., Sonkusale, S.: Thread-based multiplexed

sensor patch for real-time sweat monitoring. npj Flexible Electron. **4**(1), 1–10 (2020). https://doi.org/10.1038/s41528-020-00081-w
17. Arquilla, K., Webb, A.K., Anderson, A.P.: Textile Electrocardiogram (ECG) electrodes for wearable health monitoring. Sensors **20**, 1013 (2020). https://doi.org/10.3390/S20041013
18. Shen, S., Xiao, X., Xiao, X., Chen, J.: Wearable triboelectric nanogenerators for heart rate monitoring. Chem. Commun. **57**, 5871–5879 (2021). https://doi.org/10.1039/D1CC02091A
19. Coyle, S., Morris, D., Lau, K.-T., Diamond, D., Moyna, N.: Textile-based wearable sensors for assisting sports performance, pp. 307–311 (2009). https://doi.org/10.1109/BSN.2009.57
20. Lee, J.W., Han, D.C., Shin, H.J., Yeom, S.H., Ju, B.K., Lee, W.: PEDOT:PSS-based temperature-detection thread for wearable devices. Sensors **18**, 2996 (2018). https://doi.org/10.3390/S18092996
21. El Gharbi, M., Fernández-García, R., Gil, I.: Embroidered wearable Antenna-based sensor for real-time breath monitoring. Measurement **195**, 111080 (2022). https://doi.org/10.1016/J.MEASUREMENT.2022.111080
22. Chuang, M.C., Windmiller, J.R., Santhosh, P., Ramírez, G.V., Galik, M., Chou, T.Y., Wang, J.: Textile-based electrochemical sensing: effect of fabric substrate and detection of nitroaromatic explosives. Electroanalysis **22**, 2511–2518 (2010). https://doi.org/10.1002/ELAN.201000434
23. Kassal, P., Kim, J., Kumar, R., De Araujo, W.R., Steinberg, I.M., Steinberg, M.D., Wang, J.: Smart bandage with wireless connectivity for uric acid biosensing as an indicator of wound status. Electrochem. Commun. **56**, 6–10 (2015). https://doi.org/10.1016/J.ELECOM.2015.03.018
24. Yang, Y.L., Chuang, M.C., Lou, S.L., Wang, J.: Thick-film textile-based amperometric sensors and biosensors. Analyst **135**, 1230–1234 (2010). https://doi.org/10.1039/B926339J
25. Schazmann, B., Morris, D., Slater, C., Beirne, S., Fay, C., Reuveny, R., Moyna, N., Diamond, D.: A wearable electrochemical sensor for the real-time measurement of sweat sodium concentration. Anal. Methods **2**, 342–348 (2010). https://doi.org/10.1039/B9AY00184K
26. Choudhary, T., Rajamanickam, G.P., Dendukuri, D.: Woven electrochemical fabric-based test sensors (WEFTS): a new class of multiplexed electrochemical sensors. Lab Chip **15**, 2064–2072 (2015). https://doi.org/10.1039/C5LC00041F
27. Woven Fabrics - Google Books. https://books.google.com/books?hl=en&lr=&id=r-KdDwAAQBAJ&oi=fnd&pg=PA267&dq=B.+Gutarowska+and+A.+Michalski,+Microbial+Degradation+of+Woven+Fabrics+and+Protection+Against+Biodegradation+&ots=rqgkFves_i&sig=r_9UtVKMDWzuswPapXoF5L1Dp9Y#v=onepage&q=B.%20Gutarowska%20and%20A.%20Michalski%2C%20Microbial%20Degradation%20of%20Woven%20Fabrics%20and%20Protection%20Against%20Biodegradation&f=false
28. Murphy, A.R., John, P.S., Kaplan, D.L.: Modification of silk fibroin using diazonium coupling chemistry and the effects on hMSC proliferation and differentiation. Biomaterials **29**, 2829–2838 (2008). https://doi.org/10.1016/J.BIOMATERIALS.2008.03.039
29. Barani, H., Calvimontes, A.: Effects of oxygen plasma treatment on the physical and chemical properties of wool fiber surface. Plasma Chem. Plasma Process. **34**, 1291–1302 (2014). https://doi.org/10.1007/S11090-014-9581-X/TABLES/3
30. Barani, H., Maleki, H.: Plasma and ultrasonic process in dyeing of wool fibers with madder in presence of Lecithin **32**, 1191–1199 (2011). https://doi.org/10.1080/01932691.2010.505525
31. Tajuddin, M., Ahmad, Z., Ismail, H.: A review of natural fibers and processing operations for the production of binderless boards. Bioresources (2016). https://doi.org/10.15376/BIORES.11.2
32. Jamir, M.R.M., Majid, M.S.A., Khasri, A.: Natural lightweight hybrid composites for aircraft structural applications. In: Sustainable Composites for Aerospace Applications, pp. 155–170 (2018). https://doi.org/10.1016/B978-0-08-102131-6.00008-6
33. Delebecq, E., Pascault, J.P., Boutevin, B., Ganachaud, F.: On the versatility of urethane/urea bonds: reversibility, blocked isocyanate, and non-isocyanate polyurethane. Chem. Rev. **113**, 80–118 (2013). https://doi.org/10.1021/CR300195N/ASSET/IMAGES/LARGE/CR-2012-00195N_0014.JPEG

34. Du, Y., George, S.M.: Molecular layer deposition of nylon 66 films examined using in situ FTIR spectroscopy. J. Phys. Chem. C **111**, 8509–8517 (2007). https://doi.org/10.1021/JP067041N/ASSET/IMAGES/MEDIUM/JP067041NE00005.GIF
35. Sadeqi, A., Rezaei Nejad, H., Alaimo, F., Yun, H., Punjiya, M., Sonkusale, S.R.: Washable smart threads for strain sensing fabrics. IEEE Sens. J. **18**, 9137–9144 (2018). https://doi.org/10.1109/JSEN.2018.2870640
36. Li, W., Liu, J., Wei, J., Yang, Z., Ren, C., Li, B.: Recent progress of conductive hydrogel fibers for flexible electronics: fabrications, applications, and perspectives. Adv. Funct. Mater. **33**, 2213485 (2023). https://doi.org/10.1002/ADFM.202213485
37. De France, K., Zeng, Z., Wu, T., Nyström, G., De France, K., Zeng, Z., Wu, T., Nyström, G.: Functional materials from nanocellulose: utilizing structure-property relationships in bottom-up fabrication. Adv. Mater. **33**, 2000657 (2021). https://doi.org/10.1002/ADMA.202000657
38. Tamayol, A., Akbari, M., Annabi, N., Paul, A., Khademhosseini, A., Juncker, D.: Fiber-based tissue engineering: progress, challenges, and opportunities. Biotechnol. Adv. **31**, 669–687 (2013). https://doi.org/10.1016/J.BIOTECHADV.2012.11.007
39. Rutledge, G.C., Fridrikh, S.V.: Formation of fibers by electrospinning. Adv. Drug Deliv. Rev. **59**, 1384–1391 (2007). https://doi.org/10.1016/J.ADDR.2007.04.020
40. Liu, N., Fang, G., Wan, J., Zhou, H., Long, H., Zhao, X.: Electrospun PEDOT:PSS–PVA nanofiber based ultrahigh-strain sensors with controllable electrical conductivity. J. Mater. Chem. **21**, 18962–18966 (2011). https://doi.org/10.1039/C1JM14491J
41. Karageorgiou, V., Kaplan, D.: Porosity of 3D biomaterial scaffolds and osteogenesis. Biomaterials **26**, 5474–5491 (2005). https://doi.org/10.1016/J.BIOMATERIALS.2005.02.002
42. Kim, D.W., Lee, O.J., Kim, S.W., Ki, C.S., Chao, J.R., Yoo, H., Yoon, S. il, Lee, J.E., Park, Y.R., Kweon, H.Y., Lee, K.G., Kaplan, D.L., Park, C.H.: Novel fabrication of fluorescent silk utilized in biotechnological and medical applications. Biomaterials **70**, 48–56 (2015). https://doi.org/10.1016/J.BIOMATERIALS.2015.08.025
43. Murphy, A.R., Kaplan, D.L.: Biomedical applications of chemically-modified silk fibroin. J. Mater. Chem. **19**, 6443–6450 (2009). https://doi.org/10.1039/B905802H
44. Kim, D., Herr, A.E.: Protein immobilization techniques for microfluidic assays. Biomicrofluidics **7**, 041501 (2013). https://doi.org/10.1063/1.4816934
45. Nimse, S.B., Song, K., Sonawane, M.D., Sayyed, D.R., Kim, T.: Immobilization techniques for microarray: challenges and applications. Sensors **14**, 22208–22229 (2014). https://doi.org/10.3390/S141222208
46. Prieto-Simon, B., Campas, M., Marty, J.-L.: Biomolecule immobilization in biosensor development: tailored strategies based on affinity interactions. Protein Pept. Lett. **15**, 757–763 (2008). https://doi.org/10.2174/092986608785203791
47. Salva, M.L., Rocca, M., Niemeyer, C.M., Delamarche, E.: Methods for immobilizing receptors in microfluidic devices: a review (2021)
48. Welch, N.G., Scoble, J.A., Muir, B.W., Pigram, P.J.: Orientation and characterization of immobilized antibodies for improved immunoassays (Review). Biointerphases **12**, 02D301 (2017). https://doi.org/10.1116/1.4978435
49. Bulaj, G.: Formation of disulfide bonds in proteins and peptides. Biotechnol. Adv. **23**, 87–92 (2005). https://doi.org/10.1016/J.BIOTECHADV.2004.09.002
50. Limo, M.J., Sola-Rabada, A., Boix, E., Thota, V., Westcott, Z.C., Puddu, V., Perry, C.C.: Interactions between metal oxides and biomolecules: from fundamental understanding to applications. Chem. Rev. **118**, 11118–11193 (2018). https://doi.org/10.1021/ACS.CHEMREV.7B00660/ASSET/IMAGES/LARGE/CR-2017-006605_0023.JPEG
51. Chiodi, E., Marn, A.M., Geib, M.T., Selim Ünlü, M.: The role of surface chemistry in the efficacy of protein and DNA microarrays for label-free detection: an overview. Polymers **13**, 1026 (2021). https://doi.org/10.3390/POLYM13071026
52. Jain, P., Rauer, S.B., Möller, M., Singh, S.: Mimicking the natural basement membrane for advanced tissue engineering. Biomacromol **23**, 3081–3103 (2022). https://doi.org/10.1021/ACS.BIOMAC.2C00402/ASSET/IMAGES/LARGE/BM2C00402_0006.JPEG

53. Gharib, G., Bütün, İ., Muganlı, Z., Kozalak, G., Namlı, İ., Sarraf, S.S., Ahmadi, V.E., Toyran, E., van Wijnen, A.J., Koşar, A.: Biomedical applications of microfluidic devices: a review. Biosensors (Basel) **12** (2022). https://doi.org/10.3390/BIOS12111023/S1
54. Xia, J., Khaliliazar, S., Hamedi, M.M., Sonkusale, S.: Thread-based wearable devices. MRS Bull. **46**, 502–511 (2021). https://doi.org/10.1557/S43577-021-00116-1/FIGURES/7
55. Owyeung, R.E., Terse-Thakoor, T., Rezaei Nejad, H., Panzer, M.J., Sonkusale, S.R.: Highly flexible transistor threads for all-thread based integrated circuits and multiplexed diagnostics. ACS Appl. Mater. Interf. **11**, 31096–31104 (2019). https://doi.org/10.1021/ACSAMI.9B09522/SUPPL_FILE/AM9B09522_SI_001.PDF
56. Mousavi, M.P.S., Ainla, A., Tan, E.K.W., Abd El-Rahman, M., Yoshida, Y., Yuan, L., Sigurslid, H.H., Arkan, N., Yip, M.C., Abrahamsson, C.K., Homer-Vanniasinkam, S., Whitesides, G.M.: Ion sensing with thread-based potentiometric electrodes. Lab Chip. **18**, 2279–2290 (2018). https://doi.org/10.1039/C8LC00352A
57. Ricci, F., Palleschi, G.: Sensor and biosensor preparation, optimisation and applications of Prussian blue modified electrodes. Biosens. Bioelectron. **21**, 389–407 (2005). https://doi.org/10.1016/J.BIOS.2004.12.001
58. Renedo, O.D., Alonso-Lomillo, M.A., Martínez, M.J.A.: Recent developments in the field of screen-printed electrodes and their related applications. Talanta **73**, 202–219 (2007). https://doi.org/10.1016/J.TALANTA.2007.03.050
59. Malzahn, K., Windmiller, J.R., Valdés-Ramírez, G., Schöning, M.J., Wang, J.: Wearable electrochemical sensors for in situ analysis in marine environments. Analyst **136**, 2912–2917 (2011). https://doi.org/10.1039/C1AN15193B
60. Malon, R.S.P., Chua, K.Y., Wicaksono, D.H.B., Córcoles, E.P.: Cotton fabric-based electrochemical device for lactate measurement in saliva. Analyst **139**, 3009–3016 (2014). https://doi.org/10.1039/C4AN00201F
61. Carrilho, E., Martinez, A.W., Whitesides, G.M.: Understanding wax printing: A simple micropatterning process for paper-based microfluidics. Anal. Chem. **81**, 7091–7095 (2009). https://doi.org/10.1021/AC901071P/SUPPL_FILE/AC901071P_SI_001.PDF
62. Kinnamon, D.S., Krishnan, S., Brosler, S., Sun, E., Prasad, S.: Screen printed graphene oxide textile biosensor for applications in inexpensive and wearable point-of-exposure detection of influenza for at-risk populations. J. Electrochem. Soc. **165**, B3084–B3090 (2018). https://doi.org/10.1149/2.0131808JES/XML
63. Nomura, K.I., Horii, Y., Kanazawa, S., Kusaka, Y., Ushijima, H.: Fabrication of a textile-based wearable blood leakage sensor using screen-offset printing. Sensors **18**, 240 (2018). https://doi.org/10.3390/S18010240
64. Kincal, D., Kumar, A., Child, A.D., Reynolds, J.R.: Conductivity switching in polypyrrole-coated textile fabrics as gas sensors. Synth. Met. **92**, 53–56 (1998)
65. Collins, G.E., Buckley, L.J.: Conductive polymer-coated fabrics for chemical sensing. Synth. Met. **78**, 93–101 (1996). https://doi.org/10.1016/0379-6779(96)80108-1
66. Sekar, M., Pandiaraj, M., Bhansali, S., Ponpandian, N., Viswanathan, C.: Carbon fiber based electrochemical sensor for sweat cortisol measurement. Sci. Reports **9**(1), 1–14 (2019). https://doi.org/10.1038/s41598-018-37243-w
67. Parrilla, M., Ferrø, J., Guinovart, T., Andrade, F.J., Parrilla,] M, Ferrø, J., Guinovart, T., Andrade, F.J.: Wearable potentiometric sensors based on commercial carbon fibres for monitoring sodium in sweat. Electroanalysis **28**, 1267–1275 (2016). https://doi.org/10.1002/ELAN.201600070
68. Shim, B.S., Chen, W., Doty, C., Xu, C., Kotov, N.A.: Smart electronic yarns and wearable fabrics for human biomonitoring made by carbon nanotube coating with polyelectrolytes. Nano Lett. **8**, 4151–4157 (2008). https://doi.org/10.1021/NL801495P
69. Guinovart, T., Parrilla, M., Crespo, G.A., Rius, F.X., Andrade, F.J.: Potentiometric sensors using cotton yarns, carbon nanotubes and polymeric membranes. Analyst **138**, 5208–5215 (2013). https://doi.org/10.1039/C3AN00710C
70. Kang, M.A., Ji, S., Kim, S., Park, C.Y., Myung, S., Song, W., Lee, S.S., Lim, J., An, K.S.: Highly sensitive and wearable gas sensors consisting of chemically functionalized graphene

oxide assembled on cotton yarn. RSC Adv. **8**, 11991–11996 (2018). https://doi.org/10.1039/C8RA01184B

71. Tang, J., Wei, L., He, S., Li, J., Nan, D., Ma, L., Shen, W., Kang, F., Lv, R., Huang, Z.: A highly sensitive electrochemical glucose sensor based on room temperature exfoliated graphite-derived film decorated with dendritic copper. Materials **14** (2021). https://doi.org/10.3390/MA14175067/S1
72. Chen, C., Ran, R., Yang, Z., Lv, R., Shen, W., Kang, F., Huang, Z.H.: An efficient flexible electrochemical glucose sensor based on carbon nanotubes/carbonized silk fabrics decorated with Pt microspheres. Sens Actuators B Chem. **256**, 63–70 (2018). https://doi.org/10.1016/J.SNB.2017.10.067
73. Wang, Y., Yang, B., Hua, Z., al, Wang, D., Liu, Z., Li, J.: Review-Flexible and Stretchable Electrochemical Sensing Systems: Materials, Energy Sources, and Integrations You may also like Recent advancements in flexible and wearable sensors for biomedical and healthcare applications Thin-film transistor arrays for biological sensing systems (2020). https://doi.org/10.1149/1945-7111/ab7117
74. Liu, X., Lillehoj, P.B.: Embroidered biosensors on gauze for rapid electrochemical measurements. In: Proceedings of the IEEE International Conference on Micro Electro Mechanical Systems (MEMS), pp. 377–380 (2017). https://doi.org/10.1109/MEMSYS.2017.7863420
75. Liu, X., Lillehoj, P.B.: Embroidered electrochemical sensors for biomolecular detection. Lab Chip **16**, 2093–2098 (2016). https://doi.org/10.1039/C6LC00307A
76. Alaimo, F., Sadeqi, A., Rezaei Nejad, H., Jiang, Y., Wang, W., Demarchi, D., Sonkusale, S.: Reel-to-reel fabrication of strain sensing threads and realization of smart insole. Sens Actuators A Phys. **301**, 111741 (2020). https://doi.org/10.1016/J.SNA.2019.111741
77. Post, E.R., Orth, M., Russo, R.R., Gershenfeld, N.: E-broîdery: Design and fabrication of textile-based computing. IBM Syst. J. **39**, 840–860 (2000). https://doi.org/10.1147/SJ.393.0840
78. Sergio, M., Manaresi, N., Campi, F., Canegallo, R., Tartagni, M., Guerrieri, R.: A dynamically reconfigurable monolithic CMOS pressure sensor for smart fabric. IEEE J. Solid-State Circuits **38**, 966–975 (2003). https://doi.org/10.1109/JSSC.2003.811977
79. Meyer, J., Arnrich, B., Schumm, J., Troster, G.: Design and modeling of a textile pressure sensor for sitting posture classification. IEEE Sens. J. **10**, 1391–1398 (2010). https://doi.org/10.1109/JSEN.2009.2037330
80. Hoffmann, T., Eilebrecht, B., Leonhardt, S.: Respiratory monitoring system on the basis of capacitive textile force sensors. IEEE Sens. J. **11**, 1112–1119 (2011). https://doi.org/10.1109/JSEN.2010.2082524
81. Holleczek, T., Rüegg, A., Harms, H., Tröster, G.: Textile pressure sensors for sports applications. In: Proceedings of IEEE Sensors, pp. 732–737 (2010). https://doi.org/10.1109/ICSENS.2010.5690041
82. Catrysse, M., Puers, R., Hertleer, C., Van Langenhove, L., Van Egmond, H., Matthys, D.: Towards the integration of textile sensors in a wireless monitoring suit. Sens Actuators A Phys. **114**, 302–311 (2004). https://doi.org/10.1016/J.SNA.2003.10.071
83. Cho, G., Jeong, K., Paik, M.J., Kwun, Y., Sung, M.: Performance evaluation of textile-based electrodes and motion sensors for smart clothing. IEEE Sens. J. **11**, 3183–3193 (2011). https://doi.org/10.1109/JSEN.2011.2167508
84. Zhang, H., Tao, X., Yu, T., Wang, S.: Conductive knitted fabric as large-strain gauge under high temperature. Sens Actuators A Phys. **126**, 129–140 (2006). https://doi.org/10.1016/J.SNA.2005.10.026
85. Pacelli, M., Caldani, L., Paradiso, R.: Textile piezoresistive sensors for biomechanical variables monitoring. In: Annual International Conference of the IEEE Engineering in Medicine and Biology—Proceedings, pp. 5358–5361 (2006). https://doi.org/10.1109/IEMBS.2006.259287
86. Huang, C.T., Shen, C.L., Tang, C.F., Chang, S.H.: A wearable yarn-based piezo-resistive sensor. Sens Actuators A Phys. **141**, 396–403 (2008). https://doi.org/10.1016/J.SNA.2007.10.069

87. Liu, M., Del-Rio-Ruiz, R., Sharma, A., Asci, C., Owyeung, R., Sonkusale, S.: Preliminary results on sensing pillow to monitor head movement using strain sensing threads. In: Proceedings of IEEE Sensors. 2022-October (2022). https://doi.org/10.1109/SENSORS52175.2022.9967140
88. Pantelopoulos, A., Bourbakis, N.G.: A survey on wearable sensor-based systems for health monitoring and prognosis. IEEE Trans. Syst. Man Cybern. Part C Appl. Rev. **40**, 1–12 (2010). https://doi.org/10.1109/TSMCC.2009.2032660
89. Coyle, S., Lau, K.T., Moyna, N., O'Gorman, D., Diamond, D., Di Francesco, F., Costanzo, D., Salvo, P., Trivella, M.G., De Rossi, D.E., Taccini, N., Paradiso, R., Porchet, J.A., Ridolfi, A., Luprano, J., Chuzel, C., Lanier, T., Revol-Cavalier, F., Schoumacker, S., Mourier, V., Chartier, I., Convert, R., De-Moncuit, H., Bini, C.: BIOTEXBiosensing textiles for personalised healthcare management. IEEE Trans. Inf. Technol. Biomed. **14**, 364–370 (2010). https://doi.org/10.1109/TITB.2009.2038484
90. Sibinski, M., Jakubowska, M., Sloma, M.: Flexible temperature sensors on fibers. Sensors **10**, 7934–7946 (2010). https://doi.org/10.3390/S100907934
91. Linz, T., Gourmelon, L., Langereis, G.: Contactless EMG sensors embroidered onto textile. IFMBE Proc. **13**, 29–34 (2007). https://doi.org/10.1007/978-3-540-70994-7_5/COVER
92. Coosemans, J., Hermans, B., Puers, R.: Integrating wireless ECG monitoring in textiles. Sens Actuators A Phys. **130–131**, 48–53 (2006). https://doi.org/10.1016/J.SNA.2005.10.052
93. Kim, C., Kim, J.K., Kim, K.T., Yoon, Y.K.: Micromachined wearable/foldable super wideband (SWA) monopole antenna based on a flexible liquid crystal polymer (LCP) substrate toward imaging/sensing/health monitoring systems. In: Proceedings—Electronic Components and Technology Conference, pp. 1926–1932 (2013). https://doi.org/10.1109/ECTC.2013.6575841
94. Tronquo, A., Rogier, H., Hertleer, C., Van Langenhove, L.: Applying textile materials for the design of antennas for wireless body area networks. European Space Agency, (Special Publication) ESA SP. 626 SP (2006). https://doi.org/10.1109/EUCAP.2006.4584573
95. Kiourti, A., Volakis, J.L.: Wearable antennas using electronic textiles for RF communications and medical monitoring. In: 2016 10th European Conference on Antennas and Propagation, EuCAP 2016 (2016). https://doi.org/10.1109/EUCAP.2016.7481222
96. Ivsic, B., Bonefacic, D., Bartolic, J.: Considerations on embroidered textile antennas for wearable applications. IEEE Antennas Wirel. Propag. Lett. **12**, 1708–1711 (2013). https://doi.org/10.1109/LAWP.2013.2297698
97. Kiourti, A., Nikita, K.S.: A review of implantable patch antennas for biomedical telemetry: challenges and solutions. IEEE Antennas Propag. Mag. **54**, 210–228 (2012). https://doi.org/10.1109/MAP.2012.6293992
98. Choi, J.H., Kim, Y., Lee, K., Chung, Y.: Various wearable embroidery RFID tag antenna using electro-thread. In: 2008 IEEE International Symposium on Antennas and Propagation and USNC/URSI National Radio Science Meeting, APSURSI (2008). https://doi.org/10.1109/APS.2008.4619730
99. Amendola, S., Lodato, R., Manzari, S., Occhiuzzi, C., Marrocco, G.: RFID technology for IoT-based personal healthcare in smart spaces. IEEE Internet Things J. **1**, 144–152 (2014). https://doi.org/10.1109/JIOT.2014.2313981
100. Stoppa, M., Chiolerio, A.: Wearable electronics and smart textiles: a critical review. Sensors (Basel) **14**, 11957 (2014). https://doi.org/10.3390/S140711957
101. Krifa, M.: Electrically conductive textile materials—Application in flexible sensors and antennas. Textiles **1**, 239–257 (2021). https://doi.org/10.3390/TEXTILES1020012
102. Nurul, H.M.R., Soh, P.J., Malek, M.F.A., Vandenbosch, G.A.E.: Dual-band suspended-plate wearable textile antenna. IEEE Antennas Wirel. Propag. Lett. **12**, 583–586 (2013). https://doi.org/10.1109/LAWP.2013.2259211
103. Koski, K., Sydänheimo, L., Rahmat-Samii, Y., Ukkonen, L.: Fundamental characteristics of electro-textiles in wearable UHF RFID patch antennas for body-centric sensing systems. IEEE Trans. Antennas Propag. **62**, 6454–6462 (2014). https://doi.org/10.1109/TAP.2014.2364071

104. Abbas, B., Khamas, S.K., Ismail, A., Sali, A.: Full embroidery designed electro-textile wearable tag antenna for WBAN application. Sensors **19**, 2470 (2019). https://doi.org/10.3390/S19112470
105. Bayram, Y., Zhou, Y., Shim, B.S., Xu, S., Zhu, J., Kotov, N.A., Volakis, J.L.: E-textile conductors and polymer composites for conformal lightweight antennas. IEEE Trans. Antennas Propag. **58**, 2732–2736 (2010). https://doi.org/10.1109/TAP.2010.2050439
106. Guy, A.W., Chou, C.K., McDougall, J.A., Sorensen, C.: Measurement of shielding effectiveness of microwave-protective suits. IEEE Trans. Microw. Theory Tech. **35**, 984–994 (1987). https://doi.org/10.1109/TMTT.1987.1133796
107. Liu, G., Wang, Q.-A., Jiao, G., Dang, P., Nie, G., Liu, Z., Sun, J., Liu, G., Wang, Q.-A., Jiao, G., Dang, P., Nie, G., Liu, Z., Sun, J.: Review of wireless RFID strain sensing technology in structural health monitoring. Sensors **23**, 6925 (2023). https://doi.org/10.3390/S23156925
108. Chen, X., Ukkonen, L., Bjorninen, T.: Passive E-Textile UHF RFID-based wireless strain sensors with integrated references. IEEE Sens. J. **16**, 7835–7836 (2016). https://doi.org/10.1109/JSEN.2016.2608659
109. Kennedy, T.F., Fink, P.W., Chu, A.W., Champagne, N.J., Lin, G.Y., Khayat, M.A.: Body-worn E-textile antennas: the good, the low-mass, and the conformal. IEEE Trans. Antennas Propag. **57**, 910–918 (2009). https://doi.org/10.1109/TAP.2009.2014602
110. Liu, N., Lu, Y., Qiu, S., Li, P.: Electromagnetic properties of electro-textiles for wearable antennas applications. Front. Electr. Electron. Eng. China **6**, 563–566 (2011). https://doi.org/10.1007/S11460-011-0182-7/METRICS
111. Ouyang, Y., Chappell, W.J.: High frequency properties of electro-textiles for wearable antenna applications. IEEE Trans. Antennas Propag. **56**, 381–389 (2008). https://doi.org/10.1109/TAP.2007.915435
112. Corchia, L., Monti, G., Tarricone, L.: Wearable antennas: nontextile versus fully textile solutions. IEEE Antennas Propag. Mag. **61**, 71–83 (2019). https://doi.org/10.1109/MAP.2019.2895665
113. Amram Bengio, E., Senic, D., Taylor, L.W., Tsentalovich, D.E., Chen, P., Holloway, C.L., Babakhani, A., Long, C.J., Novotny, D.R., Booth, J.C., Orloff, N.D., Pasquali, M.: High efficiency carbon nanotube thread antennas. Appl Phys Lett. **111** (2017). https://doi.org/10.1063/1.4991822
114. Elwi, T.A., Al-Rizzo, H.M., Rucker, D.G., Dervishi, E., Li, Z., Biris, A.S.: Multi-walled carbon nanotube-based RF antennas. Nanotechnology **21**, 045301 (2009). https://doi.org/10.1088/0957-4484/21/4/045301
115. Puchades, I., Rossi, J.E., Cress, C.D., Naglich, E., Landi, B.J.: Carbon nanotube thin-film antennas. ACS Appl. Mater. Interf. **8**, 20986–20992 (2016). https://doi.org/10.1021/ACSAMI.6B05146
116. Keller, S.D., Zaghloul, A.I., Shanov, V., Schulz, M.J., Mast, D.B., Alvarez, N.T.: Radiation performance of polarization selective carbon nanotube sheet patch antennas. IEEE Trans. Antennas Propag. **62**, 48–55 (2014). https://doi.org/10.1109/TAP.2013.2287272
117. Burke, P.J., Li, S., Yu, Z.: Quantitative theory of nanowire and nanotube antenna performance. IEEE Trans. Nanotechnol. **5**, 314–334 (2006). https://doi.org/10.1109/TNANO.2006.877430
118. Plombon, J.J., O'Brien, K.P., Gstrein, F., Dubin, V.M., Jiao, Y.: High-frequency electrical properties of individual and bundled carbon nanotubes. Appl. Phys. Lett. **90**, 8 (2007). https://doi.org/10.1063/1.2437724/327159
119. Keller, S.D., Zaghloul, A.I., Shanov, V., Schulz, M.J., Mast, D.B.: Design considerations for a meshed carbon nanotube thread patch antenna. IEEE Antennas Wirel. Propag. Lett. **12**, 1192–1195 (2013). https://doi.org/10.1109/LAWP.2013.2282176
120. Jayasinghe, C., Chakrabarti, S., Schulz, M.J., Shanov, V.: Spinning yarn from long carbon nanotube arrays. J. Mater. Res. **26**, 645–651 (2011). https://doi.org/10.1557/JMR.2010.91
121. Owyeung, R., Zeng, W., Sonkusale, S.: Eutectogel electrodes for long-term biosignal monitoring. In: Proceedings of IEEE Sensors, October (2022). https://doi.org/10.1109/SENSORS52175.2022.9967321

122. Jiang, Y., Sadeqi, A., Miller, E.L., Sonkusale, S.: Head motion classification using thread-based sensor and machine learning algorithm. Sci. Reports **11**(1), 1–11 (2021). https://doi.org/10.1038/s41598-021-81284-7
123. Owyeung, R.E., Panzer, M.J., Sonkusale, S.R.: Colorimetric gas sensing washable threads for smart textiles. Sci. Reports **9**(1), 1–8 (2019). https://doi.org/10.1038/s41598-019-42054-8
124. Liu, Y., Wang, H., Zhao, W., Zhang, M., Qin, H., Xie, Y.: Flexible, Stretchable sensors for wearable health monitoring: sensing mechanisms, materials, fabrication strategies and features. Sensors **18**, 645 (2018). https://doi.org/10.3390/S18020645
125. Zhu, G.J., Ren, P.G., Guo, H., Jin, Y.L., Yan, D.X., Li, Z.M.: Highly sensitive and stretchable polyurethane fiber strain sensors with embedded silver nanowires. ACS Appl. Mater. Interf. **11**, 23649–23658 (2019). https://doi.org/10.1021/ACSAMI.9B08611/SUPPL_FILE/AM9B08611_SI_008.MP4
126. He, Z., Wang, K., Zhao, Z., Zhang, T., Li, Y., Wang, L.: A Wearable flexible acceleration sensor for monitoring human motion. Biosensors **12**, 620 (2022). https://doi.org/10.3390/BIOS12080620
127. Alrammouz, R., Podlecki, J., Abboud, P., Sorli, B., Habchi, R.: A review on flexible gas sensors: from materials to devices. Sens Actuators A Phys. **284**, 209–231 (2018). https://doi.org/10.1016/J.SNA.2018.10.036
128. Somappa, L., Malik, S., Aeron, S., Sonkusale, S., Baghini, M.S.: High resolution frequency measurement techniques for relaxation oscillator based capacitive sensors. IEEE Sens. J. **21**, 13394–13404 (2021). https://doi.org/10.1109/JSEN.2021.3068351

Continuous Monitoring of Volatile Organic Compounds in the Ear: The Development of a Headphone-Type Biosensor

Koji Toma, Kenta Iitani, Takahiro Arakawa, and Kohji Mitsubayashi

1 Introduction

Gas sensors are expected to become an essential tool in healthcare for monitoring volatile organic compounds (VOCs) emitted by various sources, including human breath and skin gas, which can be used for non-invasive assessment of metabolism and disease screening. For example, Kant et al. and Boots et al. reported the potential of VOCs in breath for diagnosing and monitoring various diseases [1, 2], including pulmonary diseases, oxidative stress, inflammation, and carcinogenesis. These compounds are important indicators of physiological and pathological states in humans, making their monitoring crucial in the diagnosis and management of diseases.

In recent years, a headphone-type gas sensor has been introduced [3]. This type of sensor offers a non-invasive and convenient method for VOC monitoring, allowing healthcare professionals to obtain real-time information about a patient's health status. In this section, we will discuss the importance of VOC monitoring in healthcare and the types of analytical instruments for VOCs and gas sensors used in healthcare, with a focus on headphone-type gas sensors for blood VOCs monitoring.

K. Toma
Department of Electronic Engineering, Shibaura Institute of Technology, 3-7-5 Toyosu, Koto-Ku, Tokyo 135-8548, Japan
e-mail: k-toma@shibaura-it.ac.jp

K. Iitani · K. Mitsubayashi (✉)
Department of Biomedical Devices and Instrumentation, Institute of Biomaterials and Bioengineering, Tokyo Medical and Dental University (TMDU), 2-3-10 Kanda-Surugadai, Chiyoda-Ku, Tokyo 101-0062, Japan
e-mail: m.bdi@tmd.ac.jp

T. Arakawa
Department of Electric and Electronic Engineering, Tokyo University of Technology, 1404-1 Katakura, Hachioji City, Tokyo 192-0982, Japan

K. Mitsubayashi (ed.), *Wearable Biosensing in Medicine and Healthcare*,
https://doi.org/10.1007/978-981-99-8122-9_14

2 Importance of VOC Monitoring in Healthcare

VOCs are organic chemicals that can vaporize at room temperature and can be found in a variety of substances, including human breath, skin gas, and blood [4–10]. These compounds have been found to be useful biomarkers for several health conditions [11], including respiratory diseases [1], cancer [12–15], diabetes [16–18], and Parkinson's disease [19]. The monitoring of VOCs in healthcare has become increasingly important in recent years, as it can provide time course information about a patient's health status [20, 21].

One of the most significant benefits of VOC monitoring in healthcare is the early detection of diseases. Many diseases, including cancer, can produce VOCs that can be detected in exhaled breath or skin gas [12–15]. By monitoring these compounds, healthcare professionals can detect diseases at an early stage, when they are more easily treatable. Early detection of cancer, for example, can significantly improve a patient's chances of survival.

In addition to early disease detection, VOC monitoring can also be used for disease management [22]. For example, in patients with respiratory diseases such as asthma [23, 24] or chronic obstructive pulmonary disease (COPD) [25, 26], VOC monitoring can help healthcare professionals evaluate disease severity and treatment effectiveness. VOCs produced during inflammation in the lungs can be detected in exhaled breath, providing a non-invasive and convenient method for monitoring disease activity [27, 28].

VOC monitoring can also be useful in diabetes management [29, 30]. Diabetic patients often experience fluctuations in blood glucose levels, which can produce VOCs such as acetone, isopropanol in breath, skin gas, and blood. By monitoring these compounds, healthcare professionals can evaluate diabetes control and adjust treatment plans accordingly. Additionally, VOC monitoring can provide early warning signs of hypoglycemia (low blood sugar), a potentially life-threatening condition that requires immediate intervention.

Another important application of VOC monitoring in healthcare is infectious disease detection [31]. Many infectious diseases produce VOCs that can be detected in breath or urine [32–34]. By monitoring these compounds, healthcare professionals can quickly and accurately diagnose infectious diseases and initiate appropriate treatment.

VOC monitoring can also provide valuable information about a patient's exposure to environmental toxins [35, 36]. Certain VOCs, such as benzene and formaldehyde, are known carcinogens and can be found in indoor air and household products. By monitoring these compounds in breath or blood, healthcare professionals can evaluate a patient's exposure and take steps to reduce their risk of exposure.

Finally, VOC monitoring can also be used for personalized medicine [4]. Each individual's VOC profile is unique, and monitoring VOCs can provide valuable information about an individual's metabolic processes and health status. By analyzing an individual's VOC profile, healthcare professionals can tailor treatment plans to their specific needs, improving treatment outcomes and reducing the risk of adverse effects.

3 VOC Detection Methods

3.1 Analytical Instruments for VOCs in Healthcare

Analytical instruments for VOCs play a critical role in healthcare by providing accurate and reliable measurement of these compounds in various biological samples. There are several types of analytical instruments used for VOC detection in healthcare, including gas chromatography-mass spectrometry (GC–MS), ion mobility spectrometry (IMS), and proton transfer reaction-mass spectrometry (PTR-MS).

GC–MS is a widely used analytical instrument for VOC detection in healthcare. GC separates the VOCs in a sample based on their chemical properties and the mass spectrometer detects and quantifies the separated compounds. GC–MS offers high sensitivity and specificity and can detect a wide range of VOCs in biological samples such as breath, blood, and urine. GC–MS has been used to detect and monitor various VOCs related to diseases such as lung cancer [14, 37], liver disease [38], kidney disease, and diabetes [39].

IMS is another analytical instrument used for VOC detection in healthcare. IMS separates and detects VOCs based on their ion mobility and charge, offering high sensitivity and fast analysis time. IMS has been used to detect various VOCs related to diseases such as lung cancer [40] and respiratory infections [41]. IMS is portable, fast, and sensitive particularly useful in point-of-care settings, where rapid and accurate diagnosis is critical [42].

PTR-MS is a specialized analytical instrument used for VOC detection in healthcare. PTR-MS uses proton transfer reactions to ionize VOCs in a sample, which are then separated and detected by a mass spectrometer. PTR-MS offers high sensitivity and fast analysis time, making it suitable for real-time monitoring of VOCs in breath and other biological samples [43, 44]. PTR-MS has been used to detect and monitor various VOCs related to diseases such as asthma [45] and lung cancer [15].

Although analytical instruments are powerful, they are not suitable for real-time and on-site monitoring of VOCs in breath or skin gas due to their large size. However, small gas sensors that are capable of real-time measurement are expected to be a technology for health management through VOC monitoring in exhaled breath and skin gas.

3.2 Gas Sensors Used in Healthcare

Gas sensors are electronic devices that detect and measure the concentration of gases in the surrounding environment. In healthcare, gas sensors are used for a variety of applications, including the monitoring of VOCs in breath, skin gas, and blood. There are several types of gas sensors commonly used in healthcare, each with their unique characteristics and applications.

Electrochemical sensors work by measuring the electrical current produced when a gas reacts with an electrode. Electrochemical sensors are highly sensitive and can detect low concentrations of gases. They are commonly used for monitoring gases such as carbon monoxide [46], oxygen [47], and VOCs [48].

Optical sensors use light to measure the concentration of gases. These sensors work by passing light through a gas sample and measuring the amount of light absorbed by the gas. Optical sensors are highly sensitive and can detect low concentrations of gases. They are commonly used for monitoring gases such as carbon dioxide and oxygen [49].

Metal oxide sensors work by measuring changes in the electrical conductivity of metal oxide materials when exposed to gases. These sensors are highly sensitive and can detect low concentrations of gases. They are commonly used for monitoring gases such as carbon monoxide [50], nitrogen dioxide [51], and VOCs [52].

Catalytic sensors work by measuring changes in the temperature of a catalytic material when exposed to gases. These sensors are highly sensitive and can detect low concentrations of gases. They are commonly used for monitoring gases such as methane [53] and hydrogen [54].

Photoionization detectors work by ionizing gases using ultraviolet light and measuring the resulting electrical current. These sensors are highly sensitive and can detect low concentrations of gases. They are commonly used for monitoring VOCs [55].

Each type of gas sensor has its unique advantages and disadvantages. Electrochemical sensors are commonly used because they are affordable and highly sensitive. Optical sensors are highly accurate but are more expensive than other types of gas sensors. Metal oxide sensors are sensitive and affordable but can be affected by humidity [56]. Catalytic sensors are highly accurate but can be affected by temperature and humidity [57].

As mentioned above, conventional gas sensors are susceptible to humidity. However, biological gases such as exhaled breath and skin gas contain a lot of moisture, making it difficult to accurately measure VOCs in these gases. On the other hand, biosensors are particularly suitable for measurements in solution systems or high-humidity environments, as the recognition elements used in biosensors are commonly derived from living organisms. Based on this point, we will now discuss biosensors for VOC measurements.

3.3 *Biosensors for Measuring VOCs*

Biosensors are analytical devices that use biological components such as enzymes, antibodies, or microorganisms to detect and measure target analytes. They offer a promising approach for the detection and monitoring of volatile organic compounds (VOCs) in various applications, including healthcare, environmental monitoring, and food safety.

One type of biosensor for measuring VOCs is the enzymatic biosensor. Enzymatic biosensors use enzymes that have been specifically designed to react with VOCs, producing a measurable signal such as a change in electrical conductivity or fluorescence. Enzymatic biosensors are highly specific and can detect low concentrations of VOCs. They have been used to detect various VOCs, such as ethanol [58], methanol [59], and acetone [60]. Enzymatic biosensors have potential to be used in a range of applications, including breath analysis for disease diagnosis [16] and monitoring, environmental monitoring [61], and food safety [62, 63].

Biosensors offer several advantages over traditional gas sensors, including higher specificity, lower detection limits, and the ability to detect a wider range of VOCs. However, biosensors are also more complex and expensive to manufacture than traditional gas sensors, and they may require more specialized training to operate. The use of biosensors for VOC detection in healthcare has gained significant attention in recent years. For example, breath analysis using biosensors could enable non-invasive diagnosis and monitoring of various diseases, including lung cancer [64] and diabetes [16].

The development of new biosensor technologies and the integration of biosensors into portable and wearable devices could enable real-time monitoring of VOCs in skin gas, providing valuable information for the diagnosis and treatment of diseases.

The following is an overview of the recently developed "headphone-type gas monitoring system" developed by combining a biosensor with an over-ear gas collection cell in the form of headphones.

4 Headphone-Type Gas Sensors for Blood VOCs Monitoring

Headphone-type gas sensors offer several advantages over traditional blood VOC monitoring methods. First, they offer a non-invasive and convenient method for VOC monitoring. Blood sampling for VOC analysis usually requires invasive procedures such as venipuncture, which can be uncomfortable and time-consuming for patients. In contrast, headphone-type sensors offer a non-invasive method for monitoring VOCs, which can reduce patient discomfort and improve compliance with monitoring protocols.

Second, headphone-type sensors offer real-time monitoring of VOCs, allowing healthcare professionals to obtain immediate feedback about a patient's health status.

Real-time monitoring is particularly important in critical care settings, where timely interventions can mean the difference between life and death.

Several studies have reported the use of headphone-type gas sensors for blood VOC monitoring in healthcare. For example, a study by Toma et al. reported the development of a headphone-type gas collection cell connected to a biosensor for monitoring ethanol levels from the ears [3]. The study demonstrated that the external ears are suitable parts for monitoring VOCs stably.

Another study by Arakawa et al. reported the development of a "headset bio-sniffer" that incorporated a biosensor to measure ethanol emitted from the ear into the headphones [65]. The study demonstrated that the sensor was capable of detecting ethanol at concentrations as low as 11 ppb, making it possible to measure ethanol from the ear after drinking alcohol.

4.1 Working Mechanism of Headphone-Type Gas Sensor

Toma et al. presented a headphone-type gas monitoring system that includes two primary components: the ethanol bio-sniffer and an over-ear gas collection cell [3] (Fig. 1a). The bio-sniffer is a gas-phase biosensor that employs biofluorometry with enzymes [66] and is comprised of excitation and detection units connected to a bifurcated optical fiber. The excitation unit features a UV light-emitting diode and a bandpass filter that sets the excitation light's peak wavelength at 340 nm, while the detection unit has a photomultiplier tube and another bandpass filter to extract fluorescence light from the reduced form of β-nicotinamide adenine dinucleotide (NADH). The flow-cell, composed of polymethyl methacrylate, is attached to an optical fiber probe end, which is then connected to the bifurcated optical fiber. An alcohol dehydrogenase (ADH) membrane is attached to the end of the flow-cell by an o-ring to work as a gas–liquid diaphragm. When ethanol vapor reaches it, the catalyzed redox reaction occurs with NAD+ in the buffer solution above the membrane, producing NADH. The NADH is excited by the UV light from the optical fiber probe, and the emitted fluorescence is collected by the same fiber probe for detection (Fig. 1b).

The over-ear gas collection cell is created by modifying a commercial earmuff with cups made of acrylonitrile-styrene-acrylate resin and urethane cushions covered by polyvinyl chloride sheets. The earmuff cups have two connectors, one for a filtered air (carrier gas) inlet and the other for a sample gas outlet containing ear gas.

To monitor external ear-derived gas, the over-ear gas collection cell collects the sample gas, which is then transported to the bio-sniffer for real-time measurement. This system provides a non-invasive and real-time approach to measure ethanol vapor in the external ear.

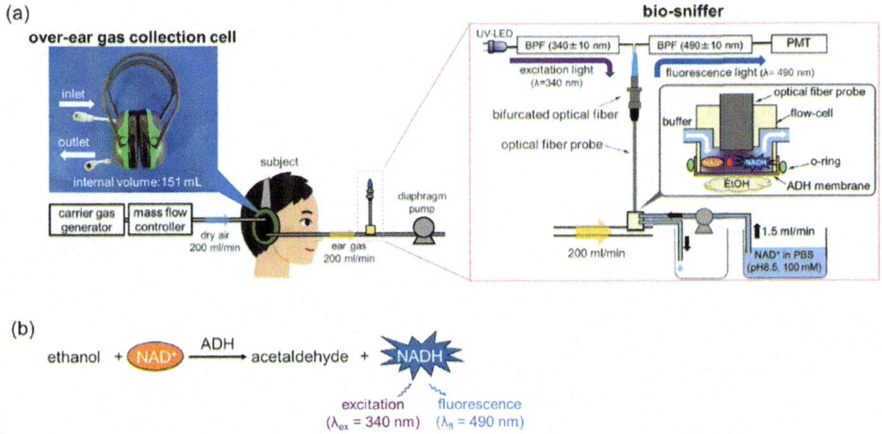

Fig. 1 **a** Headphone-type gas monitoring system that composed of the ethanol bio-sniffer and an over-ear gas collection cell for monitoring ethanol from the ear. **b** Detection principle of the ethanol bio-sniffer [3]

4.2 Characteristics of Headphone-Type Gas Sensors

Each component of the headphone-type gas sensor was characterized. The ethanol bio-sniffer was tested with various buffer solutions to study how buffer solution pH affects the activity of ADH. They measured sensor output as an increase in fluorescence intensity from the baseline and discovered that it reached a peak with pH 8.5 PB solution, which they used for further experiments. They also examined the ethanol bio-sniffer's sensitivity and reproducibility for different ethanol concentrations. The sensor output varied according to the concentration of gaseous ethanol, and the dynamic range of the ethanol bio-sniffer was determined to be 26 ppb–554 ppm. The researchers evaluated the selectivity of the ethanol bio-sniffer to gaseous ethanol by testing it with representative VOCs in breath, and the results demonstrated the bio-sniffer's high selectivity with minimal interference from other compounds in the gas mixture. The stability and repeatability of the sensor signal were also tested, and the findings suggested that the bio-sniffer could be utilized for continuous measurement of gaseous ethanol.

An Investigation was conducted on the over-ear gas collection cell to detect the presence of interfering VOCs, and the findings revealed that no such compounds were present in the gas collected from the cell (Fig. 2).

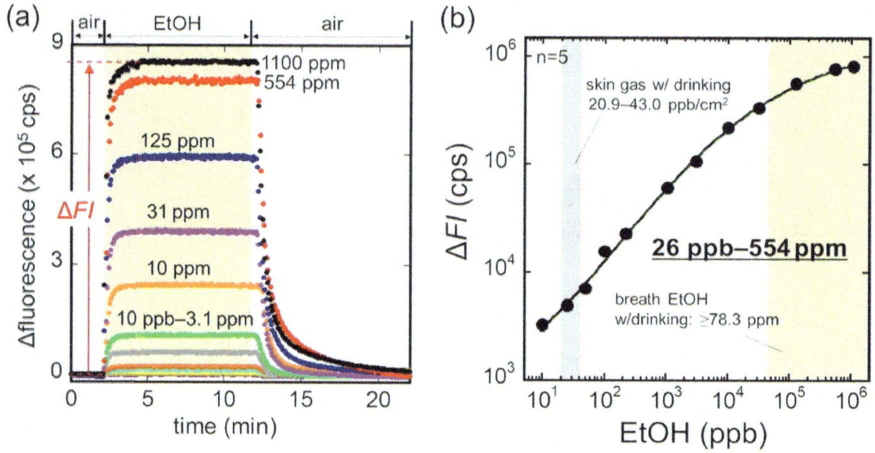

Fig. 2 **a** Sensor responses to standard gaseous ethanol at various concentrations. **b** Calibration curve of the ethanol bio-sniffer [3]

4.3 Applications of Headphone-Type Gas Sensors to Breath Ethanol

To explore the feasibility of using a headphone-type gas sensor for continuous measurement of biological gas, monitoring the release of gaseous ethanol from the ear canal, which could potentially serve as a marker of alcohol metabolism, was conducted. The study involved three healthy male subjects who had refrained from consuming alcohol or drugs for 72 h prior to the experiment. The study was approved by the ethics committee of the Faculty of Medicine at Tokyo Medical and Dental University and was conducted in accordance with guidelines and regulations after obtaining informed consent from all subjects. The sensor was attached to the external ear of each subject, and a compressor was used to deliver a carrier gas into a collection cell at a flow rate of 200 mL/min. The gas was then pumped out of the cell using a diaphragm pump located behind the bio-sniffer to maintain a constant internal pressure. The ear to be measured was chosen randomly. Initially, the baseline ethanol concentration in the external ear gas was measured for 10 min while the subjects refrained from consuming alcohol. Next, the subjects drank alcohol at a concentration of 0.4 g per kg body weight within 5 min, and the measurement continued for an additional 140 min. Ethanol concentrations in breath were also intermittently measured using detector tubes and the ethanol bio-sniffer for comparison. Breath samples were collected using the end-expiratory sampling method.

The study demonstrated that the headphone-type gas monitoring system successfully measured the change in ethanol concentration in the external ear gas, which was similar to the temporal change in breath ethanol concentration. The ethanol concentration in the external ear increased 7 min after alcohol intake, peaked at 183 ppb after 52 min, and then gradually decreased. The noise level in the signal was lower

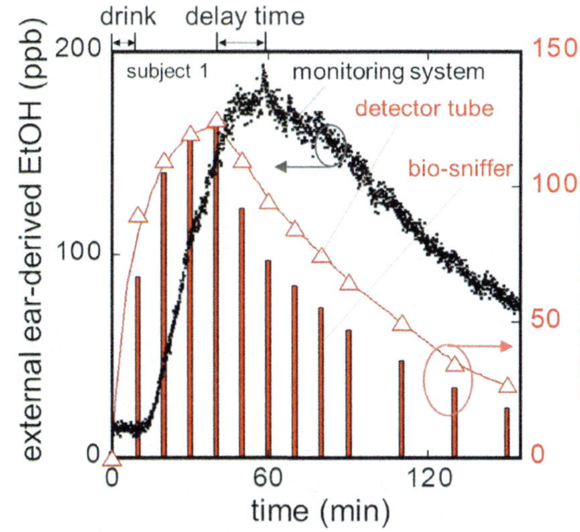

Fig. 3 Time course of ethanol concentrations from the ear (black points) measured by the headphone-type monitoring system and that in exhaled breath (red bars and triangles) measured by detector tubes and the ethanol bio-sniffer, respectively [3]

compared to that from a palm, with an average coefficient of variation (C.V.) of 6.9% over 90 min, indicating a reduction in noise by a third. The low noise level may be attributed to the low density of eccrine sweat glands in the external ear canal compared to the palm. This study's results suggest that the concentration of ethanol in the external ear is likely correlated with the concentration of ethanol in the blood (Fig. 3).

This study also measured sweat rates of four different parts of the body (palm, wrist, external ear canal, and earlobe) with a ventilated capsule-type perspiration meter. In the study, it was found that the sweat rate of the palm was in agreement with previous reports. However, the sweat rate of the external ear canal was relatively higher than those of the wrist and earlobe. This may be due to the presence of apocrine sweat glands in the outer one-third of the external ear canal, which produce a different type of sweat compared to eccrine sweat glands in the other areas of the body. Additionally, the connection of the external ear canal to the deeper part of the ear allows for sweat and moisture in the deeper region to be exhaled through the external ear canal (Fig. 4).

It is noteworthy that the sweat rate of the palm largely fluctuates, as indicated by the coefficient of variation (C.V.) of 14.9%. In contrast, the sweat rate of the external ear canal was stable, with a C.V. of 5.0%. The stable sweat rate in the external ear canal may explain the stable ethanol concentration observed in the study, as previous studies have shown that the stability of ethanol concentrations is influenced by the sweat rates. The stable ethanol concentration in the external ear canal may be due to the stability of sweating in this area.

Furthermore, to reduce the noise in the ethanol concentration derived from the external ear or reduce it in the palm-derived, the shapes of collection cells may be considered in future studies. It is possible that the shapes of collection cells may affect the collection of sweat and, consequently, the measurement of ethanol concentration.

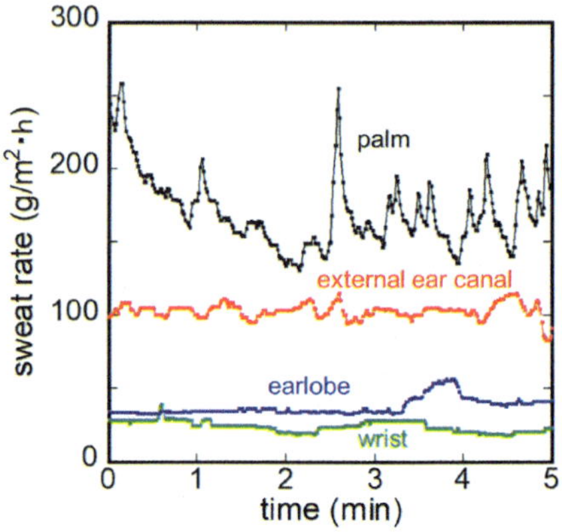

Fig. 4 Time course of sweat rate from various body parts without drinking alcohol [3]

5 Challenges, Future Directions

Healthcare applications of the headphone-type gas monitoring system have the potential to revolutionize non-invasive diagnosis and monitoring of various medical conditions. However, there are several challenges that must be overcome before this technology can be widely used in healthcare settings.

One of the major challenges is the need for more extensive clinical validation studies to establish the accuracy and reliability of the system in detecting and quantifying specific VOCs in the ear. The current research on the system's ability to detect ethanol vapor shows promise, but more research is needed to explore its potential applications in monitoring other VOCs associated with various medical conditions.

Another challenge is the need for miniaturization of the system to make it more practical for use in clinical settings. The current system is bulky and requires a trained technician to operate it, limiting its potential for widespread use in medical facilities. To make the system more accessible, efforts should be made to develop a compact and user-friendly version that can be used by healthcare professionals without extensive training.

Furthermore, there is a need for the development of robust algorithms for data analysis to enable the interpretation of complex data collected by the system. This will require collaborations between experts in signal processing, machine learning, and healthcare to develop reliable algorithms that can identify specific patterns in the data and interpret them accurately.

In terms of future directions, the headphone-type gas monitoring system has the potential to be used in various healthcare applications, including the early detection and monitoring of chronic diseases such as diabetes, cancer, and respiratory diseases. For example, VOCs such as acetone, ammonia, and nitric oxide have been found to be biomarkers of diabetes and lung cancer, and the system could be used to detect these VOCs in the ear and provide early warning signs of these conditions.

Another potential application of the system is in the monitoring of anesthesia during surgeries. Currently, the level of anesthesia is monitored using blood tests or through observation of physical signs such as heart rate and blood pressure. However, the headphone-type gas monitoring system could provide a non-invasive and real-time method of monitoring the level of anesthesia by detecting the concentration of specific anesthetic gases in the ear.

Moreover, the system could also be used in the monitoring of infectious diseases such as tuberculosis, which is known to produce specific VOCs in the breath of infected individuals. By detecting these VOCs in the ear, the system could provide a non-invasive and real-time method of screening for the disease, particularly in low-resource settings where traditional diagnostic methods may not be available.

6 Conclusion

In conclusion, the headphone-type gas monitoring system has the potential to transform non-invasive diagnosis and monitoring of various medical conditions. However, to realize its potential, there is a need for extensive clinical validation studies, miniaturization of the system, and the development of robust algorithms for data analysis. The future directions of the system include its potential use in the early detection and monitoring of chronic diseases, monitoring of anesthesia during surgeries, and screening for infectious diseases. By addressing these challenges and exploring these potential applications, the headphone-type gas monitoring system could pave the way for a new era of non-invasive medical diagnostics and monitoring.

Acknowledgements This work was partially supported by the Japan Society for the Promotion of Science (JSPS) through Kakenhi (Grant-in-Aid for Scientific Research) under grant numbers JP21H04888, JP21H01335, JP23H03864, and the Cooperative Re-search Project of the Research Center for Biomedical Engineering.

References

1. van de Kant, K.D., van der Sande, L.J., Jöbsis, Q., van Schayck, O.C., Dompeling, E.: Clinical use of exhaled volatile organic compounds in pulmonary diseases: a systematic review. Respir. Res. **13**, 117 (2012). https://doi.org/10.1186/1465-9921-13-117
2. Boots, A.W., van Berkel, J.J.B.N., Dallinga, J.W., Smolinska, A., Wouters, E.F., van Schooten, F.J.: The versatile use of exhaled volatile organic compounds in human health and disease. J. Breath Res. **6**, 027108 (2012). https://doi.org/10.1088/1752-7155/6/2/027108
3. Toma, K., Suzuki, S., Arakawa, T., Iwasaki, Y., Mitsubayashi, K.: External ears for non-invasive and stable monitoring of volatile organic compounds in human blood. Sci. Rep. **11**, 10415 (2021). https://doi.org/10.1038/s41598-021-90146-1
4. De Lacy Costello, B., Amann, A., Al-Kateb, H., Flynn, C., Filipiak, W., Khalid, T., Osborne, D., Ratcliffe, N.M.: A review of the volatiles from the healthy human body. J. Breath Res. **8** (2014). https://doi.org/10.1088/1752-7155/8/1/014001
5. Agapiou, A., Amann, A., Mochalski, P., Statheropoulos, M., Thomas, C.L.P.: Trace detection of endogenous human volatile organic compounds for search, rescue and emergency applications. TrAC – Trends Anal. Chem. **66**, 158–175 (2015). https://doi.org/10.1016/j.trac.2014.11.018

6. Mochalski, P., King, J., Klieber, M., Unterkofler, K.: Europe PMC Funders Group Blood and breath levels of selected volatile organic compounds in healthy volunteers. **138**, 2134–2145 (2016). https://doi.org/10.1039/c3an36756h.Blood
7. Amann, A., Costello, B.D.L., Miekisch, W., Schubert, J., Buszewski, B., Pleil, J., Ratcliffe, N., Risby, T.: The human volatilome: volatile organic compounds (VOCs) in exhaled breath, skin emanations, urine, feces and saliva. J. Breath Res. **8**, (2014). https://doi.org/10.1088/1752-7155/8/3/034001
8. Grabowska-Polanowska, B., Miarka, P., Skowron, M., Sułowicz, J., Wojtyna, K., Moskal, K., Śliwka, I.: Development of sampling method and chromatographic analysis of volatile organic compounds emitted from human skin. Bioanalysis **9**, 1465–1475 (2017). https://doi.org/10.4155/bio-2017-0128
9. Iitani, K., Toma, K., Arakawa, T., Mitsubayashi, K.: Transcutaneous blood VOC imaging system (skin-gas cam) with real-time bio-fluorometric device on rounded skin surface. ACS Sensors **5**, 338–345 (2020). https://doi.org/10.1021/acssensors.9b01658
10. Arakawa, T., Aota, T., Iitani, K., Toma, K., Iwasaki, Y., Mitsubayashi, K.: Skin ethanol gas measurement system with a biochemical gas sensor and gas concentrator toward monitoring of blood volatile compounds. Talanta **219**, 121187 (2020). https://doi.org/10.1016/j.talanta.2020.121187
11. Nakhleh, M.K., Amal, H., Jeries, R., Broza, Y.Y., Aboud, M., Gharra, A., Ivgi, H., Khatib, S., Badarneh, S., Har-Shai, L., Glass-Marmor, L., Lejbkowicz, I., Miller, A., Badarny, S., Winer, R., Finberg, J., Cohen-Kaminsky, S., Perros, F., Montani, D., Girerd, B., Garcia, G., Simonneau, G., Nakhoul, F., Baram, S., Salim, R., Hakim, M., Gruber, M., Ronen, O., Marshak, T., Doweck, I., Nativ, O., Bahouth, Z., Shi, D., Zhang, W., Hua, Q., Pan, Y., Tao, L., Liu, H., Karban, A., Koifman, E., Rainis, T., Skapars, R., Sivins, A., Ancans, G., Liepniece-Karele, I., Kikuste, I., Lasina, I., Tolmanis, I., Johnson, D., Millstone, S.Z., Fulton, J., Wells, J.W., Wilf, L.H., Humbert, M., Leja, M., Peled, N., Haick, H.: Diagnosis and classification of 17 diseases from 1404 subjects via pattern analysis of exhaled molecules. ACS Nano **11**, 112–125 (2017). https://doi.org/10.1021/acsnano.6b04930
12. Phillips, M., Cataneo, R.N., Ditkoff, B.A., Fisher, P., Greenberg, J., Gunawardena, R., Kwon, C.S., Tietje, O., Wong, C.: Prediction of breast cancer using volatile biomarkers in the breath. Breast Cancer Res. Treat. **99**, 19–21 (2006). https://doi.org/10.1007/s10549-006-9176-1
13. Bajtarevic, A., Ager, C., Pienz, M., Klieber, M., Schwarz, K., Ligor, M., Ligor, T., Filipiak, W., Denz, H., Fiegl, M., Hilbe, W., Weiss, W., Lukas, P., Jamnig, H., Hackl, M., Haidenberger, A., Buszewski, B., Miekisch, W., Schubert, J., Amann, A.: Noninvasive detection of lung cancer by analysis of exhaled breath. BMC Cancer **9**, 348 (2009). https://doi.org/10.1186/1471-2407-9-348
14. Buszewski, B., Ligor, T., Jezierski, T., Wenda-Piesik, A., Walczak, M., Rudnicka, J.: Identification of volatile lung cancer markers by gas chromatography-mass spectrometry: Comparison with discrimination by canines. Anal. Bioanal. Chem. **404**, 141–146 (2012). https://doi.org/10.1007/s00216-012-6102-8
15. Wehinger, A., Schmid, A., Mechtcheriakov, S., Ledochowski, M., Grabmer, C., Gastl, G.A., Amann, A.: Lung cancer detection by proton transfer reaction mass-spectrometric analysis of human breath gas. Int. J. Mass Spectrom. **265**, 49–59 (2007). https://doi.org/10.1016/j.ijms.2007.05.012
16. Chien, P.-J.J., Suzuki, T., Tsujii, M., Ye, M., Minami, I., Toda, K., Otsuka, H., Toma, K., Arakawa, T., Araki, K., Iwasaki, Y., Shinada, K., Ogawa, Y., Mitsubayashi, K.: Biochemical gas sensors (biosniffers) using forward and reverse reactions of secondary alcohol dehydrogenase for breath isopropanol and acetone as potential volatile biomarkers of diabetes mellitus. Anal. Chem. **89**, 12261–12268 (2017). https://doi.org/10.1021/acs.analchem.7b03191
17. Righettoni, M., Tricoli, A.: Toward portable breath acetone analysis for diabetes detection. J. Breath Res. **5** (2011). https://doi.org/10.1088/1752-7155/5/3/037109
18. Galassetti, P.R., Novak, B., Nemet, D., Rose-Gottron, C., Cooper, D.M., Meinardi, S., Newcomb, R., Zaldivar, F., Blake, D.R.: Breath ethanol and acetone as indicators of serum

glucose levels: an initial report. Diabetes Technol. Ther. **7**, 115–123 (2005). https://doi.org/10.1089/dia.2005.7.115

19. Trivedi, D.K., Sinclair, E., Xu, Y., Sarkar, D., Walton-Doyle, C., Liscio, C., Banks, P., Milne, J., Silverdale, M., Kunath, T., Goodacre, R., Barran, P.: Discovery of volatile biomarkers of Parkinson's disease from sebum. ACS Cent. Sci. **5**, 599–606 (2019). https://doi.org/10.1021/acscentsci.8b00879
20. Trefz, P., Schmidt, M., Oertel, P., Obermeier, J., Brock, B., Kamysek, S., Dunkl, J., Zimmermann, R., Schubert, J.K., Miekisch, W.: Continuous real time breath gas monitoring in the clinical environment by proton-transfer-reaction-time-of-flight-mass spectrometry. Anal. Chem. **85**, 10321–10329 (2013). https://doi.org/10.1021/ac402298v
21. Huang, Z., Zhang, Y., Yan, Q., Zhang, Z., Wang, X.: Real-time monitoring of respiratory absorption factors of volatile organic compounds in ambient air by proton transfer reaction time-of-flight mass spectrometry. J. Hazard. Mater. **320**, 547–555 (2016). https://doi.org/10.1016/j.jhazmat.2016.08.064
22. Williams, H.R.T., Orchard, T.R.: Editorial: volatile organic compounds in breath for monitoring IBD-longitudinal studies are essential. Aliment. Pharmacol. Ther. **46**, 371–372 (2017). https://doi.org/10.1111/apt.14135
23. Cavaleiro Rufo, J., Madureira, J., Oliveira Fernandes, E., Moreira, A.: Volatile organic compounds in asthma diagnosis: a systematic review and meta-analysis. Allergy **71**, 175–188 (2016). https://doi.org/10.1111/all.12793
24. Ibrahim, B., Basanta, M., Cadden, P., Singh, D., Douce, D., Woodcock, A., Fowler, S.J.: Non-invasive phenotyping using exhaled volatile organic compounds in asthma. Thorax **66**, 804–809 (2011). https://doi.org/10.1136/thx.2010.156695
25. Bessa, V., Darwiche, K., Teschler, H., Sommerwerck, U., Rabis, T., Baumbach, J.I., Freitag, L.: Detection of volatile organic compounds (VOCs) in exhaled breath of patients with chronic obstructive pulmonary disease (COPD) by ion mobility spectrometry. Int. J. Ion Mobil. Spectrom. **14**, 7–13 (2011). https://doi.org/10.1007/s12127-011-0060-2
26. Gaida, A., Holz, O., Nell, C., Schuchardt, S., Lavae-Mokhtari, B., Kruse, L., Boas, U., Langejuergen, J., Allers, M., Zimmermann, S., Vogelmeier, C., Koczulla, A.R., Hohlfeld, J.M.: A dual center study to compare breath volatile organic compounds from smokers and non-smokers with and without COPD. J. Breath Res. **10**, 026006 (2016). https://doi.org/10.1088/1752-7155/10/2/026006
27. Yang, H.-Y., Shie, R.-H., Chang, C.-J., Chen, P.-C.: Development of breath test for pneumoconiosis: a case-control study. Respir. Res. **18**, 178 (2017). https://doi.org/10.1186/s12931-017-0661-3
28. Buszewski, B., Grzywinski, D., Ligor, T., Stacewicz, T., Bielecki, Z., Wojtas, J.: Detection of volatile organic compounds as biomarkers in breath analysis by different analytical techniques. Bioanalysis **5**, 2287–2306 (2013). https://doi.org/10.4155/bio.13.183
29. Dalton, P., Gelperin, A., Preti, G.: Volatile metabolic monitoring of glycemic status in diabetes using electronic olfaction. Diabetes Technol. Ther. **6**, 534–544 (2004). https://doi.org/10.1089/1520915041705992
30. Smith, D., Španěl, P., Fryer, A.A., Hanna, F., Ferns, G.A.A.: Can volatile compounds in exhaled breath be used to monitor control in diabetes mellitus? J. Breath Res. **5**, 022001 (2011). https://doi.org/10.1088/1752-7155/5/2/022001
31. Sethi, S., Nanda, R., Chakraborty, T.: Clinical application of volatile organic compound analysis for detecting infectious diseases. Clin. Microbiol. Rev. **26**, 462–475 (2013). https://doi.org/10.1128/CMR.00020-13
32. Banday, K.M., Pasikanti, K.K., Chan, E.C.Y., Singla, R., Rao, K.V.S., Chauhan, V.S., Nanda, R.K.: Use of urine volatile organic compounds to discriminate tuberculosis patients from healthy subjects. Anal. Chem. **83**, 5526–5534 (2011). https://doi.org/10.1021/ac200265g
33. Nakhleh, M.K., Jeries, R., Gharra, A., Binder, A., Broza, Y.Y., Pascoe, M., Dheda, K., Haick, H.: Detecting active pulmonary tuberculosis with a breath test using nanomaterial-based sensors. Eur. Respir. J. **43**, 1522–1525 (2014). https://doi.org/10.1183/09031936.00019114

34. Beccaria, M., Bobak, C., Maitshotlo, B., Mellors, T.R., Purcaro, G., Franchina, F.A., Rees, C.A., Nasir, M., Black, A., Hill, J.E.: Exhaled human breath analysis in active pulmonary tuberculosis diagnostics by comprehensive gas chromatography-mass spectrometry and chemometric techniques. J. Breath Res. **13**, 016005 (2018). https://doi.org/10.1088/1752-7163/aae80e
35. Cao, W., Duan, Y.: Breath analysis: potential for clinical diagnosis and exposure assessment. Clin. Chem. **52**, 800–811 (2006). https://doi.org/10.1373/clinchem.2005.063545
36. Ahmed, W.M., Lawal, O., Nijsen, T.M., Goodacre, R., Fowler, S.J.: Exhaled volatile organic compounds of infection: a systematic review. ACS Infect. Dis. **3**, 695–710 (2017). https://doi.org/10.1021/acsinfecdis.7b00088
37. Rudnicka, J., Walczak, M., Kowalkowski, T., Jezierski, T., Buszewski, B.: Determination of volatile organic compounds as potential markers of lung cancer by gas chromatography-mass spectrometry versus trained dogs. Sens. Actuators B Chem. **202**, 615–621 (2014). https://doi.org/10.1016/j.snb.2014.06.006
38. Van den Velde, S., Nevens, F., Van hee, P., van Steenberghe, D., Quirynen, M.: GC-MS analysis of breath odor compounds in liver patients. J. Chromatogr. B Anal. Technol. Biomed. Life Sci. **875**, 344–348 (2008). https://doi.org/10.1016/j.jchromb.2008.08.031
39. Saidi, T., Zaim, O., Moufid, M., El Bari, N., Ionescu, R., Bouchikhi, B.: Exhaled breath analysis using electronic nose and gas chromatography–mass spectrometry for non-invasive diagnosis of chronic kidney disease, diabetes mellitus and healthy subjects. Sensors Actuators B Chem. **257**, 178–188 (2018). https://doi.org/10.1016/j.snb.2017.10.178
40. Handa, H., Usuba, A., Maddula, S., Baumbach, J.I., Mineshita, M., Miyazawa, T.: Exhaled breath analysis for lung cancer detection using ion mobility spectrometry. PLoS ONE **9**, e114555 (2014). https://doi.org/10.1371/journal.pone.0114555
41. Fink, T., Baumbach, J.I., Kreuer, S.: Ion mobility spectrometry in breath research. J. Breath Res. **8**, 027104 (2014). https://doi.org/10.1088/1752-7155/8/2/027104
42. Zhang, J.D., Baker, M.J., Liu, Z., Kabir, K.M.M., Kolachalama, V.B., Yates, D.H., Donald, W.A.: Medical diagnosis at the point-of-care by portable high-field asymmetric waveform ion mobility spectrometry: a systematic review and meta-analysis. J. Breath Res. **15**, 046002 (2021). https://doi.org/10.1088/1752-7163/ac135e
43. Pugliese, G., Trefz, P., Brock, B., Schubert, J.K., Miekisch, W.: Extending PTR based breath analysis to real-time monitoring of reactive volatile organic compounds. Analyst **144**, 7359–7367 (2019). https://doi.org/10.1039/C9AN01478K
44. King, J., Mochalski, P., Kupferthaler, A., Unterkofler, K., Koc, H., Filipiak, W., Teschl, S., Hinterhuber, H., Amann, A.: Dynamic profiles of volatile organic compounds in exhaled breath as determined by a coupled PTR-MS/GC-MS study. Physiol. Meas. **31**, 1169–1184 (2010). https://doi.org/10.1088/0967-3334/31/9/008
45. Romano, A., Doran, S., Belluomo, I., Hanna, G.B.: High-throughput breath volatile organic compound analysis using thermal desorption proton transfer reaction time-of-flight mass spectrometry. Anal. Chem. **90**, 10204–10210 (2018). https://doi.org/10.1021/acs.analchem.8b01045
46. Obermeier, J., Trefz, P., Wex, K., Sabel, B., Schubert, J.K., Miekisch, W.: Electrochemical sensor system for breath analysis of aldehydes CO and NO. J. Breath Res. **9**, 016008 (2015). https://doi.org/10.1088/1752-7155/9/1/016008
47. Xia, J., Sonkusale, S.: Flexible thread-based electrochemical sensors for oxygen monitoring. Analyst **146**, 2983–2990 (2021). https://doi.org/10.1039/D0AN02400G
48. Gębicki, J., Kloskowski, A., Chrzanowski, W.: Prototype of electrochemical sensor for measurements of volatile organic compounds in gases. Sensors Actuators B Chem. **177**, 1173–1179 (2013). https://doi.org/10.1016/j.snb.2012.12.025
49. Krabbe, S., Achatz, D.E., Nieradzik, T., Gerhardy, C., Schomburg, W.K.: Ultrasonic welding of chemical optical sensors supporting O_2, pH and CO_2 imaging in microfluidic systems. Proc. Eng. **120**, 598–601 (2015). https://doi.org/10.1016/j.proeng.2015.08.736
50. Anukunprasert, T., Saiwan, C., Traversa, E.: The development of gas sensor for carbon monoxide monitoring using nanostructure of Nb–TiO 2. Sci. Technol. Adv. Mater. **6**, 359–363 (2005). https://doi.org/10.1016/j.stam.2005.02.020

51. Park, S., Kim, S., Sun, G.-J., In Lee, W., Kim, K.K., Lee, C.: Fabrication and NO_2 gas sensing performance of TeO_2-core/CuO-shell heterostructure nanorod sensors. Nanoscale Res. Lett. **9**, 638 (2014). https://doi.org/10.1186/1556-276X-9-638
52. Ahmadipour, M., Pang, A.L., Ardani, M.R., Pung, S.-Y., Ooi, P.C., Hamzah, A.A., Mohd Razip Wee, M.F., Aniq Shazni Mohammad Haniff, M., Dee, C.F., Mahmoudi, E., Arsad, A., Ahmad, M.Z., Pal, U., Chahrour, K.M., Haddadi, S.A.: Detection of breath acetone by semiconductor metal oxide nanostructures-based gas sensors: a review. Mater. Sci. Semicond. Process. **149**, 106897 (2022). https://doi.org/10.1016/j.mssp.2022.106897
53. Karpov, E.E., Karpov, E.F., Suchkov, A., Mironov, S., Baranov, A., Sleptsov, V., Calliari, L.: Energy efficient planar catalytic sensor for methane measurement. Sensors Actuators A Phys. **194**, 176–180 (2013). https://doi.org/10.1016/j.sna.2013.01.057
54. Brauns, E., Morsbach, E., Kunz, S., Bäumer, M., Lang, W.: A fast and sensitive catalytic gas sensors for hydrogen detection based on stabilized nanoparticles as catalytic layer. Sens. Actuators B Chem. **193**, 895–903 (2014). https://doi.org/10.1016/j.snb.2013.11.048
55. Prestage, J., Day, C., Husheer, S.L.G., Winter, W.T., Ho, W.O., Saffell, J.R., Hutter, T.: Selective detection of volatile organics in a mixture using a photoionization detector and thermal desorption from a nanoporous preconcentrator. ACS Sensors **7**, 304–311 (2022). https://doi.org/10.1021/acssensors.1c02344
56. Nabil Abdullah, A., Kamarudin, K., Muhammad Mamduh, S., Hamid Adom, A., Hadi Mohd Juffry, Z.: Effect of environmental temperature and humidity on different metal oxide gas sensors at various gas concentration levels. IOP Conf. Ser. Mater. Sci. Eng. **864**, 012152 (2020). https://doi.org/10.1088/1757-899X/864/1/012152
57. Jones, M.G., Nevell, T.G.: The detection of hydrogen using catalytic flammable gas sensors. Sens. Actuators. **16**, 215–224 (1989). https://doi.org/10.1016/0250-6874(89)87003-9
58. Mitsubayashi, K., Matsunaga, H., Nishio, G., Toda, S., Nakanishi, Y., Saito, H., Ogawa, M., Otsuka, K.: Bio-sniffer sticks for breath analysis after drinking. Sens. Actuators B Chem. **108**, 660–664 (2005). https://doi.org/10.1016/j.snb.2004.11.093
59. Toma, K., Iwasaki, K., Zhang, G., Iitani, K., Arakawa, T., Iwasaki, Y., Mitsubayashi, K.: Biochemical Methanol Gas Sensor (MeOH Bio-Sniffer) for Non-Invasive Assessment of Intestinal Flora from Breath Methanol. Sensors. **21**, 4897 (2021). https://doi.org/10.3390/s21144897
60. Ye, M., Chien, P.-J., Toma, K., Arakawa, T., Mitsubayashi, K.: An acetone bio-sniffer (gas phase biosensor) enabling assessment of lipid metabolism from exhaled breath. Biosens. Bioelectron. **73**, 208–213 (2015). https://doi.org/10.1016/j.bios.2015.04.023
61. Vianello, F., Stefani, A., Paolo, M.D., Rigo, A.: Potentiometric detection of formaldehyde in air by an aldehyde dehydrogenase FET. Sens. Actuators B Chem. **37**, 49–54 (1996). https://doi.org/10.1016/S0925-4005(97)80071-8
62. Amine, A., Mohammadi, H., Bourais, I., Palleschi, G.: Enzyme inhibition-based biosensors for food safety and environmental monitoring. Biosens. Bioelectron. **21**, 1405–1423 (2006). https://doi.org/10.1016/j.bios.2005.07.012
63. Prodromidis, M.I., Karayannis, M.I.: Enzyme based amperometric biosensors for food analysis. Electroanalysis **14**, 241–261 (2002). https://doi.org/10.1002/1521-4109(200202)14:4%3c241::AID-ELAN241%3e3.0.CO;2-P
64. Khanmohammadi, A., Aghaie, A., Vahedi, E., Qazvini, A., Ghanei, M., Afkhami, A., Hajian, A., Bagheri, H.: Electrochemical biosensors for the detection of lung cancer biomarkers: a review. Talanta **206**, 120251 (2020). https://doi.org/10.1016/j.talanta.2019.120251
65. Arakawa, T., Ishikawa, R., Iitani, K., Toma, K., Iwasaki, Y., Mitsubayashi, K.: Headset biosniffer with wireless CMOS camera for percutaneous ethanol vapor from the ear canal. Biosens. Bioelectron. X. **11**, 100169 (2022). https://doi.org/10.1016/j.biosx.2022.100169
66. Kudo, H., Suzuki, Y., Gessei, T., Takahashi, D., Arakawa, T., Mitsubayashi, K.: Biochemical gas sensor (bio-sniffer) for ultrahigh-sensitive gaseous formaldehyde monitoring. Biosens. Bioelectron. **26**, 854–858 (2010). https://doi.org/10.1016/j.bios.2010.07.099

Supporting Technologies for Wearable Sensing

Design and Fabrication of Wearable Biosensors: Materials, Methods, and Prospects

Rajendra Kumar Reddy Gajjala, Sara Muñana-González, Pello Núñez-Marinero, Joseba Totoricaguena-Gorriño, Leire Ruiz-Rubio, and Francisco Javier del Campo

Objects and their manufacture are inseparable. You understand a product if you understand how it´s made.
—Jonathan Ive

1 Introduction and Background

Wearable biosensors are analytical devices intended to provide continuous and personalized information on biochemical parameters in real time. Wearable biosensors originated as tools for better health management, particularly for the monitoring of glucose in diabetes management, but have reached other areas such as sports and fitness, military applications, and environmental monitoring.

The development of wearable biosensors is a highly complex task that involves multidisciplinary teams from areas such as materials science, electronics, biomedical engineering, and information technologies. This is why, despite the recent boom and the increasing reports on wearable biosensors, very few devices (see Table 1) have made it to market.

R. K. Reddy Gajjala · P. Núñez-Marinero · J. Totoricaguena-Gorriño · L. Ruiz-Rubio · F. Javier del Campo (✉)
BCMaterials, Basque Center for Materials, Applications and Nanostructures. UPV/EHU Science Park, 48940 Leioa, Vizcaya, Spain
e-mail: Javier.delcampo@bcmaterials.net

S. Muñana-González · L. Ruiz-Rubio
Grupo de Química Macromolecular, Departamento Química Física, Universidad del País Vasco, UPV/EHU, Campus de Leioa, Vizcaya, Spain

F. Javier del Campo
IKERBASQUE, Basque Foundation for Science, 48009 Bilbao, Spain

K. Mitsubayashi (ed.), *Wearable Biosensing in Medicine and Healthcare*,
https://doi.org/10.1007/978-981-99-8122-9_15

Table 1 Summary of main commercial wearable biosensors to date

Year	Device	References
2000	Glucowatch biographer (EU)	[3, 4]
2006	Dexcom seven (US)	[5]
2014	Abbot Freestyle Libre (EU)	[6]
	Dexcom G4 Platinum	[6, 7]
2016	Senseonics Eversense (EU)	[8]
2017	Abbot Freestyle Libre (US)	[9, 10]
	Dexcom G5 mobile	[6]
	Medtronic Guardian Connect	[11]
2018	Dexcom G6 Senseonics Eversense (US)	[12–14]
2021	Supersapiens Abbot Libresense glucose sport	[15]

The most important thing about any wearable biosensor is the information they provide. Biosensor data must be accurate, specific, timely, and reliable. This has important implications in terms of sensor placement, power and data management, and device usability. These are all aspects to consider throughout device development but, in addition to usability and functionality concerns, which can be addressed through frameworks such as REASSURED [1], originated in the area of point-of-care devices as an evolution of the ASSURED criteria [2]. ASSURED criteria stressed accuracy, accessibility and affordability, and REASSURED brings this one step forward by considering real-time connectivity enabled by the Internet of things (IoT), and ease of specimen collection and environmental friendliness. The REASSURED framework thus becomes extremely important for the field of wearable biosensors, as not only the body can be regarded as a "low resource setting", but also because of the criticality of having timely information—if continuous, then even better—and also because wearable biosensor design and manufacture are also heavily conditioned by environmental and sustainability issues.

These aspects and many others need to be considered during the design stage and have an impact on materials selection and manufacturing techniques. This chapter aims to provide a broad introduction to the design, materials selection, and fabrication processes involved in wearable biosensors. We begin by looking at the evolution of wearable biosensors, with a focus on commercial devices. Next, the chapter provides an overview of product design, so that newcomers to the field of wearables have a starting point on the main issues to consider in order to make a meaningful contribution. Next, the bulk of the chapter addresses biosensor design from the perspective of wearable devices. This part is written mostly for those without a strong background in (bio)chemistry, but who need to understand the available options when it comes to the transducer and its functionalization, and the way biosensors work. Last, we provide an overview of the most common fabrication techniques used in the study and in the development of wearable devices.

2 Evolution of Wearable Biosensors

Wearable sensors are intended for the continuous monitoring of health parameters, and the development is driven mainly by glucose monitoring for diabetes management. Diabetes mellitus has a very high prevalence, and affects approximately 10% of the world population. This means millions of people worldwide, so improving their life quality therefore represents a huge challenge for scientists and technologists, and an enormous market. In addition to this, glucose monitoring is also of importance in other application domains, such as sports, work safety, and defense.

Glucowatch was the first commercial device for the monitoring of glucose [3]. However, despite gaining FDA approval, it was withdrawn from the market by 2004 because the reverse-iontophoresis method used to induce sweating caused skin rushes on some of its users. It was nevertheless a true technological milestone.

Shortly after, in 2006, Dexcom released their "Dexcom seven" in the US, which was able to monitor glucose for seven days. Abbot released their wearable "Freestyle Libre" system in 2014 in the EU, and in the US in 2017 following FDA approval. In 2021, Abbot announced a new version of the device for athletes under the commercial name of "Supersapiens system". Barring a few exceptions, the wearable biosensor market seems focused on glucose monitoring for diabetes management.

Table 1 summarizes the main commercial devices and their corresponding release dates. Except for the Glucowatch, which was withdrawn from the market, and Senseonics' Eversense, which is an implantable capsule, the rest have a very similar form factor, as wearable patches. The evolution of these devices has been toward more compact and unobtrusive designs, more reliable sensors, and higher levels of integration with other personal smart devices, particularly smartphones, through wireless communication and user-friendly apps.

All these systems have in common that they consist of a piece of instrumentation, and a disposable part. This is because biosensors typically have a shorter lifespan than the electronics and the mechanical parts supporting them. This is the same for any quantitative point-of-care diagnostic system. While the instrumentation can be used many times, the sensing part is discarded after one use. The main limitations of this model are the complexity of the instrumentation, the need to tightly control the cost of the disposable part, and the environmental impact of this disposable part. The evolution of these systems, driven or facilitated in most part by progress in miniaturization and new materials, has been toward integrating the biosensor into a small smart system that works in tandem with a smartphone or a smartwatch to facilitate user interaction and data management. The biosensing smart system, in turn, comprises a microcontroller, a potentiostat, a small memory, short-range communications, and a button battery cell to power the set. Although these products have doubtlessly improved the life quality of millions of diabetic patients worldwide, they are also extremely wasteful. Whether it is the biosensor or the adhesive that limits the lifetime of the disposable component, the fact is that these capsules are disposed of after only a few days, typically 10–14. The longer lasting device is Eversense XL from Senseonics, which can operate for 90–180 days, although it is a subcutaneous implant, which has a completely different set of issues. When the biosensing capsules are thrown away, the electronics within it and its mechanical parts are in perfectly good order, so the approach is clearly wasteful.

Turning to the academic literature, as Fig. 1 shows, Diamond et al. pioneered research on wearable biosensors in Europe. These workers proposed the integration of biological sensors in the form of wristwatches and in textiles and garments as early as 2005 [16]. Although these early works on wearables focused on textile-based biosensors, his group has also developed wrist-watch biosensor devices, and they have led the field all this time [17–20]. Other researchers have made great progress in this area, mostly in the US. Joseph Wang, at the Center for Wearable Sensors at UCSD, focused on screen-printed biosensors [21, 22] and enzymatic fuel cells [23, 24] for skin tattoos but also interacted in sportswear [25]. These seminal works led Joshua Windmiller to the creation of the spin-off company Biolinq, formerly Electrozyme. More recently, in 2014, Heikenfeld et al. reported an RFID patch for the monitoring of sweat electrolytes [26]. Heikenfeld's approach differed from the rest in that he focused on understanding the mechanism of sweat [27], modeling it [28], and developing systems able to manage the small sweat volumes available, avoiding sensor contamination from the skin itself. Heikenfeld also worked on sweat generation systems by reverse iontophoresis, similar to the Glucowatch. Around that time, Heikenfeld started Eccrine Systems Inc., a company to develop sweat sensors, initially for the military and today focusing on pharmacokinetic studies. In his latest works, Heikenfeld points that the relevant parameter to monitor dehydration is not the level of electrolytes such as Na^+, Cl^-, or K^+, but the fraction of bodyweight lost as sweat during exercise, and has presented a simple colorimetric patch for it [29]. Last, it is worth highlighting the Rogers' group at Northwestern University, whose work on sweat sensing is spectacular. In contrast to the others, Rogers' work has focused on developing stretchable electronics [30, 31]; he has presented stretchable patches integrating wireless capabilities (for power, sensor control, and data communications), improving the microfluidics to drive sweat around the device, as well as form factor design and device functionality. His systems have been demonstrated on babies and triathletes [32]. Some of the latest work by the Rogers Group features a self-powered skin patch that integrates an Mg (anode) and an Ag/AgCl (cathode) battery to power a circuit that monitors heart rate, sweat chloride, and sweat pH, plus an NFC tag to communicate with a smartphone [33]. It is also worth noting that the Rogers Group has also spun out a company, Epicore Biosystems, to commercialize skin patches. However, in contrast with the others, two out of three products currently featured in the company website seem restricted to passive microfluidic devices. One of them, named *Gx Sweat Patch*, has been developed in partnership with Gatorade to prevent dehydration, and consists of an easy-to-use patch that allows you to monitor sweat rate and (sweat) sodium in real time [34, 35]. The patch contains two microchannels impregnated in colorimetric indicators. Color can be read using a mobile phone and a dedicated app. The other two products are a passive patch for sweat collection, and a patch-like device for the monitoring of temperature, sweat rate, and sweat loss [36], with a similar form factor to the systems featured in Table 1.

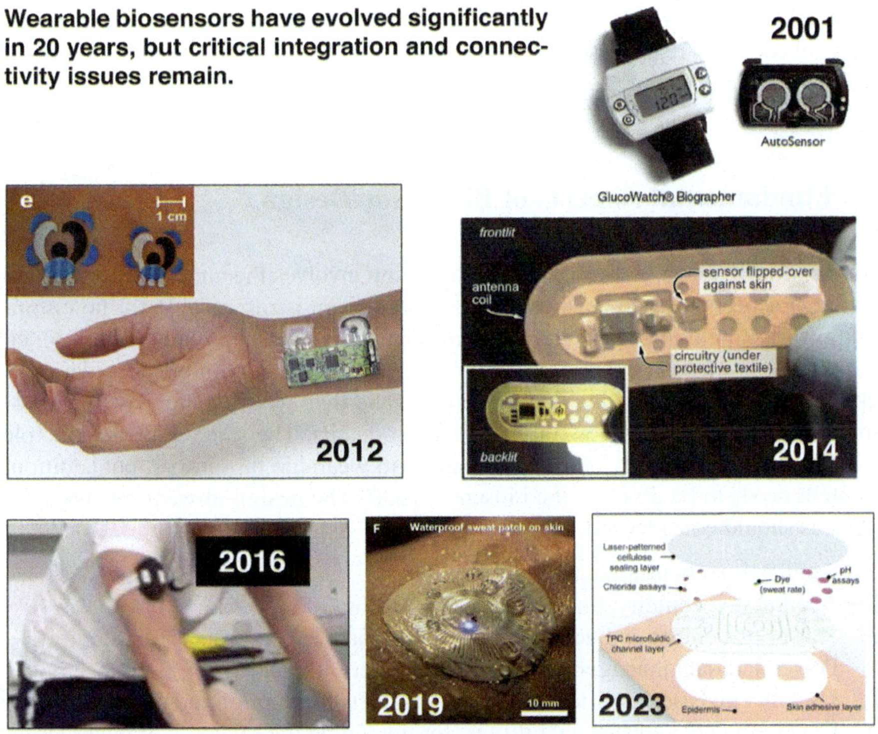

Fig. 1 Key milestones in wearable biosensors to date. 2001—Glucowatch [3]; 2012—Screen-printed biosensing tattoos [21]; 2014—RFID skin patch [26]; 2019—Waterproof skin patch [32]; 2023—Biodegradable skin patch for sweat rate and electrolyte monitoring [35]

In summary, barring the differences mentioned above, which confer a distinct identity to the works of these four leading groups, they all stumble upon the same issues, mainly related to the connectivity between the sensing part (disposable) and the control instrumentation. Wireless connectivity is superior to contact-based connection because it gives more freedom to the user, but it has two important drawbacks. First, the radiofrequency protocols typically used, such as near field communication (NFC), rely on very short-range (<3 cm) radiofrequency, which means that the two communicating nodes need to be aligned and very close to each other. Such short-range protocols can work with low information encryption levels and are in general very safe and energy efficient, but require an excellent coupling between emitter and receptor units. Second, so far, all systems reported which rely on NFC require the integration of an NFC chip and other discrete components on the sensing part. This translates into prohibitive costs of integration into flexible or even elastic substrates. On the other hand, electrochemical chipless devices require a physical connection with the instrumentation, which often makes them uncomfortable and impractical. This already suggests some areas for research to improve present-day wearable devices, regardless of whether the system mounts a biosensor or a different kind of sensor.

3 Fundamental Aspects of Biosensor Design

As mentioned above, wearable biosensor design involves the consideration of many different aspects beyond aesthetics. Industrial design is responsible for how something works, including every single detail surrounding a new product or service, and not simply how it looks. One way of approaching the subject is to start considering the application environment and the user profile, and then working the way down from the most critical aspects down to the smallest details. The same guiding principles used in the design of any other device apply to wearable biosensors, but additional thought needs to be given to the biosensor itself. The design of wearable biosensor devices should consider at least:

1. Application environment and user profile.
2. Accuracy, specificity, timeliness, and reliability of biosensor data.
3. Sensor placement and design, which can impact accuracy and user comfort.
4. Power and data management, including battery life and wireless connectivity.
5. Usability and user experience, which can affect the adoption and use of the device.
6. Environmental and sustainability issues, including the selection of materials and manufacturing processes that minimize environmental impact and the end-of-life disposal of the device.
7. Regulatory issues.

Design and manufacturability or, better, design for manufacturability, can help overcome technical challenges, comply with regulations, reduce manufacturing costs, and incorporate essential security features to ensure data privacy and protection. However, they are rarely considered in academic environments. The following aims to provide some basic ideas on the process taking wearable biosensors from concept to prototype.

3.1 The (New Device) Design Process: Useful, Usable, Beautiful

Developing a new product from scratch may seem terrifying and extremely difficult. To make it easier, setting specifications correctly is crucial. The first step is to answer a number of questions: why is this new device needed? What problem does it solve? For whom? These questions are intended to clarify the utility of the device. **The first goal has to be to make a useful product.** Additional factors may be considered to complete the specifications. On the technical side, consider application requirements such as analytical performance, durability, user profile, use environment, etc. Are there any size, weight, and form factor requirements? How about possible constraints on the physical design of the product? Based on this, what are the main features and capabilities to meet the need? Now is the time to think about materials and processes. Are there any specific materials that are necessary or desirable? Any materials that should not be used at all?

At this stage it is quite useful to consider business-related issues. Things like market size (who is the new device solving a problem for?), user profile; whether the user is actually the same person as the buyer, which is not always the case, particularly when dealing with medical products, where the user is a member of the general public, but the purchasing decision is made by the clinician. How about the competition, is there anything like it already in the market? If so, you need to decide how to compete: what will you do that makes your product better than those already in the market? How is technology evolving in this particular area? What are the competing technologies doing, and at what development point are they? These are central strategic questions to ask, although perhaps outside of the scope of the present chapter. The interested reader can find out more about strategic and business-related issues of technology in the works of Peter Drucker [37], Clayton Christensen [38], and Michael Porter [39, 40].

Once we know that the device will be useful, it is time to move to the next step: making it usable.

3.2 On Device Usability: A Heuristics Approach to Product Design

Heuristics are about experience. Heuristics comes from the Greek word meaning “to discover” or “to find”. Heuristics are based on intuition and experience, rather than a rigorous analysis. Although heuristics-based design may be prone to errors of interpretation and biases, using heuristics can be extremely helpful during the first stages of the design process. In short, it is about common sense and, in many ways, it means putting yourself in the user’s shoes with a little empathy. Above all, it is about trial and error, and how to make a product truly meaningful for the user. Having said this, while heuristics are a powerful tool, their assumptions should be checked and tested with more objective criteria and backed by data, so that the design stage can be more rigorous.

The work of Dan Norman entitled “The design of Everyday Things” [41] is an excellent source of heuristics at work in product design. This book uses great real-life examples of how to use heuristics to come up with excellent products that are both intuitive and easy to use. Another very widely used set of heuristics can be found at the Nielsen Norman Group website (https://www.nngroup.com/articles/ten-usability-heuristics/). In addition to this, it is worth considering also Dieter Rams’ ten

Table 2 Dieter Rams' design principles and Norman Nielsen's ten heuristics for new product design

Dieter Rams' design principles	Nielsen's ten heuristics
Good design is innovative	
Good design makes a product useful	Flexibility and efficiency of use
Good design is aesthetic	Aesthetic and minimalist design
Good design makes a product understandable	Match between system and real world Recognition rather than recall Help users recognize, diagnose, and recover from errors
Good design is unobtrusive	User control and freedom
Good design is honest	Visibility of system status
Good design is long lasting	Consistency and standards
Good design is thorough down to the last detail	Error prevention
Good design is environmentally friendly–	
Good design is as little design as possible	Aesthetic and minimalist design

principles of good design.[1] Table 2 lists Dieter Rams' principles along Nielsen's ten heuristics. While there may be slight differences in wording, both approaches have very much in common. Dieter Rams' principles were conceived with physical objects in mind, whereas Nielsen's heuristic finds more application in software user interfaces. Be as it may, they both can be extremely helpful when applied to the design of new wearable biosensors.

Only after usefulness and usability have been addressed is time to work on the device aesthetics. Prototyping offers a systematic way to move from concept to production.

3.3 Prototyping: From Design to Product

At this stage of the work, prototyping techniques play an important role to bridge the gap between the proof of concept and the production line. As Fig. 2 shows, prototyping is an iterative process whose goal is to bring out problems so that they do not appear later, during, or after production. It is generally useful to focus on one or two very specific aspects of the device at a time, and gradually integrate the lessons learned into subsequent prototyping cycles. One round may focus on how process conditions affect materials and functionality. This will help fine-tune the production sequence, to check that the processes are done in the right order, and that the conditions are adequate. It may be possible to identify alternative processes that will facilitate overall yield, or to find that certain design assumptions need to

[1] https://www.vitsoe.com/gb/about/good-design.

be revised. Both hardware and software prototypes can and should be developed. Hardware prototypes should address all aspects of the device: mechanical, optical, electric, etc. Last, these parts need to be integrated into a coherent whole, which can also be addressed through prototypes. Complete functional prototypes can be used to assess user experience and usability: how intuitive and clear is the device? How can we make it more "user-proof"? Generally speaking, how can we improve the user experience without starting all over? In sum, there are many ways in which prototyping is helpful, but in our case we ought to be concerned with materials processability and device functionality. It is important to choose prototyping techniques that are scalable, and relevant to the industrial processes that will be used later but, more importantly, prototyping methods should allow for an easy work cycle. It is important that prototypes can be produced in a short period of time with reasonable resources. The more prototyping cycles that can be performed in a given time, the more robust the final device will be. Although the later sections of this chapter provide an overview of the techniques and processes, prototyping techniques should be industrially relevant, versatile, and able to process different materials in different ways, and easily scaled and combined with other processes.

The following provides an overview of the main fabrication techniques used in the development of new wearable biosensor devices.

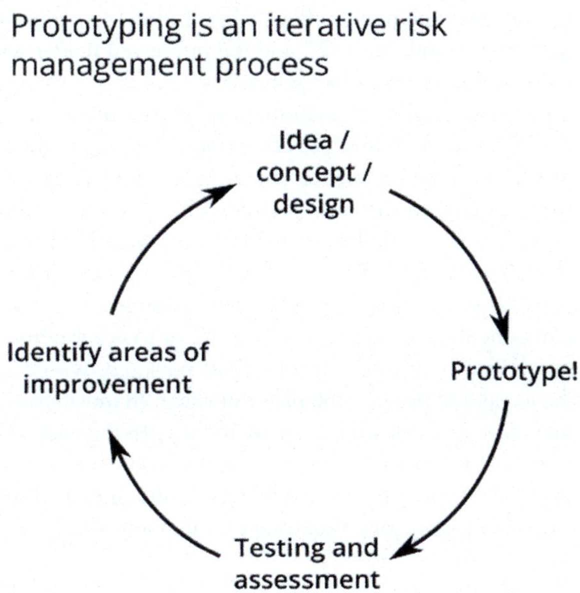

Fig. 2 Prototyping is a key iterative process in product development

4 Fabrication Processes

The production of wearable biosensors involves the combination of several manufacturing processes, from sensor fabrication, device assembly, and packaging. Based on the process results, fabrication techniques or processes could be divided into four main types of processes:

1. Additive processes.
2. Subtractive processes.
3. Pattering techniques.
4. Forming processes.

Additive processes include deposition methods such as thin-film processes like sputtering and other physical vapor deposition methods, chemical vapor deposition methods, electrodeposition, and thick-film processes such as coating methods, printing methods, and replica molding. Subtracting methods include etching methods and cutting and engraving methods. Patterning methods whose function is to transfer functional structures to the device by means of additive or subtractive techniques are mainly based on lithography. This chapter will cover the first three types. Forming processes are those where material is neither added nor removed, but given a particular shape by means of deformation, pressing, calendering, extrusion, thermoforming, and many others. Chris Lefteri provides an excellent overview of manufacturing techniques [42], and the interested reader will find a comprehensive list of industrial manufacturing processes.

The fabrication of wearable biosensors involves a combination of processes and the integration of multiple components, typically through heterogeneous integration processes. Heterogeneous integration consists in the combination of components from a range of different sources into a single working microsystem. The integration process should be oriented toward ensuring higher performance levels, smaller overall size, and low power consumption. Among the processes involved in heterogeneous integration, we can find pick and place techniques, various ways of bonding components to a common substrate, or to other components. Following integration, the device is inserted into its final package, whose main mission is to protect the device and to provide the user interface. In most cases, the case is made of plastic by injection, and assembled mechanically. In the case of wearables, special care needs to be taken to make the device waterproof, for which the IP ratings[2] provide guidance. The following sections describe the main techniques used in the production of wearable biosensors, illustrated with examples from the literature.

[2] https://www.iec.ch/ip-ratings.

4.1 Additive Manufacturing Techniques

Additive manufacturing involves the addition of material on a substrate [43]. There are many ways in which material may be added. Thin-film methods involve the deposition of material layers up to a few hundred nanometres thick. The main thin-film processes are physical and chemical vapor deposition. An important case is that of metal parts, typically gold tracks, pads, and electrodes, which are deposited by physical deposition methods, typically sputtering or evaporation. Thick-film techniques, on the other hand, provide means to deposit layers and structures more than 1 micron-thick. Among thick-layer techniques we find various coating and printing methods. Screen-printing and 3D printing are the most widely adopted.

4.1.1 Coating and Printing Techniques

Broadly, coating refers to covering of a substrate with another material. The coating process is carried out on multiple surfaces or substrates depending on the application. Although there are numerous coating processes, we will focus on those with the greatest interest for the manufacture of wearable biosensors. The coating techniques to be used in each development will vary depending on parameters such as the nature of the inks or coatings, and the substrates, and how these interact. Figure 3 shows a diagrammatic representation of some of the main coating techniques, also used in wearable fabrication.

4.2 Drop Casting

Drop casting is the simplest of the coating techniques, and perhaps the most widely spread. It is used to build the selective layer over a transducer, effectively transforming it into a biosensor. The technique consists of adding a drop of solution, containing the necessary reagents, over the surface to be modified, usually a transducer. After the solvent has evaporated, the solid residue left on the transducer becomes an active part. Thus, enzymes, cross-linking agents, and (semi)permeable membranes can be deposited by drop casting using a micropipette or a spotter. For instance, in a recent work describing a self-powered skin-patch biosensor, glucose oxidase was drop cast over the surface of a screen-printed graphite transducer [44].

4.3 Dip Coating

In Dip coating, the substrate is dipped and lifted out of a cuvette filled with the coating material. The substrate is immersed in the coating material at a constant speed, and

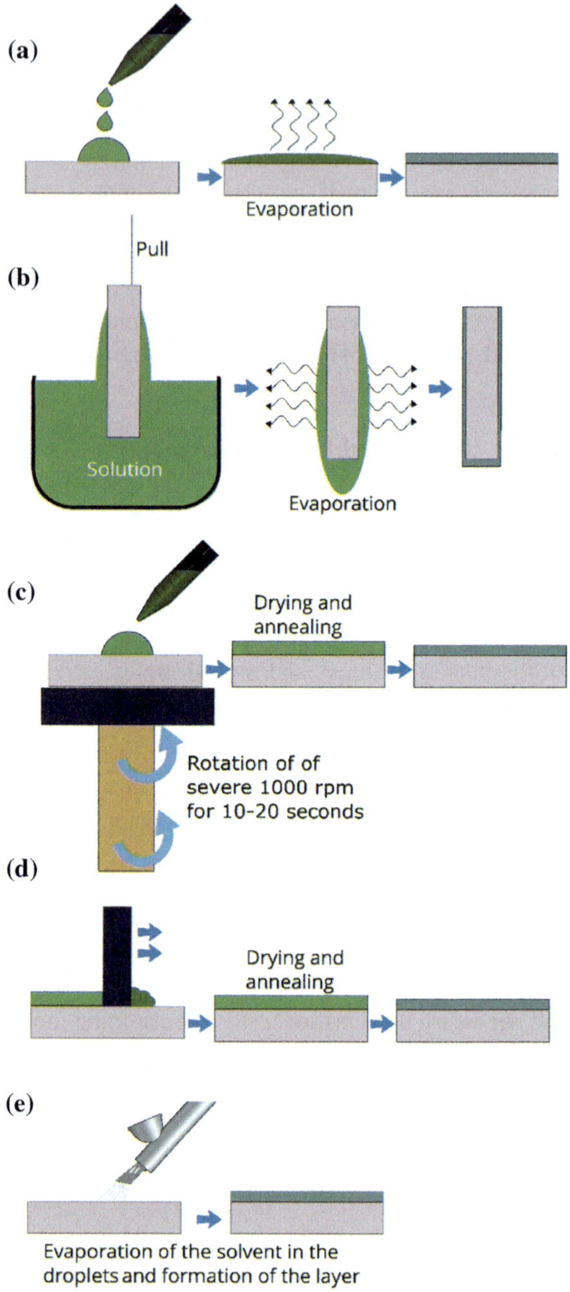

Fig. 3 Diagrammatic representation of the main coating techniques. **a** Drop casting; **b** Dip coating; **c** Spin coating; **d** Blade and Bar coating; **e** Spray coating

after some time the material is lifted carefully, so it does not jolt. The film starts depositing as the substrate exits the solution, and the extraction speed determines the resulting film thickness. Dip coating is used when spin coating or spray coating are not possible.

Dip coating can be used to create different types of functional materials. For example, it was used for the fabrication of a non-enzymatic sweat biosensor [45, 46] a polymeric humidity sensor was fabricated by coating a PDLLA fiber coated in a CNC/PEDOT:PSS. In the two articles different solutions were made to coat the material with the wanted layer, and in both a previous coating was made with a different material to make sure that the wanted layer wouldn't have any problem to stick, this shows a common problem of dip coating, sometimes an extra layer of other material is need.

4.4 *Spin Coating*

This technique consists of pouring or dispensing a few mL of coating material on top of a substrate that will be spun at a high rotational speed of over several 1000 rpm, and excess material is spun off the substrate. Due to centrifugal force, the dispersed material spreads into a uniform film of desired thickness, typically in the micron range; this technique is used to coat the wanted material with a suspension of other material. This is mostly used in microfabrication of electronics and in lithography creating uniform thin layer with nanoscale thickness.

On the one hand, we have a very fast method, a few seconds, and we can control the thickness of the layer applied and the obtained layers are very smooth. On the other hand, the shape of the substrate is an important factor in this method, square or rectangular shapes have edge beads at the corners of the substrate, and this also can happen with thick films. In addition, the surface of the substrate can affect the homogeneity of the films.

Spin coating is used in different fabrication processes. In [47] they used spin coating to add a layer of RuS2 nanoparticles in the PMDS substrate, this way they obtain a RuS2/PMDS electrode, as they use spin coating to coat the electrode, they could put a smooth layer in a very fast way spinning at 5000 rpm, as the shape is kind of a limiting factor for this technique, previous or subsequent work must have been carried out to give the wanted shape [48].

4.5 *Blade and Bar Coating*

Blade coating is another very popular method. The coating material is cast on the substrate, which rests on a flat surface, typically a glass tile. Then a bar that is kept parallel to the substrate at a controlled distance, slides over the substrate and spreads the coating material over it, leaving behind a coat of same thickness as the distance between the bar and the substrate surface. When the solvent evaporates, the remaining film is usually slightly thinner than the height of the blade, but it is possible to predict the final thickness by accounting for the volume of solvent present in the original coating mixture. One way in which bar and blade coating may be very useful is in the elaboration of flexible and elastomeric substrates several hundred microns thick. This gives the freedom to prepare substrates in materials other than the usual PET, polyimide (Kapton), and PDMS. Also, it allows the possibility to prepare functional substrates, if the polymer is combined with a functional component.

4.6 Spray Coating

In spray coating the material is deposited in μm size droplets and lands on the surface of the substrate. If the substrate surface or size does not allow for spin coating with required homogeneity spray coating can be used, because it offers the potential to coat arbitrary shaped substrates. To form the droplet ultrasonic atomization or nitrogen filed nozzle can be used, most of the droplet's land outside of the substrate and due to Micro-turbulences, the center of the substrate is thicker than in the edges. A particular advantage of spray coating is the ability to produce thinner coatings than other methods, such as screen-printing or blade coating. One particular application is in the fabrication of transparent electrodes based on carbon nanotubes. Nanotube suspensions can be sprayed and, on evaporation of the solvent, leave a conducting mesh behind that can be used to produce electrochromic or other electro-optical devices, of high potential interest also in wearables. Asaduzzaman et al. made a 3D graphene-based epidermal patch for glucose and lactate analysis in sweat [49] and Zahid and co-workers made a Strain-responsive wearable based on PEDOT:PSS/ graphene [50] using spray coating.

As a final word, spray coating can also be considered one of the non-contact printing methods, as depicted in Fig. 4. Recently, aerosol jet-printing has gained attention because it allows the deposition of very fine and thin structures over 3D objects [51].

4.7 Screen-Printing

Screen-printing or serigraphy is a printing method with a long history and tradition in the graphic arts. In screen-printing, ink is pressed through a mesh screen onto the substrate, transferring the patterns defined on the screen or stencil, as shown in Fig. 5. This technique is one of the most commonly used printing techniques, and possibly the most used technique in the fabrication of wearable devices. Screen-printing is a straightforward and cost-effective way of reproducing patterns. In contrast to inkjet-printing, screen-printing is a very robust technique. It can print materials (inks) with particle size several microns in size (typically less than 10 microns), and widely

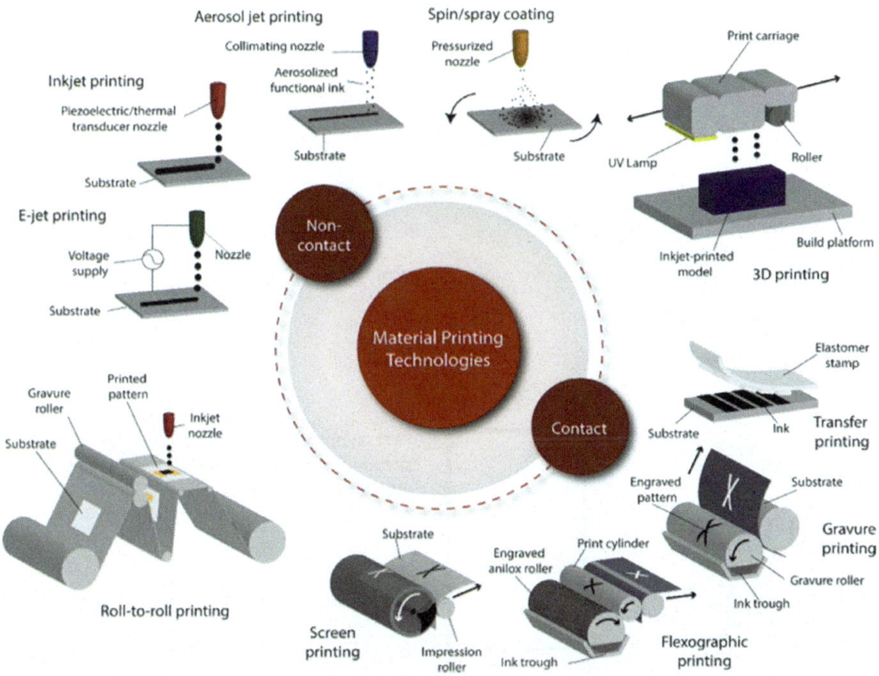

Fig. 4 Overview of contact and non-contact printing techniques [43]

different viscosity and rheological properties. By changing mesh count and squeegee profile, it is possible to control the amount of material transferred, and hence the lateral size and thickness of the features. However, the larger the particle, the larger the necessary opening in the mesh and hence the lower the resolution. Although the state of the art allows the printing of structures with critical dimensions down to few tens of microns, such screens are very costly and are only used in high-end applications and in research environments. Most screens consist of nylon meshes, and mesh counts between 77 and 120 thread cm^{-1} are typical in the area of electrochemical sensors and wearable biosensors. Such mesh counts allow the printing of structures down to 100–150 microns, which is enough in most cases.

Although screen-printing is typically used to print electrodes and dielectric structures, it is in principle possible to print structures up to 100 microns in height, which could lead to printing gaskets, fluidic channels, and membranes. Screen-printing also allows printing virtually any material imaginable, and it enables the fabrication of complete devices if suitable inks are available [44].

Figure 5 depicts the process of screen-printing an electrochemical sensor using three screen levels. First, the conducting pads and tracks are printed on the substrate, followed by the transducer and auxiliary electrodes, which are typically graphite or other carbon allotrope and, last, a dielectric structure is printed to protect conducting

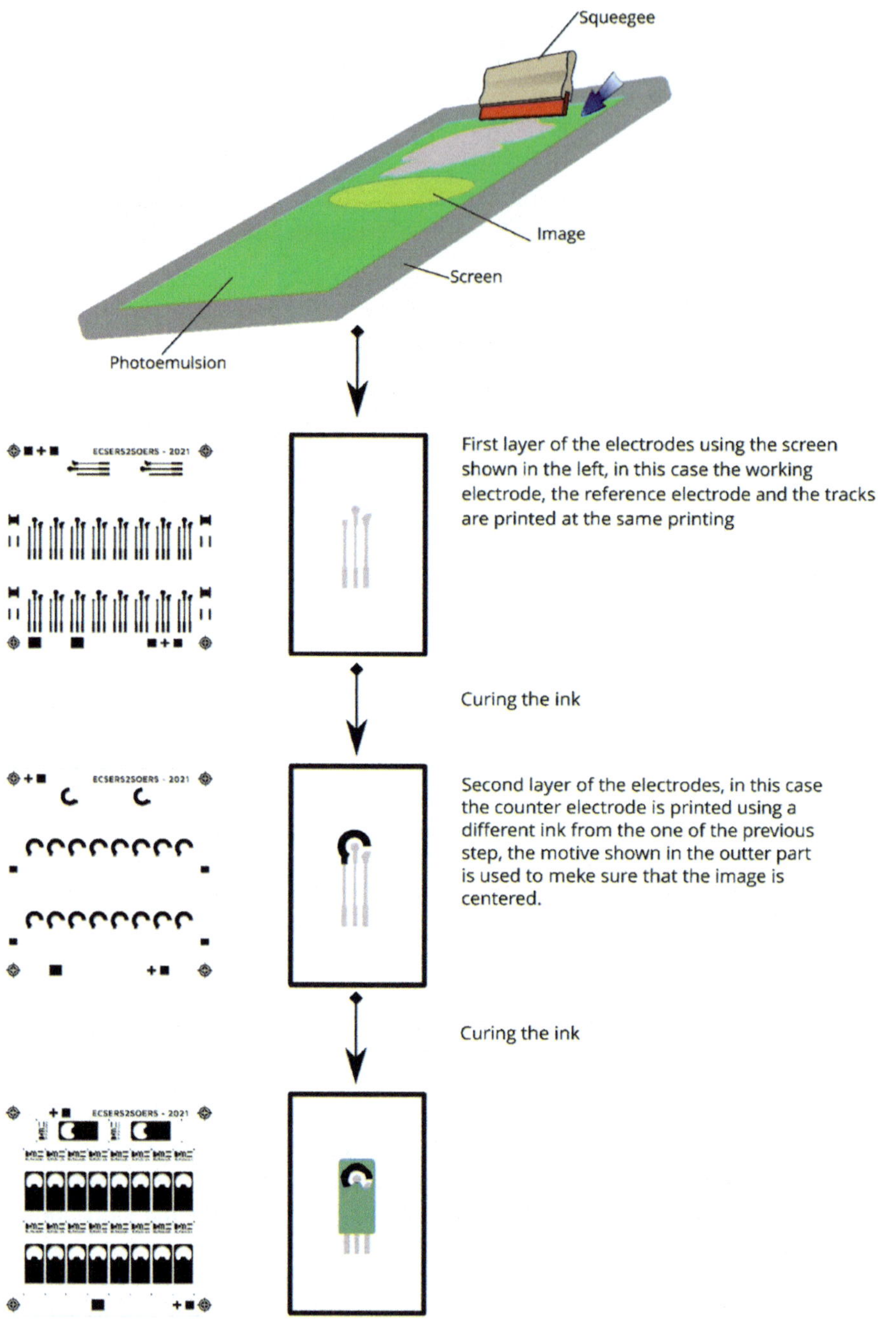

Fig. 5 Representation of the screen-printing process to produce electrodes for electroanalytical devices

tracks and to define the working areas of the electrodes. On the side of each step, film positives are shown.

This technique has a lot of versatility and can be used to fabricate wearables, but often the required functional inks have to be developed specifically. For example, a screen-printed GOx/NQ/MWCNT-based bioanode and a screen-printed GOx/PB/MWCNT-based biocathode were prepared to obtain a self-powered and energy-harvesting biosensing device, these inks were created to function as a single-enzyme-based energy-harvesting device and a self-powered biosensor driven by glucose on bioanode and biocathode [52]. This is a good example of how screen printing allows to obtain different types of devices with diverse types of functions depending on what ink we use. On the other hand, we see that this technique is very useful and very fast, as [53] it replicates a lot of biosensors as a flexible wearable biosensor using screen-printing as a technique. So, this shows that the future of this technique looks like is going to be creating different types of inks with new kinds of materials to obtain different effects in the wearables.

4.8 3D Printing

3D printing is a fabrication process that involves creating three-dimensional objects by adding successive material layers. The different ways to deposit the material, or to grow layer upon layer of material, define 3D printing techniques. Fused filament deposition modeling (FDM) works by extruding melted material through a heated nozzle, and each subsequent layer is printed or deposited on top of the previous one until object is formed. This technique allows the creation of complex objects in a fast and cost-effective process. 3D printing could be used in biosensor fabrication allowing the rapid prototyping, tailor-made sensor designs, and miniaturization. Stereolithography, SLA, is becoming even more popular than FDM because it allows for higher resolutions than the latter, and typical photoresins used are thermally and mechanically more stable than PLA used by filament deposition techniques. In stereolithography, an ultraviolet light source is selectively applied to cure a photocurable resin with a high degree of accuracy. Selective laser sintering, SLS, is a third stereolithographic technique where a laser beam is used to selectively cure a powder into a solid shape. Polymers such as polyamide, and even metal powders can be sintered, which opens a vast number of possibilities.

3D printing has transformed many of the manufacturing sectors as it allows rapid prototyping of devices for many different applications. In this sense, and despite the fact that 3D printing has not been as widely adopted as other techniques in the area of wearable biosensors, even if things may be starting to change [54]. Until now, many of the developments have focused on pressure sensors, in which highly stretchable sensors had created by 3D printing a resistive carbon-based ink onto an elastomeric matrix [55–57]. However, recently Kim et al. presented a wearable sweat sensor by means of 3D microprinting [58]. The authors describe the development of a low-cost, mechanically flexible, all-inclusive integrated wearable patch (AIIW) containing 3D-printed flexible sensors, which are shown in Fig. 6. In addition, the developed system integrates a microfluidic system. Studies successfully demonstrate the ability to obtain ex situ and in situ measurements of multiple electrolyte levels (Na^+, K^+, and Ca^{2+}) in sweat.

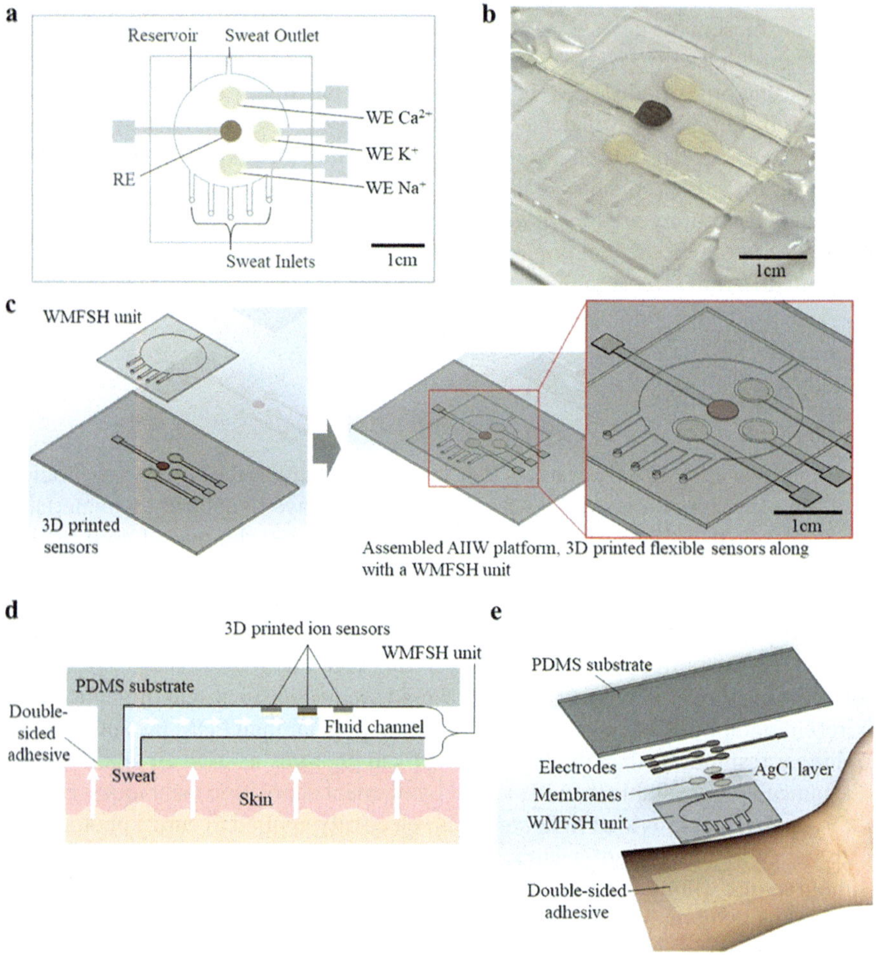

Fig. 6 Scheme of the multisensing integrated flexible device developed by Kim et al. [58] (Reproduced with permission of Wiley)

5 Subtractive Manufacturing Techniques

Subtractive techniques involve the selective removal of material from a substrate. Etching processes are very widely used subtractive techniques in micro-, nanomachining, integrated circuits, and when fabricating wearables. Etching processes may be wet or dry, depending on the phase of the etchant. Wet etching is the chemical removal by the immersion in a solution, and dry etching is the removal of the material by a reactive ion plasma. Etching may be isotropic if the same amount of material is lost vertically and horizontally, or anisotropic, if the etch rate is direction dependent. This section will describe different types of etching processes used in the fabrication of wearable biosensors.

5.1 Etching

Etching is a type of subtractive technique where the material is removed; this removal can happen in a bath with an etchant (wet etching) or in a reactive ion plasma (dry etching).

The most important parameters in etching processes are:

1. Etch rate: The rate at which the material is removed from the substrate. It is typically expressed in depth, *D*, Å, nm or μm per unit of time, *t*, and can be calculated using the following formula $ER = \frac{D}{t}$
2. Uniformity: It is defined as the etch rate constancy across the substrate. It may consider other post-etch characteristics such as selectivity and profile.
3. Throughput: Amount of material etched during one process cycle.
4. Directional control: Control of the horizontal and vertical etch rate.

    i. isotropic: the same amount of material from the substrate is removed horizontally and vertically
    ii. Anisotropic: different amount of material is removed from the substrate vertically and horizontally.

Anisotropy can be calculated using this formula $A = 1 - \frac{r_{hor}}{r_{vert}}$ where *rhor* is the horizontal etch rate and *rvert* is the vertical etch rate.

5. Etch selectivity: It is the ratio between two different etch rates, normally the mask and the material being etched. Selectivity can be adjusted through the selection or thickness of the mask, in dry etching we work with plasma and during the etching process part of the mask is removed, understanding the selectivity in this example lets to choose a perfect mask thickness and control the etching results.

5.2 Wet Etching

Wet etching is the chemical removal of the material by immersion of a substrate in a bath with a suitable etchant. Wet etching can be used in microfabrication of pattern metals like aluminum, gold, nickel, and chromium, among others. In addition glass and silicon can be etched using this technique. Due to diffusion controlling of the etch reaction, this type of technique is in most cases isotropic. This has important implications in mask design and the process of critical dimensions, due to the fact that structure with aspect ratio greater than 1:1 is difficult to obtain using wet etching. On the other hand, wet etching can be very selective, uniform, and is used in the removal of thin layers and for cleaning purposes. Wet etching or chemical etching is used a lot in the synthesis of biosensors based in MXenes [59, 60] and Nanoporous metals [61, 62].

5.3 Dry Etching

Dry etching is the removal of the material in a reactive ion plasma, without using wet chemicals or rising, they enter and leave the etching cycle in dry state. In dry etching, a solid surface is etched in the gas or vapor phase physically by ion bombardment. There are 3 different types of etching methods:

1. Physical sputter/ion etching and ion-beam milling etching occurs because of a purely physical effect.
2. In chemical plasma etching, reactive species such as chlorine or fluorine atoms generated in the plasma diffuse to the substrate, where they react to form volatile products with the layer to be removed.
3. In the case of physical/chemical etching (RIE), Line of sight impacting ions damage the surface, inducing highly anisotropic chemical reactions on the surface with plasma neutrals.

This technique is anisotropic because the high directionality of the electric field inside the reaction chamber makes the etch rate of the direction perpendicular to the substrate surface much higher than etching in the lateral. Structures with very high aspect ratios can be obtained by controlling the plasma conditions.

Figure 7 shows the etching of a substrate using a mask. High aspect-ratio structures typically involve a series of steps of etching and masking, as in the Bosch process, to ensure verticality of the structures, as there is always some degree of lateral etching [63].

Fig. 7 Selective etching using a mask

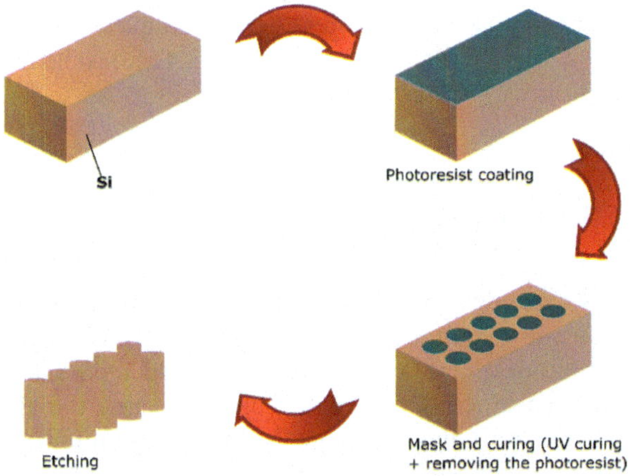

5.4 *Milling*

In milling, a rotatory cutting tool like a milling cutter or drill is used to carve structures from a block of material, chipping it off. Milling can be used to make components ranging from fluidic channels to molds to cast different types of materials such as PDMS. The most important parameters during milling processes are the spindle rotation speed and the material feed rate. Although tool manufacturers provide approximate work conditions, these values are generally a starting point, and conditions need to be adapted as each material will behave differently.

To calculate them you can use the following formulas in decimal metric system.

Cutting speed (m/min):

$$Vc = \frac{\left(\pi \times DC_{ap} \times n\right)}{1000}$$

where *DCap* is Cutting diameter at cutting depth ap (mm) and *n* Spindle speed in rpm.

Feed rate (per revolution mm/r):

$$F_n = \frac{V_f}{n}$$

where *Vf* is the table feed (mm/min) and *n* Spindle speed rpm.

Milling is used to produce objects from a few hundred microns up to meters. In the case of wearable devices, milling can be used to produce mm-size objects with sub-millimetric features, such as molds and casings. Although the versatility and increasingly high resolution of stereolithographic 3D printing methods offer an easy alternative, the fact is that milling still provides far superior finish quality. A smart approach is to combine both techniques. Using a 3D-printed part as the starting point, and ending the part by milling. This results in less wasted material, and very high-quality finished parts.

5.5 Laser Cutting

Laser cutting and etching works by ablating the material; that is, by sublimating the material where the laser energy is focused. An advantage of laser cutting compared to milling or other mechanical cutting techniques is that the material does not suffer mechanical stress during the cutting because this is a "contactless" technique. On the one hand, this technique is faster than milling, but on the other hand, the high temperatures during the cutting can lead to problems such as the cracking of brittle materials like ceramics or the formations of burr in polymeric materials. In addition, as with milling the optimal conditions for the cutting has to be obtained with experience.

In addition to cutting, laser can also be used to transform materials. For instance, it is possible to produce graphene structures from Kapton, a polyimide, simply by using the right laser conditions [64].

5.6 Patterning Methods: Lithography

It is difficult to classify lithographic techniques as either additive (a photoresist material is added onto a substrate) or subtractive (photoresist is selectively removed during development following UV insolation). However, it is easy to agree that the main function of lithography is to transfer patterns over a substrate. This is the reason why we have decided to separate it from the other two big families of techniques.

The term lithography comes from the Greek words Lithos (stone) and Graphein (write), it was discovered by Aloys Senefelder when he found that the limestone properly inked and treated with chemicals was able to transfer a carved image onto paper. Lithographic methods, developed in the microelectronics industry, are widely used in biosensor production, since complex micro/nanoscale structures with high-resolution topography can be produced from it. Lithography offers the capability to fabricate sensors in a microscale, reaching sub-micrometric resolution.

Lithography encompasses a wide range of surface fabrication methods (photolithography, electron/ion-beam lithography, soft lithography, etc.) and selecting the most appropriate technique for a specific application involves a compromise between resolution, performance, and cost considerations [65]. In the area of wearable or flexible devices, lithographic techniques are also limited due to the resolution that can be achieved on flexible substrates compared to silicon substrates.

The most used lithography method is photolithography. In photolithography, patterns transfer from masks onto substrates using photomasking and chemical processing, as schematized in Fig. 1. Photolithography relies on three key components: a photoresist, a suitable light source, and a photomask. However, direct writing and projection methods are gradually gaining ground as means to insolate the resist and transfer the desired patterns to the photoresist.

A photoresist is a light-sensitive polymeric material that protects the underlying substrate, providing a selective barrier to the processes following lithography. The chain length of the resin molecule determines the properties of the photoresist; long chains have better stability against thermal rounding while shorter chains improve adhesion to the substrate. Also, depending on how they react when exposed to UV light, we can differentiate two types of photoresists, positive resists and negative resists. In positive resists, the insolated areas are removed during the development stage. When the photoresist is exposed to the UV light, the chemical structure changes and becomes more soluble in the solvent, these exposed areas are washed away with the photoresist developer solvent, leaving the underlying material. In contrast, in the case of negative resists, insolated are not removed in the development stage. With negative resists, exposure to UV light causes the chemical structure of the photoresist to polymerize, instead of becoming more soluble, negative photoresists become extremely difficult to dissolve. As a result, the UV exposed negative resist remains on the surface while the photoresist developer solution works to remove the unexposed areas.

Conventionally, photoresists are insolated through photomasks, although direct-write laser systems are becoming more popular recently as they facilitate prototyping and changing designs on the fly. Photomasks are typically chromium or soda lime glass featuring a patterned chromium layer that block light during substrate exposure. If photoresists were either positive or negative, photomasks may be bright-field or dark-field depending on whether chromium patterns are the positive or the negative images of the desired structures.

The set of process steps to transfer a pattern in photolithography are the following, summarized in Fig. 1:

1. A wafer is coated with a photoresist layer (typically between 0.5 and 10 μm thickness).
2. Insolation, using UV light of suitable energy, through a photomask or by direct writing to activate exposed resist areas.
3. Resist development to remove resist from unwanted areas.
4. Deposition or etching step.
5. Resist removal and substrate clean-up.

This technique can be used to fabricate different types of electronics [66]. Photolithography is widely used in industry for the fabrication of devices that play an important role in the miniaturization of next-generation electronics. While photolithography patterns are commonly performed on rigid semiconductors such as silicon and glass wafers, the demand for micropatterned flexible substrates has significantly raised, leading to the emergence of new photolithographic techniques. As an example, Ma et al. [1] has developed a new photolithographic method to create micropatterned metal arrays on flexible substrates. In this work, they successfully fabricated a flexible glucose sensor with an LOD of 1.2 μM, a dynamic range from 0.05 to 0.30 mM, and good stability and reproducibility. Photolithography can be an interesting technique for manufacturing self-healing sensors. Thus, for example, Kim et al. fabricated pressure sensors via photolithography using polyurethane substrate. This sensor presents a chemical resistance and self-healing capability due to the polymeric substrate and the photolithography patterning allowed for the precise and controlled fabrication of the sensor arrays, demonstrating high potential application in the development of skin-attachable flexible devices [67].

6 Materials in Wearables

Materials play a crucial role in the functionality, durability, and overall performance of biosensors. As described so far, only the biosensor incorporates several different materials, ranging from the substrate on which they are placed, the transducer, the biorecognition element, any protective coating to preserve the biosensor functionality, and to provide an effective interface with the body.

Manufacturing processes not only define the device architecture. They also have a strong influence on the final properties that materials display in the device. Materials and fabrication processes are intimately related. Once the desired function has been identified, materials are selected that can meet the desired requirements in terms of functionality but biocompatibility, cost, and sustainability issues. These materials then need to be processed by suitable manufacturing methods that will transform them into their final forms in the device. However, devices display the properties of the processed materials, so it is important to know the effects of the manufacturing process on the desired material properties involved. For instance, high process temperatures, pressures, or harsh chemical conditions can be either beneficial or damaging. For instance, polyvinylidene fluoride, PVDF, is a fluoropolymer, it is elastic, and it is also piezoelectric. It can be processed in a number of different ways, such as by using from solution, for instance using casting and printing methods to produce coatings or specific geometries selectively, it can be turned into fibers and non-woven fabrics by electrospinning, or it can be turned into membranes and films that can subsequently be cut by different means. The precise technique of choice will be dictated by the function that PVDF needs to fulfill in a device. Thus, if PVDF is to be used as a mere substrate, casting into films and subsequent slot-die or laser cutting can be perfectly fine, and then other criteria will be needed to select the most adequate process. Typically, one looks at the processes upstream and downstream in the production line to ensure that the process does not spoil any of the work from the previous processes, but also that any subsequent processes will be able to deal with the processed parts.

Going back to the PVDF example, if the reason to use this material is its piezoelectricity, then it is important that the processes involved result in the correct orientation of its polymer chains and that the thickness of the material will allow the observation of the piezoelectricity. This sets important temperature constraints for any process downstream in the chain, as temperatures above PVDF Curie temperature will result in loss of piezoelectricity. One last possible application of PVDF in a wearable biosensor is as matrix of solid electrolytes, in combination with organic salts and ionic liquids. In this case, high transparency may be the desired property, and the process of choosing a printing technique such as screen-printing [44]. To produce an ink, PVDF needs to be added to a suitable solvent, and heated while stirred. However, most PVFD forms will turn yellow if heated above 70 ºC, so the temperature of any processes downstream ink formulation needs to be kept below this temperature.

This is only an example of how the material influences the choice of manufacturing process and process conditions.

The substrate material is critical in wearable biosensors because it supports the sensor and provides mechanical flexibility. Flexible substrates, such as silicon and polyurethane, facilitate the conformability of the device to the user's body. Flexible substrates enhance comfort and facilitate the necessary contact to ensure that the sample, typically sweat or interstitial fluid, reaches the sensor in optimal conditions.

The materials used in the fabrication of wearable biosensors also include adhesives, sealants, and coatings. These materials are used to ensure that the biosensor remains securely attached to the skin and to protect the sensor and substrate from damage due to exposure to the environment or sweat. Although adhesives are carefully chosen, cases of allergies and skin irritation are occasionally reported [68].

One significant issue with wearable biosensors is their disposable nature, which raises sustainability issues. Wearable biosensors are designed to be used once, or for a few days at best, and then discarded. This generates a significant amount of waste. There are two main strategies to prevent this. One is to develop more durable wearable biosensors, which involves biorecognition element engineering. The other way consists of using more sustainable materials and manufacturing processes. Using biodegradable or compostable materials in the fabrication of wearable biosensors can alleviate the environmental impact. In addition to this, exploring ways to recycle or repurpose used biosensors is also needed.

All this has become very important in recent years and sustainability has become one of the most important issues in today's society. Moreover, since 2015, the member states of the United Nations and the United Nations Framework Convention on Climate Change (UNFCCC) adopted the Sustainable Development Goals (SDGs), also known as Agenda 2030. The SDGs consist of 17 goals and 169 targets that aim to end poverty, protect the planet, and ensure prosperity for all [69]. As a result, the materials used in the manufacture of wearables have evolved over the last few years and more advanced materials and generally manufacturing methods are being used that have less impact on the environment. Personalized medicine is also of great importance in the SDGs, so wearable biosensors help to better monitor health but have to be environmentally friendly to follow the SDGs.

The wearable-body interface, sensors and actuators, substrates, electronics, and encapsulation and packaging (Fig. 2).

Poitout et al. showed a blood glucose sensor implanted in the subcutaneous area with a wearable unit control [70]. In this case, the sensor consists of a Teflon coated platinum anode except for the end which is functionalized with glucose oxidase. This glucose oxidase in turn is coated with polyurethane. Around the Teflon is an Ag/AgCl cathode. The size of the control unit is $6 \times 12 \times 18$ cm and it has a memory which is able to store the information as well as a display which shows it. The information in the memory can be transferred to a PC so that it can be read out. Although it was one of the first wearables to monitor glucose in humans, it was a very large device and it was an invasive monitoring system because it involved subcutaneous measurements.

In 2006, Kudo et al. [71] presented a flexible and functional wearable glucose sensor based on polymers. Soft-microelectromechanical systems were used to fabricate the sensor. In this case it is a two-electrode system as it is possible to see in Fig. 3. The working electrode is a Pt electrode with two poly phospholipid polymer membranes copolymerized with dodecyl methacrylate (PDM) and one of them is functionalized with glucose oxidase. The auxiliary/reference electrode is Ag/AgCl coated with a PDM layer. The substrate is also polymeric because it is hydrophobic polydimethyl siloxane (PDMS). The Pt and Ag electrode were deposited by ion-beam sputtering obtaining a thickness of 200 nm and 300 nm respectively. Subsequently, the Ag was electrochemically chlorinated and the whole was coated with PDM by dip coating. Glucose oxidase was immobilized by solution casting of PDM.

Mannoor et al. [72] showed a graphene biosensor printed on silk fibroin. In this case the biosensor is placed in the patient's mouth but the silk fibroin is 100% biocompatible. This biosensor is used to detect bacteria by bio-functionalizing antimicrobial peptides in graphene-modified silk tattoos. These measurements are made by changing the resistance of the sensor. One of the major advantages of this sensor is that it is wireless, which means better use of energy and no batteries and less environmental impact.

Kim et al. in 2014 reported one of the first oral biosensor for continuous lactate monitoring [25], This wearable is non-invasive because it is a mouthguard where the three-electrode electrochemical cell has been printed on a polyethylene terephthalate (PET) film and is therefore flexible. These materials, being flexible, adapt perfectly to the shape of the person's teeth and at the same time take up very little space in the mouth, making it a comfortable device for the end user. The reference electrode is made of Ag/AgCl while the auxiliary electrode and the working electrode are made of Prussian blue-graphite ink. The electrode was attached to the mouthguard by using double-sided tape. To make the working electrode lactate selective it was functionalized by means of a lactate oxidase immobilized with an electropolymeric entrapment coating in a poly(*o*-phenylenediamine) (PPD).

In 2015, Rose et al. demonstrated an RFID adhesive patch which could be used for the co-concentration of sodium ions in sweat. In this case, they claim it can be used to measure other ionic solutes in sweat, Cl^-, K^+ and others [26]. Although this device can only measure ions, it is battery-free which makes it interesting because of its energy efficiency. They made two different sizes of devices, one 25 mm $\times$ 60 mm for the arm and one 40×70 mm for the leg. This device uses an RFID tag which makes the device very interesting because the sensor is powered wirelessly and without a battery. For this purpose, the printed circuit board (PCB) is a combination of flexible, conformal polyimide and a thin copper foil. For the adhesion to the body, they use a double-sided medical adhesive and a medical textile for the encapsulation. The working and reference electrodes have Pd and Ag electrodeposited on top of the copper. The reference electrode has had the Ag layer chlorinated. To detect the Na^+ ions, an ISE layer has been applied, in this case the ionophore membrane establishes a difference in potential across the electrode-ionophore barrier corresponding to the Na^+ concentration.

In 2016, Gao et al. presented a device with an array of 5 sensors which were able to detect lactate, glucose, temperature, Na^+, and K^+ in sweat [73]. The substrate of this wearable device is PET on which a flexible printed circuit board (FPCB) is printed. After that, a parylene layer is used to passivate the circuit. The auxiliary electrode is Ag/AgCl, so to obtain it, first the parylene is removed from the electrode by O_2 etching and Ag patterning and then chlorinating it. Finally, the working electrodes have been modified. For the glucose sensor, Prussian blue was deposited on the gold electrode. Finally, by drop casting the sensor was functionalized with a solution of glucose oxidase, chitosan, and carbon nanotubes. For the $Na^+$and K^+ sensors, the reference electrode was coated with polyvinyl butyral in order to stabilize the potential. The working electrodes, in this case, were made of poly(3,4-ethylenedioxythiophene) polystyrene sulfonate, PEDOT:PSS, and ISE membranes were applied by drop casting. The temperature sensor is a Cr/Au nanowire resistor.

In 2022, Huang et al. demonstrated an ultra-thin, flexible, cotton-based microelectronic device powered by sweat-activated battery for real-time monitoring of sweat [74]. This wearable device, represented in Fig. 4, is a big step forward because it does not use any conventional batteries or power transmission as it is powered by sweat. Instead, a sweat-activated battery generates a maximum power density of 3.17 mW cm^{-2}. The battery size is 3.5 cm $\times$ 1 cm so it produces a maximum power of 11,095 mW. This wearable can measure Na^+ concentration, pH, and skin impedance. In this case the electronics are mounted on a polyimide substrate. The conducting material is a 10 cm copper coated with 50 nm gold. The impedance sensors are two gold electrodes. For the Na^+ sensor PEDOT:PSS was electrodeposited and then an Na^+ selective layer was applied. The working electrode of the pH sensor was obtained by electrodeposition of a polyaniline layer (PANI). The pseudo-reference electrodes for the Na^+ and pH sensors are made of Ag/AgCl obtained by Ag deposition and subsequent chlorination. For the Na^+ sensor a PVB layer was applied on the Ag/AgCl electrode. The SAB has been fabricated by homogeneously dispersing graphene on absorbent paper and Mg on each side of the electrode. Next, by putting obtained graphene/paper and Mg sheet side by side on the air-permeable cloth and finally, attaching a thin layer of cotton with KCl powders on it to the electrodes.

As it has been possible to observe over the years there has been an evolution in materials. Nowadays, more miniaturized devices are being used and they are based more on energy efficiency, avoiding the use of conventional batteries. This also comes from the evolution of electronics which has allowed electronic components to be printed on flexible substrates, with these complex circuits feeding and obtaining information from sensors.

7 Biosensors and Wearables

7.1 Overview

Biosensors are miniaturized analytical devices. Their fundamental feature, in contrast to chemical sensors, is that a biological recognition element (also known as a bioreceptor) is intimately coupled to a transducer, enabling the selective detection of a particular target analyte [75]. Bioreceptors are functionally active molecules, and this functionality is usually maintained over a predefined range of certain parameters such as pH, temperature, and ion concentration, which presents some limitations from the design point of view that must be considered when developing wearable biosensing devices [76].

Figure 5 is a schematic representation of the basic structure of a biosensor. Typically, biomolecules are attached or immobilized onto a biocompatible surface in the body-wearable interface. A transducer coupled to that biocompatible layer converts the biochemical responses coming from the interaction between analytes and bioreceptors into a readable/measurable signal (optical, electrical, thermal…) that can be amplified and processed for analyte concentration monitoring.

When designing a wearable biosensing device, it is also important to consider biosensor placement, and the fluid or matrix where the target analyte is measured, as that will determine the required sensitivity of the biosensor as well as possible interferences or incompatibilities. For example, as skin covers most of the body surface, it seems reasonable to consider sweat as an easily available biological fluid to measure various analytes of interest. Biomarkers available to be measured in sweat are well documented, and some analytes exhibit strong correlation between sweat and blood concentrations. This is usually the case for small lipophilic analytes such as hormones (cortisol, testosterone…) and various drugs (ethanol, methylxanthine, levodopa…) that are known to travel transcellularly through the lipophilic cell membranes [77]. However, bigger, or more hydrophilic compounds that enter the sweat through alternative routes (paracellular, vesicular…) experience a considerable dilution during this process, decreasing their measurable concentration. This is notably the case for glucose, which is found in sweat over 50 times more diluted than in blood at concentration values around 0.06 to 0.2 mM, corresponding to 3.3 and 17.3 mM of blood glucose respectively [78]. Other interesting analytes available in sweat that lack blood-sweat correlation are lactate and urea [79]. This underlines the need for highly sensitive and selective detection systems, and the use of bioreceptors such as enzymes, antibodies, aptamers, or DNA (that bind selectively to specific target analytes), combined with more efficient sampling methods, could be a way to solve some of these problems and could play a key role in the evolution of wearable monitorization devices.

This section will mostly focus on the recognition sites of the biosensing device, looking at the various types of bioreceptors and their incorporation into wearables. Some examples of recent advances in this field will help understanding the role of biomolecules in fully functional wearable devices and provide an overview of the main advantages and disadvantages of each biomolecule and the techniques employed to immobilize them for repeated usage.

7.2 *Enzymatic Biosensors*

Enzymatic biosensors have been extensively used for the fabrication of wearables and monitorization devices for direct detection of small target analytes (glucose, urea, catechol, glutamate…) or inhibition-based indirect detection of heavy-metals and other pollutants. These types of sensors are based on the biocatalytic ability of enzymes to convert target analytes at a high rate, which allows for the measurement of changes caused by that reaction (H^+ concentration, electron transfer, heat, or light emission…) [80]. The main advantage of enzymes is their great selectivity due to their unique 3D structure. However, this high selectivity comes at the expense of relatively high production costs and structural instability under certain pH and temperature conditions that can lead to loss of activity, limiting their repeatable usage. For that reason, immobilization of enzymes on electrode surfaces is a prominent area of research nowadays, as it is a way to extend the lifespan and reduce the degradation of these bioreceptors [81]. The focus is on developing strategies that allow for the enzyme mobility required for catalysis (folding and conformational changes), while keeping them uniformly immobilized on the substrate, avoiding denaturing and aggregation to allow for catalytic activity under various reaction conditions. Figure 6 summarizes the main approaches for enzyme immobilization: physical adsorption, covalent binding, entrapment, and cross-linking. These techniques have been used for the fabrication of enzymatic biosensors as well as for devices that use other types of bioreceptors (e.g., antibodies, DNA, or aptamers).

7.2.1 Physical Adsorption

Physical adsorption is a simple and quick method for manufacturing enzymatic biosensors. Although this method has the benefit of speed and simplicity, unfavorable orientations and decreased functionality are likely. Usually, this method is carried out by some sort of prior modification of the surface of the electrode in order to provide the functionalization that allows for non-covalent interactions (Van der Waals forces, electrostatic interactions, H-bonding…). After a proper functionalization of the working surface, enzymes can be adsorbed by simply dipping the material into a solution containing the bioreceptor. An example of this could be the reduction of gold nanoparticles (AuNPs) with a negatively charged ligand like citrate, allowing for electrostatic interactions between the positively charged amino residues of enzymes and the now negatively charged surface of AuNPs [82].

Nanomaterials seem to be the preferred option for surface functionalization. Nanoparticles (NPs), nanotubes (NTs), and nanoarrays (NAs) have distinct properties due to their reduced size and unique morphology [83]. Some nanomaterials, such as the aforementioned AuNPs and carbon nanotubes (CNTs) are popular choices for biosensor electrode functionalization, as their great chemical stability, relative non-toxicity, high surface area, and good electron transfer capabilities, make them compatible for biomolecule immobilization, and at the same time, act as the electrochemical or optical (in case of AuNPs) transducer of the biosensor [84]. For example, Yanyan Niu et al. used star-shaped AuNPs and horseradish peroxidase (HRP) for biosensor electrode modification with enhanced electrocatalytic activity [85]. Fang et al. combined 3D nanoarchitecture ZnO with gold nanoparticles for successful immobilization of GOx enzyme on a high-performance glucose-sensing electrode, achieving satisfactory low detection limits (0.02 mM) and an acceptable linear range (1–20 mM) [86].

Hydrophobic domains of proteins and enzymes have a natural affinity toward carbon nanotubes, which allows for spontaneous adsorption of the biomolecules [82]. In some cases, is also possible to modify the enzyme as well as the transducer surface to include functional groups for non-covalent interactions. Holzinger et al. immobilized cyclodextrin-tagged glucose oxidase (CD-GOx) and histidine-tagged glucose oxidase (His-GOx) via coordination and host–guest interactions on adamantane-modified single-wall carbon nanotubes (SWCNT) [87]. However, despite advances on surface modification, enzyme immobilization through physical adsorption still faces issues with durability and repeated usage. For that reason, even if it has been used in non-wearable or single use biosensing devices, there is a lack of reported wearable biosensors for continuous analysis and monitorization.

In contrast to the soft approach of physical adsorption, **covalent bonding** provides a strong and effective binding of the enzyme to the substrate or transducer surface. It is one of the most used enzyme immobilization techniques, creating a usually uniform and stable biomolecule layer, that allows for increased lifetime and reusability of the biosensor [88]. Side chain amino acids (e.g., lysine, cysteine, aspartic acid, glutamic acid…) of enzymes and other proteins have reactive functional groups such as -OH, -NH_2, -COOH, or -SH that can be used for covalent bond formation [89]. It is crucial, however, that the amino acids involved in the immobilization process are different from the ones on the active site of the enzyme, as mistakenly binding the active site amino acids to the substrate can lead to severe conformational changes that result in denaturization of the enzyme and its loss of catalytic activity [75]. Despite these drawbacks, and the large amount of bioreagent required, the high uniformity of the covalently bonded monolayer and the good control over the amount of immobilized enzyme are solid advantages of this immobilization method [89].

7.2.2 Cross-Linking

This is another enzyme immobilization technique that has the advantage of providing strong chemical binding between the biomolecules using relatively simple and fast methods such as drop casting and dip coating. Enzymes are cross-linked to each other using bifunctional reagents (glutaraldehyde, glyoxal, or hexamethylenediamine, for example). This allows for greater stability for repeated usage, but it comes at the expense of possible distortion of the enzyme 3D structure during cross-linking reaction and its consequential loss of activity, although inert proteins such as gelatine or bovine serum albumin (BSA) are often added during the process to minimize enzyme damage [90]. Cross-linking of enzymes is sometimes combined with electropolymerization of biocompatible polymers used for protein stabilization such as polyethyleneimine (PEI) [91].

Following this approach, enzymatic biosensors have successfully been incorporated into wearable devices for the detection of small analytes in different matrices during the last decade. For example, in 2015, Kim et al. integrated cross-linked uricase enzyme on a screen-printed electrode system for a mouthguard wearable device for the detection of uric acid in saliva [92] and in 2017 Tur-García et al. prepared an enzyme laminate of cross-linked lactate oxidase/BSA placed between two polycarbonate membranes to fabricate a flexible wearable sensor for lactate analysis in sweat [93].

A widely used enzyme immobilization method in biosensor fabrication is the **entrapment** of the enzymes in a semipermeable or porous system where ions, electrons, and small molecules can freely flow. This is usually achieved by the use of membranes, polymeric or inorganic 3D networks, and hydrogels.

The first biosensor using a semipermeable membrane for enzyme entrapment was developed in 1962 by Leland Clark and Champ Lyons for a glucose detection electrode immobilizing glucose oxidase (GOx) enzyme [94]. In recent times, polymeric films in biosensors are used for enzyme immobilization as well as protection of the transducer material. This becomes relevant following the recent advances in electrochemical sensing materials that convert common enzyme-catalyzed reaction products (e.g., H_2O_2) into measurable electrical signals. Some examples of these materials are conductive carbon inks and Prussian Blue inks, which exhibit excellent electrochemical performance with relatively low toxicity, but are unstable over extended periods of time and that limits their use in wearable devices [95].

Hydrogels are a popular choice of material for the entrapment approach. Being an insoluble 3D network that can hold fluids provides a few advantages. Hydrogels establish a liquid contact between the biosensor and the body surface, which allows for analytes to flow through. They also create a viscous environment that protects the three-dimensional conformation of the enzyme, minimizing the loss of activity. They can also serve as skin contact material and extraction site, as a reservoir, or as an electrolyte for the transducing system [96]. Some biopolymers extracted from natural resources, such as chitosan or alginate, can form gels by cross-linking or complexation, and have the advantage of being biocompatible and inert, making them suitable for direct contact with the body [97].

Hydrogels were used as enzyme entrapment matrices for biosensing purposes all the way back to the mid-60 s, with Updike and Hicks' glucose-sensing electrode using GOx immobilized in an acrylamide gel [98]. A famous example of the use of hydrogels for this purpose has already been mentioned in this chapter: the Glucowatch, the first commercial glucose monitorization device [3]. Nowadays there is still a prominent use of hydrogels in wearable enzymatic biosensors, now taking advantage of modern manufacturing techniques and materials. Gun Jin Kim and Kyu Oh

Kim used electrospinning to create hydrogel nanofibers immobilizing GOx enzyme for the fabrication of a flexible and transparent glucose-responsive sweat patch [99]. Jayoung Kim et al. created a flexible tattoo-like alcohol biosensor via screen-printing of a Prussian Blue electrode immobilizing alcohol-oxidase enzyme in a chitosan matrix with an agarose gel cover [100].

A drawback of the entrapment approach is that, even though analytes can flow through the hydrogel matrix, diffusion is still restricted, which affects the sensitivity and response time of the biosensor [101]. Nagamine et al. developed a non-invasive enzymatic sweat-lactate biosensor employing an agarose gel touch pad as enzyme immobilization and sweat extraction site [102]. They pointed out that the signal detected did not reflect the actual lactate concentration in sweat, but the concentration extracted in the agarose gel. On top on that, since analytes diffuse directly into the gel, there is no control over the volume of sweat extracted, so the amount of lactate detected depends not only on the concentration in sweat but also on the perspiration rate.

An alternative to conventional cross-linked polymer hydrogel entrapment could be inorganic sol–gel matrices. These porous materials are chemically inert and can be formed under mild conditions and maintain enzyme structure. The key differences when comparing to their organic counterparts are the mechanical stability and neglectable swelling of the sol–gel matrices. This restricts undesired movement and conformational changes of the trapped enzymes and results in good preservation of the chemical and bioactive properties. However, the rigidity of the matrix makes its use limited for wearable applications that require flexibility, and even though the synthesis conditions are mild, and the method is relatively simple, sol–gel process can be exceedingly long sometimes [103].

8 Immunosensors

Antibodies are specialized immunoglobulins, a type of glycoproteins produced by lymphocyte cells in response to the presence of a foreign species known as antigen. Among immunoglobulins, the most widely used for biosensing (IgG) have a Y-shaped structure where two identical pairs of light chains (25,000 Da) and heavy chains (50,000 Da) are linked together by disulphide and hydrogen bonds [104]. Antibodies have been used as bioreceptors for a long time due to their broad application range, strong antigen–antibody interactions, and high specificity and sensitivity. Biosensors based on this type of interactions are called immunosensors [88].

In a similar manner to other biomolecule-based sensing devices (enzymatic, DNA…), immobilization of the bioreceptor on the surface of the transducer material is a crucial step in the fabrication if immunosensors. Surface coverage, binding stability, or orientation of the immobilized antibodies influence the detection performance of the system and need to be considered during the manufacturing process.

Table 3 DNA-based wearable biosensors

S. No	Probe	Immobilization method	Transduction method	Mechanism	Analyte	Limit of detection	Ref.
1	Aptamer/ SiO_2microneedle arrays-based skin patch	EDC/NHS-based Covalent immobilization	Optical/ Colorimetric	SiO_2 nanoparticle hydrogen swells upon the aptamer analyte interaction and changes its color due to the photonic nanocrystals	Histamine in rat muscle	2 μg/mL	[183]
2	dCAS9/Graphene/ Gold microneedle skin patch	EDC/NHS-based Covalent immobilization	Electrochemical	CRISPR/dCAS9-based detection for continuous measurement	Universal cell-free DNA	1.1 fM	[148]
3	5'NH-ssDNA/P-Si electrode	APTES/ Glutaraldehyde cross-linking	Optical	Hybridization of 5´NH-ssDNA with HPV complimentary DNA	Human papilloma virus	6,700 copies	[152]
4	Thiolated MB-modified DNA aptamer/Gold microelectrode	Thiolated aptamer immobilization on gold electrode surface	Electrochemical	Aptamers undergo reversible binding induced conformational changes upon interaction with the analyte molecules that affect electron transfer (eT) between the reporter and the gold needles	Tobramycin, Irinotecan, and Doxorubicin	Tobramycin = 1 μM, Irinotecan = 10 nM Doxorubicin = 10 nM	[120]
5	6-FAM-Biotin-SS DNA probe	Freeze dried in the μ-PAD system integrated into face mask	Optical	CRISPR-Cas12a-SHERLOCK-based 6-FAM-Biotin-SS DNA cleavage	SARS-CoV-2	17aM/500 copies	[182]
6	6-FAM-5' Iowa Black® FQ quenched ssDNA fluorescent reporter	Quickly deposited in-fabric to be snap-frozen and then lyophilized for 4–8 h within the device	Optical	CRISPR-Cas12a-based 6-FAM-5' Iowa Black® FQ quenched ssDNA cleavage	*mec*A/*spa*/ *erm*A genes	2.7 fM	[182]

(continued)

Table 3 (continued)

S. No	Probe	Immobilization method	Transduction method	Mechanism	Analyte	Limit of detection	Ref.
7	Methylene blue tagged aptamer	Thiolated aptamer immobilized on gold nano particle modified surgical microneedle patch implanted in rat for the measurement of administered drugs in blood	Electrochemical	Aptamer undergoes conformational change from linear upright to bent position upon interaction with analyte	Tobramycin and vancomycin	20 mg/kg of Tobramycin and 40 mg/kg of vancomycin	[184]
8	MB tagged poly-thymine (T-12) pseudoknot aptamer	MB tagged thiolated aptamer self-assembled on gold screen-printed electrode	Electrochemical	Pseudoknot unwinds bringing the MB redox probe close to the gold electrode surface upon cortisol interaction	Cortisol in sweat	0.2 pM	[169]
9	Guanine-rich VR11 aptamer	Aptamer was cross-linked to graphene surface via 1-Pyrenebutanoic acid succinimidyl ester (PBASE)	Field effect transistor	Change in charge carrier concentration due to confirmation switch of the aptamer to G-quadraplex upon interaction with the analyte	TNF-α	26 pM	[173]

Physical adsorption is a fast and simple strategy, as it was for enzyme immobilization. Functionalized surfaces and nanomaterials are used to create a self-assembled monolayer (SAM) of antibodies via non-covalent interactions such as H-bonding and electrostatic interactions. Due to the difference in the functional groups present in the F_c section of the bioreceptor (abundance of $-COOH$ groups) and the F_{ab} sections, where the active antigen binding sites are found (abundance of $-NH2$ groups), changing some parameters such as pH can favor certain biomolecule orientations when forming the monolayer. This phenomenon was studied, for example, to immobilize anti-horseradish peroxidase antibodies (anti-HRP) on gold nanoparticles (AuNPs), finding that pH 7.5 provided more favorable orientations, with the active sites of the F_{ab} sections facing outwards, than pH 8 or higher [105]. However, physical adsorption still faces some limitations, as, because of the weak interactions, the final concentration of the immobilized antibodies is usually low. Some other factors that can affect the quality of an adsorption-based immunosensor are surface contamination at a low concentration of antibody, blocking active sites when the adsorption takes place near to the substrate surface, leading to loss of binding capacity, or partial denaturation of the immunoglobulin [104].

Covalent bonding is a more popular approach for antibody immobilization, and it has been used for immunosensors that have been successfully incorporated in working wearable devices. Jingwei et al. developed a multifunctional biosensing patch for simultaneous detection of various biomarkers. Enzymatic detection of ascorbic acid via hydrogel-entrapped ascorbic acid oxidase (AAOx) was complemented with the detection of neuropeptide Y (an important biomarker related to cardiovascular diseases) using its complementary antibody NPY-Ab. Immobilization of the antibody on the surface-modified Au transducer was carried out via amide-bond formation from the reaction of a self-assembled monolayer of carboxylic acid groups and the amine groups of the antibody. This covalent approach, as we have previously seen for enzyme immobilization ensures strong binding and proper orientation of the bioreceptor [106].

A popular strategy is to first form a SAM on the electrode surface and then covalently bind the antibody to the monolayer. SAM-forming alkanethiols are commonly used for covalently immobilization of antibodies on the surface of electrodes since they offer good surface coverage, stable covalent binding, and the possibility of controlling the orientation and distribution, of the sensing element while reducing non-specific interactions [107]. Functionalization of gold surfaces with highly ordered monolayer of alkyl thiols can be efficiently achieve via strong S–Au bond [107]. This combined approach allowed Han-Byeol et al. (2020) to build a stretchable lab-on-patch immunosensor that used an Au-ZnO nanostructured electrode with a di(N-succinimidyl)3,3'-dithiodipropionate (DSP) monolayer to covalently bind an antibody probe biomolecule for selective cortisol detection [108]. Alternatively, antibodies can be engineered to contain modifications at site-specific locations to control orientation upon immobilization. However, even though it is effective for that purpose, simpler modification methods need to be developed, as they currently involve complex conjugate synthesis with additional steps or require advanced protein engineering [105].

8.1 DNA-Based Biosensors

DNA-based biosensors have become increasingly popular due to their ability to detect and quantify biological and chemical analytes with high sensitivity and specificity. One of the most significant applications of DNA-based biosensors is the detection of DNA association with small molecules, such as drugs and chemicals [109]. These biosensors can be utilized in various fields, including pharmaceuticals and environmental monitoring, to determine the presence and concentration of small molecules of interest. Another application of DNA-based biosensors is the investigation of DNA interactions with proteins [110]. Biosensors can detect catalytic activities of DNA-processing enzymes and study affinity interactions between DNA and proteins, providing valuable insights into various cellular processes [111].

Aptamers are single-stranded DNA or RNA molecules that can bind to specific target molecules with high affinity and specificity. They are often referred to as "chemical antibodies" due to their ability to recognize and bind to targets with high specificity and affinity, similar to antibody-antigen interactions. Aptamers are selected from a large pool of oligonucleotides through a process called SELEX (Systematic Evolution of Ligands by Exponential Enrichment). During SELEX, a random pool of oligonucleotides is incubated with the target molecule of interest, and the oligonucleotides that bind to the target are isolated and amplified by polymerase chain reaction (PCR). This process is repeated several times to enrich for oligonucleotides with high affinity and specificity for the target [112]. There have been significant development over microfluidic electrochemical aptamer-based sensors. These sensors offer a promising approach for non-invasive and real-time monitoring of biomarkers in different biofluids such as blood, sweat, and interstitial fluid [113]. As an example, Ferguson et al. developed an ex vivo microchip that accesses the bloodstream via catheters, allowing continuous and sensitive detection of drug concentrations. By integrating an electrochemical aptamer-based (E-AB) sensor with a continuous-flow diffusion filter, the system effectively prevents biofouling and interference from blood-borne substances. The sensor achieved a low limit of detection (LOD) for the chemotherapy drug doxorubicin and successfully monitored its concentration over an extended period in live animal models [114]. Another area of focus is the use of wearable sweat sensors for continuous non-invasive monitoring of biochemical biomarkers. Human sweat, which contains valuable diagnostic information, serves as an attractive medium for capturing molecular data [115, 116]. Researchers have developed wearable, flexible sensing platforms that can detect multiple sweat metabolites, electrolytes, and even drug concentrations. For example, a study where a wearable sweat sensor successfully monitored caffeine levels in real time during exercise [117, 118]. This concept has been expanded to continuously collect other drugs, such as levodopa, in sweat for personalized management of conditions like Parkinson's disease [119]. Another important application of wearable microneedle aptasensors for continuous monitoring of chemical biomarkers in interstitial fluid. Microneedle arrays penetrate the outermost layer of the skin, allowing minimally invasive and rapid sensing [120].

Lastly, DNA-based biosensors have been used for in vitro DNA damage detection, which can provide crucial information for assessing the genotoxicity of various compounds [121]. There is wide spectrum of transducing methods such as optical, acoustic, gravimetric, electrical, and electrochemical techniques have been employed in DNA-based biosensors, making them versatile tools for wearable sensor applications [122–126].

Integration of nucleic acid amplification methods with wearable biosensors for improved sensitivity in DNA-based sensing. Traditional amplification techniques like PCR, which require thermal cycling, are not suitable for wearable devices. Therefore, isothermal nucleic acid amplification includes Nucleic Acid Sequence-based Amplification, Loop-mediated Isothermal Amplification, Strand Displacement Amplification, Rolling Circle Amplification, and Recombinase Polymerase Amplification (RPA) [127]. RPA, in particular, stands out due to its fast reaction time and near-physiological temperature (37 °C). It has been successfully employed in wearable devices as it is highly sensitive and can utilize the heat generated by the human body as a heat source. The use of such techniques enables point-of-care testing with improved detection capabilities, making wearable biosensors valuable tools in various diagnostic and monitoring applications [128].

This section is focused to discuss the different immobilization methods which are suitable for the wearable biosensor with examples and the changes in DNA which leads to signal generation after binding to the analyte of interest. This includes DNA hybridization, DNA conformational changes, and CRISPR-CAS (clustered regularly interspaced short palindromic repeats—with CRISPR-associated Protein)-based detection methods.

Figure 7 is schematic diagram which shows a wearable biosensor with plausible mechanisms and components of wearable DNA. There are several wearable biosensors mentioned in the Table 1, which covers detection mechanism, immobilization technique for the probe molecule on the transducer surface and detection limits.

8.1.1 Immobilization Methods

Functionalization of transducer surface with the DNA probe depends on the choice of application, which in turn decides the sensitivity, stability, and minimizes the cross reactivity of the sensor. There were different types of immobilizations of DNA probes on the functional surfaces such as adsorption, covalent immobilization, avidin–biotin interactions, and entrapment.

8.1.2 Adsorption

DNA oligomers, aptamers, and SS-DNA can be immobilized on transducer surfaces by adsorption. Negatively charged DNA strands can be adsorbed onto positively charged surfaces via electrostatic interactions. The adsorbed DNA is oriented randomly, presenting different directions without uniformity. Cationic polymers such as Poly(L-Lysine), chitosan, polyethyleneimine, and poly(3-(3′-N,N,N-triethylamino-1′-propyloxy)-4-methyl-2,5-thiophene) (PMNT) were used to demonstrate for the DNA immobilization for biosensing applications [129–132]. FAST™ slides are a kind of glass slides with microporous nitrocellulose surface was used as microarray technology. The FAST™ surface can be modified with DNA molecules, which adsorbs on the polymer surface irreversibly. These slides were employed for fluorescent and chemiluminescence-based assay [133]. DNA can be immobilized via physical adsorption on surface such as pyrolytic graphite electrode, glassy carbon electrode, and gold electrode [134]. FAM (fluorescein) labeled DNA oligonucleotide was immobilized on 13 nm sized gold nanoparticle. The FAM labeled DNA incubated with gold nanoparticles at lower pH for a duration of 5 to 8 min followed by centrifuging and washing steps in PBS at neutral pH condition [135]. Wang et al. demonstrated metal–organic frameworks (MOFs) with different metal compositions such as Zr, Cr, Fe, and Al were successfully adsorbed with DNA oligonucleotides with rich phosphorous moieties for the successful demonstration of measuring and manipulating intracellular processes [136]. Sun et al. successfully developed colorimetric sensor array by adsorbing thiolate SS-DNA gold nanoparticles on zirconium metal–organic frameworks (Zr-MOFs) for the determination of semen quality of humans [137]. Xiong et al. developed a MOF, heme-like ligand FeTCPP into commonly used MOFs (UiO-66) (FeTCPP $\subset$ UiO-66) and modified by incubating 6-carboxy fluorescein (FAM)-labeled SS-DNA for a duration of 10 min. DNA physically adsorb on the MOF surface with one among the electrostatic, π-stacking, hydrogen-bonding, and coordination interactions and was used to perform dual fluorescence resonance energy transfer (FRET)-based fluorescence quenching assay and peroxide mimicking colorimetric assay for the determination of AFB1 (Aflatoxin B1) [138].

Though several biosensors were designed based on DNA physical adsorption on the substrates such as gold nanoparticles, polymers, MOFs, gold and graphite electrodes their use in the design of real-time biosensor will be limited due to non-specific, disoriented immobilization on the surfaces [139, 140]. Moreover the adsorption of DNA is susceptible to varying conditions such as pH, salt concentrations, non-specific desorption, etc. [141].

8.1.3 Covalent Immobilization

Covalent immobilization methods are based on oligonucleotide covalently binding to the immobilizing surfaces, which offers good stability, flexibility, binding strength, and provides a suitable vertical orientation to the detecting probe [142].

The oligonucleotide typically binds covalently to the amine or carboxy terminated self-assembled monolayers on the gold surfaces at the 3′ or 5′ end. Thiols show a strong affinity to the gold surface, also referred chemisorption, attaching covalently forming a self-assembled layer.

Glenn et al. conjugated oligonucleotide with gold nanoparticles by mixing 1:20 volume ratio of 100 μM thiolated oligonucleotide with gold nanoparticles and incubated at 37 °C for 30 min, followed by addition of 150 μL of 1 M NaCl/100 mM phosphate buffer of pH 7.0 and incubated at 37 °C for aging. Excess oligonucleotides were removed by centrifuging the volume, and resuspending into 0.3 M NaCl/ phosphate buffer at pH 7.0 for several cycles [143]. Ahmadi et al. prepared thiolated oligonucleotides by adding nucleotide to 0.1 N dithiothreitol (DTT) and incubated for 15 min. Excess DTT was removed by washing thrice with ethyl acetate. The thiolated DNA-gold nanoparticle mixture was used as colorimetric sensor for the detection of Klebsiella pneumoniae [144]. Self-assembly of the 32-mer-5′ oligonucleotide on Nano-Au electrode surface achieved by adding thiolated DNA to the electrode surface and incubating at –4 °C for 6 h. The developed DNA-based electrochemical biosensor is used for DNA-based bioassay of legionella pneumonia pathogen [145].

Covalent attachment of modified probe on functionalized surface can be achieved by self-assembled monolayers on the transducer surface. As it is well known thiol show strong affinity toward gold surfaces, several alkane thiols, dithiols, and thiol containing amino acids [146, 147]. These self-assembled monolayers mostly terminated with carboxyl functional groups which can be attached to the amine terminated oligonucleotide probe via carbodiimide chemistry, which also referred as 1-ethyl-3-(3-dimethylaminopropyl) carbodiimide hydrochloride and N-hydroxy succinimide coupling (EDC-NHS coupling). EDC enables coupling carboxylic groups to primary amines of the linking molecules by forming and amine-reactive O-acylisourea intermediate. NHS mediates to form active amine ester which will efficiently couples the EDC mediated reaction.

Yang et al. developed graphene oxide decorated gold micro needle electrode to demonstrate CRISPR-dCAS9-based skin-patch biosensor for long-term capture and real-time monitoring of universal cell-free DNA. 1-pyrenebutanoic acid (PBA), binds to graphene surface with π–π interaction was used to establish EDC-NHS coupling between COOH of PBA and dCAS9. 1:1 ratio of 4 mM EDC and 11 mM NHS mixture in 50 mM 2-morpholinoethanesulfonic acid buffer was added and incubated for a duration of 60 min. 1% of BSA was used as blocking agent to avoid non-specific binding [148]. 1-pyrenebutanoic acid succinimidyl ester (PBASE) is analogue to PBA and can bind to graphene-modified surfaces in the similar fashion. The advantage of using PBASE is EDC-HNS activation is no longer required, since succinimide ester already exist on PBASE. Kusku et al. modified graphene surface with PBASE by adding 0.2 mM PBASE/N, N Dimethylformamide (DMF). The PBASE modified graphene surface was directly immobilized with probe DNA and incubated at 4 °C overnight. The developed sensor was used as DNA-graphene field effect transistor (GFET) as a proof of concept for the microfluidic-graphene-based molecular communication receiver for Internet of Nano Things (IoNT) [149].

Glutaraldehyde is a widely used alternative as a fixative and cross-linking agent in biological assays. It is a dialdehyde whose aldehydic groups are highly reactive and can form covalent bonds with functional groups such as amines, thiols, phenols, hydroxyl, and imidazoles.

Most commonly glutaraldehyde was employed in coupling reaction between two amine terminated chemical species via its two aldehyde groups, forming water molecule as a by-product, also referred as aldehyde-ammonia condensation reaction. Chitosan-multiwalled carbon nanotunes (CS-MWCNT) modified glassy carbon electrode (GC) was incubated with 1% glutaraldehyde for 2 h followed by cross-linking amine terminated aptamer for the electrochemical detection of tetracycline with detection limit of 5.6 fM [150]. ~ 60 nm thick YbTixOy on silica wafer with ~ 400 nm-thickness Al film at the bottom was used as transducer for the electrochemical biosensor. 3-aminopropyl triethoxysilane (APTES) was immobilized on sensor surface which provide hydroxyl groups. Amine groups pf APTES SAM layer were modified with 2.5% glutaraldehyde solution overnight followed by addition of ssDNA probe for the electrochemical detection of KRAS and BRAF gene mutations in colorectal cancer [151]. 5'Amine C12 modified ssDNA was used as probe molecule and APTES was used as SAM layer on electrochemically etched boron doped porous silica substrate. Glutaraldehyde cross-linking agent was used to bridge the ssDNA probe and APTES molecules. The developed sensor was used as optical sensor utilizing the porous structure of silica for the detection of human papilloma virus (HPV) detection [152]. Owing to the excellent cross-linking ability of the glutaraldehyde as across-linking agent it also offers high molecular weight and hydrophobicity due to the replacement of amino group with aldehyde group. This may affect the confirmations of the probe/analyte molecule confirmation, in turn affecting the sensitivity of the detection system [153].

8.1.4 Avidin (Streptavidin)–Biotin Interactions

Biotin, a vitamin B7 derivative, exhibits exceptionally strong and specific binding to avidin and streptavidin proteins. This characteristic has made the avidin–biotin and streptavidin–biotin interactions a cornerstone of molecular biology and biochemistry research, enabling the development of a wide range of applications such as protein purification, cell labeling, and cross-linking. In cross-linking applications, biotinylated molecules are attached to either avidin or streptavidin, which in turn bind specifically to the other protein or molecule of interest. The resulting complex forms a stable, non-covalent interaction that can be utilized to link enzymes, antibodies, or DNA. The tetrameric avidin protein, isolated from egg whites, and the homotetrameric streptavidin protein, derived from the bacterium Streptomyces avidinii, each have four high-affinity binding sites for biotin. These strong and specific interactions have dissociation constants in the femtomolar range, making them ideal for applications requiring high affinity and specificity. Streptavidin with an isoelectric point (pI) equal to 5.0 is thus preferably used over avidin, which has a pI of 10.5, to avoid non-specific interactions.

5′ biotinlyted DNA probe was immobilized on glassy carbon electrode by cross-linking the DNA probe to the avidin immobilized GCE. The avidin was immobilized on GC electrode by modifying the GC electrode with COOH terminated 4-carboxyphenyl diazonium salt. COOH groups of the GC electrode were activated by carbodiimide chemistry. The avidin–biotin interaction was achieved by immersing electrode in 0.1 M PBS at pH 7.0, which contains 10 pM biotinylated DNA probe molecules and incubated for one hour at room temperature followed by washing the electrode with 0.1/ SDS to remove unbound DNA probes [154].

Liu et al. designed biotinylated gold electrode sensor array by forming a mixed self-assembling monolayer (SAM) with 11-mercaptoundecanoic acid (MUA) and 11-mercapto-1-undecanol (MU) and activating the COOH with EDC-NHS chemistry to cross-link with biotin. A dual labeled the stem-loop probe oligonucleotide (oligos 1 and 6), has a 5′-digoxigenin (DIG) affinity label and a 3′-biotin was cross-linked to biotinylated electrode surface via biotin-avidin–biotin cross-linking. As a result, the DIG label becomes available by the anti-DIG-HRP, and the target hybridization event can be efficiently transduced via the enzymatically amplified faradaic electrochemical electron transfer phenomenon to sensitively detect femtomolar detection of sequence specific DNA [155]. Pan et al. conducted similar studies, where gold coated Indium tin oixide glass substrate was passivated with mixed thiols of 2-mercaptoethanol (2-ME) and 11-mercaptoundecanoic acid (11-MUA) and activated carboxylic group of 11-MUA by carbodiimide chemistry. The activated COOH group was covalently bound with streptavidin, was meant to bind with 5′ biotinylated probe sequence SS-DNA. The detection event was performed by exposing the mixture of DNA samples to the electrode surface, the hybridization event was transduced into electrochemical event in terms of increment in the charge transfer resistance (Rct). The detection limit was noted up to picomolar of concentration of DNA [156].

8.1.5 DNA Probe-Analyte Interactions

DNA undergoes physical changes which leads to signal generation after binding to the analyte of interest. Based on the changes post to the DNA probe and analyte interaction DNA biosensor can be broadly divided into two types (1) DNA biosensors based on hybridization and (2) DNA biosensors based on conformational changes.

8.1.6 DNA Biosensor Based on Hybridization

DNA biosensor based on the hybridization were studied extensively. Hybridization-based detection mechanism can be broadly divided into two types (1) Label free DNA detection and (2) Labeled DNA detection.

Label free detection of DNA is mostly based on electrochemical detection utilizing the redox properties of DNA base pairs. Purines such as Adenine and Guanine can undergo electrooxidation at lower potentials than that of pyrimidines, moreover purine attracted the focus due to their ability to bind with various labeling molecules such as redox probes, fluorescent probes and dyes, etc. [141, 157]. The principle behind the label free electrochemical DNA biosensors is interaction of guanine with its complementary molecule cytosine and adenine with thiamine during DNA hybridization process leaving few free adenine/guanine moieties for electrochemical oxidation. As a result, the faradaic current response due to adenine/guanine decreases. Jalit et al. reported electrooxidation of nucleic acids at Poly-L-lysine-MWCNT modified glassy carbon electrode. Single-stranded electrodes were deposited electrostatically on the Poly-L-lysine modified multiwalled carbon nanotubes by immersing in a PBS solution with 50 mg/mL DNA. In this study guanine was monitored for the electrooxidation studies using adaptive stripping voltammetry. The results showed free guanine undergo electrooxidation resulting high current response, whereas, current response decreased as hybridization process increases [158]. The main disadvantage of label free DNA detection is sluggish electron transfer rates and poor current response. To circumvent the problem, redox probe-based electron mediators were introduced to improve the electron transfer kinetics and improved current responses. These redox mediators can either labeled on the ssDNA/dsDNA or can be used without labeling [159]. $Ru(bpy)_3^{2+}$ metal ion complex was widely used as label free redox indicator for the electrochemical detection of DNA hybridization. It was shown in Fig. 8 Guanine can undergo oxidation by reducing the oxidized ruthenium bipyridine $Ru(bpy)_3^{3+}$ to $Ru(bpy)_3^{2+}$[160] also used as probe molecule for the DNA hybridization detection by generating electrochemiluminescence (ECL) signal. Electrochemically oxidation of $Ru(bpy)_3^{2+}$ to $Ru(bpy)_3^{3+}$ can be reduced by co-reactant guanine nucleotide forming excited $Ru(bpy)_3^{2+*}$. ECL signal was generated when $Ru(bpy)_3^{2+*}$ relaxed to ground state form $Ru(bpy)_3^{2+}$. Wei et al. reported MWCNT/Nafion modified glassy carbon electrode for the detection of DNA hybridization process in P53 gene using label free $Ru(bpy)_3^{2+}$-based ECL detection. It was found that the ECL signal generation is sensitive up to 39.3 nM concentration and able to change the signal with single base pair mismatch during the hybridization process [159]. Another redox probe for label free DNA redox mediator is the widely studied $Fe(CN)_6^{3-/4-}$, which do not bind to the DNA due to the electrostatic repulsion of DNA, (depicted in Fig. 8) can be used to quantify the amount of DNA present before and after the hybridization process.

Labeled DNA hybridization-based electrochemical biosensors were reported. The labels are redox responsive which were attached to the single-stranded DNA probe molecules to monitor the DNA hybridization process [161]. The redox labels can bind to the DNA with simple immersion method for a certain period of time. Interaction event can take place between DNA and probe molecule via different modes such as electrostatic, affinity toward a specific nucleotide base, groove binding and intercalation, etc. For example, $Ru(NH_3)_6^{3+}$, a metal complex, can bind electrostatically to negatively charged phosphate of hybridized DNA, was widely used to evaluate the DNA modified surface to understand the surface coverage [121, 162]. Other redox probes such as doxorubicin, an anticancer drug also was used as a labeled indicator for the detection of DNA hybridization. Ferrocene another important redox probe binds specifically to major grove of dsDNA and perform redox reaction on the electrode surface [163, 164]. Ethidium bromide is also one of the important redox indicators for DNA hybridization, show affinity to the guanine-cytosine pair and intercalate between them.

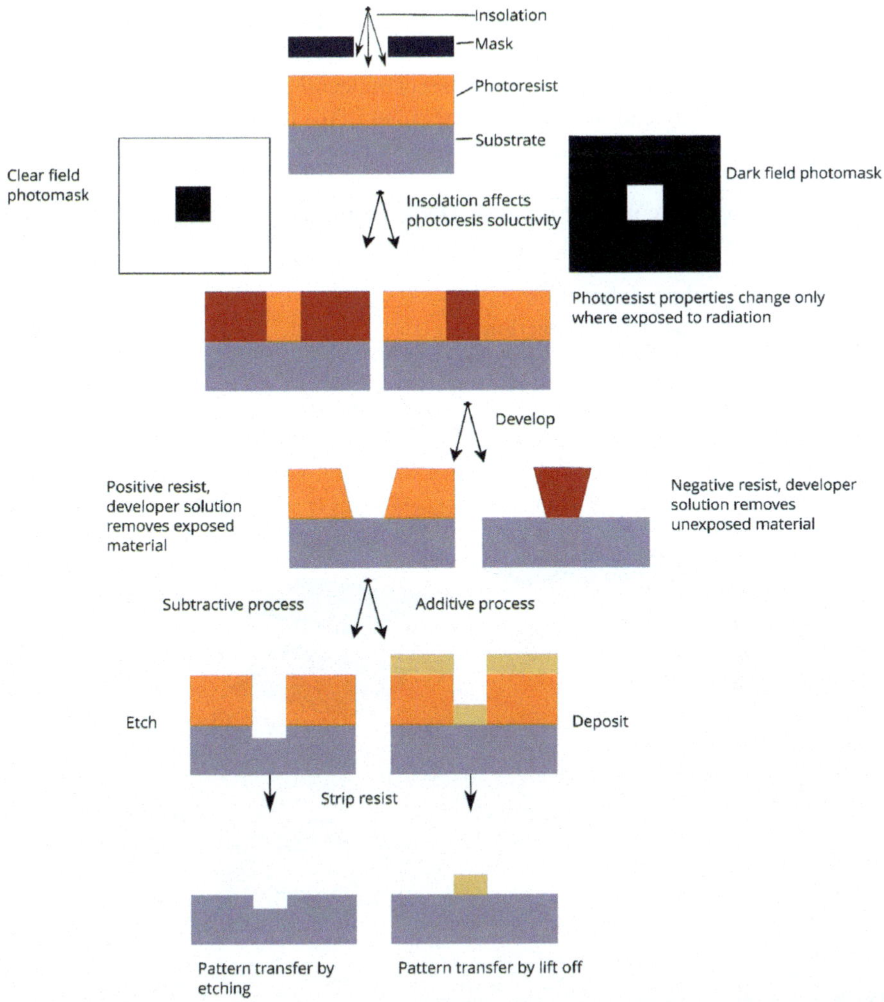

Fig. 8 Schematics summary of the photolithographic process from the wafer with photoresist to the final product

8.1.7 DNA Biosensor Based on Conformational Changes

A DNA biosensor based on conformational change leverages the inherent structural dynamics of DNA molecules to detect and analyze target molecules. The biosensor comprises a DNA probe engineered to selectively interact with the target molecule of interest. Binding or interaction between the target molecule and the DNA probe induces a conformational rearrangement within the DNA structure, which can be exploited for detection and quantification. These conformational changes can be interrogated through various sensing methods such as electrochemical, fluorescence, and surface plasmonic resonance (SPR) techniques. The detection technique purely depends on the type of detection probe tangled to the DNA and detection event happens once the DNA probe bind to the target molecule, changing the DNA's confirmation which in turn leads to change in the either of the fluorescent/electrochemical/SPR signals providing the qualitative or quantitative information of the analyte of interest.

There are several common types of DNA conformational changes widely used for detection purposes (Fig. 9). These include:

1. DNA Hairpin-based detection
2. DNA pseudoknot-based detection
3. DNA G-quadruplex-based detection.

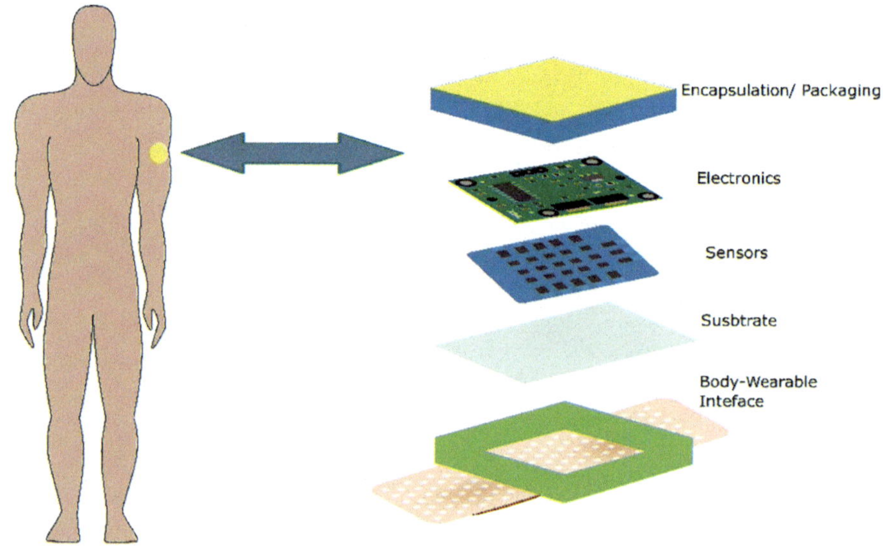

Fig. 9 Schematic representation of a wearable device

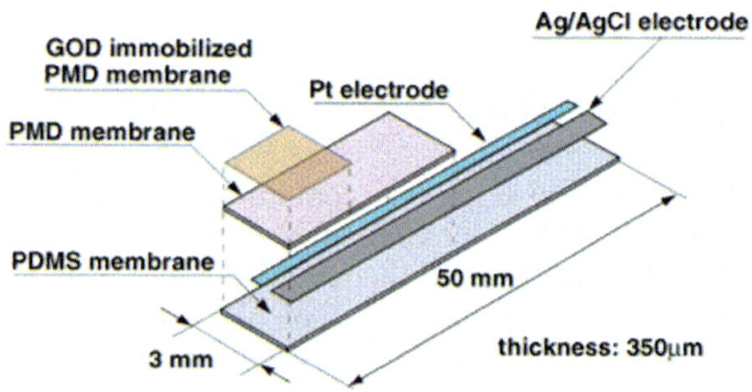

Fig. 10 Schematic representation of the flexible glucose sensor based on polymers by Kudo et al. [71]

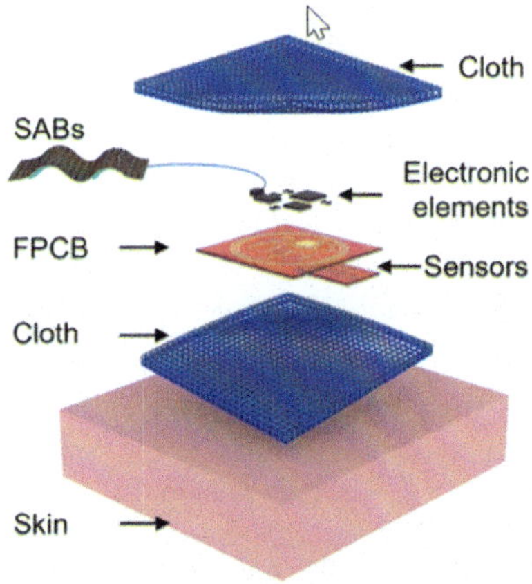

Fig. 11 Schematic illustration of a cotton-fabric-based wearable device [74]

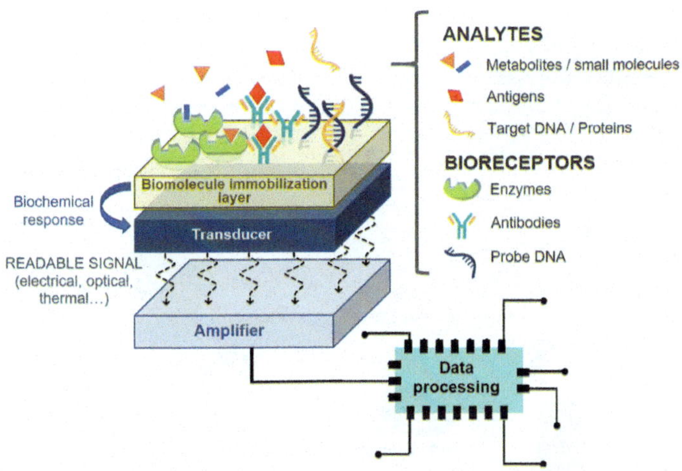

Fig. 12 Schematic representation of the components of a biosensor

8.1.8 DNA Hairpin-Based Detection

A hairpin bend DNA-based biosensor utilizes the conformational changes that occur in hairpin-structured DNA molecules for biosensing applications. Hairpin DNA structures consist of a single DNA strand that folds back on itself, forming a stem-loop structure. The stem region is composed of complementary base pairs, while the loop region contains non-complementary bases.

In a hairpin bend DNA-based biosensor, the DNA probe is designed with a specific recognition sequence for the target molecule of interest, which is typically located within the loop region. The binding or interaction of the target molecule with the

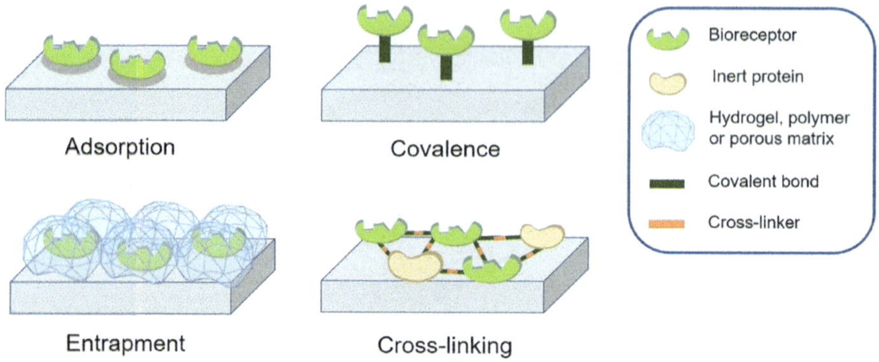

Fig. 13 Diagrammatic representation of the four main enzyme immobilization strategies for transducer functionalization

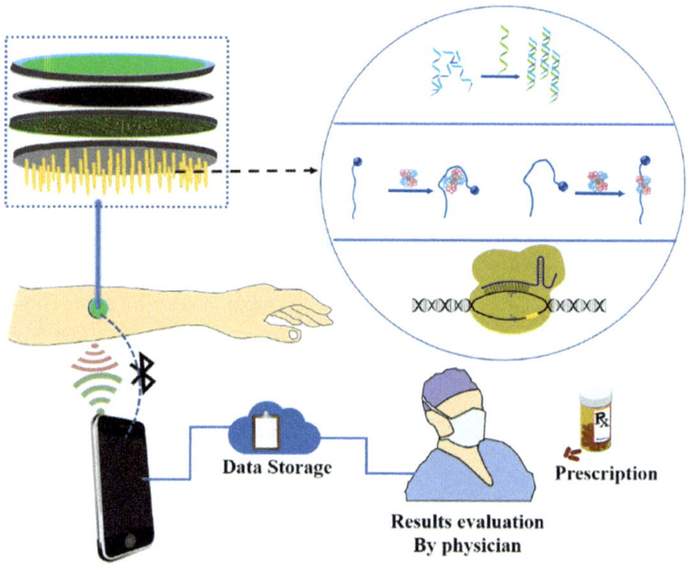

Fig. 14 Scheme of DNA biosensor depicting the different types of DNA detection mechanisms plausible for the wearable DNA biosensors and cell phone-based readout, storage, results evaluation, and prescription by a physician

DNA probe induces a conformational change in the hairpin structure, leading to the bending or distortion of the stem region.

The detection of the hairpin bend conformational change can be achieved through various techniques such as electrochemical and fluorescence techniques. In fluorescent-based detection method, the fluorescent labels or dyes can be attached

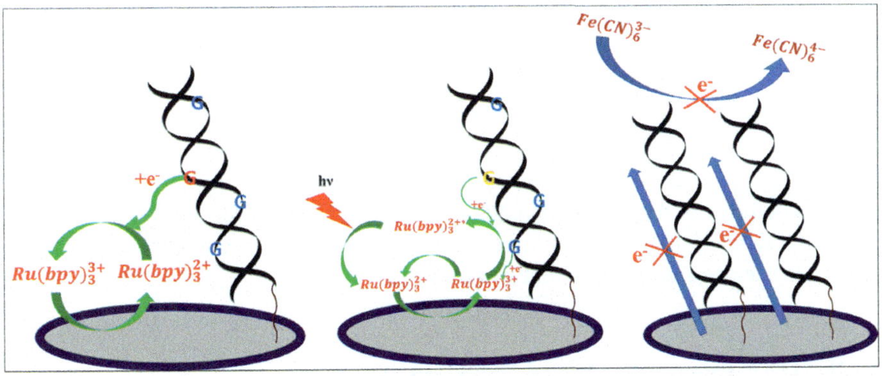

Fig. 15 Mechanism of electron exchange between the redox probes such as $Ru(bpy)_3^{2+}$ and $Fe(CN)_6^{3-/4-}$ with the DNA on the electrode surface

Fig. 16 Three dimensional conformational changes of aptamers. Reprinted (adapted) with permission from [179]. Copyright (2016) American Chemical Society

to the ends of the hairpin structure, such that in the folded or unbent state, the fluorescence signal is quenched due to close proximity or efficient energy transfer between the labels. Upon target molecule binding, the hairpin structure undergoes a conformational change, resulting in the separation or alteration of the fluorophores, leading to an increase in fluorescence signal, which can be measured and quantified [165]. Farjami et al. reported DNA beacons modified gold electrodes with their hair pin folded states through the alkanethiol linker at the 3′ end, while the 5′ end was labeled with a methylene blue (MB) redox probe. The MB lies proximity with the electrode surface enabling the faster electron transfer in the absence of analyte, whereas during the biorecognition event, the electron transfer rate was lowered due to loop unfolding upon hybridization process moving the redox probe away from the electrode surface. This mechanism was successfully employed to construct DNA-based electrochemical detection of cancer biomarker-TP53 gene [166].

Gao et al. reported a flexible multiplexed aptamer biosensor for the insitu wound monitoring of mice model. The sensor designed to detect inflammatory mediators such as inflammatory mediators [tumor necrosis factor-α (TNF-α), interleukin-6 (IL-6), IL-8, and transforming growth factor-β1] and microbial load of Staphylococcus aureus. Hair pin bend aptamer was designed with thiol group for the self-assemble on the gold electrode surface and methylene blue was tagged for the faradaic reaction. In the absence of the analyte the hairpin structure of aptamer keeps the methylene blue redox probe close to the electrode surface resulting high faradaic response, whereas in the presence of the analyte, upon aptamer bind to the analyte and undergo conformational change to upright position. This makes redox move away from the electrode surface lowering the current response [113].

8.1.9 DNA Pseudoknot-Based Detection

A DNA pseudoknot-based biosensor utilizes the formation and disruption of pseudoknot structures within DNA molecules for sensing applications. Pseudoknots are intricate three-dimensional structures formed by specific folding patterns of nucleic acids. They consist of regions where a single-stranded DNA segment forms a loop by base-pairing with another region of the same strand, creating a structure resembling a knot. In a DNA pseudoknot-based biosensor, the DNA probe is designed to contain a target recognition sequence, a loop region, and complementary regions that can form the pseudoknot structure. The presence of the target molecule induces the formation of the pseudoknot structure by bringing the complementary regions together. This conformational change results in a detectable signal that can be measured to indicate the presence or concentration of the target molecule [167].

The detection of the pseudoknot formation can be achieved through various techniques. For instance, fluorescence-based methods involve incorporating fluorophores or quenchers within the DNA probe. The pseudoknot formation alters the proximity or accessibility of these fluorophores or quenchers, leading to changes in the fluorescence signal. Electrochemical methods rely on changes in the electrical properties of the DNA probe upon pseudoknot formation, which can be measured as an electrical signal. Other techniques, such as surface plasmon resonance (SPR) or quartz crystal microbalance (QCM), can also be utilized to monitor the mass or refractive index changes associated with the pseudoknot formation.

The advantage of DNA pseudoknot-based biosensors lies in their ability to achieve highly specific and sensitive target detection. The conformational changes resulting from pseudoknot formation offer a robust and tunable mechanism for signal transduction. These biosensors have been utilized in various applications, including the detection of nucleic acids, proteins, small molecules, and pathogens.

Heeger et al. reported an electrochemical pseudoknot-based DNA sensor for the detection of DNA in blood serum. Pseudoknot DNA structure attached with methylene blue (MB) responds to the analyte and undergoes conformational change, bringing it near to the electrode surface, which supports electron transfer. The electrochemical DNA biosensor was reported to be able to detect the analyte as low as 2 nM with a dynamic linear concentration range between 2 and 200 nM [168].

Singh et al. developed aptamer wearable biosensor for the continuous measurement of stress levels by estimating the cortisol concentrations in human sweat. The developed device consists of a poly-thymine (T-12) pseudoknot aptamer with methylene blue tag immobilized on gold screen-printed electrode and a pH sensor to measure and correct the generated signal accordingly. Upon the cortisol interaction with the aptamer the pseudoknot switches its shape enabling the methylene blue redox probe reach the electrode surface to enhance the faradaic current response. The lower detection was reported to be 0.2 pM with the linear concentration range of 1 pM–1 μM [169].

8.1.10 DNA G-Quadruplex-Based Detection

G-quadruplexes are stable secondary structures formed by guanine-rich DNA sequences. These structures consist of stacked planar arrangements of guanine tetrads stabilized by Hoogsteen base-pairing [170]. The formation of G-quadruplexes can induce substantial conformational changes in DNA. Biosensors have exploited the conformational changes associated with G-quadruplex formation for the detection of specific DNA sequences, as well as the binding of small molecules or metal ions. The design of a DNA G-quadruplex biosensor typically involves incorporating a G-quadruplex-forming DNA sequence into a sensing platform, which can be a DNA probe, aptamer, or other DNA-based recognition element. The biosensor takes advantage of the specific binding affinity of G-quadruplexes to various target molecules, such as small molecules, proteins, or nucleic acids, for detection purposes.

Tan et al. reported response of a G-quadruplex of human telomere sequence $(TTAGGG)_4$ to ionnic concentrations of K^+ and Na^+. During their study, they observed that the association and dissociation rate of G-quadruplex with its complimentary DNA were lowered with the concentrations of K^+ and Na^+, while in the presence of K^+ ion kinetics of association and dissociation was lowered compared to Na^+ ion. This study was useful to understand telomere length homeostasis in telomerase-positive cells, such as germ line and cancer cells [171]. AS114q is a G-rich quadraplex 3D structured aptamer with unique property, and has selective affinity to the Cu^{+2} ions. This property of AS114q was employed to design AS114q dependent fluorescent aptasensor for the detection of Cu^{+2}, in which gel red was used as fluorescent dye. Cu^{+2} ions showed the high affinity, supressing the complimentary DNA and bind to the AS114q resulting in weaker fluorescence [172].

Graphene-modified flexible SiO_2 coated polyethylene napthalate (PEN)-based graphene field effect transistor (GFET) was designed by Hao et al. for the wearable biosensor application to detect cytokines biomarker-tumor necrosis factor (TNF-α) from human sweat. TNF-α-specific aptamer, a guanine-rich VR11 upon binding to the TNF-α, changes its confirmation to a compact G-quadruplex structure. The charge carrier concentrations were altered due to interaction dependent conformational change to close proximity of the negatively charged base pairs to the graphene surface. Lower detection limit was reported to be 26 pM with a detection time of 5 min [173].

Travascio et. al. first reported a G-quafraplex-hemen complex showed enhanced peroxidase activity than the individual iron porphyrin containing pristine hemin [174]. This mechanism was widely used for the detection of nucleases, screening G-quadruplex ligands (potential anticancer reagents), heterocyclic compounds such as pentamethine, symmetric trimethine cyanines, micro-RNA, etc. [175–178].

8.1.11 CRISPR-Cas-Based Biosensors

Clustered regularly interspaced short palindromic repeats (CRISPR) are short DNA sequences found in bacterial and archaeal genomes that are part of an adaptive immune system. Cas (CRISPR-associated) proteins are enzymes that use CRISPR sequences to target and cleave foreign DNA, such as viral genomes [180]. A CRISPR-Cas biosensor is a molecular tool that detects specific DNA sequences using the CRISPR-Cas system.

In a CRISPR-Cas biosensor, the CRISPR-Cas system is used to detect specific DNA sequences in a sample. The biosensor consists of a CRISPR RNA (crRNA) that is designed to recognize the target DNA sequence and a Cas enzyme that cleaves the target DNA in the presence of the crRNA. The biosensor also includes a reporter molecule, such as a fluorescent, colorimetric dye, or a redox probe that is activated when the Cas enzyme cleaves the target DNA [181].

To use a CRISPR-Cas biosensor, the crRNA is first designed to specifically target the DNA sequence of interest. The biosensor is then introduced to a sample containing the target DNA, and if the target DNA is present, the crRNA will hybridize to the target DNA and activate the Cas enzyme to cleave the DNA. The reporter molecule will then signal the presence of the target DNA by generating a detectable signal. CRISPR-Cas biosensors have numerous applications in medical diagnostics. They offer a rapid and sensitive method for detecting specific DNA sequences with high specificity and accuracy.

Peter et al. reported face mask integrated sensor with microfluidic paper-based analytical device (μ-PAD) containing lateral flow assay (LFA) strip. The sensor is incorporated with freeze-dried lysis and detection molecules. The molecules get activated upon interaction with sample containing fluid by wicking action and show the response. The sensor is designed based on CRISPR-Cas12a mechanism, in the presence of sample, i.e., SARS-CoV-2, the virus undergoes lysis in the presence of lysis reagent, exposing the SARS-CoV-2 S gene. The SARS-CoV-2 S gene triggers the function of CAS-12a enzyme, which is already paired reverse transcription-recombinase polymerase amplification (RPA) to produce amplicons. The presence of amplicons triggers the cleavage of 6-FAM-(TTATTATT)-Biotin ssDNA probe, leading to emission of fluorescence at 495 nm. This phenomenon is called CRISPR-based specific high-sensitivity enzymatic reporter unlocking (SHERLOCK). The sensor can give results in 1 h 30 min with a limit of detection of 500 copies/17aM. A similar mechanism was integrated into a multi-sensor wearable freeze-dried cell-free (wFDCF)-CRISPR-based sensor for the detection of *mecA/spa/ermA* genes of methicillin-resistant *Staphylococcus aureus* (MRSA). The sensor works based on the wicking liquid sample of the fiber optic embedded fabric which is integrated in a jacket. The fiber optic embedded sensor functions upon rehydration of the fabric and measures the changes in the excitation event due to the changes in the molecular interaction cascade during the CRISPR-Cas mechanism [182].

Table 2 shows that the Relation of Various DNA-Based Wearable Biosensors

9 Challenges and Opportunities for Wearable Biosensors

Wearable biosensors are becoming increasingly important not only in healthcare, but also in other areas such as sports, work safety, and defense. The main reason why one uses wearable devices, and particularly wearable biosensors, beyond the ability to obtain physiological information in real time, is their convenience. Wearable devices should push Dieter Rams' idea of unobtrusiveness to the extreme where users become unaware that they are being monitored. This is extremely difficult to accomplish, given the relatively short lifespan of chemical sensors and biosensors, which requires the sensing part to be replaced with certain frequency. Wearable biosensors face other challenges in common to other wearable devices, such as accuracy, reliability. Power management, data security, and user acceptance. In this sense, something worth remembering about wearables in general, is that the users will find it very hard to wear anything they are not used to wearing already. The exceptions to this, of course, stem out of need. Obviously, the patient suffering from a disease such as diabetes mellitus will wear a skin patch if this improves their life quality, or a worker will wear a certain garment or accessory if it improves their safety. But designing wearables for the broader population can pose significant cultural challenges but represents enormous opportunities.

This chapter has focused on the design and fabrication of wearable biosensors, and has provided many examples of different wearable biosensors taken from the academic literature. The design should begin with the user and the use environment, and then work its way down to the smallest detail. This includes not only the perspective of the user, but also of manufacturing. In this sense, fabrication technologies and materials available today make almost anything possible. In the case of wearable biosensors, the disposable nature of the sensing part has driven many a worker to explore the construction of biosensors using biodegradable or absorbable materials that can reduce or eliminate the environmental impact of disposing of a packaged device with relatively high frequency (every 2 to 3 weeks at best in the case of state-of-the-art commercial devices). Another interesting direction for future development is not to make short-lived devices less environmentally damaging, but to make them much longer lasting. However, this not only involves developing new materials or new synthetic receptors, but also new ways to measure. Self-calibrating or calibration-less systems are perhaps a Holy Grail in electroanalytical science, but one that will eventually be found.

References

1. Land, K.J., Boeras, D.I., Chen, X.-S., Ramsay, A.R., Peeling, R.W.: REASSURED diagnostics to inform disease control strategies, strengthen health systems and improve patient outcomes. Nat. Microbiol. **4**, 46–54 (2019). https://doi.org/10.1038/s41564-018-0295-3
2. Mabey, D., Peeling, R.W., Ustianowski, A., Perkins, M.D.: Diagnostics for the developing world. Nat. Rev. Microbiol. **2**, 231–240 (2004). https://doi.org/10.1038/nrmicro841
3. Tierney, M.J., Tamada, J.A., Potts, R.O., Jovanovic, L., Garg, S.: Clinical evaluation of the GlucoWatch® biographer: a continual, non-invasive glucose monitor for patients with diabetes. Biosens. Bioelectron. **16**, 621–629 (2001). https://doi.org/10.1016/S0956-5663(01)00189-0
4. Tierney, M.J., Tamada, J.A., Potts, R.O., Eastman, R.C., Pitzer, K., Ackerman, N.R., Fermi, S.J.: The GlucoWatch ® biographer: a frequent, automatic and noninvasive glucose monitor. Ann. Med. **32**, 632–641 (2000). https://doi.org/10.3109/07853890009002034
5. Ellis, S., Naik, R., Gemperline, K., Garg, S.: Use of Continuous Glucose Monitoring in Patients with Type 1 Diabetes. Curr. Diabetes Rev. **4**, 207–217 (2008). https://doi.org/10.2174/157339908785294370
6. Christiansen, M., Bailey, T., Watkins, E., Liljenquist, D., Price, D., Nakamura, K., Boock, R., Peyser, T.: A New-Generation Continuous Glucose Monitoring System: Improved Accuracy and Reliability Compared with a Previous-Generation System. Diabetes Technol. Ther. **15**, 881–888 (2013). https://doi.org/10.1089/dia.2013.0077
7. Garcia, A., Rack-Gomer, A.L., Bhavaraju, N.C., Hampapuram, H., Kamath, A., Peyser, T., Facchinetti, A., Zecchin, C., Sparacino, G., Cobelli, C.: Dexcom G4AP: An Advanced Continuous Glucose Monitor for the Artificial Pancreas. J. Diabetes Sci. Technol. **7**, 1436–1445 (2013). https://doi.org/10.1177/193229681300700604
8. Kropff, J., Choudhary, P., Neupane, S., Barnard, K., Bain, S.C., Kapitza, C., Forst, T., Link, M., Dehennis, A., DeVries, J.H.: Accuracy and Longevity of an Implantable Continuous Glucose Sensor in the PRECISE Study: A 180-Day, Prospective, Multicenter. Pivotal Trial. Diabetes Care. **40**, 63–68 (2017). https://doi.org/10.2337/dc16-1525
9. Laffel, L.M., Aleppo, G., Buckingham, B.A., Forlenza, G.P., Rasbach, L.E., Tsalikian, E., Weinzimer, S.A., Harris, D.R.: A Practical Approach to Using Trend Arrows on the Dexcom G5 CGM System to Manage Children and Adolescents With Diabetes. J. Endocr. Soc. **1**, 1461–1476 (2017). https://doi.org/10.1210/js.2017-00389
10. Aleppo, G., Laffel, L.M., Ahmann, A.J., Hirsch, I.B., Kruger, D.F., Peters, A., Weinstock, R.S., Harris, D.R.: A Practical Approach to Using Trend Arrows on the Dexcom G5 CGM System for the Management of Adults With Diabetes. J. Endocr. Soc. **1**, 1445–1460 (2017). https://doi.org/10.1210/js.2017-00388
11. Yeoh, E., Png, D., Khoo, J., Chee, Y.J., Sharda, P., Low, S., Lim, S.C., Subramaniam, T.: A head-to-head comparison between Guardian Connect and FreeStyle Libre systems and an evaluation of user acceptability of sensors in patients with type 1 diabetes. Diabetes Metab. Res. Rev. 38, (2022). https://doi.org/10.1002/dmrr.3560
12. Guillot, F.H., Jacobs, P.G., Wilson, L.M., Youssef, J.E., Gabo, V.B., Branigan, D.L., Tyler, N.S., Ramsey, K., Riddell, M.C., Castle, J.R.: Accuracy of the Dexcom G6 Glucose Sensor during Aerobic, Resistance, and Interval Exercise in Adults with Type 1 Diabetes. Biosensors **10**, 138 (2020). https://doi.org/10.3390/bios10100138
13. Davis, G.M., Spanakis, E.K., Migdal, A.L., Singh, L.G., Albury, B., Urrutia, M.A., Zamudio-Coronado, K.W., Scott, W.H., Doerfler, R., Lizama, S., Satyarengga, M., Munir, K., Galindo, R.J., Vellanki, P., Cardona, S., Pasquel, F.J., Peng, L., Umpierrez, G.E.: Accuracy of Dexcom G6 Continuous Glucose Monitoring in Non-Critically Ill Hospitalized Patients With Diabetes. Diabetes Care **44**, 1641–1646 (2021). https://doi.org/10.2337/dc20-2856
14. Joseph, J.I.: Review of the Long-Term Implantable Senseonics Continuous Glucose Monitoring System and Other Continuous Glucose Monitoring Systems. J. Diabetes Sci. Technol. **15**, 167–173 (2021). https://doi.org/10.1177/1932296820911919

15. Abbot: Perform Stronger and Recover Faster with CGM, https://www.supersapiens.com
16. Brady, S., Dunne, L.E.E., Lynch, A., Smyth, B., Diamond, D.: Wearable sensors? what is there to sense? IOS Press, Adaptive Information Cluster, National Centre for Sensor Research, Dublin City University, Ireland (2005)
17. Coyle, S., Morris, D., Lau, K.-T., Diamond, D., Moyna, N.: Textile-based wearable sensors for assisting sports performance. Presented at the , Clarity: The Centre for Sensor Web Technologies, National Centre for Sensor Research, Dublin City University, Dublin 9, Ireland (2009)
18. Coyle, S., Diamond, D.: Medical applications of smart textiles. In: Multidisciplinary Know-How for Smart-Textiles Developers. pp. 420–443. Elsevier Ltd, CLARITY: Centre for Sensor Web Technologies, National Centre for Sensor Research, Dublin City University, Glasnevin, Dublin 9, Ireland (2013)
19. Morris, D., Coyle, S., Wu, Y., Lau, K.T., Wallace, G., Diamond, D.: Bio-sensing textile based patch with integrated optical detection system for sweat monitoring. Sens. Actuators B Chem. **139**, 231–236 (2009). https://doi.org/10.1016/j.snb.2009.02.032
20. Glennon, T., O'Quigley, C., McCaul, M., Matzeu, G., Beirne, S., Wallace, G.G.G., Stroiescu, F., O'Mahoney, N., White, P., Diamond, D.: 'SWEATCH': A Wearable Platform for Harvesting and Analysing Sweat Sodium Content. Electroanalysis **28**, 1283–1289 (2016). https://doi.org/10.1002/elan.201600106
21. Windmiller, J.R., Bandodkar, A.J., Valdés-Ramírez, G., Parkhomovsky, S., Martinez, A.G., Wang, J.: Electrochemical sensing based on printable temporary transfer tattoos. Chem. Commun. **48**, 6794 (2012). https://doi.org/10.1039/c2cc32839a
22. Bandodkar, A.J., Jia, W., Wang, J.: Tattoo-Based Wearable Electrochemical Devices: A Review. Electroanalysis **27**, 562–572 (2015). https://doi.org/10.1002/elan.201400537
23. Wang, J., Windmiller, J.R., Jia, W.: Printed biofuel cells, (2012)
24. Jia, W., Valdés-Ramírez, G., Bandodkar, A.J., Windmiller, J.R., Wang, J.: Epidermal biofuel cells: Energy harvesting from human perspiration. Angew. Chem. - Int. Ed. **52**, 7233–7236 (2013). https://doi.org/10.1002/anie.201302922
25. Kim, J., Valdés-Ramírez, G., Bandodkar, A.J., Jia, W., Martinez, A.G., Ramírez, J., Mercier, P., Wang, J.: Non-invasive mouthguard biosensor for continuous salivary monitoring of metabolites. Analyst **139**, 1632–1636 (2014). https://doi.org/10.1039/c3an02359a
26. Rose, D., Ratterman, M., Griffin, D., Hou, L., Kelley-Loughnane, N., Naik, R., Hagen, J., Papautsky, I., Heikenfeld, J.: Adhesive RFID Sensor Patch for Monitoring of Sweat Electrolytes. IEEE Trans. Biomed. Eng. **9294**, 1–1 (2014). https://doi.org/10.1109/TBME.2014.2369991
27. Hou, L., Hagen, J., Wang, X., Papautsky, I., Naik, R., Kelley-Loughnane, N., Heikenfeld, J.: Artificial microfluidic skin for in vitro perspiration simulation and testing. Lab Chip **13**, 1868 (2013). https://doi.org/10.1039/c3lc41231h
28. La Count, T.D., Jajack, A., Heikenfeld, J., Kasting, G.B.: Modeling Glucose Transport From Systemic Circulation to Sweat. J. Pharm. Sci. **108**, 364–371 (2019). https://doi.org/10.1016/j.xphs.2018.09.026
29. Zhao, F.J.J., Bonmarin, M., Chen, Z.C.C., Larson, M., Fay, D., Runnoe, D., Heikenfeld, J.: Ultra-simple wearable local sweat volume monitoring patch based on swellable hydrogels. Lab Chip **20**, 168–174 (2020). https://doi.org/10.1039/c9lc00911f
30. Kim, D.-H., Lu, N., Ma, R., Kim, Y.-S., Kim, R.-H., Wang, S., Wu, J., Won, S.M., Tao, H., Islam, A., Yu, K.J., Kim, T., Chowdhury, R., Ying, M., Xu, L., Li, M., Chung, H.-J., Keum, H., McCormick, M., Liu, P., Zhang, Y.-W., Omenetto, F.G., Huang, Y., Coleman, T., Rogers, J.A.: Epidermal Electronics. Science **333**, 838–843 (2011). https://doi.org/10.1126/science.1206157
31. Kim, J.U., Seo, S.G., Rogers, J.A.: Compound semiconductor devices for the skin. Nat. Mater. (2022). https://doi.org/10.1038/s41563-022-01441-9
32. Reeder, J.T., Choi, J., Xue, Y., Gutruf, P., Hanson, J., Liu, M., Ray, T., Bandodkar, A.J., Avila, R., Xia, W., Krishnan, S., Xu, S., Barnes, K., Pahnke, M., Ghaffari, R., Huang, Y., Rogers, J.A.: Waterproof, electronics-enabled, epidermal microfluidic devices for sweat collection,

biomarker analysis, and thermography in aquatic settings. Sci. Adv. 5, eaau6356 (2019). https://doi.org/10.1126/sciadv.aau6356

33. Bandodkar, A.J.J., Lee, S.P.P., Huang, I., Li, W., Wang, S., Su, C.-J., Jeang, W.J.J., Hang, T., Mehta, S., Nyberg, N., Gutruf, P., Choi, J., Koo, J., Reeder, J.T., Tseng, R., Ghaffari, R., Rogers, J.A.A.: Sweat-activated biocompatible batteries for epidermal electronic and microfluidic systems. Nat. Electron. **3**, 554–562 (2020). https://doi.org/10.1038/s41928-020-0443-7
34. Baker, L.B., Model, J.B., Barnes, K.A., Anderson, M.L., Lee, S.P., Lee, K.A., Brown, S.D., Reimel, A.J., Roberts, T.J., Nuccio, R.P., Bonsignore, J.L., Ungaro, C.T., Carter, J.M., Li, W., Seib, M.S., Reeder, J.T., Aranyosi, A.J., Rogers, J.A., Ghaffari, R.: Skin-interfaced microfluidic system with personalized sweating rate and sweat chloride analytics for sports science applications. Sci. Adv. 6, eabe3929 (2020). https://doi.org/10.1126/sciadv.abe3929
35. Liu, S., Yang, D.S., Wang, S., Luan, H., Sekine, Y., Model, J.B., Aranyosi, A.J., Ghaffari, R., Rogers, J.A.: Soft, environmentally degradable microfluidic devices for measurement of sweat rate and total sweat loss and for colorimetric analysis of sweat biomarkers. EcoMat. 5, (2023). https://doi.org/10.1002/eom2.12270
36. Kwon, K., Kim, J.U., Deng, Y., Krishnan, S.R., Choi, J., Jang, H., Lee, K., Su, C.-J., Yoo, I., Wu, Y., Lipschultz, L., Kim, J.-H., Chung, T.S., Wu, D., Park, Y., Kim, T., Ghaffari, R., Lee, S., Huang, Y., Rogers, J.A.: An on-skin platform for wireless monitoring of flow rate, cumulative loss and temperature of sweat in real time. Nat. Electron. **4**, 302–312 (2021). https://doi.org/10.1038/s41928-021-00556-2
37. Drucker, P.F.: Innovation and entrepreneurship: practice and principles. HarperBusiness, New York, NY (2006)
38. Christensen, C.M.: The innovator's dilemma: the revolutionary book that will change the way you do business ; [with a new preface]. Harper Business, New York, NY (2011)
39. Porter, M.E.: Competitive strategy: techniques for analyzing industries and competitors. Free Press, New York London Toronto Sydney (2004)
40. Porter, M.E.: Competitive advantage: creating and sustaining superior performance. Free Press, New York, NY (2004)
41. Norman, D.A.: The design of everyday things. Tantor Media Inc., Old Saybrook, Ct (2011)
42. Lefteri, C.: Making It: Manufacturing techniques for product design. Laurence King Publishing, London (2019)
43. Sui, X., Downing, J.R., Hersam, M.C., Chen, J.: Additive manufacturing and applications of nanomaterial-based sensors. Mater. Today **48**, 135–154 (2021). https://doi.org/10.1016/j.mattod.2021.02.001
44. Santiago-Malagón, S., Río-Colín, D., Azizkhani, H., Aller-Pellitero, M., Guirado, G., del Campo, F.J.: A self-powered skin-patch electrochromic biosensor. Biosens. Bioelectron. **175**, 112879 (2021). https://doi.org/10.1016/j.bios.2020.112879
45. Wang, Z., Gui, M., Asif, M., Yu, Y., Dong, S., Wang, H., Wang, W., Wang, F., Xiao, F., Liu, H.: A facile modular approach to the 2D oriented assembly MOF electrode for non-enzymatic sweat biosensors. Nanoscale **10**, 6629–6638 (2018). https://doi.org/10.1039/C8NR00798E
46. Cheng, Y.-T., Chen, L.-C., Wang, W.-C.: Development of a fiber shape polymeric humidity sensor. Presented at the SPIE Smart Structures and Materials + Nondestructive Evaluation and Health Monitoring , Portland, Oregon, United States April 17 (2017)
47. Veeralingam, S., Khandelwal, S., Badhulika, S.: AI/ML-Enabled 2-D - RuS $_2$ Nanomaterial-Based Multifunctional, Low Cost, Wearable Sensor Platform for Non-Invasive Point of Care Diagnostics. IEEE Sens. J. **20**, 8437–8444 (2020). https://doi.org/10.1109/JSEN.2020.2984807
48. Jiang, D., Xu, C., Zhang, Q., Ye, Y., Cai, Y., Li, K., Li, Y., Huang, X., Wang, Y.: In-situ preparation of lactate-sensing membrane for the noninvasive and wearable analysis of sweat. Biosens. Bioelectron. **210**, 114303 (2022). https://doi.org/10.1016/j.bios.2022.114303
49. Asaduzzaman, M., Zahed, M.A., Sharifuzzaman, M., Reza, M.S., Hui, X., Sharma, S., Shin, Y.D., Park, J.Y.: A hybridized nano-porous carbon reinforced 3D graphene-based epidermal patch for precise sweat glucose and lactate analysis. Biosens. Bioelectron. **219**, 114846 (2023). https://doi.org/10.1016/j.bios.2022.114846

50. Zahid, M., Papadopoulou, E.L., Athanassiou, A., Bayer, I.S.: Strain-responsive mercerized conductive cotton fabrics based on PEDOT:PSS/graphene. Mater. Des. **135**, 213–222 (2017). https://doi.org/10.1016/j.matdes.2017.09.026
51. Fisher, C., Skolrood, L.N., Li, K., Joshi, P.C., Aytug, T.: Aerosol-Jet Printed Sensors for Environmental, Safety, and Health Monitoring: A Review. Adv. Mater. Technol. n/a, 2300030. https://doi.org/10.1002/admt.202300030
52. Veenuttranon, K., Kaewpradub, K., Jeerapan, I.: Screen-Printable Functional Nanomaterials for Flexible and Wearable Single-Enzyme-Based Energy-Harvesting and Self-Powered Biosensing Devices. Nano-Micro Lett. **15**, 85 (2023). https://doi.org/10.1007/s40820-023-01045-1
53. Smith, A.A., Li, R., Tse, Z.T.H.: Reshaping healthcare with wearable biosensors. Sci. Rep. **13**, 4998 (2023). https://doi.org/10.1038/s41598-022-26951-z
54. Kalkal, A., Kumar, S., Kumar, P., Pradhan, R., Willander, M., Packirisamy, G., Kumar, S., Malhotra, B.D.: Recent advances in 3D printing technologies for wearable (bio)sensors. Addit. Manuf. **46**, 102088 (2021). https://doi.org/10.1016/j.addma.2021.102088
55. Muth, J.T., Vogt, D.M., Truby, R.L., Mengüç, Y., Kolesky, D.B., Wood, R.J., Lewis, J.A.: Embedded 3D Printing of Strain Sensors within Highly Stretchable Elastomers. Adv. Mater. **26**, 6307–6312 (2014). https://doi.org/10.1002/adma.201400334
56. Yi, Q., Najafikhoshnoo, S., Das, P., Noh, S., Hoang, E., Kim, T., Esfandyarpour, R.: All-3D-Printed, Flexible, and Hybrid Wearable Bioelectronic Tactile Sensors Using Biocompatible Nanocomposites for Health Monitoring. Adv. Mater. Technol. **7**, 2101034 (2022). https://doi.org/10.1002/admt.202101034
57. Zhou, L., Gao, Q., Fu, J., Chen, Q., Zhu, J., Sun, Y., He, Y.: Multimaterial 3D Printing of Highly Stretchable Silicone Elastomers. ACS Appl. Mater. Interfaces **11**, 23573–23583 (2019). https://doi.org/10.1021/acsami.9b04873
58. Kim, T., Yi, Q., Hoang, E., Esfandyarpour, R.: A 3D Printed Wearable Bioelectronic Patch for Multi-Sensing and In Situ Sweat Electrolyte Monitoring. Adv. Mater. Technol. **6**, 2001021 (2021). https://doi.org/10.1002/admt.202001021
59. Xin, M., Li, J., Ma, Z., Pan, L., Shi, Y.: MXenes and Their Applications in Wearable Sensors. Front. Chem. **8**, 297 (2020). https://doi.org/10.3389/fchem.2020.00297
60. Bhardwaj, S.K., Singh, H., Khatri, M., Kim, K.-H., Bhardwaj, N.: Advances in MXenes-based optical biosensors: A review. Biosens. Bioelectron. **202**, 113995 (2022). https://doi.org/10.1016/j.bios.2022.113995
61. Hondred, J.A., Johnson, Z.T., Claussen, J.C.: Nanoporous gold peel-and-stick biosensors created with etching inkjet maskless lithography for electrochemical pesticide monitoring with microfluidics. J. Mater. Chem. C. **8**, 11376–11388 (2020). https://doi.org/10.1039/D0TC01423K
62. Wang, C., Zhu, S., Liang, Y., Cui, Z., Wu, S., Qin, C., Luo, S., Inoue, A.: Understanding the macroscopical flexibility/fragility of nanoporous Ag: Depending on network connectivity and micro-defects. J. Mater. Sci. Technol. **53**, 91–101 (2020). https://doi.org/10.1016/j.jmst.2020.04.010
63. Sánchez-Molas, D., Esquivel, J.P., Sabaté, N., Muñoz, F.X., del Campo, F.J., Esquivel, J.P.: High Aspect-Ratio, Fully Conducting Gold Micropillar Array Electrodes: Silicon Micromachining and Electrochemical Characterization. J. Phys. Chem. C **116**, 18831–18846 (2012). https://doi.org/10.1021/jp305339k
64. Chyan, Y., Ye, R., Li, Y., Singh, S.P., Arnusch, C.J., Tour, J.M.: Laser-Induced Graphene by Multiple Lasing: Toward Electronics on Cloth, Paper, and Food. ACS Nano **12**, 2176–2183 (2018). https://doi.org/10.1021/acsnano.7b08539
65. Fruncillo, S., Su, X., Liu, H., Wong, L.S.: Lithographic Processes for the Scalable Fabrication of Micro- and Nanostructures for Biochips and Biosensors. ACS Sens. **6**, 2002–2024 (2021). https://doi.org/10.1021/acssensors.0c02704
66. Cao, R., Pu, X., Du, X., Yang, W., Wang, J., Guo, H., Zhao, S., Yuan, Z., Zhang, C., Li, C., Lin Wang, Z.: Screen-Printed Washable Electronic Textiles as Self-Powered Touch/Gesture Tribo-Sensors for Intelligent Human–Machine Interaction Article. ACS Nano **12**, 38 (2018). https://doi.org/10.1021/acsnano.8b02477

67. Ma, D., Chon, S., Cho, S., Lee, Y., Yoo, M., Kim, D., Lee, D.Y., Lim, J.K.: A novel photolithographic method for fabrication of flexible micro-patterned glucose sensors. J. Electroanal. Chem. **876**, 114720 (2020). https://doi.org/10.1016/j.jelechem.2020.114720
68. Oppel, E., Högg, C., Oschmann, A., Summer, B., Kamann, S.: Contact allergy to the Dexcom G6 glucose monitoring system—Role of 2,2′-methylenebis(6-tert-butyl-4-methylphenol) monoacrylate in the new adhesive. Contact Dermatitis **87**, 258–264 (2022). https://doi.org/10.1111/cod.14141
69. Jones, P., Wynn, M., Hillier, D., Comfort, D.: The Sustainable Development Goals and Information and Communication Technologies. Indones. J. Sustain. Account. Manag. 1, 1 (2017). https://doi.org/10.28992/ijsam.v1i1.22
70. Poitout, V., Moatti-Sirat, D., Reach, G., Zhang, Y., Wilson, G.S., Lemonnier, F., Klein, J.C.: A glucose monitoring system for on line estimation in man of blood glucose concentration using a miniaturized glucose sensor implanted in the subcutaneous tissue and a wearable control unit. Diabetologia **36**, 658–663 (1993). https://doi.org/10.1007/BF00404077
71. Kudo, H., Sawada, T., Kazawa, E., Yoshida, H., Iwasaki, Y., Mitsubayashi, K.: A flexible and wearable glucose sensor based on functional polymers with Soft-MEMS techniques. Biosens. Bioelectron. **22**, 558–562 (2006). https://doi.org/10.1016/j.bios.2006.05.006
72. Mannoor, M.S., Tao, H., Clayton, J.D., Sengupta, A., Kaplan, D.L., Naik, R.R., Verma, N., Omenetto, F.G., McAlpine, M.C.: Graphene-based wireless bacteria detection on tooth enamel. Nat. Commun. **3**, 763 (2012). https://doi.org/10.1038/ncomms1767
73. Gao, W., Emaminejad, S., Nyein, H.Y.Y., Challa, S., Chen, K., Peck, A., Fahad, H.M., Ota, H., Shiraki, H., Kiriya, D., Lien, D.-H., Brooks, G.A., Davis, R.W., Javey, A.: Fully integrated wearable sensor arrays for multiplexed in situ perspiration analysis. Nature **529**, 509–514 (2016). https://doi.org/10.1038/nature16521
74. Huang, X., Liu, Y., Zhou, J., Nejad, S.K., Wong, T.H., Huang, Y., Li, H., Yiu, C.K., Park, W., Li, J., Su, J., Zhao, L., Yao, K., Wu, M., Gao, Z., Li, D., Li, J., Shi, R., Yu, X.: Garment embedded sweat-activated batteries in wearable electronics for continuous sweat monitoring. Npj Flex. Electron. **6**, 10 (2022). https://doi.org/10.1038/s41528-022-00144-0
75. Rasitanon, N., Ittisoponpisan, S., Kaewpradub, K., Jeerapan, I.: Wearable Electrodes for Lactate: Applications in Enzyme-Based Sensors and Energy Biodevices. Anal. Sens. (2023). https://doi.org/10.1002/anse.202200066
76. Sadani, K., Nag, P., Thian, X.Y., Mukherji, S.: Enzymatic optical biosensors for healthcare applications. Biosens. Bioelectron. X. **12**, 100278 (2022). https://doi.org/10.1016/j.biosx.2022.100278
77. Heikenfeld, J., Jajack, A., Feldman, B., Granger, S.W., Gaitonde, S., Begtrup, G., Katchman, B.A.: Accessing analytes in biofluids for peripheral biochemical monitoring. Nat. Biotechnol. **37**, 407–419 (2019). https://doi.org/10.1038/s41587-019-0040-3
78. Zafar, H., Channa, A., Jeoti, V., Stojanović, G.M.: Comprehensive Review on Wearable Sweat-Glucose Sensors for Continuous Glucose Monitoring. Sensors. **22**, 638 (2022). https://doi.org/10.3390/s22020638
79. Chung, M., Fortunato, G., Radacsi, N.: Wearable flexible sweat sensors for healthcare monitoring: a review. J. R. Soc. Interface. **16**, 20190217 (2019). https://doi.org/10.1098/rsif.2019.0217
80. Bucur, B., Purcarea, C., Andreescu, S., Vasilescu, A.: Addressing the Selectivity of Enzyme Biosensors: Solutions and Perspectives. Sensors. **21**, 3038 (2021). https://doi.org/10.3390/s21093038
81. Lipińska, W., Grochowska, K., Siuzdak, K.: Enzyme Immobilization on Gold Nanoparticles for Electrochemical Glucose Biosensors. Nanomaterials **11**, 1156 (2021). https://doi.org/10.3390/nano11051156
82. Cabaj, J., Sołoducho, J.: Nano-Sized Elements in Electrochemical Biosensors. Mater. Sci. Appl. **05**, 752–766 (2014). https://doi.org/10.4236/msa.2014.510076
83. Skaria, E., Patel, B.A., Flint, M.S., Ng, K.W.: Poly(lactic acid)/Carbon Nanotube Composite Microneedle Arrays for Dermal Biosensing. Anal. Chem. **91**, 4436–4443 (2019). https://doi.org/10.1021/acs.analchem.8b04980

84. Lawal, A.T.: Synthesis and utilization of carbon nanotubes for fabrication of electrochemical biosensors. Mater. Res. Bull. **73**, 308–350 (2016). https://doi.org/10.1016/j.materresbull.2015.08.037
85. Niu, Y., Liu, J., Chen, W., Yin, C., Weng, W., Li, X., Wang, X., Li, G., Sun, W.: A direct electron transfer biosensor based on a horseradish peroxidase and gold nanotriangle modified electrode and electrocatalysis. Anal. Methods **10**, 5297–5304 (2018). https://doi.org/10.1039/C8AY01980K
86. Fang, L., Liu, B., Liu, L., Li, Y., Huang, K., Zhang, Q.: Direct electrochemistry of glucose oxidase immobilized on Au nanoparticles-functionalized 3D hierarchically ZnO nanostructures and its application to bioelectrochemical glucose sensor. Sens. Actuators B Chem. **222**, 1096–1102 (2016). https://doi.org/10.1016/j.snb.2015.08.032
87. Holzinger, M., Baur, J., Haddad, R., Wang, X., Cosnier, S.: Multiple functionalization of single-walled carbon nanotubes by dip coating. Chem. Commun. **47**, 2450–2452 (2011). https://doi.org/10.1039/C0CC03928D
88. Faculty of Military Health Sciences, University of Defence, Trebesska 1575, 50001 Hradec Kralove, Czech Republic, Martinkova, P.: Main streams in the Construction of Biosensors and Their Applications. Int. J. Electrochem. Sci. 7386–7403 (2017). https://doi.org/10.20964/2017.08.02
89. Nguyen, H.H., Lee, S.H., Lee, U.J., Fermin, C.D., Kim, M.: Immobilized Enzymes in Biosensor Applications. Materials. **12**, 121 (2019). https://doi.org/10.3390/ma12010121
90. Sassolas, A., Blum, L.J., Leca-Bouvier, B.D.: Immobilization strategies to develop enzymatic biosensors. Biotechnol. Adv. **30**, 489–511 (2012). https://doi.org/10.1016/j.biotechadv.2011.09.003
91. Salazar, P., Martín, M., O'Neill, R.D., González-Mora, J.L.: Glutamate microbiosensors based on Prussian Blue modified carbon fiber electrodes for neuroscience applications: In-vitro characterization. Sens. Actuators B Chem. **235**, 117–125 (2016). https://doi.org/10.1016/j.snb.2016.05.057
92. Kim, J., Imani, S., de Araujo, W.R., Warchall, J., Valdés-Ramírez, G., Paixão, T.R.L.C., Mercier, P.P., Wang, J.: Wearable salivary uric acid mouthguard biosensor with integrated wireless electronics. Biosens. Bioelectron. **74**, 1061–1068 (2015). https://doi.org/10.1016/j.bios.2015.07.039
93. Tur-García, E.L., Davis, F., Collyer, S.D., Holmes, J.L., Barr, H., Higson, S.P.J.: Novel flexible enzyme laminate-based sensor for analysis of lactate in sweat. Sens. Actuators B Chem. **242**, 502–510 (2017). https://doi.org/10.1016/j.snb.2016.11.040
94. Clark, L.C., Lyons, C.: ELECTRODE SYSTEMS FOR CONTINUOUS MONITORING IN CARDIOVASCULAR SURGERY. Ann. N. Y. Acad. Sci. **102**, 29–45 (2006). https://doi.org/10.1111/j.1749-6632.1962.tb13623.x
95. Jiang, Y., Yang, Y., Shen, L., Ma, J., Ma, H., Zhu, N.: Recent Advances of Prussian Blue-Based Wearable Biosensors for Healthcare. Anal. Chem. **94**, 297–311 (2022). https://doi.org/10.1021/acs.analchem.1c04420
96. Herrmann, A., Haag, R., Schedler, U.: Hydrogels and Their Role in Biosensing Applications. Adv. Healthc. Mater. **10**, 2100062 (2021). https://doi.org/10.1002/adhm.202100062
97. Lipińska, W., Siuzdak, K., Karczewski, J., Dołęga, A., Grochowska, K.: Electrochemical glucose sensor based on the glucose oxidase entrapped in chitosan immobilized onto laser-processed Au-Ti electrode. Sens. Actuators B Chem. **330**, 129409 (2021). https://doi.org/10.1016/j.snb.2020.129409
98. Updike, S.J., Hicks, G.P.: The Enzyme Electrode. Nature **214**, 986–988 (1967). https://doi.org/10.1038/214986a0
99. Kim, G.J., Kim, K.O.: Novel glucose-responsive of the transparent nanofiber hydrogel patches as a wearable biosensor via electrospinning. Sci. Rep. **10**, 18858 (2020). https://doi.org/10.1038/s41598-020-75906-9
100. Kim, J., Jeerapan, I., Imani, S., Cho, T.N., Bandodkar, A., Cinti, S., Mercier, P.P., Wang, J.: Noninvasive Alcohol Monitoring Using a Wearable Tattoo-Based Iontophoretic-Biosensing System. ACS Sens. **1**, 1011–1019 (2016). https://doi.org/10.1021/acssensors.6b00356

101. Kemp, E., Palomäki, T., Ruuth, I.A., Boeva, Z.A., Nurminen, T.A., Vänskä, R.T., Zschaechner, L.K., Pérez, A.G., Hakala, T.A., Wardale, M., Haeggström, E., Bobacka, J.: Influence of enzyme immobilization and skin-sensor interface on non-invasive glucose determination from interstitial fluid obtained by magnetohydrodynamic extraction. Biosens. Bioelectron. **206**, 114123 (2022). https://doi.org/10.1016/j.bios.2022.114123
102. Nagamine, K., Mano, T., Nomura, A., Ichimura, Y., Izawa, R., Furusawa, H., Matsui, H., Kumaki, D., Tokito, S.: Noninvasive Sweat-Lactate Biosensor Emplsoying a Hydrogel-Based Touch Pad. Sci. Rep. **9**, 10102 (2019). https://doi.org/10.1038/s41598-019-46611-z
103. Liu, J., Zhang, L., Fu, C.: Os-complex-based amperometric bienzyme biosensor for continuous determination of lactate in saliva. Anal. Methods **7**, 6158–6164 (2015). https://doi.org/10.1039/C5AY01110H
104. Mollarasouli, K.: Ozkan: The Role of Electrochemical Immunosensors in Clinical Analysis. Biosensors **9**, 86 (2019). https://doi.org/10.3390/bios9030086
105. Ruiz, G., Tripathi, K., Okyem, S., Driskell, J.D.: PH Impacts the Orientation of Antibody Adsorbed onto Gold Nanoparticles. Bioconjug. Chem. **30**, 1182–1191 (2019). https://doi.org/10.1021/acs.bioconjchem.9b00123
106. Wei, J., Zhang, X., Mugo, S.M., Zhang, Q.: A Portable Sweat Sensor Based on Carbon Quantum Dots for Multiplex Detection of Cardiovascular Health Biomarkers. Anal. Chem. **94**, 12772–12780 (2022). https://doi.org/10.1021/acs.analchem.2c02587
107. Haji-Hashemi, H., Norouzi, P., Safarnejad, M.R., Ganjali, M.R.: Label-free electrochemical immunosensor for direct detection of Citrus tristeza virus using modified gold electrode. Sens. Actuators B Chem. **244**, 211–216 (2017). https://doi.org/10.1016/j.snb.2016.12.135
108. Lee, H.-B., Meeseepong, M., Trung, T.Q., Kim, B.-Y., Lee, N.-E.: A wearable lab-on-a-patch platform with stretchable nanostructured biosensor for non-invasive immunodetection of biomarker in sweat. Biosens. Bioelectron. **156**, 112133 (2020). https://doi.org/10.1016/j.bios.2020.112133
109. Bagni, G., Osella, D., Sturchio, E., Mascini, M.: Deoxyribonucleic acid (DNA) biosensors for environmental risk assessment and drug studies. Instrum. Methods Anal. -IMA **2005**(573–574), 81–89 (2006). https://doi.org/10.1016/j.aca.2006.03.085
110. Lymperopoulos, K., Crawford, R., Torella, J.P., Heilemann, M., Hwang, L.C., Holden, S.J., Kapanidis, A.N.: Single-Molecule DNA Biosensors for Protein and Ligand Detection. Angew. Chem. Int. Ed. **49**, 1316–1320 (2010). https://doi.org/10.1002/anie.200904597
111. Choi, J.-H., Lim, J., Shin, M., Paek, S.-H., Choi, J.-W.: CRISPR-Cas12a-Based Nucleic Acid Amplification-Free DNA Biosensor via Au Nanoparticle-Assisted Metal-Enhanced Fluorescence and Colorimetric Analysis. Nano Lett. **21**, 693–699 (2021). https://doi.org/10.1021/acs.nanolett.0c04303
112. Mukherjee, M., Gajjala, R.K.R., Gade, P.S., Bhatt, P.: 3.42 - Aptasensors: Paradigm Shift for Detection of Food Toxins. In: Knoerzer, K. and Muthukumarappan, K. (eds.) Innovative Food Processing Technologies. pp. 712–730. Elsevier, Oxford (2021)
113. Gao, Y., Nguyen, D.T., Yeo, T., Lim, S.B., Tan, W.X., Madden, L.E., Jin, L., Long, J.Y.K., Aloweni, F.A.B., Liew, Y.J.A., Tan, M.L.L., Ang, S.Y., Maniya, S.D., Abdelwahab, I., Loh, K.P., Chen, C.-H., Becker, D.L., Leavesley, D., Ho, J.S., Lim, C.T.: A flexible multiplexed immunosensor for point-of-care in situ wound monitoring. Sci. Adv. 7, eabg9614. https://doi.org/10.1126/sciadv.abg9614
114. Ferguson, B.S., Hoggarth, D.A., Maliniak, D., Ploense, K., White, R.J., Woodward, N., Hsieh, K., Bonham, A.J., Eisenstein, M., Kippin, T.E., Plaxco, K.W., Soh, H.T.: Real-Time, Aptamer-Based Tracking of Circulating Therapeutic Agents in Living Animals. Sci. Transl. Med. 5, 213ra165–213ra165 (2013). https://doi.org/10.1126/scitranslmed.3007095
115. Raju, K.S.R., Taneja, I., Singh, S.P.: Wahajuddin: Utility of noninvasive biomatrices in pharmacokinetic studies. Biomed. Chromatogr. **27**, 1354–1366 (2013). https://doi.org/10.1002/bmc.2996
116. Tsunoda, M., Hirayama, M., Tsuda, T., Ohno, K.: Noninvasive monitoring of plasma l-dopa concentrations using sweat samples in Parkinson's disease. Clin. Chim. Acta **442**, 52–55 (2015). https://doi.org/10.1016/j.cca.2014.12.032

117. Tai, L.-C., Gao, W., Chao, M., Bariya, M., Ngo, Q.P., Shahpar, Z., Nyein, H.Y.Y., Park, H., Sun, J., Jung, Y., Wu, E., Fahad, H.M., Lien, D.-H., Ota, H., Cho, G., Javey, A.: Methylxanthine Drug Monitoring with Wearable Sweat Sensors. Adv. Mater. **30**, 1707442 (2018). https://doi.org/10.1002/adma.201707442
118. Gal, P.: Caffeine Therapeutic Drug Monitoring Is Necessary and Cost-effective. J. Pediatr. Pharmacol. Ther. **12**, 212–215 (2007). https://doi.org/10.5863/1551-6776-12.4.212
119. Tai, L.-C., Liaw, T.S., Lin, Y., Nyein, H.Y.Y., Bariya, M., Ji, W., Hettick, M., Zhao, C., Zhao, J., Hou, L., Yuan, Z., Fan, Z., Javey, A.: Wearable Sweat Band for Noninvasive Levodopa Monitoring. Nano Lett. **19**, 6346–6351 (2019). https://doi.org/10.1021/acs.nanolett.9b02478
120. Wu, Y., Tehrani, F., Teymourian, H., Mack, J., Shaver, A., Reynoso, M., Kavner, J., Huang, N., Furmidge, A., Duvvuri, A., Nie, Y., Laffel, L.M., Doyle, F.J.I., Patti, M.-E., Dassau, E., Wang, J., Arroyo-Currás, N.: Microneedle Aptamer-Based Sensors for Continuous, Real-Time Therapeutic Drug Monitoring. Anal. Chem. **94**, 8335–8345 (2022). https://doi.org/10.1021/acs.analchem.2c00829
121. Mousavisani, S.Z., Raoof, J.B., Ojani, R., Bagheryan, Z.: An impedimetric biosensor for DNA damage detection and study of the protective effect of deferoxamine against DNA damage. Bioelectrochemistry **122**, 142–148 (2018). https://doi.org/10.1016/j.bioelechem.2018.03.012
122. Parab, H.J., Jung, C., Lee, J.-H., Park, H.G.: A gold nanorod-based optical DNA biosensor for the diagnosis of pathogens. Biosens. Bioelectron. **26**, 667–673 (2010). https://doi.org/10.1016/j.bios.2010.06.067
123. Zhang, H., Wang, R., Tan, H., Nie, L., Yao, S.: Bovine serum albumin as a means to immobilize DNA on a silver-plated bulk acoustic wave DNA biosensor. Talanta **46**, 171–178 (1998). https://doi.org/10.1016/S0039-9140(97)00271-3
124. Afzal, A., Mujahid, A., Schirhagl, R., Bajwa, S.Z., Latif, U., Feroz, S.: Gravimetric viral diagnostics: QCM based biosensors for early detection of viruses. Chemosensors. **5**, 7 (2017)
125. Hu, L., Hu, S., Guo, L., Shen, C., Yang, M., Rasooly, A.: DNA Generated Electric Current Biosensor. Anal. Chem. **89**, 2547–2552 (2017). https://doi.org/10.1021/acs.analchem.6b04756
126. Han, S., Liu, W., Zheng, M., Wang, R.: Label-Free and Ultrasensitive Electrochemical DNA Biosensor Based on Urchinlike Carbon Nanotube-Gold Nanoparticle Nanoclusters. Anal. Chem. **92**, 4780–4787 (2020). https://doi.org/10.1021/acs.analchem.9b03520
127. Kim, E.R., Joe, C., Mitchell, R.J., Gu, M.B.: Biosensors for healthcare: Current and future perspectives. Trends Biotechnol. (2022)
128. Yang, B., Kong, J., Fang, X.: Bandage-like wearable flexible microfluidic recombinase polymerase amplification sensor for the rapid visual detection of nucleic acids. Talanta **204**, 685–692 (2019). https://doi.org/10.1016/j.talanta.2019.06.031
129. Galandová, J., Ovádeková, R., Ferancová, A., Labuda, J.: Disposable DNA biosensor with the carbon nanotubes–polyethyleneimine interface at a screen-printed carbon electrode for tests of DNA layer damage by quinazolines. Anal. Bioanal. Chem. **394**, 855–861 (2009). https://doi.org/10.1007/s00216-009-2740-x
130. Suginta, W., Khunkaewla, P., Schulte, A.: Electrochemical Biosensor Applications of Polysaccharides Chitin and Chitosan. Chem. Rev. **113**, 5458–5479 (2013). https://doi.org/10.1021/cr300325r
131. Di Iorio, D., Marti, A., Koeman, S., Huskens, J.: Clickable poly-l-lysine for the formation of biorecognition surfaces. RSC Adv. **9**, 35608–35613 (2019). https://doi.org/10.1039/C9RA08714A
132. Zhang, P., Lu, C., Niu, C., Wang, X., Li, Z., Liu, J.: Binding Studies of Cationic Conjugated Polymers and DNA for Label-Free Fluorescent Biosensors. ACS Appl. Polym. Mater. **4**, 6211–6218 (2022). https://doi.org/10.1021/acsapm.2c00986
133. Stillman, B.A., Tonkinson, J.L.: FASTTM Slides: A Novel Surface for Microarrays. Biotechniques **29**, 630–635 (2000). https://doi.org/10.2144/00293pf01
134. Sassolas, A., Leca-Bouvier, B.D., Blum, L.J.: DNA Biosensors and Microarrays. Chem. Rev. **108**, 109–139 (2008). https://doi.org/10.1021/cr0684467

135. Zhang, F., Wang, S., Liu, J.: Gold Nanoparticles Adsorb DNA and Aptamer Probes Too Strongly and a Comparison with Graphene Oxide for Biosensing. Anal. Chem. **91**, 14743–14750 (2019). https://doi.org/10.1021/acs.analchem.9b04142
136. Wang, S., McGuirk, C.M., Ross, M.B., Wang, S., Chen, P., Xing, H., Liu, Y., Mirkin, C.A.: General and Direct Method for Preparing Oligonucleotide-Functionalized Metal-Organic Framework Nanoparticles. J. Am. Chem. Soc. **139**, 9827–9830 (2017). https://doi.org/10.1021/jacs.7b05633
137. Sun, Z., Wu, S., Ma, J., Shi, H., Wang, L., Sheng, A., Yin, T., Sun, L., Li, G.: Colorimetric Sensor Array for Human Semen Identification Designed by Coupling Zirconium Metal-Organic Frameworks with DNA-Modified Gold Nanoparticles. ACS Appl. Mater. Interfaces **11**, 36316–36323 (2019). https://doi.org/10.1021/acsami.9b10729
138. Xiong, D., Cheng, J., Ai, F., Wang, X., Xiao, J., Zhu, F., Zeng, K., Wang, K., Zhang, Z.: Insight into the Sensing Behavior of DNA Probes Based on MOF–Nucleic Acid Interaction for Bioanalysis. Anal. Chem. **95**, 5470–5478 (2023). https://doi.org/10.1021/acs.analchem.3c00832
139. Weizmann, Y., Chenoweth, D.M., Swager, T.M.: Addressable Terminally Linked DNA−CNT Nanowires. J. Am. Chem. Soc. **132**, 14009–14011 (2010). https://doi.org/10.1021/ja106352y
140. Vittala, S.K., Han, D.: DNA-Guided Assemblies toward Nanoelectronic Applications. ACS Appl. Bio Mater. **3**, 2702–2722 (2020). https://doi.org/10.1021/acsabm.9b01178
141. Rashid, J.I.A., Yusof, N.A.: The strategies of DNA immobilization and hybridization detection mechanism in the construction of electrochemical DNA sensor: A review. Sens. Bio-Sens. Res. **16**, 19–31 (2017). https://doi.org/10.1016/j.sbsr.2017.09.001
142. Papadopoulou, E., Gale, N., Thompson, J.F., Fleming, T.A., Brown, T., Bartlett, P.N.: Specifically horizontally tethered DNA probes on Au surfaces allow labelled and label-free DNA detection using SERS and electrochemically driven melting. Chem. Sci. **7**, 386–393 (2016). https://doi.org/10.1039/C5SC03185K
143. Goodrich, G.P., Helfrich, M.R., Overberg, J.J., Keating, C.D.: Effect of Macromolecular Crowding on DNA: Au Nanoparticle Bioconjugate Assembly. Langmuir **20**, 10246–10251 (2004). https://doi.org/10.1021/la0484341
144. Ahmadi, S., Kamaladini, H., Haddadi, F., Sharifmoghadam, M.R.: Thiol-Capped Gold Nanoparticle Biosensors for Rapid and Sensitive Visual Colorimetric Detection of Klebsiella pneumoniae. J. Fluoresc. **28**, 987–998 (2018). https://doi.org/10.1007/s10895-018-2262-z
145. Mobed, A., Hasanzadeh, M., Babaie, P., Agazadeh, M., Mokhtarzadeh, A., Rezaee, M.A.: DNA-based bioassay of legionella pneumonia pathogen using gold nanostructure: A new platform for diagnosis of legionellosis. Int. J. Biol. Macromol. **128**, 692–699 (2019). https://doi.org/10.1016/j.ijbiomac.2019.01.125
146. Tour, J.M., Jones, L.I., Pearson, D.L., Lamba, J.J.S., Burgin, T.P., Whitesides, G.M., Allara, D.L., Parikh, A.N., Atre, S.: Self-Assembled Monolayers and Multilayers of Conjugated Thiols, .alpha.,.omega.-Dithiols, and Thioacetyl-Containing Adsorbates. Understanding Attachments between Potential Molecular Wires and Gold Surfaces. J. Am. Chem. Soc. 117, 9529–9534 (1995). https://doi.org/10.1021/ja00142a021
147. Reddy Gajjala, R.K., Gade, P.S., Bhatt, P., Vishwakarma, N., Singh, S.: Enzyme decorated dendritic bimetallic nanocomposite biosensor for detection of HCHO. Talanta **238**, 123054 (2022). https://doi.org/10.1016/j.talanta.2021.123054
148. Yang, B., Kong, J., Fang, X.: Programmable CRISPR-Cas9 microneedle patch for long-term capture and real-time monitoring of universal cell-free DNA. Nat. Commun. **13**, 3999 (2022). https://doi.org/10.1038/s41467-022-31740-3
149. Kuscu, M., Ramezani, H., Dinc, E., Akhavan, S., Akan, O.B.: Fabrication and microfluidic analysis of graphene-based molecular communication receiver for Internet of Nano Things (IoNT). Sci. Rep. **11**, 19600 (2021). https://doi.org/10.1038/s41598-021-98609-1
150. Guo, Y., Shen, G., Sun, X., Wang, X.: Electrochemical Aptasensor Based on Multiwalled Carbon Nanotubes and Graphene for Tetracycline Detection. IEEE Sens. J. **15**, 1951–1958 (2015). https://doi.org/10.1109/JSEN.2014.2370051

151. Pan, T.-M., Liao, P.-Y.: High sensitivity and rapid detection of KRAS and BRAF gene mutations in colorectal cancer using YbTixOy electrolyte-insulator-semiconductor biosensors. Mater. Today Chem. **25**, 100979 (2022). https://doi.org/10.1016/j.mtchem.2022.100979
152. Rodríguez-Montelongo, S.A., Moreno-Gutiérrez, D.S., Terán-Figueroa, Y., Gómez-Durán, C.F.A., Bañuelos-Frías, A., Palestino, G.: Porous Silicon-Based DNA Biosensor for Human Papillomavirus Detection: Towards the Design of Fast and Portable Test. SILICON **15**, 2371–2383 (2023). https://doi.org/10.1007/s12633-022-02179-4
153. Capo, C., Bongrand, P., Benoliel, A., Depieds, R.: Non-specific recognition in phagocytosis: ingestion of aldehyde-treated erythrocytes by rat peritoneal macrophages. Immunology **36**, 501 (1979)
154. Chung, D.-J., Kim, K.-C., Choi, S.-H.: Electrochemical DNA biosensor based on avidin–biotin conjugation for influenza virus (type A) detection. Appl. Surf. Sci. **257**, 9390–9396 (2011). https://doi.org/10.1016/j.apsusc.2011.06.015
155. Liu, G., Wan, Y., Gau, V., Zhang, J., Wang, L., Song, S., Fan, C.: An Enzyme-Based E-DNA Sensor for Sequence-Specific Detection of Femtomolar DNA Targets. J. Am. Chem. Soc. **130**, 6820–6825 (2008). https://doi.org/10.1021/ja800554t
156. Pan, S., Rothberg, L.: Chemical Control of Electrode Functionalization for Detection of DNA Hybridization by Electrochemical Impedance Spectroscopy. Langmuir **21**, 1022–1027 (2005). https://doi.org/10.1021/la048083a
157. Deng, C., Xia, Y., Xiao, C., Nie, Z., Yang, M., Si, S.: Electrochemical oxidation of purine and pyrimidine bases based on the boron-doped nanotubes modified electrode. Biosens. Bioelectron. **31**, 469–474 (2012). https://doi.org/10.1016/j.bios.2011.11.018
158. Jalit, Y., Moreno, M., Gutierrez, F.A., Sanchez Arribas, A., Chicharro, M., Bermejo, E., Zapardiel, A., Parrado, C., Rivas, G.A., Rodríguez, M.C.: Adsorption and Electrooxidation of Nucleic Acids at Glassy Carbon Electrodes Modified with Multiwalled Carbon Nanotubes Dispersed In Polylysine. Electroanalysis **25**, 1116–1121 (2013). https://doi.org/10.1002/elan.201200622
159. Zhang, S., Ding, Y., Wei, H.: Ruthenium Polypyridine Complexes Combined with Oligonucleotides for Bioanalysis: A Review. Molecules **19**, 11933–11987 (2014). https://doi.org/10.3390/molecules190811933
160. Drummond, T.G., Hill, M.G., Barton, J.K.: Electrochemical DNA sensors. Nat. Biotechnol. **21**, 1192–1199 (2003). https://doi.org/10.1038/nbt873
161. Wu, N., Gao, W., He, X., Chang, Z., Xu, M.: Direct electrochemical sensor for label-free DNA detection based on zero current potentiometry. Biosens. Bioelectron. **39**, 210–214 (2013). https://doi.org/10.1016/j.bios.2012.07.038
162. Yu, H.-Z., Luo, C.-Y., Sankar, C.G., Sen, D.: Voltammetric Procedure for Examining DNA-Modified Surfaces: Quantitation, Cationic Binding Activity, and Electron-Transfer Kinetics. Anal. Chem. **75**, 3902–3907 (2003). https://doi.org/10.1021/ac034318w
163. Ilkhani, H., Hughes, T., Li, J., Zhong, C.J., Hepel, M.: Nanostructured SERS-electrochemical biosensors for testing of anticancer drug interactions with DNA. Biosens. Bioelectron. **80**, 257–264 (2016). https://doi.org/10.1016/j.bios.2016.01.068
164. Ribeiro Teles, F.R., França dos Prazeres, D.M., De Lima-Filho, J.L.: Electrochemical Detection of a Dengue-related Oligonucleotide Sequence Using Ferrocenium as a Hybridization Indicator. Sensors. 7, 2510–2518 (2007). https://doi.org/10.3390/s7112510
165. Huang, J., Wu, J., Li, Z.: Biosensing using hairpin DNA probes. **34**, 1–27 (2015). https://doi.org/10.1515/revac-2015-0010
166. Farjami, E., Clima, L., Gothelf, K., Ferapontova, E.E.: "Off–On" Electrochemical Hairpin-DNA-Based Genosensor for Cancer Diagnostics. Anal. Chem. **83**, 1594–1602 (2011). https://doi.org/10.1021/ac1032929
167. Fu, L., Zhuang, J., Tang, D., Que, X., Lai, W., Chen, G.: DNA pseudoknot-functionalized sensing platform for chemoselective analysis of mercury ions. Analyst **137**, 4425–4427 (2012). https://doi.org/10.1039/C2AN35662G
168. Xiao, Y., Qu, X., Plaxco, K.W., Heeger, A.J.: Label-Free Electrochemical Detection of DNA in Blood Serum via Target-Induced Resolution of an Electrode-Bound DNA Pseudoknot. J. Am. Chem. Soc. **129**, 11896–11897 (2007). https://doi.org/10.1021/ja074218y

169. Singh, N.K., Chung, S., Chang, A.-Y., Wang, J., Hall, D.A.: A non-invasive wearable stress patch for real-time cortisol monitoring using a pseudoknot-assisted aptamer. Biosens. Bioelectron. **227**, 115097 (2023). https://doi.org/10.1016/j.bios.2023.115097
170. Majee, P., Kumar Mishra, S., Pandya, N., Shankar, U., Pasadi, S., Muniyappa, K., Nayak, D., Kumar, A.: Sci. Rep. **10**, 1477 (2020). https://doi.org/10.1038/s41598-020-58406-8
171. Zhao, Y., Kan, Z., Zeng, Z., Hao, Y., Chen, H., Tan, Z.: Determining the Folding and Unfolding Rate Constants of Nucleic Acids by Biosensor. Application to Telomere G-Quadruplex. J. Am. Chem. Soc. 126, 13255–13264 (2004). https://doi.org/10.1021/ja048398c
172. Bahreyni, A., Ramezani, M., Alibolandi, M., Hassanzadeh, P., Abnous, K., Taghdisi, S.M.: High affinity of AS1411 toward copper; its application in a sensitive aptasensor for copper detection. Anal. Biochem. **575**, 1–9 (2019). https://doi.org/10.1016/j.ab.2019.03.016
173. Hao, Z., Wang, Z., Li, Y., Zhu, Y., Wang, X., De Moraes, C.G., Pan, Y., Zhao, X., Lin, Q.: Measurement of cytokine biomarkers using an aptamer-based affinity graphene nanosensor on a flexible substrate toward wearable applications. Nanoscale **10**, 21681–21688 (2018). https://doi.org/10.1039/C8NR04315A
174. Travascio, P., Witting, P.K., Mauk, A.G., Sen, D.: The Peroxidase Activity of a Hemin−DNA Oligonucleotide Complex: Free Radical Damage to Specific Guanine Bases of the DNA. J. Am. Chem. Soc. **123**, 1337–1348 (2001). https://doi.org/10.1021/ja0023534
175. Wang, Z., Zhao, J., Bao, J., Dai, Z.: Construction of Metal-Ion-Free G-quadruplex-Hemin DNAzyme and Its Application in S1 Nuclease Detection. ACS Appl. Mater. Interfaces **8**, 827–833 (2016). https://doi.org/10.1021/acsami.5b10165
176. Cheng, X., Liu, X., Bing, T., Cao, Z., Shangguan, D.: General Peroxidase Activity of G-Quadruplex−Hemin Complexes and Its Application in Ligand Screening. Biochemistry **48**, 7817–7823 (2009). https://doi.org/10.1021/bi9006786
177. Xu, J., Yan, C., Wang, X., Yao, B., Lu, J., Liu, G., Chen, W.: Ingenious Design of DNA Concatamers and G-Quadruplex Wires Assisted Assembly of Multibranched DNA Nanoarchitectures for Ultrasensitive Biosensing of miRNA. Anal. Chem. **91**, 9747–9753 (2019). https://doi.org/10.1021/acs.analchem.9b01353
178. Owens, E.A., Huynh, H.T., Stroeva, E.M., Barman, A., Ziabrev, K., Paul, A., Nguyen, S.V., Laramie, M., Hamelberg, D., Germann, M.W., Wilson, W.D., Henary, M.: Second Generation G-Quadruplex Stabilizing Trimethine Cyanines. Bioconjug. Chem. **30**, 2647–2663 (2019). https://doi.org/10.1021/acs.bioconjchem.9b00571
179. Tello, A., Cao, R., Marchant, M.J., Gomez, H.: Conformational Changes of Enzymes and Aptamers Immobilized on Electrodes. Bioconjug. Chem. **27**, 2581–2591 (2016). https://doi.org/10.1021/acs.bioconjchem.6b00553
180. Sorek, R., Lawrence, C.M., Wiedenheft, B.: CRISPR-Mediated Adaptive Immune Systems in Bacteria and Archaea. Annu. Rev. Biochem. **82**, 237–266 (2013). https://doi.org/10.1146/annurev-biochem-072911-172315
181. Wan, Y., Zong, C., Li, X., Wang, A., Li, Y., Yang, T., Bao, Q., Dubow, M., Yang, M., Rodrigo, L.-A., Mao, C.: New Insights for Biosensing: Lessons from Microbial Defense Systems. Chem. Rev. **122**, 8126–8180 (2022). https://doi.org/10.1021/acs.chemrev.1c01063
182. Nguyen, P.Q., Soenksen, L.R., Donghia, N.M., Angenent-Mari, N.M., de Puig, H., Huang, A., Lee, R., Slomovic, S., Galbersanini, T., Lansberry, G., Sallum, H.M., Zhao, E.M., Niemi, J.B., Collins, J.J.: Wearable materials with embedded synthetic biology sensors for biomolecule detection. Nat. Biotechnol. **39**, 1366–1374 (2021). https://doi.org/10.1038/s41587-021-00950-3
183. Lu, M., Zhang, X., Xu, D., Li, N., Zhao, Y.: Encoded Structural Color Microneedle Patches for Multiple Screening of Wound Small Molecules. Adv. Mater. n/a, 2211330 (2023). https://doi.org/10.1002/adma.202211330
184. Lin, S., Cheng, X., Zhu, J., Wang, B., Jelinek, D., Zhao, Y., Wu, T.-Y., Horrillo, A., Tan, J., Yeung, J., Yan, W., Forman, S., Coller, H.A., Milla, C., Emaminejad, S.: Wearable microneedle-based electrochemical aptamer biosensing for precision dosing of drugs with narrow therapeutic windows. Sci. Adv. 8, eabq4539. https://doi.org/10.1126/sciadv.abq4539

Printable Wearable Self-Powered Biosensing System Based on Paper-Based Biofuel Cells Using Porous Carbon Material

Isao Shitanda, Noya Loew, and Seiya Tsujimura

1 Introduction

Recently, wearable biosensors that detect physiological parameters in body fluids, such as sweat, saliva, and tears have been developed for the early detection and prevention of diseases [1–10].

Because these devices can diagnose exercise efficiency and health status when worn, they have the potential to be applied for the measurement of physical and physiological parameters related to health and health monitoring during daily life and exercise. Furthermore, these devices are attracting attention in a wide range of fields such as medicine and health. For example, they can facilitate outpatient healthcare in conjunction with drug delivery modules [11, 12].

Conventional wearable devices are generally powered by coin-type lithium-ion or metal-air batteries. The miniaturization of these devices is desirable, as it improves comfort. However, miniaturization reduces the battery capacity, which means that frequent recharging in case of a secondary battery or replacement in case of a primary battery is required. Therefore, the miniaturization of wearable devices necessitates the consideration of new power sources, as well as safety and manufacturing costs. Biofuel cells (BFCs) have attracted significant research interest as energy harvesters for wearable devices [13–18].

I. Shitanda (✉) · N. Loew
Department of Pure and Applied Chemistry, Faculty of Science and Technology, Tokyo University of Science, Noda 2641, Yamazaki, Japan
e-mail: shitanda@rs.noda.tus.ac.jp

N. Loew · S. Tsujimura
Research Institute for Science and Technology, Tokyo University of Science, Yamazaki, Noda, Chiba, Japan

S. Tsujimura
Division of Material Science, Faculty of Pure and Applied Science, University of Tsukuba, 1-1-1, Tennodai, Tsukuba 305-8573, Ibaraki, Japan

K. Mitsubayashi (ed.), *Wearable Biosensing in Medicine and Healthcare*,
https://doi.org/10.1007/978-981-99-8122-9_16

BFCs are power sources that use biocatalysts as electrocatalysts; BFCs that use enzymes (redox enzymes) as biocatalysts are often referred to as enzymatic BFCs, and those that use microorganisms are called microbial BFCs [19–22].

Biocatalysts have the advantage of being able to generate electricity from stable, safe, and energy-dense organic materials (especially sugars, organic acids, and alcohols) in a mild environment similar to that of living organisms (near neutral pH, room to body temperature, and atmospheric pressure). BFCs can generate electricity without a separator between the anode and cathode fuels because of the high substrate selectivity of the enzyme. BFCs can be constructed from components that have a low environmental impact, and the total cost from manufacturing to disposal can be lower than that of conventional portable power generators and energy storage devices. As the enzymes are proteins, they can be mass-produced through bioengineering without the need for limited resources, such as rare metals.

BFCs can also be fueled by body fluids, and the development of self-driven biosensors that generate electricity from body fluid components and measure their concentrations is highly desirable. For example, self-driven biosensors operate by converting chemical energy into electric power when the glucose in body fluid reacts with the enzymes in a BFC and then uses the power to send signals from a transmitter (Fig. 1).

Because the output of BFCs depends on the glucose concentration, the glucose concentration can be determined from the output value of the BFC, which combines the functions of the power source and sensor. Self-driven biosensors are being investigated worldwide because they are an essential component of next-generation wearable devices that are lightweight and comfortable [23–25].

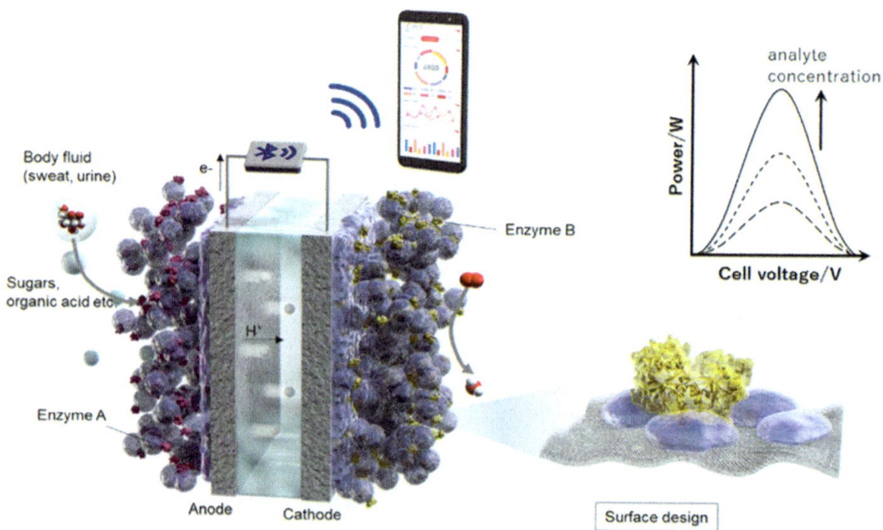

Fig. 1 Conceptual diagram of a self-driven biosensor that uses body fluid as fuel

Here, we focus on self-driven biosensors using BFCs. Recently, a contact-lens-type BFC device [26] that generates electricity from tears to measure sugar concentration, a mouthpiece-type device [27] that generates electricity from saliva to monitor stress markers, and a patch-type device [28] that monitors lactate concentration in sweat have been reported. However, to power wireless communication devices sufficiently without primary or secondary batteries, the output power must be increased.

We aimed to develop porous carbon materials with nanointerfaces specifically designed for enzyme electrodes to realize wearable-type BFCs with high power output and operational stability [29–34].

In enzymatic BFCs, the current density per geometric surface area in the presence of sufficient fuel is determined by the amount of enzyme loaded and the electrocatalytic activity of the enzyme. Enzymes generally have active sites embedded in the protein shell, making it difficult to achieve direct electron transfer with the electrode. This results in low output. Additionally, enzymes lose their activity due to structural changes or degradation. The aggregation and desorption of enzyme on the electrode surface also contribute to the durability of the BFC.

To improve the output power (current × voltage), it is necessary to not only search for and develop enzymes with high electrocatalytic activity (through screening and protein engineering, etc.) but also develop electrode materials with controlled structures and surfaces that can fully demonstrate the enzyme activity. The interactions between the nanostructure and enzyme may prevent the enzyme electrode from losing its activity. The design of three-dimensional structures must also be considered to effectively increase the number of electrochemically active enzymes per unit geometric area. Therefore, the presence of both nano- and mesopores is important to enhance enzyme stability and activity, and the presence of macropores enables fuel transport to the enzyme, which means that the design of a hierarchical nano-, meso-, and macropore structure is important.

In this chapter, we describe the development of porous carbon materials with controlled nano-, meso-, and macropores. In particular, carbon materials with controlled pore structures that use magnesium oxide (MgO) as a template (hereafter referred to as MgOCs) are introduced. Next, we describe an attempt to further improve the stability of the enzymes via surface modification with MgOC. Finally, biosensors and BFCs based on MgOCs or surface-modified MgOCs are introduced, especially sensors for monitoring body fluid components such as glucose in urine [29–31, 35] and lactate in sweat [32, 36].

2 Control of Nano-/Meso-/Macropores in Carbon Materials

To obtain a high enzyme catalytic current, the amount of enzyme loaded per unit geometric area should be increased, which effectively increases the enzyme utilization efficiency, that is, the amount of electrochemically active enzyme. The conditions that must be met to achieve this are listed below:

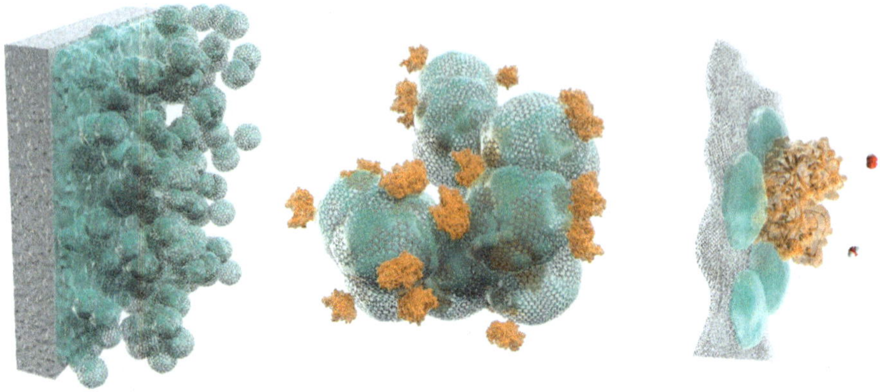

Fig. 2 Nano/meso/macropore control of carbon/enzyme/reactant to obtain high enzyme catalytic current

(1) The enzyme should be immobilized on the electrode.
(2) An electrode with a large specific surface area should be used.
(3) Reactants should be supplied to the enzyme (Fig. 2).

Regarding (1): When immobilizing the enzyme on the electrode surface, the enzyme should be localized near the electrode surface to improve the efficiency because enzyme-electrode reactions occur near the electrode surface. In direct electron transfer enzyme reactions, only the enzymes in contact with the electrode surface react; therefore, a monolayer is desirable. In the mediator enzyme reaction, the effective thickness of the enzyme layer can be determined by the reaction rate of the enzyme and the charge transport rate by the mediator [37]. The thickness of the enzyme layer can be easily controlled by changing the pore size.

Regarding (2): The number of electrochemically active enzymes can be increased by increasing the specific surface area and loading enzymes onto the surface. Carbon materials with nanostructures, such as carbon nanotubes and carbon black can also be used [38, 39]. However, when such nanomaterials are used, it is difficult to control the pore structure of the electrode; therefore, the enzyme-immobilized layer may form non-uniformly within the porous carbon layer, and the electro-enzymatic reaction may not proceed as expected [40]. Considering that enzymes are approximately 10 nm in size, the number of enzymes with electrochemical activity can be controlled on the scale of tens of nanometers. MgOC is synthesized as follows: MgO with a controlled crystallite size as a template is mixed with a carbon precursor, such as polyvinyl alcohol, to form carbon, followed by removal of the template with dilute acid (Fig. 3) [41–43]. This material has a large specific surface area with interconnected pores. The use of mesoporous carbon as electrode material offers the following advantages. In the case of direct-electron transfer enzyme reactions, the active center is more likely to be located near the electrode surface because the

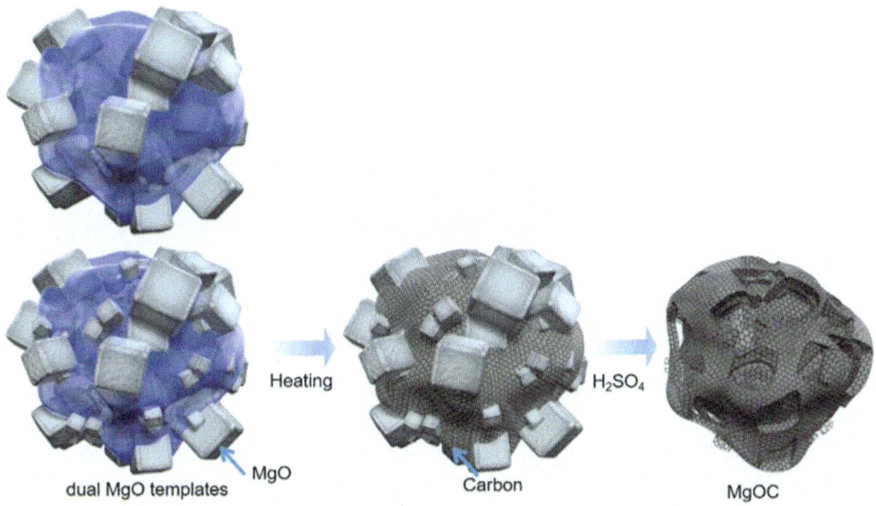

Fig. 3 Fabrication of MgO-templated carbon

enzyme reaction occurs in mesopores with similar sizes to that of the enzyme. In addition, for mediator-type enzyme electrode reactions, although the reactions are not affected by the orientation of the enzyme, the electron transport distance through the mediator can be reduced, which is advantageous as it can accelerate the net bioelectrocatalytic reaction. In particular, when the mediator is immobilized on a polymer, the electron transport rate decreases [44]. However, packing the mesopores with the enzyme and polymerized mediator can overcome the disadvantage of the slow electron transport rate, leading to a high-performance enzyme electrode [45].

Notably, if the carbon materials contain only mesopores with pore sizes comparable to the enzyme size, the enzyme will be present only near the surface of the porous carbon material and not inside it, resulting in a lower utilization efficiency per volume of the porous material. The use of an electrode with a controlled hierarchical structure comprising both mesopores and macropores improves enzyme dispersion and increases the number of active enzymes. In the case of MgOC, a material with a hierarchical structure can be obtained by mixing MgO templates of different sizes [46].

Regarding (3): If the supply of fuels to the enzymes immobilized on the carbon surface and the transport of electrolytes (protons) is not smooth and mass transport becomes a rate-limiting process, a large current cannot be obtained even if the number of electrochemically active enzymes is increased [47, 48]. Smaller pore sizes lead to larger surface areas, but greater resistance to mass transport and a more pronounced depletion of the fuel inside. In particular, when glucose is used as a fuel, its solubility in water is limited and its viscosity increases at high concentrations, decreasing the diffusion rate. The appropriate arrangement of the macropores is necessary to secure mass-transport pathways. However, an increase in the number of macropores

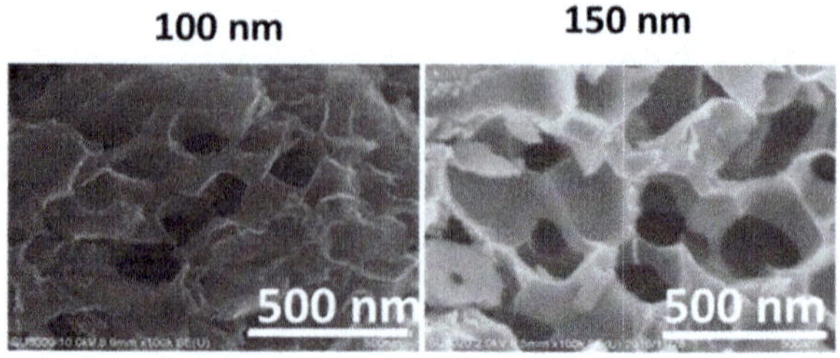

Fig. 4 Scanning electron microscopy image of MgO-templated carbon

decreased the specific surface area and number of immobilized enzymes. The ideal structure of the macropores and mesopores should be optimized, and although it may be possible to find the ideal structure by calculation, it is not easy to realize in practice.

As a practical solution to the trade-off between electrochemical enzyme activity and mass transport in the meso- and macroporous regions, the authors have proposed the use of macroporous carbon with pores of 100–200 nm as the enzyme electrode (Fig. 4). Macroporous carbons are not expected to enclose enzymes with a molecular size of approximately 10 nm [49]. However, by localizing the enzymes on the electrode surface, it is possible to control the meso-region involved in the reaction and the macro-region involved in mass transport. In direct electron transfer enzyme reactions, the use of technology to control the orientation of enzymes on the surface of carbon electrodes can lead to more efficient and stable enzyme utilization, and in mediator enzyme reactions, the formation of an enzyme-immobilized layer near the surface of macroporous carbon allows the construction of a reaction system that does not interfere with mass transport [50].

3 Surface Modification of MgOC

As mentioned previously, the power output of a BFC is highly dependent on the maximum current density and electron transfer rate at the enzyme-electrode interface. The maximum current density depends on the amount of immobilized enzyme per unit geometric area, the activity of each enzyme, the hierarchical structure of porous carbon, and the enzyme dispersion. In contrast, the electron transfer rate at the interface is determined by the electron transfer distance between the active center and electrode. The structure and molecular orientation of the enzyme are also important factors that determine the electron transfer distance. For electron transfer mediators, the interfacial electron transfer kinetics, and the molecular diffusion rate of the mediator (or the apparent electron diffusion rate in the case of the immobilized mediator), must be considered during the design process. In this section, we introduce the molecular modification of carbon surfaces to enhance the electron transfer rate between the carbon surface and the enzyme or to improve the stability.

3.1 Electrochemical Modification

For enzymes capable of direct electron transfer, controlling the molecular orientation of the enzyme makes it possible to minimize the electron transfer distance and construct an efficient electron transfer system. For example, a hydrophobic anthracene promotes the direct electron transfer of laccases, which have hydrophobic active centers. In the case of bilirubin oxidase (BOD), direct electron transfer is promoted by modifying naphthoic acid, which has π-electrons and a negative charge similar to bilirubin, the natural substrate of BOD [51]. One method for modifying such molecules on carbon surfaces is electrochemical modification which utilizes electrode reactions and the subsequent reaction associated with the modification [52].

The simplest modification method involves the electrochemical oxidation of amines. The nitrogen of the amine is activated and reacts directly with the neighboring electrode, forming a C-N covalent bond and modifying the molecule on the carbon surface. In the case of aromatic amines, the amine is converted to a diazonium salt, which is subsequently modified on the electrode surface by electrochemical reduction. The carbon bonded to the diazonium group is activated to form a C–C bond. However, during these reactions, not only do the electrochemically activated molecules react with the carbon surface to form bonds, but they may also react with other molecules near the electrode surface, thereby linking multiple molecules. Controlling the reactions on the surface is difficult, and the molecular size of the modified molecules can vary widely.

Azure-A and toluidine blue are derivatives of phenothiazine [53], which can be a good mediator for FAD-type glucose dehydrogenase (FAD-GDH). They also contain an -NH_2 group, which can be used for the modification of mediators in anodic systems that require mediators. The electrochemical oxidation of amines allows the formation of C-N covalent bonding. When modified with methylene blue, electron transfer from the active center of FAD-GDH has been observed [54].

These phenothiazines can also be modified on a carbon electrode via conversion to a diazonium salt and the subsequent electrochemical reduction. For Azure-A, the redox potential shifted significantly due to the formation of covalent bonds. Relative to the adsorbed Azure-A molecule, the covalently modified Azure-A electrode demonstrated a better electron transfer reaction with FAD-GDH [55, 56]. As mentioned above, the control of the molecular size of a modified molecule by this method is difficult, and clarification of the electron transfer mechanism and interface design based on this mechanism are required. The monolayers of these redox molecules can be modified on carbon surfaces using organic chemical methods instead of electrochemical methods. For example, the carbon surface can be oxidized to form carboxyl groups and phenothiazines can be modified on the carbon surface using carbodiimide and *N*-hydroxysuccinimide or acid chlorides with thionyl chloride. In this case, the electron transfer rate at the interface can be increased with the aid of a surface-grafted mediator. The mediator molecule on the carbon surface fits well enough into the binding site of the enzyme active center to allow electron transfer.

3.2 Modification by Electron-Beam Graft Polymerization

Graft polymerization is one promising approach for improving the hydrophilicity, anion or cation adsorption properties, biocompatibility, and electrical conductivity of polymers. Graft polymerization creates covalent bonds that link the grafted polymer chains and the carbon surface tightly, which is expected to prevent desorption.

Surface graft polymerization can be classified as "grafting-to," in which pre-synthesized polymers are joined, or "grafting-from," in which polymers are grafted by polymerizing monomers from the carbon surface. Obtaining a large quantity of grafted polymer using the "grafting-to" method is difficult due to the low reactivity of macromolecules [57]. In contrast, "grafting-from" is an efficient and controllable graft polymerization method [58]. Here, we describe a method for graft polymerization on a porous carbon surface using a "grafting-from" process and electron-beam irradiation.

When the carbon material is irradiated with electron beams, the bonds between carbon and hydrogen undergo homolysis, and radicals are generated. Subsequently, when the radicals are brought into contact with a vinyl monomer with a double bond, graft polymerization is initiated, the graft chain grows, and polymerization eventually stops. In contrast with other polymerization methods, electron-beam irradiation graft polymerization has the following advantages: no initiator is required, radicals are generated efficiently, the radicals can be maintained in a radical state, large quantities of substrate can be irradiated simultaneously, and the selectivity of the monomers is broad. However, the need for radiation facilities is a disadvantage. However, irradiation can be outsourced to specialist companies by universities that do not have their own irradiation facilities, if required.

Electron-beam graft polymerization can be classified into two main categories; pre-irradiation and simultaneous irradiation. The pre-irradiation graft polymerization method, in which vinyl monomers are brought into contact with an irradiated substrate as a gas or liquid after radicals are generated on the substrate in a nitrogen or argon atmosphere, is known as pre-irradiation graft polymerization. Another method, the simultaneous irradiation graft polymerization method involves placing the substrate in a gaseous or liquid vinyl monomer and then irradiating with electron beams to promote graft polymerization. In our study, a pre-irradiation graft polymerization method was adopted. The pre-irradiation method has the advantage of allowing a large amount of time and space by creating and storing radicals after irradiation. Retaining the radicals formed on the substrate after irradiation is necessary. However, the radicals will not be reduced if the temperature of the irradiated substrate is kept lower than the glass transition temperature of the substrate [59].

Glycidyl methacrylate (GMA) is a useful monomer for adding functional groups onto hydrophobic substrates, and a variety of compounds containing amino, thiol, or alcoholic hydroxyl groups can be bound to the epoxy groups. Various functional compounds can be connected to the grafted chains using epoxy groups on the grafted chains [59]. Electron-beam irradiation graft polymerization has been used on plastic surfaces such as polyethylene and polypropylene, but its application to porous carbon modification has not been previously reported. The advantage of this technique is that not only the surface of the carbon material but also the surface of the inner pores can be activated. Poly(GMA) grafted onto the porous carbon surface can react with the amino functional groups on the enzyme surface. Therefore, the enzyme can be easily and stably immobilized on a carbon surface by simply depositing the enzyme solution onto the carbon surface with grafted poly(GMA).

First, we investigated the stability of the immobilized enzyme on the epoxy group of poly(GMA) [62]. The MgOC modified with poly(GMA) is, hereafter, referred to as GMgOC. A mediator (1,2-naphthoquinone; 1,2-NQ) was added to the GMgOC, and FAD-GDH solution was deposited onto the GMgOC. The FAD-GDH could be immobilized by creating a stable bond between the amino group of the surface lysin and the epoxy groups on the GMA-grafted chains. In the FT-IR spectrum of GMgOC, a $C = O$ stretching vibration originating from the carbonyl group was observed near 1735 cm^{-1} and a C–O–C symmetric stretching vibration originating from the epoxy group was observed near 1150 cm^{-1}, confirming that both peaks are attributable to the formation of poly(GMA). Thermogravimetric analysis revealed that the grafting percentage was 10.0%. A clear difference in the stability of the electrodes was observed in the cyclic voltammograms after 25 cycles. For the GMgOC electrode with immobilized FAD-GDH, the normalized current density was maintained at 95% even after 25 cycles. In contrast, the FAD-GDH-immobilized MgOC electrode retained only 51% of its current density. As there was no change in the redox peak of 12NQ in the absence of glucose, the decrease in the catalytic current on the MgOC electrode was attributed to the elution of the enzyme, which was suppressed using the GMgOC electrode. In addition, the capping effect of the enzyme might have suppressed the elution of the mediator from the MgOC. This technique dramatically improved the stability of the FAD-GDH-modified MgOC electrode for long-term storage and continuous operation.

We also recently succeeded in simultaneously immobilizing the mediators aminoferrocene [63], Azure-A [35], and FAD-GDH using poly(GMA) graft chains on MgOC (Fig. 5). We are currently developing biosensors, wearable BFCs, and self-driven biosensors that use these materials, which are described in detail in the following section.

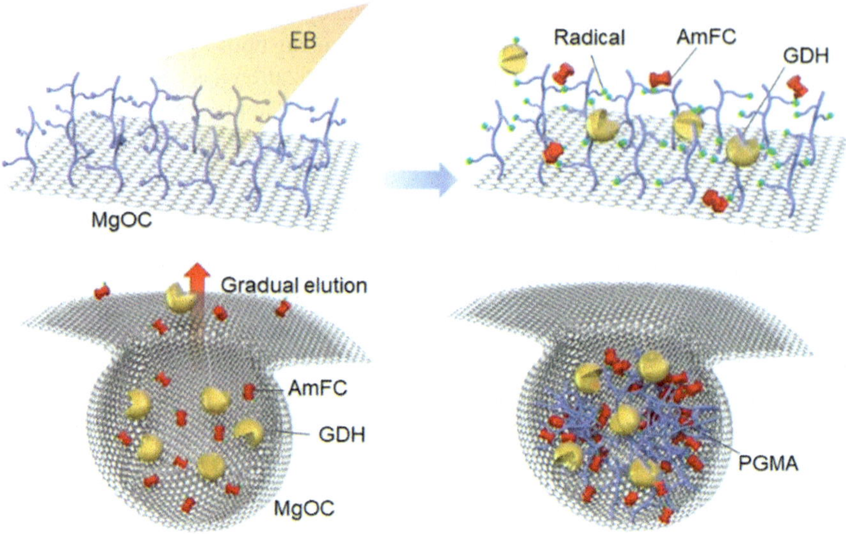

Fig. 5 Poly(GMA) modification of MgOC by electron-beam graft polymerization, and immobilization of enzymes and mediators [4]

4 Biosensors and BFCs Based on MgO-Templated Carbon

4.1 BFCs Combining MgOC and Carbon Cloth

The authors developed a glucose/oxygen enzyme BFC consisting of carbon cloth electrodes coated with MgOC [47]. Carbon cloth is highly electroconductive and serves as a lightweight and flexible MgOC carrier; MgOC with a pore size of 40 nm was added to the carbon cloth. Polyvinylidene fluoride was used as the anode binder, and polytetrafluoroethylene (PTFE) was used as the cathode binder. FAD-GDH was selected as the anode enzyme, and 1,4-naphthoquinone (1,4-NQ) was selected as the redox mediator. 1,4-NQ is highly reactive with FAD-GDH [53] and can be easily adsorbed onto the surface of porous carbons. BOD and 2,2'-azino-bis(3-ethylbenzothiazoline-6-sulfonic acid) (ABTS) were used as the oxygen reduction enzyme and mediator, respectively, for the cathode. The gas diffusion cathode was optimized for hydrophilicity and hydrophobicity by adjusting the ratio of the PTFE binder to carbon to maximize the O_2 supply from the atmosphere without inhibiting the proton supply from the solution. The open-circuit voltage potential of the BFC was 0.75 V and the maximum power density was 2 mW cm^{-2} at 0.4 V (room temperature and air atmosphere). The BFC showed no output power decrease even when bent, clearly demonstrating that the MgOC-loaded carbon clothes are very promising candidates for wearable BFCs. As the anode has high water absorbency, it can also be used to store aqueous glucose solution, and it demonstrates a discharge behavior similar to that of a metal-air battery. When discharged at a constant current, the current flows until the glucose is completely consumed, and the estimated cell capacity depends on the concentration of glucose. This result suggests that the BFC can be used as a self-driven sensor.

This platform was also applied to a lactate/oxygen BFC [32]. Lactate is a useful biomarker for evaluating health conditions, and its high theoretical energy density as a fuel is attractive for BFCs. A composite electrode consisting of a carbon cloth as the current collector, MgOC as the enzyme immobilization and mediator reaction site, lactate oxidase (LOx) as the enzyme, and 1-methoxy-5-methylphenazine methylsulfate (1-MeO-mPMS) as the redox mediator was used as the anode. BOD was used as the electrocatalyst and ABTS as the mediator for the oxygen reduction cathode. One side of the carbon cloth was coated with PTFE to improve the gas diffusion. The lactate/oxygen BFC demonstrated a maximum power density of 4.3 mW cm^{-2} at 300 mM lactate. The maximum power output is proportional to the lactate concentration. This is because the anode determines the overall performance of the BFC by improving the performance of the cathode. The lactate concentration could be evaluated from the maximum power output and the BFC could be used as a self-driven lactate sensor.

4.2 Screen-Printed Self-Driven Biosensor

Recently, we developed a self-driven wearable biosensing system based on a paper-substrate BFC (PBFC) by screen-printing using a MgOC ink. In this section, a new biosensing device that can generate electricity from biomarkers in urine and sweat using PBFCs is described. This device allows the measurement of biomarker concentrations from the amount of generated electricity. Blood and sweat are known to be rich in lactate, and the real-time detection of lactate in sweat has garnered considerable attention for the noninvasive monitoring of exercise or training. We developed a self-driven wearable lactate-sensing device that can communicate wirelessly and is fueled by the lactate in sweat (Fig. 6) [36].

The anode was modified with 1-MeO-mPMS, and LOx. The cathode was modified with BOD. The device achieves a high-power output via the arraying of the BFCs. One of the advantages of screen-printed PBFCs is that the required output can be adjusted according to the application by connecting (arraying) multiple cells in series when a high voltage is required and in parallel when a higher current is required. Even if the output from a single cell is low, the required power can be obtained by connecting the cells in series or parallel. For example, by printing a PBFC system with multiple cells arranged in series, an operating voltage sufficient to operate commercial devices, such as activity meters and thermometers, can be obtained without a voltage booster circuit. The output current can also be increased by increasing the surface area of the electrode. The voltage–power correlations of the arrayed PBFC with (1 × 1), (6 × 1), (4 × 4), and (6 × 6) electrodes in series and parallel, respectively, revealed that the maximum power output values were 0.11, 0.51, 2.55, and 4.30 mW, which

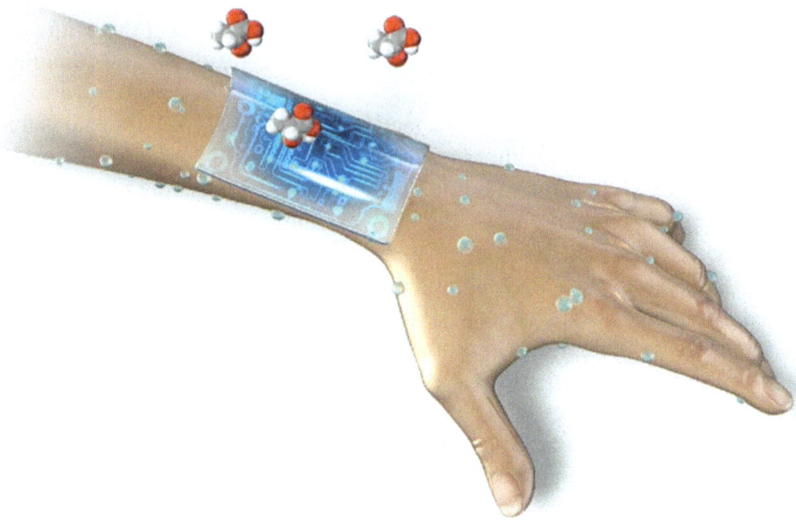

Fig. 6 Self-driven lactate biosensor fueled by lactate in sweat

were higher than those of previously reported thin film-based lactate BFCs. The results also show that the number of arrays can be adjusted to achieve the required power output. Each cell in the array is covered with Japanese paper (washi), which prevents the electrodes from directly touching the skin, thus eliminating irritation. The absorbed sweat permeates the washi paper and is evenly distributed to each electrode. The relationship between the lactate concentration in the buffer solution and the output was found to be linear. When a 100 mM lactate solution was used in the (6 × 6) PBFC, the open-circuit voltage was approximately 3.4 V, and the maximum output power was 4.3 mW, which enabled the device to drive a Bluetooth low-energy wireless communication device without the need for a voltage booster circuit and wirelessly transmit the lactate concentration to a smartphone in real time. Activity meters (health management devices that monitor calorie consumption) can also be driven simultaneously using the output of the PBFC instead of a coin-type lithium battery. Flexibility and comfort are attractive qualities for activity meters and other devices, which ideally would not be felt at all by the wearer. However, the thickness and rigidity of commercially available batteries have led to a bottleneck for wearable applications. Extremely thin and lightweight PBFC arrays are required for the design of ultralight activity meters.

The power density of PBFC increases proportionally to the lactate concentration in the range of 2.5 to 25 mM, which covers the range of lactate concentration in normal sweat. This array-type PBFC is expected to be used as a self-driven wearable tool for the continuous monitoring of the physiological state of athletes during exercise and the prevention of heatstroke.

4.3 Diaper Fuel Cell that Generates Electricity from Urine

Urinary glucose is detected when blood glucose is excreted in urine. Blood glucose is typically reabsorbed into the body through the kidneys; however, when blood glucose levels remain abnormally high, extra glucose is excreted in the urine. In general, when urinary glucose is detected, blood glucose levels are considered high, and a complete check-up at a hospital is required. However, in patients with postprandial hyperglycemia, blood glucose levels rise rapidly after a meal and return to normal over time. Therefore, abnormal blood glucose levels cannot be detected using normal fasting blood tests, and by the time symptoms are noticed, diabetes mellitus may have progressed and become severe. A sensing device that can detect urinary glucose and monitor its concentration at any time in a simple and timely manner will lead to early detection and prevention of diabetes mellitus. We developed a paper-based self-driven glucose biosensor system that can be integrated into a diaper (i.e., wearable) to detect urinary glucose in diabetic patients [35]. This system generates electricity from urinary glucose, detects when urine is voided, and monitors urinary glucose (Fig. 7).

The (1 × 1) glucose PBFC consists of the FAD-GDH anode with Azure-A as the mediator on the GMgOC printed on the substrate and BOD cathode. Reactions were performed with glucose solutions (1, 3, 5, 7, and 10 mM) to evaluate the correlation between PBFC power density and glucose concentration. The results revealed a linear correlation between the power density and glucose concentration. The sensor device consists of a PBFC and a wireless transmitter. When glucose is detected, the PBFC generates electricity, which is stored in a capacitor in the radio transmitter

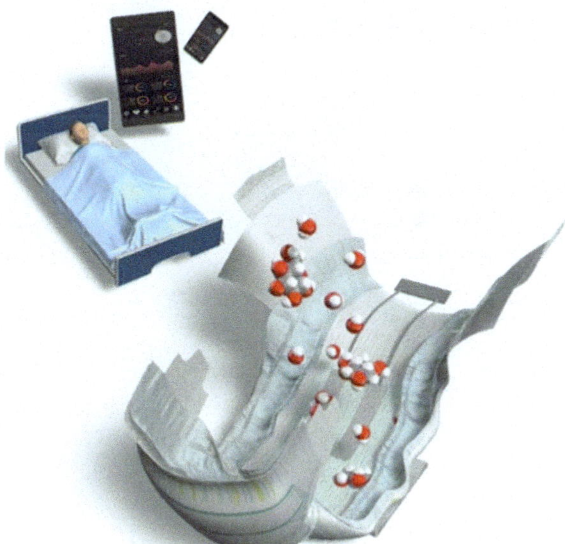

Fig. 7 Diaper fuel cell that generates electricity from urinary glucose and monitors urinary glucose levels

and is used to transmit radio waves. The transmitted radio waves are received by an external device such as a smartphone. The frequency of the radio waves sent from the radio wave transmitter to the receiver, i.e., the inverse of the response interval, can be converted into glucose concentration. The glucose concentration can be determined from the frequency of the transmitted radio waves. The PBFC is also able to monitor the timing of urination and urine volume. This sensing system is expected to reduce the mental and physical stress of patients and caregivers in homes, aged care facilities, and hospitals.

4.4 Screen-Printed Lactate Biosensor Based on GMgOC

Finally, we would like to discuss a lactate biosensor based on GMgOC (Fig. 8) [64]. This lactate biosensor was fabricated by immobilizing LOx and an appropriate mediator on a GMgOC electrode. A sensing system was developed for the continuous monitoring of lactate in sweat. The device consists of a GMgOC electrode modified with LOx and an integrated microfluidic sweat collection and delivery system fabricated using polydimethylsiloxane. The developed lactate-sensing system utilizes a simple immobilization method that takes advantage of the high surface area of GMgOC, which is advantageous because it can be easily mass-produced, and the parts that come into direct contact with the skin are fabricated from non-irritant materials. The detection limit of lactate in this biosensor is 0.3 mM, and the dynamic range is up to approximately 50 mM. This makes it suitable for detecting the lactate threshold, which indicates the change from aerobic to anaerobic metabolism. We are currently evaluating the lactate levels during training using a lactate biosensor and a self-driven lactate sensor.

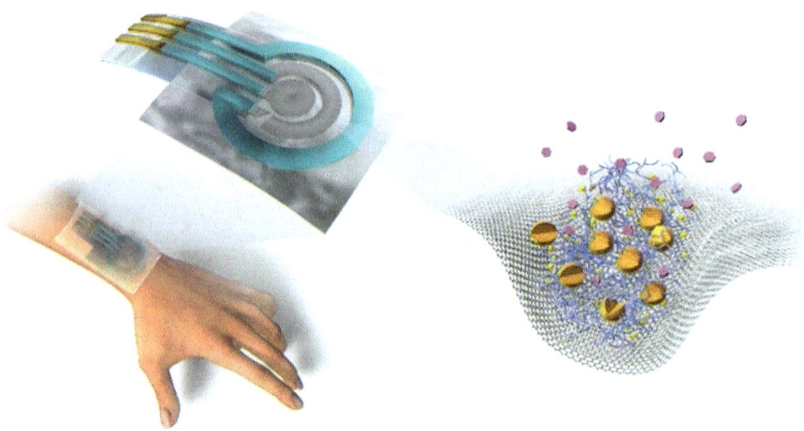

Fig. 8 Wearable lactate biosensor based on GMgOC

5 Conclusion

In this chapter, BFCs and biosensors using MgOC or GMgOC as porous carbons are described. BFCs are lightweight, compact, flexible, low-cost, environmentally friendly, and safe. They are also indispensable next-generation power sources for energy conservation and a low-carbon society. BFCs are also promising candidates for self-driven biosensors. However, owing to setbacks, such as poor output, short lifetime, and high cost, practical application has not yet been achieved. Remarkable progress has been made across multiple disciplines in this field. However, to facilitate practical application, the development of enzymes and design and engineering within nano-reaction fields are required to fully demonstrate the potential activity of enzymes. Close collaboration will become increasingly important in the future not only within the fields of enzymes and electrochemistry, but also within the field of carbon materials.

References

1. Nagamine, K., Mano, T., Nomura, A., Ichimura, Y., Izawa, R., Furusawa, H., Matsui, H., Kumaki, D., Tokito, S.: Noninvasive Sweat-Lactate Biosensor Emplsoying a Hydrogel-Based Touch Pad. Sci. Rep. **9**, 10102 (2019). https://doi.org/10.1038/s41598-019-46611-z
2. Salim, A., Lim, S.: Recent advances in noninvasive flexible and wearable wireless biosensors. Biosens. Bioelectron. **141**, 111422 (2019). https://doi.org/10.1016/j.bios.2019.111422
3. Dang, W., Manjakkal, L., Navaraj, W.T., Lorenzelli, L., Vinciguerra, V., Dahiya, R.: Stretchable wireless system for sweat pH monitoring. Biosens. Bioelectron. **107**, 192–202 (2018). https://doi.org/10.1016/j.bios.2018.02.025
4. Gao, W., Emaminejad, S., Nyein, H.Y.Y., Challa, S., Chen, K., Peck, A., Fahad, H.M., Ota, H., Shiraki, H., Kiriya, D., Lien, D.H., Brooks, G.A., Davis, R.W., Javey, A.: Fully integrated wearable sensor arrays for multiplexed in situ perspiration analysis. Nature **529**, 509–514 (2016). https://doi.org/10.1038/nature16521
5. Windmiller, J.R., Wang, J.: Wearable Electrochemical Sensors and Biosensors: A Review. Electroanalysis **25**, 29–46 (2013). https://doi.org/10.1002/elan.201200349
6. Iguchi, S., Kudo, H., Saito, T., Ogawa, M., Saito, H., Otsuka, K., Funakubo, A., Mitsubayashi, K.: A flexible and wearable biosensor for tear glucose measurement. Biomed. Microdevices **9**, 603–609 (2007). https://doi.org/10.1007/s10544-007-9073-3
7. Bandodkar, A.J., Molinnus, D., Mirza, O., Guinovart, T., Windmiller, J.R., Valdés-Ramírez, G., Andrade, F.J., Schöning, M.J., Wang, J.: Epidermal tattoo potentiometric sodium sensors with wireless signal transduction for continuous non-invasive sweat monitoring. Biosens. Bioelectron. **54**, 603–609 (2014). https://doi.org/10.1016/j.bios.2013.11.039
8. Song, Y., Min, J., Yu, Y., Wang, H., Yang, Y., Zhang, H., Gao, W.: Wireless battery-free wearable sweat sensor powered by human motion. Sci. Adv. 6, eaay9842 (2020). https://doi.org/10.1126/sciadv.aay9842
9. Yu, Y., Nassar, J., Xu, C., Min, J., Yang, Y., Dai, A., Doshi, R., Huang, A., Song, Y., Gehlhar, R., Ames, A.D., Gao, W.: Biofuel-powered soft electronic skin with multiplexed and wireless sensing for human-machine interfaces. Sci. Robot 5, eaaz7946 (2020). https://doi.org/10.1126/scirobotics.aaz7946
10. Reid, R.C., Mahbub, I.: Wearable self-powered biosensors. Curr. Opin. Electrochem. **19**, 55–62 (2020). https://doi.org/10.1016/j.coelec.2019.10.002

11. Lee, H., Song, C., Hong, Y.S., Kim, M.S., Cho, H.R., Kang, T., Shin, K., Choi, S.H., Hyeon, T., Kim, D.H.: Wearable/disposable sweat-based glucose monitoring device with multistage transdermal drug delivery module. Sci. Adv. **3**, e1601314 (2017). https://doi.org/10.1126/sciadv.1601314
12. Tan, E.K.W., Au, Y.Z., Moghaddam, G.K., Occhipinti, L.G., Lowe, C.R.: Towards Closed-Loop Integration of Point-of-Care Technologies. Trends Biotechnol. **37**, 775–788 (2019). https://doi.org/10.1016/j.tibtech.2018.12.004
13. Barton, S.C., Gallaway, J., Atanassov, P.: Enzymatic biofuel cells for implantable and microscale devices. Chem. Rev. **104**, 4867–4886 (2004). https://doi.org/10.1021/cr020719k
14. Grattieri, M., Minteer, S.D.: Self-Powered Biosensors. ACS Sens. **3**, 44–53 (2018). https://doi.org/10.1021/acssensors.7b00818
15. Miyake, T., Haneda, K., Nagai, N., Yatagawa, Y., Onami, H., Yoshino, S., Abe, T., Nishizawa, M.: Enzymatic biofuel cells designed for direct power generation from biofluids in living organisms. Energy Environ. Sci. **4**, 5008 (2011). https://doi.org/10.1039/c1ee02200h
16. Togo, M., Takamura, A., Asai, T., Kaji, H., Nishizawa, M.: An enzyme-based microfluidic biofuel cell using vitamin K3-mediated glucose oxidation. Electrochim. Acta **52**, 4669–4674 (2007). https://doi.org/10.1016/j.electacta.2007.01.067
17. Sakai, H., Mita, H., Sugiyama, T., Tokita, Y., Shirai, O., Kano, K.: Construction of a Multi-stacked Sheet-type Enzymatic Biofuel Cell. Electrochemistry **82**, 156–161 (2014). https://doi.org/10.5796/electrochemistry.82.156
18. Tsujimura, S., Kano, K., Ikeda, T.: Glucose/O_2 Biofuel Cell Operating at Physiological Conditions. Electrochemistry **70**, 940–942 (2002). https://doi.org/10.5796/electrochemistry.70.940
19. Zebda, A., Alcaraz, J.P., Vadgama, P., Shleev, S., Minteer, S.D., Boucher, F., Cinquin, P., Martin, D.K.: Challenges for successful implantation of biofuel cells. Bioelectrochemistry **124**, 57–72 (2018). https://doi.org/10.1016/j.bioelechem.2018.05.011
20. Aston, W.J., Turner, A.P.F.: Biosensors and Biofuel Cells. Biotechnol. Genet. Eng. Rev. **1**, 89–120 (1984). https://doi.org/10.1080/02648725.1984.10647782
21. Willner, I., Yan, Y.-M., Willner, B., Tel-Vered, R.: Integrated Enzyme-Based Biofuel Cells-A Review. Fuel Cells **9**, 7–24 (2009). https://doi.org/10.1002/fuce.200800115
22. Bandodkar, A.J.: Review—Wearable Biofuel Cells: Past, Present and Future. J. Electrochem. Soc. **164**, H3007–H3014 (2017). https://doi.org/10.1149/2.0031703jes
23. Nasar, A., Perveen, R.: Applications of enzymatic biofuel cells in bioelectronic devices – A review. Int. J. Hydrog. Energy **44**, 15287–15312 (2019). https://doi.org/10.1016/j.ijhydene.2019.04.182
24. Fu, L.Y., Liu, J.J., Hu, Z.Q., Zhou, M.: Recent Advances in the Construction of Biofuel Cells Based Self-powered Electrochemical Biosensors: A Review. Electroanalysis **30**, 2535–2550 (2018). https://doi.org/10.1002/elan.201800487
25. Sode, K., Yamazaki, T., Lee, I., Hanashi, T., Tsugawa, W.: BioCapacitor: A novel principle for biosensors. Biosens. Bioelectron. **76**, 20–28 (2016). https://doi.org/10.1016/j.bios.2015.07.065
26. Sánchez, I., Laukhin, V., Moya, A., Martin, R., Ussa, F., Laukhina, E., Guimera, A., Villa, R., Rovira, C., Aguiló, J., Veciana, J., Pastor, J.C.: Prototype of a nanostructured sensing contact lens for noninvasive intraocular pressure monitoring. Invest. Ophthalmol. Vis. Sci. **52**, 8310–8315 (2011). https://doi.org/10.1167/iovs.10-7064
27. Bandodkar, A.J., Wang, J.: Non-invasive wearable electrochemical sensors: a review. Trends Biotechnol. **32**, 363–371 (2014). https://doi.org/10.1016/j.tibtech.2014.04.005
28. Kai, H., Suda, W., Yoshida, S., Nishizawa, M.: Organic electrochromic timer for enzymatic skin patches. Biosens. Bioelectron. **123**, 108–113 (2019). https://doi.org/10.1016/j.bios.2018.07.013
29. Shitanda, I., Kato, S., Tsujimura, S., Hoshi, Y., Itagaki, M.: Screen-printed, Paper-based, Array-type. Origami Biofuel Cell. Chem. Lett. **46**, 726–728 (2017). https://doi.org/10.1246/cl.170047
30. Shitanda, I., Nohara, S., Hoshi, Y., Itagaki, M., Tsujimura, S.: A screen-printed circular-type paper-based glucose/O2 biofuel cell. J. Power. Sources **360**, 516–519 (2017). https://doi.org/10.1016/j.jpowsour.2017.06.043

31. Shitanda, I., Fujimura, Y., Nohara, S., Hoshi, Y., Itagaki, M., Tsujimura, S.: Paper-Based Disk-Type Self-Powered Glucose Biosensor Based on Screen-Printed Biofuel Cell Array. J. Electrochem. Soc. **166**, B1063–B1068 (2019). https://doi.org/10.1149/2.1501912jes
32. Shitanda, I., Takamatsu, K., Niiyama, A., Mikawa, T., Hoshi, Y., Itagaki, M., Tsujimura, S.: High-power lactate/O2 enzymatic biofuel cell based on carbon cloth electrodes modified with MgO-templated carbon. J. Power. Sources **436**, 226844 (2019). https://doi.org/10.1016/j.jpowsour.2019.226844
33. Shitanda, I., Momiyama, M., Watanabe, N., Tanaka, T., Tsujimura, S., Hoshi, Y., Itagaki, M.: Toward Wearable Energy Storage Devices: Paper-Based Biofuel Cells based on a Screen-Printing Array Structure. ChemElectroChem **4**, 2460–2463 (2017). https://doi.org/10.1002/celc.201700561
34. Shitanda, I., Kato, S., Hoshi, Y., Itagaki, M., Tsujimura, S.: Flexible and high-performance paper-based biofuel cells using printed porous carbon electrodes. Chem. Commun. (Camb.) **49**, 11110–11112 (2013). https://doi.org/10.1039/c3cc46644b
35. Shitanda, I., Fujimura, Y., Takarada, T., Suzuki, R., Aikawa, T., Itagaki, M., Tsujimura, S.: Self-Powered Diaper Sensor with Wireless Transmitter Powered by Paper-Based Biofuel Cell with Urine Glucose as Fuel. ACS Sens. **6**, 3409–3415 (2021). https://doi.org/10.1021/acssensors.1c01266
36. Shitanda, I., Morigayama, Y., Iwashita, R., Goto, H., Aikawa, T., Mikawa, T., Hoshi, Y., Itagaki, M., Matsui, H., Tokito, S., Tsujimura, S.: Paper-based lactate biofuel cell array with high power output. J. Power. Sources **489**, 229533 (2021). https://doi.org/10.1016/j.jpowsour.2021.229533
37. Albery, W.J., Cass, A.E., Shu, Z.X.: Inhibited enzyme electrodes. Part 1: Theoretical model. Biosens. Bioelectron. 5, 367–378 (1990). https://doi.org/10.1016/0956-5663(90)80016-7
38. Mazurenko, I., de Poulpiquet, A., Lojou, E.: Recent developments in high surface area bioelectrodes for enzymatic fuel cells. Curr. Opin. Electrochem. **5**, 74–84 (2017). https://doi.org/10.1016/j.coelec.2017.07.001
39. Holzinger, M., Le Goff, A., Cosnier, S.: Carbon nanotube/enzyme biofuel cells. Electrochim. Acta **82**, 179–190 (2012). https://doi.org/10.1016/j.electacta.2011.12.135
40. Suraniti, E., Vivès, S., Tsujimura, S., Mano, N.: Designing Thin Films of Redox Hydrogel for Highly Efficient Enzymatic Anodes. J. Electrochem. Soc. **160**, G79–G82 (2013). https://doi.org/10.1149/2.072306jes
41. Tsujimura, S., Nishina, A., Hamano, Y., Kano, K., Shiraishi, S.: Electrochemical reaction of fructose dehydrogenase on carbon cryogel electrodes with controlled pore sizes. Electrochem. Commun. **12**, 446–449 (2010). https://doi.org/10.1016/j.elecom.2010.01.016
42. Funabashi, H., Murata, K., Tsujimura, S.: Effect of Pore Size of MgO-templated Carbon on the Direct Electrochemistry of D-fructose Dehydrogenase. Electrochemistry **83**, 372–375 (2015). https://doi.org/10.5796/electrochemistry.83.372
43. Inagaki, M., Toyoda, M., Soneda, Y., Tsujimura, S., Morishita, T.: Templated mesoporous carbons: Synthesis and applications. Carbon **107**, 448–473 (2016). https://doi.org/10.1016/j.carbon.2016.06.003
44. Mao, F., Mano, N., Heller, A.: Long tethers binding redox centers to polymer backbones enhance electron transport in enzyme "Wiring" hydrogels. J. Am. Chem. Soc. **125**, 4951–4957 (2003). https://doi.org/10.1021/ja029510e
45. Tsujimura, S., Murata, K., Akatsuka, W.: Exceptionally high glucose current on a hierarchically structured porous carbon electrode with "wired" flavin adenine dinucleotide-dependent glucose dehydrogenase. J. Am. Chem. Soc. **136**, 14432–14437 (2014). https://doi.org/10.1021/ja5053736
46. Funabashi, H., Takeuchi, S., Tsujimura, S.: Hierarchical meso/macro-porous carbon fabricated from dual MgO templates for direct electron transfer enzymatic electrodes. Sci. Rep. **7**, 45147 (2017). https://doi.org/10.1038/srep45147
47. Niiyama, A., Murata, K., Shigemori, Y., Zebda, A., Tsujimura, S.: High-performance enzymatic biofuel cell based on flexible carbon cloth modified with MgO-templated porous carbon. J. Power. Sources **427**, 49–55 (2019). https://doi.org/10.1016/j.jpowsour.2019.04.064

48. Hamano, Y., Tsujimura, S., Shirai, O., Kano, K.: Micro-cubic monolithic carbon cryogel electrode for direct electron transfer reaction of fructose dehydrogenase. Bioelectrochemistry **88**, 114–117 (2012). https://doi.org/10.1016/j.bioelechem.2012.07.005
49. Mazurenko, I., Clément, R., Byrne-Kodjabachian, D., de Poulpiquet, A., Tsujimura, S., Lojou, E.: Pore size effect of MgO-templated carbon on enzymatic H2 oxidation by the hyperthermophilic hydrogenase from Aquifex aeolicus. J. Electroanal. Chem. **812**, 221–226 (2018). https://doi.org/10.1016/j.jelechem.2017.12.041
50. Tsujimura, S., Takeuchi, S.: Toward an ideal platform structure based on MgO-templated carbon for flavin adenine dinucleotide-dependent glucose dehydrogenase-Os polymer-hydrogel electrodes. Electrochim. Acta **343**, 136110 (2020). https://doi.org/10.1016/j.electacta.2020.136110
51. Tsujimura, S., Oyama, M., Funabashi, H., Ishii, S.: Effects of pore size and surface properties of MgO-templated carbon on the performance of bilirubin oxidase–modified oxygen reduction reaction cathode. Electrochim. Acta **322**, 134744 (2019). https://doi.org/10.1016/j.electacta.2019.134744
52. Downard, A.J.: Electrochemically Assisted Covalent Modification of Carbon Electrodes. Electroanalysis **12**, 1085–1096 (2000). https://doi.org/10.1002/1521-4109(200010)12:14%3c1085::AID-ELAN1085%3e3.0.CO;2-A
53. Tsuruoka, N., Sadakane, T., Hayashi, R., Tsujimura, S.: Bimolecular Rate Constants for FAD-Dependent Glucose Dehydrogenase from Aspergillus terreus and Organic Electron Acceptors. Int. J. Mol. Sci. **18**(3), 604 (2017). https://doi.org/10.3390/ijms18030604
54. Tsuruoka, N., Soto, S.S., Tahar, A.B., Zebda, A., Tsujimura, S.: Mediated electrochemical oxidation of glucose via poly(methylene green) grafted on the carbon surface catalyzed by flavin adenine dinucleotide-dependent glucose dehydrogenase. Colloids Surf. B Biointerfaces **192**, 111065 (2020). https://doi.org/10.1016/j.colsurfb.2020.111065
55. Gross, A.J., Tanaka, S., Colomies, C., Giroud, F., Nishina, Y., Cosnier, S., Tsujimura, S., Holzinger, M.: Diazonium Electrografting vs. Physical Adsorption of Azure A at Carbon Nanotubes for Mediated Glucose Oxidation with FAD-GDH. ChemElectroChem 7, 4543–4549 (2020). https://doi.org/10.1002/celc.202000953
56. Tsujimura, S., Tanaka, S., Gross, A., Holzinger, M.: Electrochemical modification at multi-walled carbon nanotube electrodes with Azure A for FAD- glucose dehydrogenase wiring: structural optimization to enhance catalytic activity and stability. J. Phys. Energy **3**, 024004 (2021). https://doi.org/10.1088/2515-7655/abd298
57. Cao, L., Yang, W., Yang, J., Wang, C., Fu, S.: Hyperbranched Poly(amidoamine)-modified Multi-walled Carbon Nanotubes via Grafting-from Method. Chem. Lett. **33**, 490–491 (2004). https://doi.org/10.1246/cl.2004.490
58. Gao, C., Vo, C.D., Jin, Y.Z., Li, W., Armes, S.P.: Multihydroxy Polymer-Functionalized Carbon Nanotubes: Synthesis, Derivatization, and Metal Loading. Macromolecules **38**, 8634–8648 (2005). https://doi.org/10.1021/ma050823e
59. Saito, K. Gurafutojūgo ni yoru Kōbunnshi-Kyūtyakuzai-Kakumei (A Revolution in Polymer Modifications by Graft Polymerization). Maruzen-Yushodo, Tokyo (2014). (in Japanese)
60. Rattan, S., Sehgal, T.: Stimuli-responsive membranes through peroxidation radiation-induced grafting of 2-hydroxyethyl methacrylate (2-HEMA) onto isotactic polypropylene film (IPP). J. Radioanal. Nucl. Chem. **293**, 107–118 (2012). https://doi.org/10.1007/s10967-012-1728-8
61. López-Saucedo, F., Flores-Rojas, G.G., Bucio, E., Alvarez-Lorenzo, C., Concheiro, A., González-Antonio, O.: Achieving antimicrobial activity through poly(N-methylvinylimidazolium) iodide brushes on binary-grafted polypropylene suture threads. MRS Commun **7**, 938–946 (2017). https://doi.org/10.1557/mrc.2017.121
62. Shitanda, I., Kato, T., Suzuki, R., Aikawa, T., Hoshi, Y., Itagaki, M., Tsujimura, S.: Stable Immobilization of Enzyme on Pendant Glycidyl Group-Modified Mesoporous Carbon by Graft Polymerization of Poly(glycidyl methacrylate). Bull. Chem. Soc. Jpn **93**, 32–36 (2020). https://doi.org/10.1246/bcsj.20190212
63. Suzuki, R., Shitanda, I., Aikawa, T., Tojo, T., Kondo, T., Tsujimura, S., Itagaki, M., Yuasa, M.: Wearable glucose/oxygen biofuel cell fabricated using modified aminoferrocene and

flavin adenine dinucleotide-dependent glucose dehydrogenase on poly(glycidyl methacrylate)-grafted MgO-templated carbon. J. Power. Sources **479**, 228807 (2020). https://doi.org/10.1016/j.jpowsour.2020.228807

64. Shitanda, I., Mitsumoto, M., Loew, N., Yoshihara, Y., Watanabe, H., Mikawa, T., Tsujimura, S., Itagaki, M., Motosuke, M.: Continuous sweat lactate monitoring system with integrated screen-printed MgO-templated carbon-lactate oxidase biosensor and microfluidic sweat collector. Electrochim. Acta **368**, 137620 (2021). https://doi.org/10.1016/j.electacta.2020.137620

Energy Harvesting from Bite Force Using Electret Sheet

Wataru Hijikata and Kenta Ichikawa

1 Introduction

The technology surrounding the Internet of Things (IoT) has developed rapidly over the past few decades. Wireless sensor networks (WSNs) are the core technology for IoT and are used in industrial and consumer applications. One promising application of IoT is wearable biosensing for healthcare, which is used to obtain health-related information from the human body. The vital data measured using these devices can be used for disease prevention and proactive prediction [1]. In the oral cavity, various biological activities such as breathing, speaking, salivation, occlusion, swallowing, and other actions associated with eating are performed, which provide a large amount of medically and dentally important information, such as nutrient intake, saliva composition, and bite force.

Mouthguard-type sensors have been developed to measure this information continuously. Arakawa et al. [2, 3] developed a mouthguard-type device with a sensor embedded within a double-layered mouthguard that measures the concentration of glucose in saliva by detecting a chemical reaction of glucose catalyzed by glucose oxidase immobilized on an electrode. Kim et al. [4] developed a mouthguard-type sensor to measure uric acid levels in saliva by detecting the chemical reaction of uric acid catalyzed by uricase immobilized on an electrode. Because glucose concentrations and uric acid in saliva measured by mouthguard-type sensors correlate with blood concentrations [5], they are expected to be used as non-invasive biomarkers that can substitute blood glucose levels and blood uric acid. These wearable devices enable continuous and non-invasive measurement of these indices, which are difficult to measure using conventional technology and enable us to understand the diurnal

W. Hijikata (✉)
School of Engineering, Tokyo Institute of Technology, Tokyo, Japan
e-mail: hijikata.w.aa@m.titech.ac.jp

K. Ichikawa
Institute of Biomaterials and Bioengineering, Tokyo Medical and Dental University, Tokyo, Japan

K. Mitsubayashi (ed.), *Wearable Biosensing in Medicine and Healthcare*,
https://doi.org/10.1007/978-981-99-8122-9_17

variation of these indices. Compared to conventional invasive methods involving puncture, this method is expected to be useful for the early diagnosis of hidden diabetes mellitus.

Although mouthguard-type sensors have the advantage that they can be easily attached to and detached from the oral cavity when compared to the sensors implanted in the body, it is difficult to use batteries for powering the sensors because of the risk of accidental ingestion [6], which is a bottleneck to the realization of long-term health monitoring. Energy harvesting from the body is a promising method for realizing battery-free mouthguard sensors. Energy harvesting is a technology that generates electricity using tiny energy sources that have not yet been utilized. By integrating a power generation system based on this technology into an intraoral wearable device, a battery-free intraoral health monitoring system can be realized. Several potential energy sources are available in the oral cavity. Organic matter in saliva can drive microbial fuel cells (MFC) [7]. Among the various daily activities involving mechanical energy, such as swallowing, breathing, and speaking in the oral cavity, biting is one of the most energy-intensive. Harvesters use processes such as the piezoelectric effect [8, 9], electromagnetic induction [10, 11], and electrostatic induction [12]. This chapter introduces electrostatic inductive energy harvesters that use electret sheets and their application in bite-force sensors.

2 Principle

An overview of the proposed intraoral power generation system [13] is shown in Fig. 1. The system is a mouthguard that can be attached to the occlusal surface of the dentition, where the bite force is exerted. A thin sheet-type harvester is embedded in a double-layered mouthguard. The system generates electrical power from the bite force applied to the harvester during mastication. The power generated is then fed to a sensor embedded in the mouthguard to realize self-powered intraoral health monitoring without the use of chemical batteries. Because the bite force is an intermittent biological activity, the sensor is equipped with a capacitor to store the generated power temporarily. This capacitor has a lower capacity than a chemical battery, does not retain high energy continuously, and is sequentially stored and discharged during sensing, thereby significantly reducing health risks in the event of accidental ingestion.

To realize such an intraoral power generation system, it is necessary to develop a sheet-type harvester that meets the following requirements.

(i) Ultrathin:

The harvester should be ultrathin, less than 200 μm in thickness, so as not to interfere with occlusal movement, since it is embedded in the occlusal surface of the mouthguard dentition.

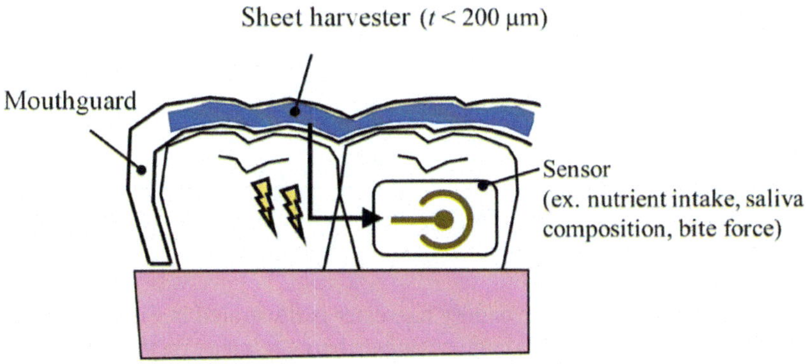

Fig. 1 Schematic of the proposed intra-oral energy system

(ii) Low-frequency drive:

The harvester uses a low-frequency (approximately 1 Hz) bite force, acting only in the compressive direction, as an energy source.

(iii) High power:

The harvester should achieve power generation on the order of microwatts, which is sufficient to power a mouthguard-type sensor.

For the above requirements, electromagnetic induction is not suitable for this system because of the limitation of miniaturization of coils and permanent magnets owing to size restrictions and the difficulty in the relative movement of coils and permanent magnets inside the mouthguard. Sheet piezoelectric materials or electrostatic inductive elements can be considered ultrathin sheet-type harvesters. In the case of piezoelectric materials, it is required to drive the materials at their resonant frequency to generate high power. However, it is difficult to achieve a resonance drive in a very small space inside the mouthguard against a bite force of 1 Hz because the space for the harvester mechanism is limited. If the harvesters are driven without resonance, the output characteristic is generally much lower than that of the resonant drive, making it difficult to achieve the target high-power generation performance in the order of microwatts. In the case of the electrostatic induction type, a power source is required to induce electrical charges on the electrodes, which leads to a larger mechanism. Therefore, an ultrathin, low-frequency, high-power harvester that meets the requirements of the proposed intraoral power generation system has not yet been realized. In this chapter, a newly developed energy harvester based on electrostatic induction consisting of a laminated electret and a dielectric elastomer is introduced.

A schematic of the proposed energy harvester is shown in Fig. 2. The power-generating sheet consists of an electrode, an electret, and a dielectric elastomer. The surface of the electret on the dielectric elastomer side semi-permanently holds a negative charge. Both electrodes are connected to a load (sensor devices, etc.) to form an electrically closed circuit.

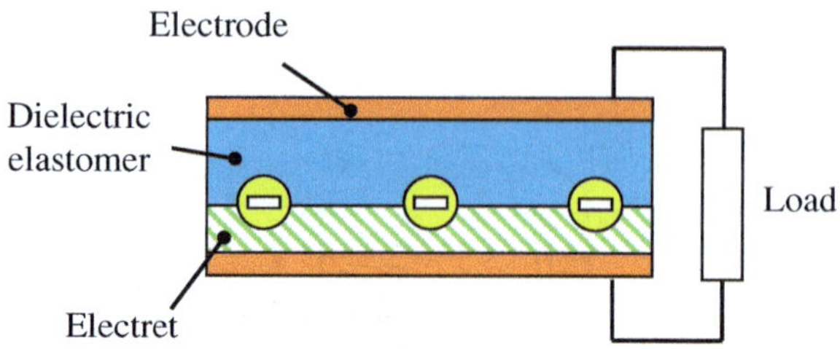

Fig. 2 Schematic of the proposed electrostatic energy harvester

The driving principle of the energy harvester is illustrated in Fig. 3. In the initial state with no bite force (Fig. 3a), the negative charge retained on the electret surface induces positive charges on both electrodes. At this point, the entire system is in a state of electrical equilibrium; however, each electrode is positively charged. The total amount of charge induced in both electrodes is equal to the amount of charge held in the electret; this ratio is determined by the capacitance from the electret surface to both electrodes.

When a bite force is applied to the power-generating sheet, each layer of the sheet is compressively deformed. In this deformation process (Fig. 3b), the displacement of the dielectric elastomer layer, which is significantly deformed by the rubber elasticity, is dominant. As the dielectric elastomer is compressed and deformed, the capacitance of the elastomer layer increases as its thickness decreases. Because the displacement of the electret layer can be ignored and the initial capacitance is maintained, the induced charge changes with the ratio of the capacitance to that of the elastomer layer.

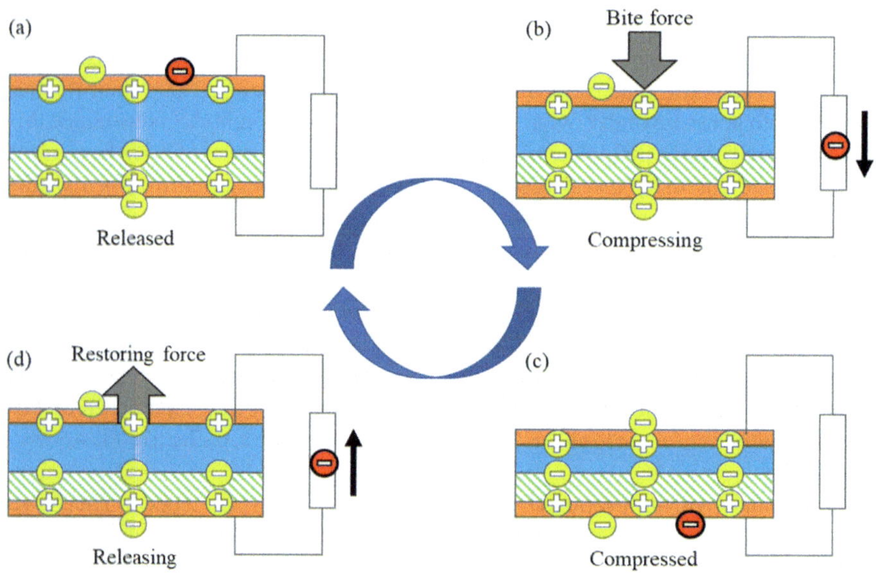

Fig. 3 Power generation principle of the proposed harvester

In this process, charge is transferred from the electrode on the dielectric elastomer side to the electrode on the electret side through a connected electrical load.

In the fully compressed state (Fig. 3c), the induced charge on the dielectric elastomer side increases, that on the electret side decreases to a stable state, and no charge transfer occurred between the electrodes. When the upper and lower dental arches are separated and the bite force is removed, the dielectric elastomer, which had been compressed and deformed, is elongated to its initial thickness due to the restoring force (Fig. 3d). In contrast to the compression process, the capacitance of the dielectric elastomer layer decreases, and the ratio of induced charges to the two electrodes returns to its initial state; thus, charges are transferred from the electrode on the electret side to the electrode on the dielectric elastomer side in this process.

When the dielectric elastomer layer is restored to its initial thickness by the restoring force, the state of charges on both electrodes becomes the same as the initial state, and this cycle is repeated whenever the bite force is applied again. In this cycle, the charge passes through the load during the compression and restoration processes, which are called the induced currents. The entire cycle realizes energy conversion, in which the mechanical energy resulting from the bite forces is stored as electrostatic energy in the dielectric elastomer layer and output as electrical energy. The large deformation achieved using dielectric elastomers allows a large amount of electrostatic energy to be utilized for power generation. Thus, high-output characteristics are expected to be achieved even in low-frequency drives. By utilizing the change in capacitance of the electret layer, the harvester can be driven without a switching circuit, which is required in conventional dielectric elastomer harvesters. Moreover, using a simple structure that utilizes the restoring force of the dielectric elastomer, a periodic drive can be realized in a very small space inside the mouthguard, even when the movement acts only in the compressive direction. This system is expected to meet the above-mentioned requirements for an intraoral power generation system.

3 Mathematical Model

To optimize the dimensions of the proposed harvester, a mathematical model was derived and the amount of power generated was simulated. An equivalent circuit of the electrostatic induction power generation sheet is shown in Fig. 4.

The dielectric elastomer and electret layers were modeled as capacitors C_{DE} and C_{EL}, respectively, and are represented by the following equations:

$$C_{DE} = \frac{\varepsilon_0 \varepsilon_{DE} S}{d_{DE}} \tag{1}$$

$$C_{EL} = \frac{\varepsilon_0 \varepsilon_{EL} S}{d_{EL}} \tag{2}$$

where ε_{DE} and ε_{EL} are the relative permittivities of the dielectric elastomer and electret, respectively, and d_{DE} and d_{EL} are the thicknesses of the dielectric elastomer

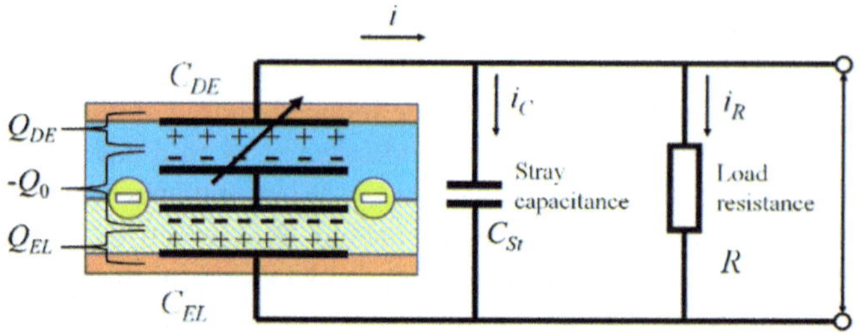

Fig. 4 Equivalent circuit diagram of the proposed electrostatic harvester

and electret, respectively. Capacitor C_{DE} in the dielectric elastomer layer is modeled as a variable capacitance because its capacitance varies according to the compression and recovery processes. Because the internal impedance of the electrostatic induction harvester is high and the loss due to parasitic capacitance is larger than that of other power generation principles, the parasitic capacitance C_{St} is considered in addition to the load resistance R as an external circuit.

According to the conservation of charge law, the charges Q_{DE} and Q_{EL} stored in the capacitors of the dielectric and electret layers, respectively, satisfy the following relationship with charge Q_0 held in the electret:

$$Q_{DE} + Q_{EL} = -Q_0 \tag{3}$$

Based on Kirchhoff's law for the equivalent circuit of a generator sheet, the following equation is derived for the output voltage v.

$$v = Ri_R = \frac{Q_{DE}}{C_{DE}} - \frac{Q_{EL}}{C_{EL}}$$

The output current i of the generator sheet is divided into current i_R flowing through the load resistance and current i_C flowing to the parasitic capacitance, satisfying the following equation:

$$i = -\frac{dQ_{DE}}{dt} = \frac{dQ_{EL}}{dt} = i_R + i_C \tag{5}$$

The current i_C flowing to the parasitic capacitor can be expressed by the following equation using the output voltage v based on the relationship between the voltage and current of the capacitor:

$$i_C = C_{St}\frac{dv}{dt} \tag{6}$$

By combining Eqs. (3)–(6) and rearranging them for Q_{DE}, the circuit equation for an electrostatic induction-type generator sheet is obtained as follows:

$$\alpha\frac{dQ_{DE}}{dt} + \beta Q_{DE} + \gamma Q_0 = 0 \tag{7}$$

where α, β, and γ are given as follows:

$$\alpha = 1 + C_{St}\left(\frac{1}{C_{DE}} + \frac{1}{C_{EL}}\right) \tag{8}$$

$$\beta = \frac{1}{R}\left(\frac{1}{C_{DE}} + \frac{1}{C_{EL}}\right) + C_{St}\frac{d}{dt}\left(\frac{1}{C_{DE}} + \frac{1}{C_{EL}}\right) \tag{9}$$

$$\gamma = \frac{1}{RC_{EL}} + C_{St}\frac{d}{dt}\left(\frac{1}{C_{EL}}\right) \tag{10}$$

Because the capacitance C_{DE} of the dielectric elastomer layer changes with the bite force, the coefficients of the circuit equations (Eqs. (8) and (9)) also change accordingly. For the capacitance of the dielectric elastomer layer shown in Eq. (1), the thickness of the dielectric elastomer, d_{DE}, was varied during the power generation process.

The mechanical properties of dielectric elastomers are expressed as rubber elasticity, which is characterized by the ability to achieve large deformations compared to metallic materials, strong nonlinearity in the stress–strain relationship, and return to their original shape when unloaded. The hyperelastic model is widely used as a model of rubber elasticity in which the material properties are expressed by the strain energy density function. Various models have been proposed for the strain energy density function [14], and the Mooney-Rivlin model [15] [16], which is widely used in the finite element analysis of rubber materials, was applied in this study. In the Mooney–Rivlin model, the strain energy density function W is formulated as follows:

$$W = \sum_{i+j=1}^{n} C_{ij}(I_1 - 3)^i (I_2 - 3)^j \tag{11}$$

where n is the order of the strain energy density function, C_{ij} is a material-specific parameter, and I_1 and I_2 are the first and second invariants of the strain, respectively.

Assuming that the dielectric elastomer used in the generator sheet is an incompressible material that is compressed uniaxially only in the direction of the bite force, the first and second invariants of strain are expressed by the following equations using the elongation ratio λ in the compression direction, respectively.

$$I_1 = \lambda^2 + 2\lambda^{-1} \tag{12}$$

$$I_2 = 2\lambda + \lambda^{-2} \tag{13}$$

Because the Cauchy stress of the material is obtained by partial differentiation of the strain energy density function by the elongation ratio, the Cauchy stress σ_t in the compressive direction can be expressed as

$$\sigma_t = 2\left(\lambda^2 - \lambda^{-1}\right)\left(\frac{\partial W}{\partial I_1} + \frac{1}{\lambda}\frac{\partial W}{\partial I_2}\right) \tag{14}$$

From the assumptions of axial compression and material incompressibility, the Cauchy stress σ_t in the compressive direction obtained in Eq. (17) can be converted to nominal stress σ to obtain the following equation.

$$\sigma = \frac{\sigma_t}{\lambda} = 2\left(\lambda - \lambda^{-2}\right)\left(\frac{\partial W}{\partial I_1} + \frac{1}{\lambda}\frac{\partial W}{\partial I_2}\right) \tag{15}$$

For simplicity, the power generation was estimated with $n = 1$ in the order of the Mooney-Rivlin strain energy density function. In this case, Eqs. (11) and (18) are simplified to the following equations.

$$W = C_{10}(I_1 - 3) + C_{01}(I_2 - 3) \tag{16}$$

$$\sigma = 2\left(\lambda - \lambda^{-2}\right)\left(C_{10} + \frac{C_{01}}{\lambda}\right) \tag{17}$$

Using the stress–strain relationship obtained in Eq. (20), the input stress can be converted to the elongation ratio λ in the compression direction of the dielectric elastomer. The thickness d_{DE} of the dielectric elastomer is expressed by the following equation using thickness d_{DE0} in the unloaded state:

$$d_{DE} = \lambda d_{DE0} \tag{18}$$

Substituting Eq. (18) into Eqs. (1) and (7) through (10) and calculating the current i_R numerically, it is possible to calculate the power generated Ri_R^2 for the proposed harvester.

4 Design

Design optimization of the proposed harvester using a mathematical model to maximize the output power is introduced in this section. In this study, the design parameters are the thickness of the electret d_{EL} [m], dielectric elastomer d_{DE} [m], and the load resistance R [Ω]. Table 1 lists the physical properties of the materials used in the design. Silicone rubber (Dragon Skin 10 FAST, Smooth-On, Inc.) was used as the dielectric elastomer. For the electret, the physical properties of CYTOP (809-M, AGC Inc.), an amorphous fluoropolymer known for its high charge retention performance as an electret material, were used. Sheet area S was defined as the area equivalent to the four pairs of molars where the bite forces occurred. The values measured by Putra et al. [17] were used for the coefficients of the Mooney-Rivlin strain energy density function of the dielectric elastomer.

The surface potential V_S, which is an indicator of the charge retention performance of the electret, was set to -1000 V, which has been reported by Suzuki [18]. The surface potential V_S satisfies the following relationship with the electret-retained charge Q_0.

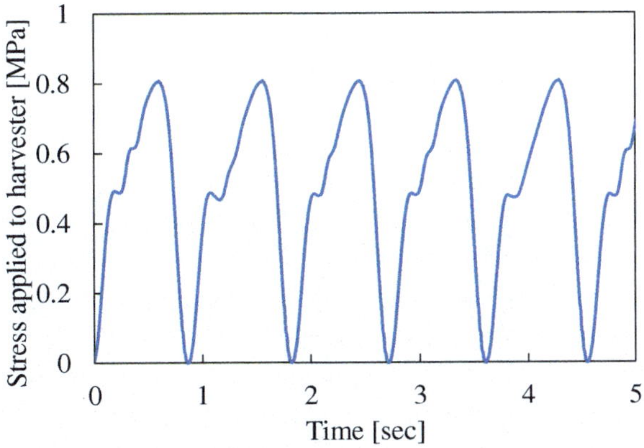

Fig. 5 Stress data simulating biting force

$$V_S = \frac{Q_0}{C_{EL}} = \frac{d_{EL} Q_0}{\varepsilon_0 \varepsilon_{EL} S} \tag{19}$$

Electrets require a certain thickness to hold the charge. According to the preliminary tests, the lower limit of the thickness of the electret fabricated by the proposed process was 2.5 μm. In this optimized design, the thickness of the electret was considered to be 2.5 μm or more. The total thickness of the harvester sheet containing the two electrodes was set to within 0.2 mm. Cu was used as the electrode sheet.

The stress σ input to the harvester was from the stress data simulating the bite force during mastication at a maximum of 0.8 MPa and a frequency of 1.1 Hz, as shown in Fig. 5.

Figure 6 shows the results of the power generation maximization design. In this figure, the thicknesses of the dielectric elastomer and electret layers are taken as the axes, and the amount of power generated when load resistance is connected, which maximizes the amount of power generated for each thickness, is shown. As indicated by the red dashed line in the figure, an optimal thickness ratio exists for the thickness of the dielectric elastomer and electret layers. Within the thickness constraints of this design, the maximum power generation is achieved when the thickness of the electret layer is 2.5 μm and the thickness of the dielectric elastomer layer is 10 μm shown as point A in Fig. 6.

However, with this design value, the total thickness of the electret layer and dielectric elastomer layer is very thin (12.5 μm), and there is a risk of rupture when embedding in a mouthguard with a complicated curved surface. Furthermore, the dielectric breakdown voltage of silicone rubber, which is assumed to be the material of the dielectric elastomer layer, is about 20 kV/mm, which is 200 V for the 10 μm thickness of this design value. Therefore, the design values shown as point B in Fig. 6, with an electret layer thickness of 12 μm and a dielectric elastomer layer thickness of 50 μm, were adopted. The characteristics of the generated power with respect to the load resistance are shown in Fig. 7. When the load resistance is 1000 MΩ, the power generated is expected to be maximized to 2.5 μW, indicating that the

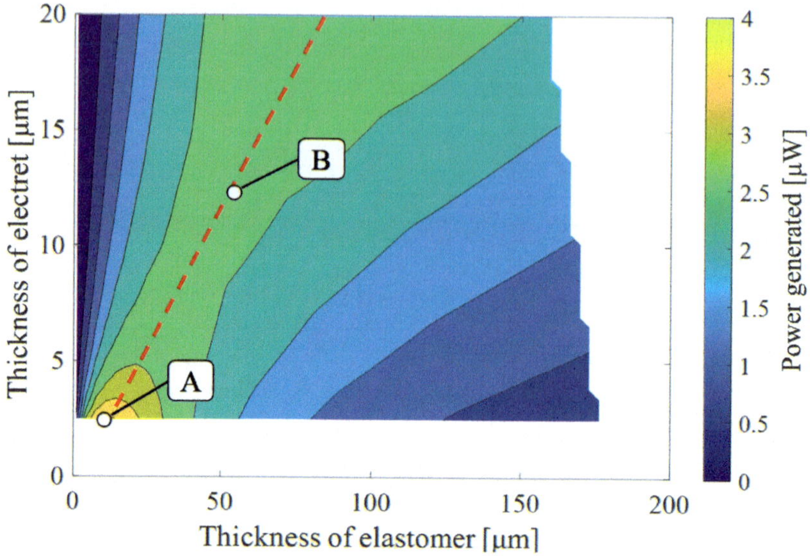

Fig. 6 Optimization results of the electrostatic harvester

proposed harvester can achieve a high-power generation performance in the order of μW.

Fig. 7 Load resistance characteristics of the optimized harvester designed at point B in Fig. 6

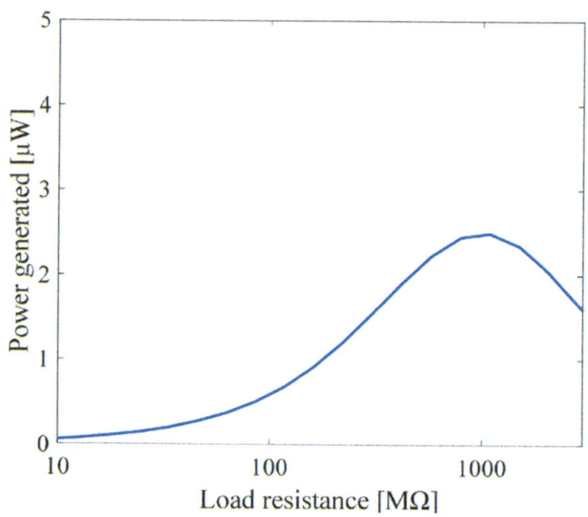

5 Fabrication

Based on the optimal design described in the previous section, a harvester sheet that could be embedded inside the mouthguard was fabricated. The CYTOP used for the electret in this prototype is generally deposited on the electrode by spin coating; however, it is difficult to install a thin electrode film flat on a spin coater. Therefore, CYTOP was applied to the copper electrodes using a manual film applicator (see each experimental condition described below) and dried for 1 h at 100 °C, and another was applied and dried for 1 h at 230 °C. After drying, the specimens were charged using a corona discharge to form an electret. A schematic of the corona discharge process is shown in Fig. 8.

The corona discharge device had wire and grid electrodes that were placed 35 and 17.5 mm from the specimens, respectively. Voltages of −9 and −3 kV were applied to the wire and grid electrodes, respectively. During this process, the specimens were heated to 130 °C and exposed to a voltage for 10 min.

The electrical performance of an electret is determined based on its surface voltage and charge density. The surface voltages of the fabricated electrets were measured using a surface electrometer (AS-20, ACHILLES CORPORATION), and the surface charge density was calculated from the measured surface voltage using the following relationship:

$$\sigma_s = \frac{\varepsilon_0 \varepsilon_{EL} V_S}{d_{EL}}, \tag{20}$$

where σ_s is the surface charge density of the electret, ε_0 is the permittivity of vacuum, V_s is the surface voltage of the electret, and d_{EL} is the electret thickness.

The relationship between the thickness of the fabricated electret and the experimentally measured surface voltage and charge density is shown in Fig. 9. The surface voltage and charge density were directly and inversely proportional to the thickness, respectively. The electrical performance of the electret was maintained quasi-permanently after corona discharge. Figure 10 shows the surface voltage of a 10-μm-thick electret 1500 days after corona discharge. The surface voltage significantly dropped in the first few weeks because of the removal of charges attached to

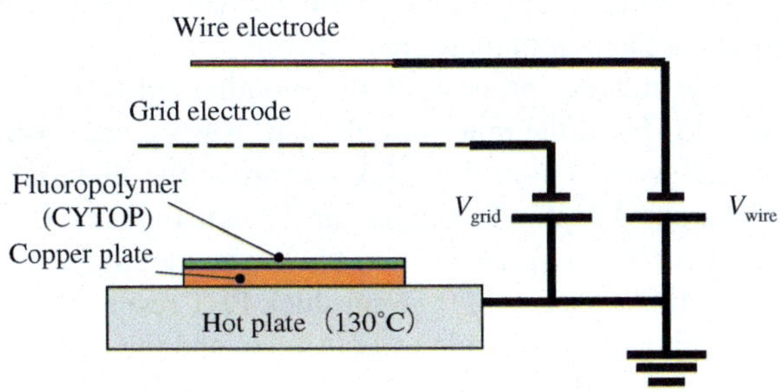

Fig. 8 Schematic of the corona discharge system

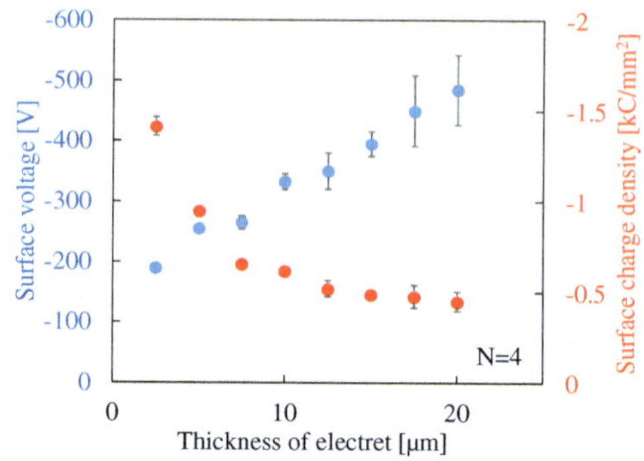

Fig. 9 Surface voltage and charge density of electret

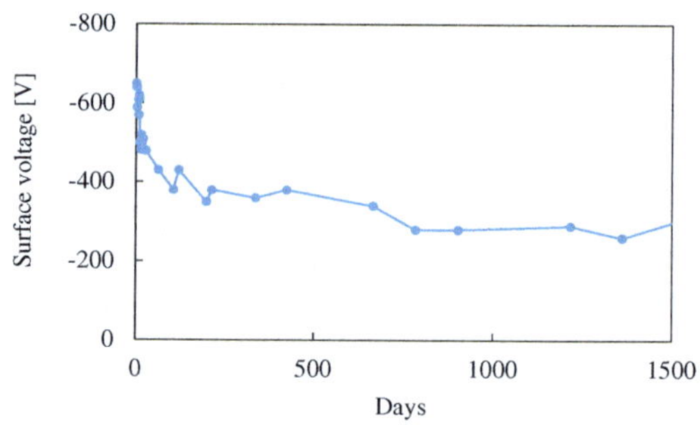

Fig. 10 Surface voltage of electret

the surface of the materials, after which the voltage was stable under -200 V. Dielectric elastomers were also deposited using a film applicator. The specimens were dried for 2 h at room temperature.

Because the occlusal surface has a complex shape, the biting force was not uniformly distributed. When the mouthguard material is too soft, it deforms during occlusion, and only a small occlusal contact area of the harvester is compressed. To ensure that the harvester was uniformly compressed between the double-structured mouthguards, even with a small contact area, a stiffer material that did not follow occlusion was utilized for the outer mouthguard. Therefore, a double-structured mouthguard was fabricated using two different materials. Ethylene vinyl acetate (EVA) and polyethylene terephthalate (PET) resins were used for the inner and outer parts of the mouthguard, respectively.

The making process of the mouthguard-type harvester prototype is shown in Fig. 11. First, the inner mouthguard was vacuum-aspirated and molded on a dental model using 1 mm-thick EVA resin as the material. A vacuum-forming machine (Vacuum Adaptor I, Yamahachi Dental Industry Co., Ltd.) shown in Fig. 12 was used for vacuum aspiration. The outer mouthguard was then formed by vacuum-aspiration molding of 0.5 mm-thick PET resin on top of the inner mouthguard. The inner mouthguard made of EVA resin and the outer mouthguard made of PET resin were removed from the tooth mold and peeled off. The outer PET resin mouthguard was bonded to the copper foil side of the dielectric elastomer sheet and the inner

EVA resin mouthguard was bonded to the copper foil side of the electret sheet using a cyanoacrylate instant adhesive (Aron Alpha, Toa Gosei Co., Ltd.); the two were laminated. The prototype mouthguard harvester is illustrated in Fig. 13. The mouthguard is sufficiently thick compared to the power generation sheet; therefore, even when the harvester sheet is embedded, it is possible to achieve the same wearing comfort as a normal mouthguard.

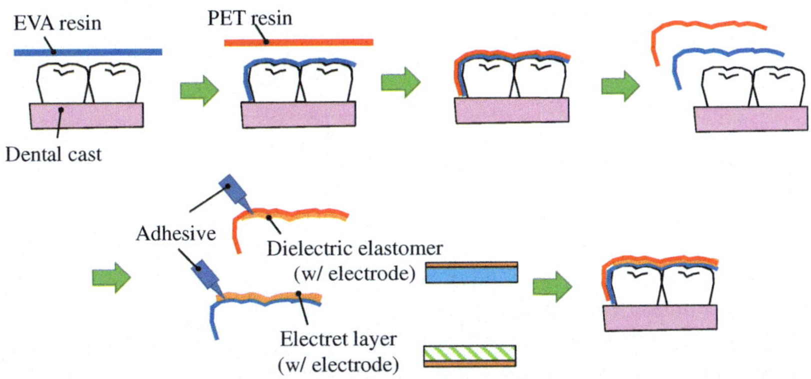

Fig. 11 Fabrication process for mouthguard embedded harvester

Fig. 12 Fabrication of mouthguard

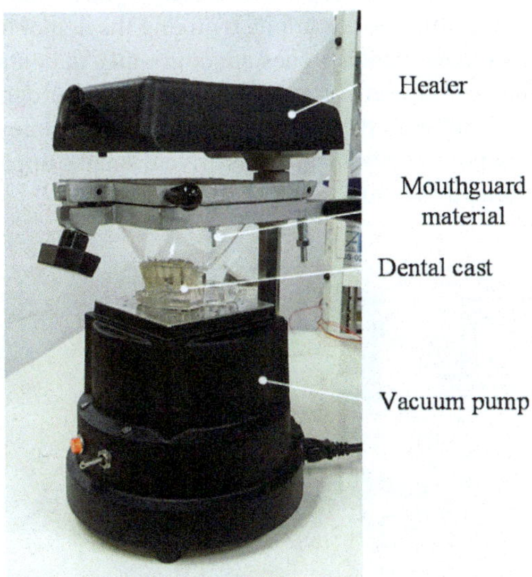

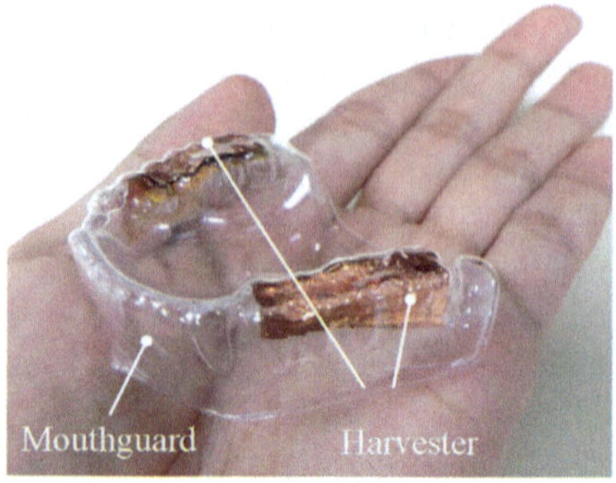

Fig. 13 Mouthguard embedded harvester

6 Experimental Setup

The generated power was evaluated by applying a compressive load equivalent to the bite force. The bite-force simulator shown in Fig. 14 was designed and fabricated for the experiments. The simulator uses a cam and spring to realize a compressive load that regenerates the occlusion force of the human. The spring stiffness and cam shape were designed so that the maximum compression pressure is 0.85 MPa, which is equivalent to the pressure generated during mastication. Using this setup, various compression conditions could be realized by adjusting the rotational speed of the motor, fixing the position of the dental model, and adjusting the stiffness of the springs. For example, by fixing the dental model at a position such that the upper and lower dental arches are constantly in contact with a jig, clenching behavior can be reproduced. In addition, by fixing the dental model in a position where there is a gap with the jig, it is possible to simulate mastication movements in which the temporomandibular joint is driven, and the upper and lower dental arches periodically come into contact with each other.

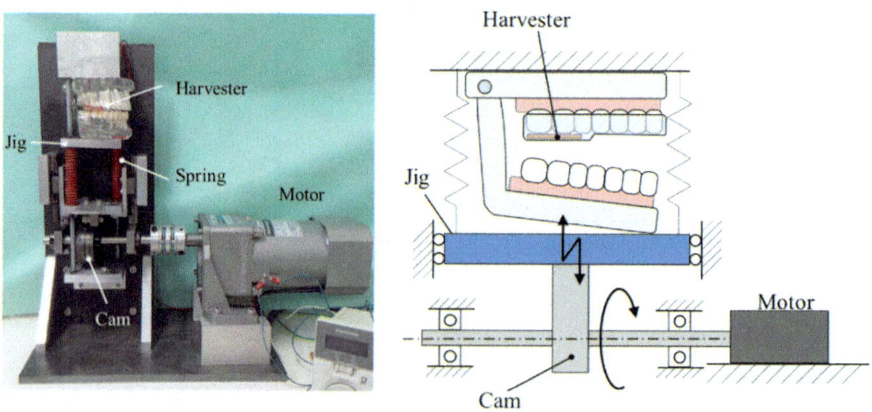

Fig. 14 Bite force simulator

7 Results of Power Generation

Figure 15 shows the results of the power generation experiments with different load resistances. The maximum power of 2.5 μW was generated when the load resistance was 1010 MΩ. The relationship between the power generated and load resistance obtained in this experiment agrees well with the simulation results obtained using the mathematical model, confirming the validity of the model. Figure 16 shows the output voltage of the harvester measured when connected to a load resistance of 1010 MΩ, and the output current calculated from the output voltage and load resistance values. The peak-to-peak output voltage of the harvester was approximately 210 V and the peak-to-peak output current was approximately 210 nA.

To compare the performance of the proposed harvester, a piezoelectric material was embedded in the molar area of the mouthguard in the same manner, and its power generation performance was evaluated. The piezoelectric material was made of 110 μm PVDF, which is the same total thickness as the proposed harvester, and was fabricated using the mouthguard embedding procedure shown in Fig. 11. The prototype piezoelectric harvester is shown in Fig. 17. Subsequently, the piezoelectric harvester was installed in a bite-force simulator, similar to the proposed electrostatic harvester, and compression experiments were conducted. The experimental results are presented in Fig. 18. In the case of the piezoelectric harvester, the maximum amount of power generated was 0.0208 μW when the load resistance was 110 MΩ. Compared with the proposed electrostatic harvester made of an electret and a dielectric elastomer, the electrostatic induction-type mouse guard harvester can realize 120 times higher power generation than the piezoelectric type harvester.

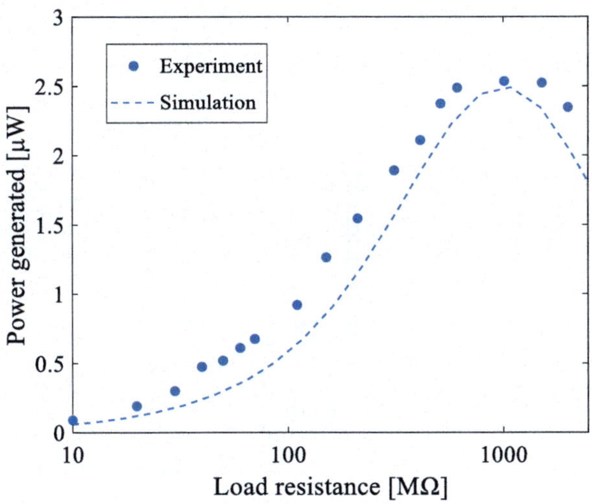

Fig. 15 Experimental results of mouthguard embedded harvester

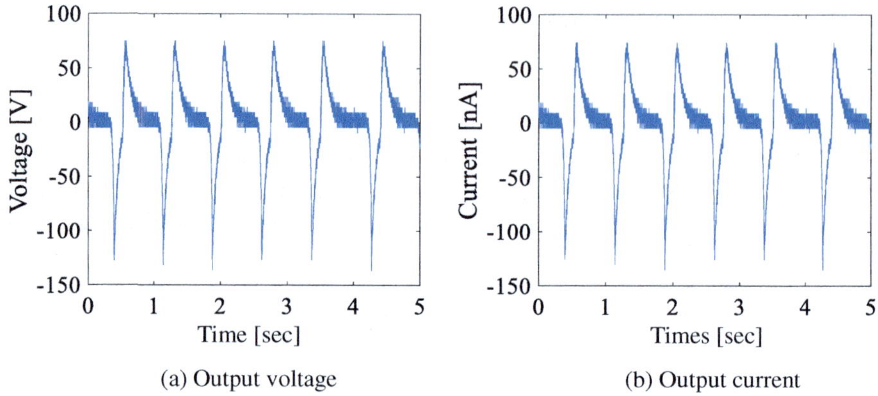

Fig. 16 Output voltage and current of the mouthguard embedded harvester

Fig. 17 Mouthguard embedding piezoelectric harvester

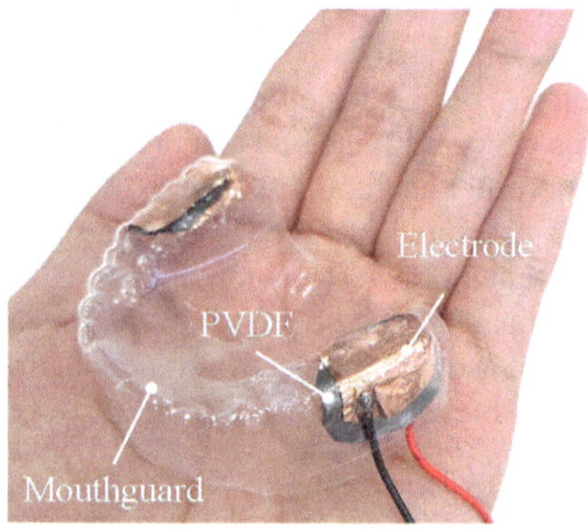

Fig. 18 Experimental results of the mouthguard-type harvester

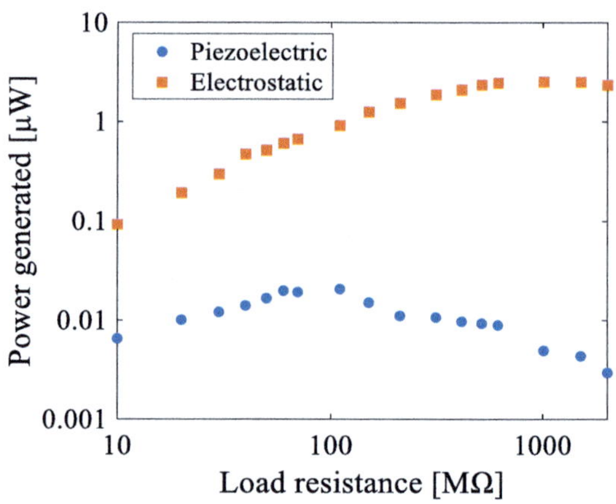

8 Applications of a Bite Force Sensor

Vital information on bite force is also used as a diagnostic indicator of oral function and is expected to be utilized in the prevention of flailing and the treatment of bruxism. However, methods for measuring bite force are generally based on pressure-sensitive films or small load cells, and bite-force sensors applicable to wearable monitoring have not yet been developed. In this section, a mouthguard-type energy harvester is applied to bite-force sensors [19]. To measure the bite force, an inverse model of the proposed harvester was developed, as shown in Fig. 19, in which the input (occlusal force) and output (voltage) of the mathematical model described in the principal section are replaced.

The relationship between the output voltage v of the generator seat, which is the input of the inverse model for occlusion force estimation, and the variable Q_{DE} in the circuit equation can be expressed as follows based on Eqs. (5) and (6)

$$\frac{dQ_{DE}}{dt} = -\frac{v}{R} - C_S\frac{dv}{dt}. \tag{21}$$

By rearranging the circuit equations from Eqs. (7) to (10) for the dielectric elastomer layer capacitanceC_{DE}, the following equation is obtained :

$$\tilde{\alpha}\,\frac{d}{dt}\left(\frac{1}{C_{DE}}\right)+\tilde{\beta}\left(\frac{1}{C_{DE}}\right)+\tilde{\gamma}=0, \tag{22}$$

where $\tilde{\alpha}$, $\tilde{\beta}$, and $\tilde{\gamma}$ are given as follows:

$$\tilde{\alpha}= C_S Q_{DE}, \tag{23}$$

$$\tilde{\beta}= \frac{Q_{DE}}{R} + C_S\frac{dQ_{DE}}{dt}, \tag{24}$$

$$\tilde{\gamma}= \left(1+\frac{C_S}{C_{EL}}\right)\frac{dQ_{DE}}{dt} + \frac{1}{RC_{EL}}(Q_0+Q_{DE}). \tag{25}$$

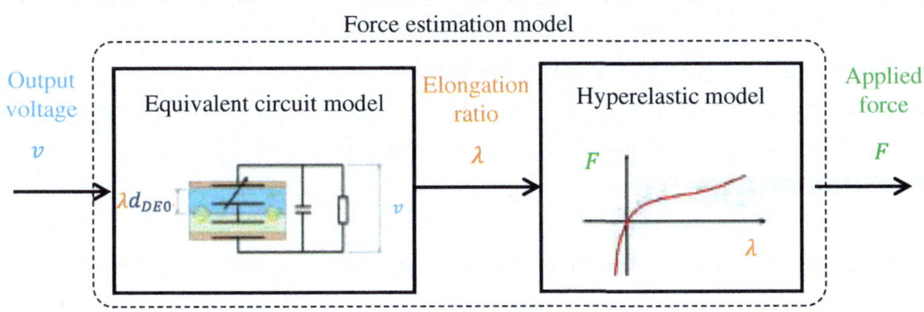

Fig. 19 Concept and flow of the force estimation model

Equations (22)–(25) are differential equations for the dielectric elastomer layer capacitance C_{DE}, which can be solved as a sequential problem because the output voltage v, which is known, is included in the coefficient of each term. The elongation ratio λ of the dielectric elastomer layer in the compression direction was calculated from Eqs. (1) and (18) using the capacitance of the dielectric elastomer layer obtained from the equivalent circuit model as follows:

$$\lambda = \frac{\varepsilon_0 \varepsilon_{DE} S}{d_{DE0} C_{DE}}. \tag{26}$$

The subsequent hyperelastic model estimates the compressive force applied to the generator sheet using the compressive elongation ratio of the dielectric elastomer layer, which is estimated using the equivalent circuit model. The stress–strain characteristics of the dielectric elastomer were modeled as a third-order Mooney-Rivlin elastic body using Eq. (11) to improve sensing accuracy. The strain energy density function W of the dielectric elastomer is expressed using a total of nine material-specific parameters from Eq. (11), and the relationship between the nominal stress σ and the elongation ratio λ follows Eq. (15). The nominal stress was calculated by substituting the elongation ratio in the compression direction estimated by the equivalent circuit model into Eq. (15) and then multiplying by the area of the generator sheet to estimate the compressive force applied to the generator sheet.

The accuracy of the bite-force estimation model was experimentally evaluated. As shown in Fig. 20, a prototype mouthguard harvester was attached to the dental model and fixed to a bite-force simulator. A load cell was placed at the lower part of the dental model, in contact with the mandible, to compare the compressive forces. To ensure that all the compressive forces measured by the load cell were applied to the generator sheet, all teeth except the molars that were in contact with the generator sheet were removed from the dental model in advance.

First, the model parameters were identified, as shown in Table 2 so that the estimated bite force and load cell values would match. The results of the bite-force estimation using these parameters are shown in Fig. 21. The output voltage of the generator sheet was approximately 10 V at both amplitudes for the maximum compression force of 100 N and the maximum compression force of 200 N, respectively. The compression force estimated from the voltage data using the inverse model for estimating the occlusion force agreed well with the values measured using the load cell. These results confirm that the proposed harvester can be applied as a wearable bite-force sensor with a self-powering function.

9 Summary

The oral cavity contains a wealth of biometric information, such as saliva and occlusion, which is considered important from medical and dental points of view. An intraoral wearable health monitoring system is expected to contribute significantly to the development of preventive medicine in the future. This chapter describes a

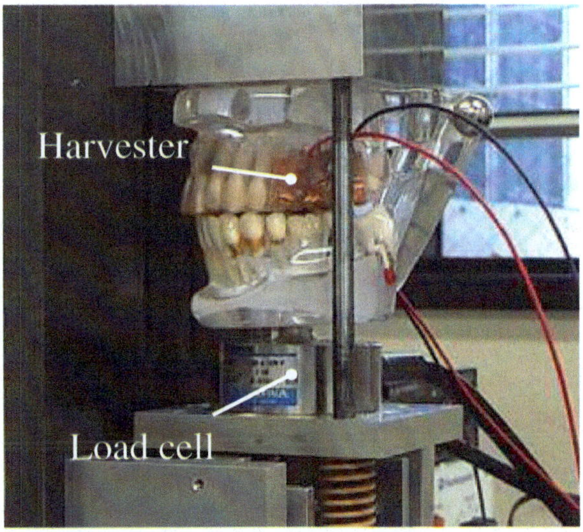

Fig. 20 Experimental setup of the mouthguard embedded harvester with a load cell

mouthguard-type energy harvester using electret sheets. Unlike piezoelectric devices, which show a high-power output at high frequencies, this system has a high-power generation performance for low-frequency forces. Experiments confirmed an output that is 120 times higher than that of piezoelectric devices, making it a suitable power generation solution for mastication at approximately 1 Hz. However, as with the piezoelectric element, the electret itself has a high impedance; therefore, the storage circuit must also receive power with a high impedance, and an electrical circuit must be developed to improve the power generation efficiency of the entire system. The newly developed power-generating sheet has features such as ultrathinness, flexibility, and high-output power generation at low frequencies and is expected to be applied to movements of the human body other than the oral cavity, such as breathing, walking, and pulsation.

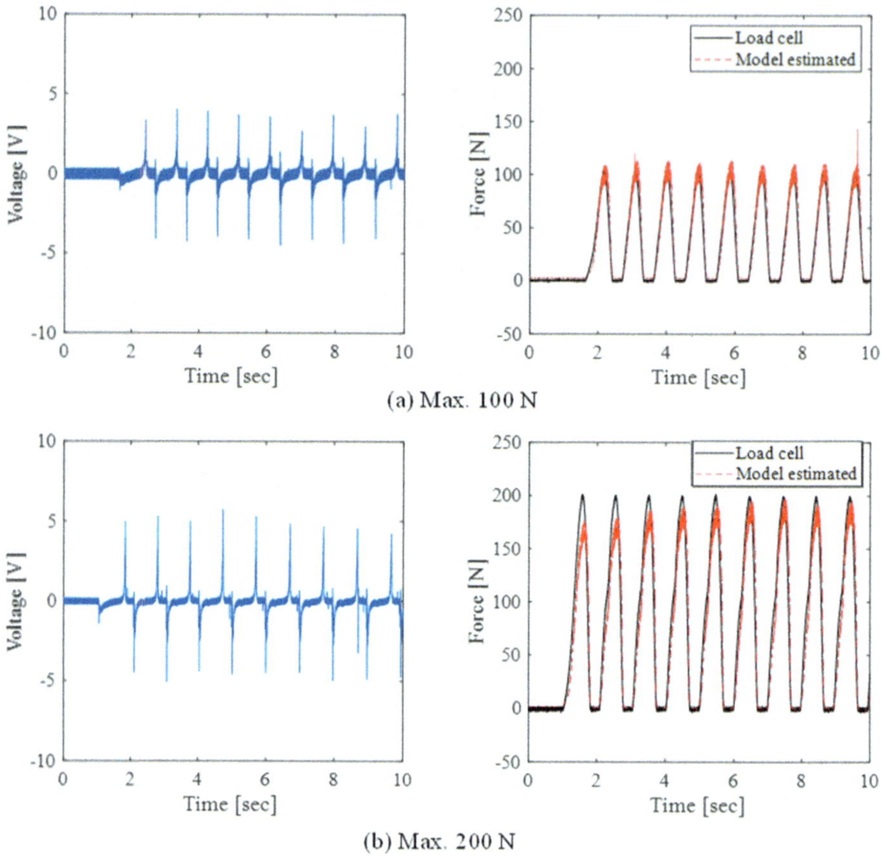

Fig. 21 Results of bite-force estimation using the mouthguard embedded harvester

Table 1 Parameters of the electrostatic harvester for simulation

Parameters	Value	Unit
Surface area S	400×10^{-6}	m^2
Dielectric of dielectric elastomer ε_{DE}	3.00	–
Dielectric of electret ε_{EL}	2.10	–
Mooney-Rivlin constant C_{10}	0.132	MPa
Mooney-Rivlin constant C_{01}	–0.0128	MPa
Stray capacitance C_{St}	10×10^{-12}	F
Surface voltage of electret V_S	–1000	V

Table 2 Identified parameters of the mouthguard embedded harvester

Parameters	Value	Unit
Surface area S	406×10^{-6}	m^2
Thickness of dielectric elastomer d_{DE0}	20.4×10^{-6}	m
Thickness of electret d_{EL}	1.02×10^{-6}	m
Dielectric of dielectric elastomer ε_{DE}	2.72	–
Dielectric of electret ε_{EL}	2.17	–
Surface voltage of electret V_S	-364	V
Load resistance R	9.75×10^{6}	Ω
Stray capacitance C_S	18.5×10^{-12}	F
Mooney-Rivlin constant C_{10}	1.26×10^{-6}	MPa
Mooney-Rivlin constant C_{01}	0.0681	MPa
Mooney-Rivlin constant C_{20}	2.14×10^{3}	MPa
Mooney-Rivlin constant C_{11}	0.443	MPa
Mooney-Rivlin constant C_{02}	-2.005×10^{3}	MPa
Mooney-Rivlin constant C_{30}	0.963	MPa
Mooney-Rivlin constant C_{21}	8.40×10^{3}	MPa
Mooney-Rivlin constant C_{12}	35.3	MPa
Mooney-Rivlin constant C_{03}	1.38×10^{3}	MPa

References

1. Tricoli, A., Nasiri, N., De, S.: Wearable and Miniaturized Sensor Technologies for Personalized and Preventive Medicine. Adv. Funct. Mater. **27**, 1–19 (2017). https://doi.org/10.1002/adfm.201605271
2. Arakawa, T., Kuroki, Y., Nitta, H., Chouhan, P., Toma, K., Sawada, S. ichi, Takeuchi, S., Sekita, T., Akiyoshi, K., Minakuchi, S., Mitsubayashi, K.: Mouthguard biosensor with telemetry system for monitoring of saliva glucose: A novel cavitas sensor. Biosens. Bioelectron. 84, 106–111 (2016). https://doi.org/10.1016/j.bios.2015.12.014
3. Arakawa, T., Tomoto, K., Nitta, H., Toma, K., Takeuchi, S., Sekita, T., Minakuchi, S., Mitsubayashi, K.: A Wearable Cellulose Acetate-Coated Mouthguard Biosensor for in Vivo Salivary Glucose Measurement. Anal. Chem. **92**, 12201–12207 (2020). https://doi.org/10.1021/acs.analchem.0c01201
4. Kim, J., Imani, S., de Araujo, W.R., Warchall, J., Valdés-Ramírez, G., Paixão, T.R.L.C., Mercier, P.P., Wang, J.: Wearable salivary uric acid mouthguard biosensor with integrated wireless electronics. Biosens. Bioelectron. **74**, 1061–1068 (2015). https://doi.org/10.1016/j.bios.2015.07.039
5. Gupta, S., Nayak, M., Sunitha, J., Dawar, G., Sinha, N., Rallan, N.: Correlation of salivary glucose level with blood glucose level in diabetes mellitus. J. Oral Maxillofac. Pathol. **21**, 334 (2017). https://doi.org/10.4103/jomfp.JOMFP_222_15
6. Litovitz, T., Whitaker, N., Clark, L., White, N.C., Marsolek, M.: Emerging Battery-Ingestion Hazard: Clinical Implications. Pediatrics **125**, 1168–1177 (2010). https://doi.org/10.1542/PEDS.2009-3037
7. Ramya, M., Senthil Kumar, P.: A review on recent advancements in bioenergy production using microbial fuel cells. Chemosphere **288**, 132512 (2022). https://doi.org/10.1016/j.chemosphere.2021.132512

8. Joung, Y.H.: Development of implantable medical devices: From an engineering perspective. Int. Neurourol. J. **17**, 98–106 (2013). https://doi.org/10.5213/inj.2013.17.3.98
9. Kim, H.S., Kim, J.H., Kim, J.: A review of piezoelectric energy harvesting based on vibration. Int. J. Precis. Eng. Manuf. **12**, 1129–1141 (2011). https://doi.org/10.1007/s12541-011-0151-3
10. Beeby, S.P., Torah, R.N., Tudor, M.J., Glynne-Jones, P., O'Donnell, T., Saha, C.R., Roy, S.: A micro electromagnetic generator for vibration energy harvesting. J. Micromechanics Microengineering. **17**, 1257–1265 (2007). https://doi.org/10.1088/0960-1317/17/7/007
11. Tan, Y., Dong, Y., Wang, X.: Review of MEMS electromagnetic vibration energy harvester. J. Microelectromechanical Syst. **26**, 1–16 (2017). https://doi.org/10.1109/JMEMS.2016.2611677
12. Khan, F.U., Qadir, M.U.: State-of-the-art in vibration-based electrostatic energy harvesting. J. Micromechanics Microengineering. 26, (2016). https://doi.org/10.1088/0960-1317/26/10/103001
13. Ichikawa, K., Hijikata, W.: Energy harvesting from biting force with thin sheet harvester based on electret and dielectric elastomer. Nano Energy **99**, 107357 (2022). https://doi.org/10.1016/j.nanoen.2022.107357
14. Haines, D.W., Wilson, W.D.: Strain-energy density function for rubberlike materials. J. Mech. Phys. Solids **27**, 345–360 (1979). https://doi.org/10.1016/0022-5096(79)90034-6
15. Rivlin, R.S.: Large elastic deformations of isotropic materials IV. further developments of the general theory. Philos. Trans. R. Soc. London. Ser. A, Math. Phys. Sci. 241, 379–397 (1948). https://doi.org/10.1098/rsta.1948.0024
16. Mooney, M.: A theory of large elastic deformation. J. Appl. Phys. **11**, 582–592 (1940). https://doi.org/10.1063/1.1712836
17. Putra, K.B., Plott, J., Shih, A.J.: Biaxial Mooney-Rivlin Coefficient of Silicone Sheet by Additive Manufacturing. Procedia CIRP. **65**, 189–195 (2017). https://doi.org/10.1016/j.procir.2017.04.049
18. Suzuki, Y.: Recent progress in MEMS electret generator for energy harvesting. IEEJ Trans. Electr. Electron. Eng. **6**, 101–111 (2011). https://doi.org/10.1002/tee.20631
19. Ichikawa, K., Hijikata, W.: Novel Self-powered Flexible Thin Bite Force Sensor with Electret and Dielectric Elastomer. Sensors Mater. 34, 4237–4245 (2022). https://doi.org/10.18494/SAM4134

Wireless Power Transfer Application for Healthcare and Treatments

Yuji Tanabe

1 Introduction

Standardization of fifth-generation (5G) communication has been completed, and 5G networks have been commercialized. Consequently, planning and development for the sixth generation (6G) have already begun [1]. While the first generation (1G) and second generation (2G) used analog and digital communication methods respectively, the third generation introduced an internationally standardized digital method that enabled high-speed data communication. The fourth generation (4G) replaced mobile phones with smartphones as the mainstream device and made high-quality image and large-volume data communication at high speeds possible. The fifth generation (5G) enables ultra-high-speed communication of up to 20 Gbps, 1 million simultaneous connections in a kilometer square, and ultra-low latency of 1us. For the world of 6G (Beyond 5G), no specific requirements or international standards have been established yet. However, it is anticipated that ultra-high-speed communications of up to 100 Gbps, much faster than 5G, and an exponential increase in the number of low-power wireless devices due to the construction of smart wireless sensor networks will be among the expected applications. These low-power wireless devices, including Bluetooth devices, beacons, cameras, and industrial sensors, will no longer be powered by wires or batteries as countless devices will be scattered around the world and connected to a huge network, leading to a singularity in which it will become prohibitively expensive and difficult to power them by wiring or batteries (Fig. 1). It is, therefore, essential to develop a non-contact method of transferring power over medium distances. In this chapter, we introduce a medium-range wireless power transfer method that uses radio waves. The distance from the transmitter ranges from several meters to several tens of meters, making it more efficient and suitable for a higher number of devices. Wireless power transfer technology using

Y. Tanabe (✉)
Aeterlink, Corp, 4-17-1, Kinshi, Sumida, Tokyo 130-0013, Japan
e-mail: ytanabe@aeterlink.com

K. Mitsubayashi (ed.), *Wearable Biosensing in Medicine and Healthcare*,
https://doi.org/10.1007/978-981-99-8122-9_18

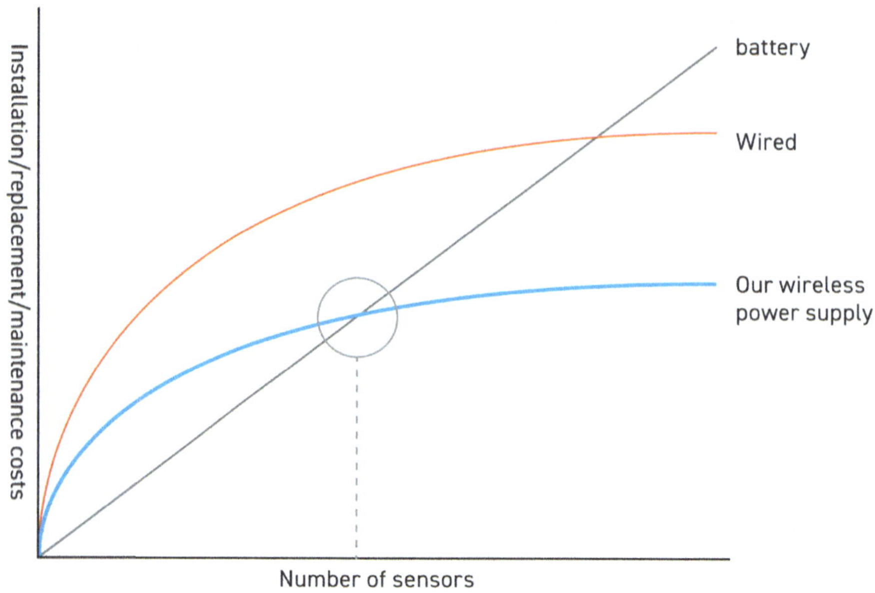

Fig. 1 Comparison of the number of sensors installed and operating costs by wireless power transfer method

radio waves dates back over 100 years, with Nicola Tesla demonstrating the first wireless power transfer in 1899. In 1981, Panasonic Corporation commercialized a wirelessly powered electric toothbrush using electromagnetic induction. The Wireless Power Consortium was established in 2008 to standardize international standards and establish a market for wireless power transfer, and various industry associations have since been established for different wireless power transfer methods.

Although various wireless power transfer methods have been studied for over 100 years, and some have recently been commercialized, electric toothbrushes, shavers, and wireless earbuds remain the only wireless-powered products that have truly penetrated the market [2]. To successfully commercialize wireless power transfer technology, it is important to develop methods that offer added value commensurate with the cost. For example, wireless power transfer can offer advantages such as not requiring waterproofing of power transfer terminals and other parts, which is essential for electric toothbrushes used on a washbasin. Another potential application of wireless power transfer is in powering implanted devices that cannot be easily removed from the outside, such as cardiac pacemakers. Therefore, it is necessary to improve the power transfer efficiency of wireless power transfer devices that require large amounts of power, such as cell phones, drones, and electric vehicles, as well as to explore and develop the potential applications for wireless power transfer technology.

The challenge lies in improving power transfer efficiency, particularly for devices that require large amounts of power such as cell phones, drones, and electric vehicles.

Even under ideal positioning and other conditions, power feed loss is fundamentally large due to the structure, resulting in longer charging times. As such, wireless power charging is more expensive than wired charging and battery charging alone. It is crucial to identify applications that offer benefits commensurate with the price. For example, powering implantable devices, such as cardiac pacemakers, eliminates the need for invasive battery replacement surgeries. Other applications include devices that are placed in numerous locations, such as sensors, and devices that operate within a certain range. Wireless powering is advantageous in these scenarios, as wired or battery-operated power can be costly. Thus, selecting applications that are suitable for wireless power transfer is key to practical application.

In this chapter, we focus on the application of wireless power transfer to sensor devices and implantable devices, which is expected to expand in the 6G world. In the first half of this chapter, we discuss the use of wireless power transfer systems in office applications. Specifically, by installing multiple transmitters on the ceiling and placing receivers around people, we demonstrate that by sensing the environment, controlling air conditioning, and providing a comfortable space, the stress level on people can be reduced as measured by the activity value of amylase. In particular, we show that air conditioning control can save electricity costs and can be applied to healthcare (Fig. 2). On the other hand, we also describe the use of wireless power transfer as a medical device. Recent advancements in embedding electronic systems in the human body have led to significant improvements in medical care [3]. Examples include cardiac pacemakers, vagus nerve stimulation, spinal cord stimulation, and cavitas technology, and ultimately, devices can be implanted directly into the brain to treat Alzheimer's disease. While advances in semiconductor technology have allowed for the realization of implantable devices (micro-implants) less than one millimeter in size, miniaturizing power supplies remains a challenge. In the latter half of the chapter, we discuss implantable nano-bioelectronics that can selectively and accurately treat deep brain tumors through wireless power transfer. After intracerebral administration, gold nanostars (GNS) can be diffused within the tumor tissue, and the tumor can be treated by exclusively heating the tumor area using LEDs based on the power received from wireless power feeding. This study could help in the accurate treatment of deep brain tumors through wireless power transfer.

2 Building Management Application and Healthcare

The use of IoE (Internet of Everything) sensors to control building environments and enhance occupant satisfaction has been a long-standing goal, but one major obstacle has been the complexity of periodically replacing batteries and managing power supply [4]. However, the introduction of wireless power transfer for IoE sensors is a revolutionary solution that eliminates the need for battery replacements, greatly reducing maintenance costs. With constant network connectivity, wireless-powered IoE sensors enable seamless building control, providing the flexibility to perform

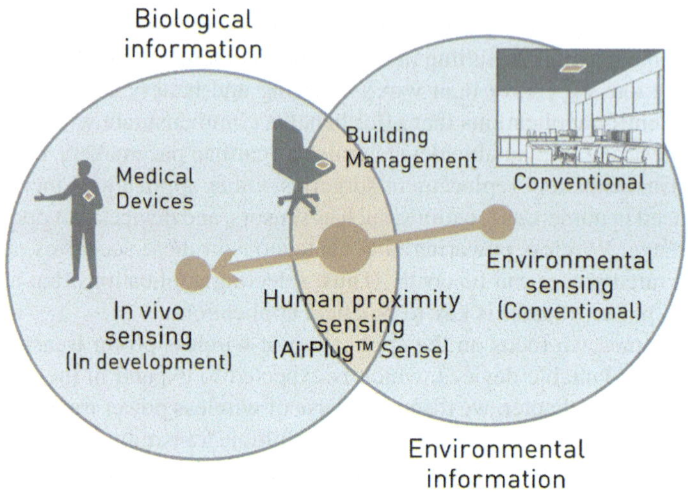

Fig. 2 Wireless power transfer facilitates environmental sensing in the vicinity of people and collects environmental data close to persons

desired tasks anytime. Aeterlink Corp's AirPlug™ (Fig. 3) is the world's only practically available power transmission solution that uses a 1 W radio wave to wirelessly transmit power over long distances in an inhabited environment. In this chapter, we report on an application of a building management solution using AirPlug™ that can reduce power consumption by more than 5% through air conditioning control by manipulating wirelessly operated temperature sensors. In addition, we examine the amylase activity values to reveal that this solution provides a comfortable space and alleviates human stress levels.

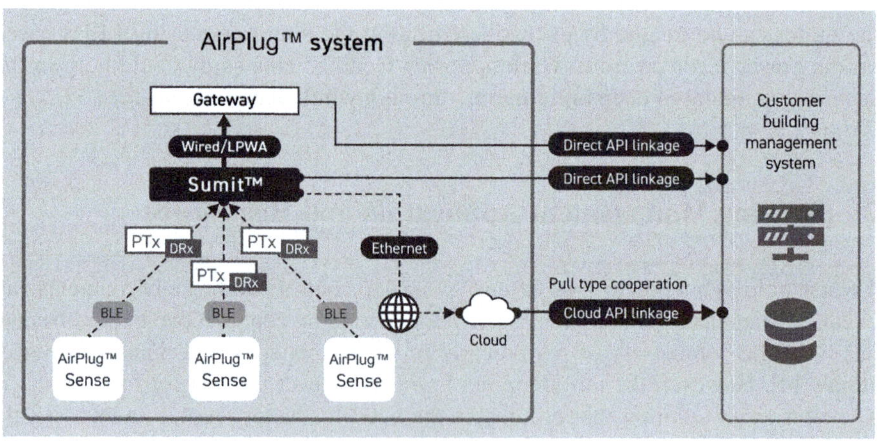

Fig. 3 System configuration of AirPlug™

The AirPlug™ PowerTx is a wireless power transmitter that delivers power to AirPlug™ Sense, which are wirelessly powered environmental sensors (Fig. 4). The Power transmitter transmits radio waves at 918 MHz and also acquires sensor data at 2.4 GHz, which is then transmitted to the AirPlug™ Sumit™, a data organizer that aggregates the data received from multiple transmitters and sends it to the cloud or a building management system at the appropriate size and timing. With the data aggregation system Sumit™, the AirPlug™ system can aggregate IoE sensing data and connect to the building management system through the cloud or gateway. Wireless power transmitters transmit 1 W radio waves. Compared to typical lighting, which consumes 20–30 W, the AirPlug™ system can drive and sense multiple sensors with extremely low power. By sensing spatial information near people, such as temperature, humidity, and CO_2, the connection between space and building control can be strengthened (Fig. 5). By linking the control of air conditioning, lighting, and other systems based on sensing data, energy consumption in buildings can be optimized, leading to a reduction in CO_2 emissions. The corresponding AirPlug™ Beacon in the AirPlug™ spot makes it possible to obtain the location information of things and people, enabling the provision of services according to the location and attributes of people.

We investigated the wireless charging capability of the AirPlug™ system for batteries, and Fig. 6 illustrates the amount of charge delivered to the capacitor or supercapacitor at varying distances between the transmitter and receiver. Our evaluation used a patch antenna with a transmitter power of 800 mW and an antenna gain of 9.5 dBi. As demonstrated in Fig. 6, our system can deliver more than 1 mW of charge to the capacitor from a distance of 4.5 m. Using a beacon type of smaller antenna (40 × 15 mm) allows for wireless charging at distances of 1 m or greater. The power generated from wireless power transfer is used to charge batteries or capacitors in the system, which then operate the sensors, microcontroller, and small wireless communication device. The total power consumption of these devices is approximately 1 mW, with sensor information transmitted to the hub every 5 ms. Environmental sensors require less power (approximately 300 μW) due to their lack of high-speed data communication. However, it is important to select batteries that

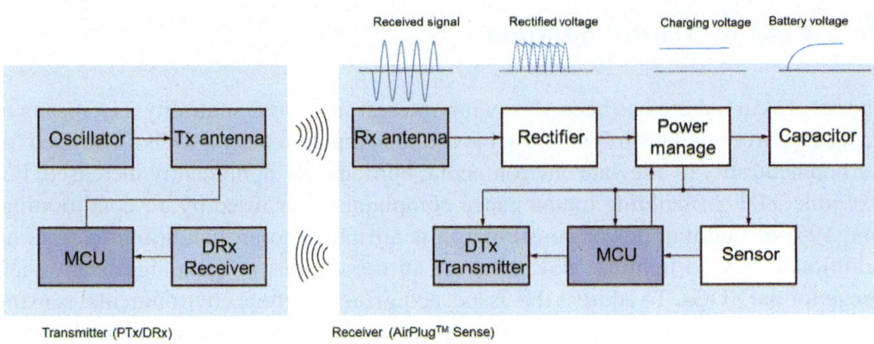

Fig. 4 System block diagram of transmitter/receiver

Fig. 5 System installation image of AirPlug™

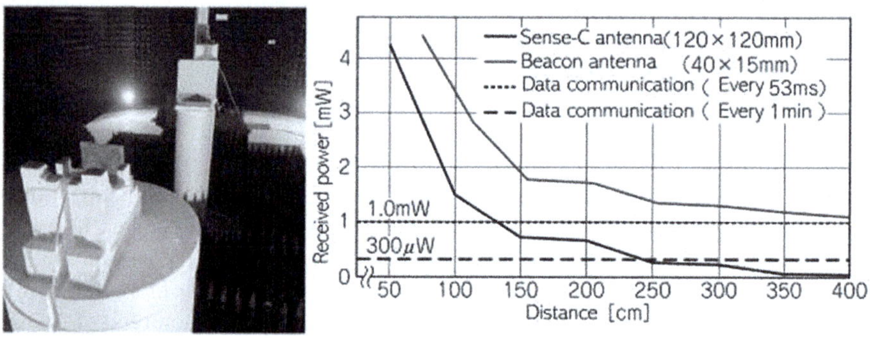

Fig. 6 Measurement setup (left) and distance between transmitter and receiver vs received power (right)

can handle their instantaneous power consumption (current). This approach enables a high-speed data communication and long-distance wireless power transfer system.

2.1 Carbon Neutral Solution

In recent years, there has been a growing interest in carbon neutrality as a means to reduce environmental burdens and achieve a sustainable society (SDGs). The need for carbon neutrality to alleviate environmental burdens has significantly increased. For example, 80% of building maintenance complaints are caused by air conditioning, and 49% of building power consumption is attributed to air conditioning, with an additional 23% to lighting. Power saving in these areas poses a significant challenge for the SDGs. To address this issue, acquiring real-time environmental sensing data, such as temperature information, at multiple points in the vicinity of people in an office is crucial. This data can be fed back to air conditioners via a building network. When multiple sensors operate simultaneously in a wirelessly powered space, a substantial amount of time cross-sectional data can be obtained, forming the basis for real-time data acquisition. Utilizing this data, a learning model can be constructed, typically with lower computational costs than traditional cloud-based models. Figure 7 illustrates the collection of data from a temperature sensor integrated into an air conditioner and a wireless power supply. A comparison between the temperature sensor in the air conditioner and the wireless-powered temperature sensor reveals significant fluctuations. Figure 8 further demonstrates temperature variations near a person, emphasizing the effectiveness of wireless-powered sensors in maintaining a consistent temperature close to the set temperature. Figure 9 shows that when the wireless power-providing temperature sensor is installed near a person, the temperature difference between the set temperature and the temperature under the desk is minimized. Results indicate that wireless sensor control and algorithms can reduce room temperature by 0.5 to 1.0°C, leading to a remarkable 23% to 26% average annual reduction in power consumption compared to conventional air conditioning control, as depicted in Fig. 10.

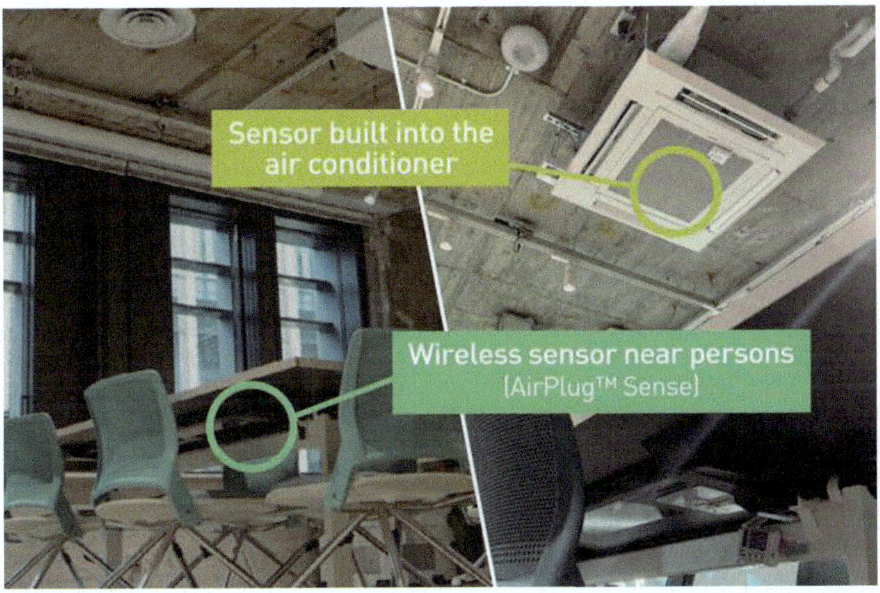

Fig. 7 Validation test setup of wirelessly powered temperature sensors deployed underneath the desk

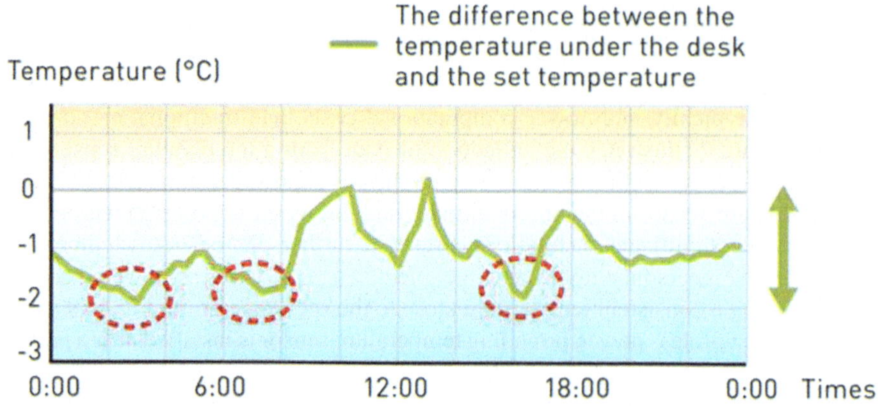

Fig. 8 Temperature trends by conventional AC control. (*This measurement was conducted by Mitsubishi Electric Corp.)

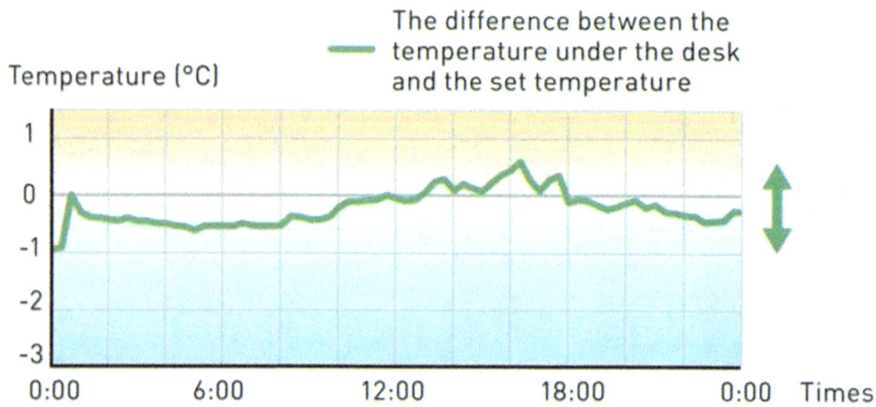

Fig. 9 Temperature trends by our wireless sensor control. (*This measurement was conducted by Mitsubishi Electric Corp.)

2.2 *Integrated Health Sciences Solution*

Acute stress felt by humans can be assessed by monitoring amylase activity in saliva shown in Fig. 11. The amylase activity level in saliva increases with acute stress, which allows an estimation of the stress level immediately prior to two hours before the assessment. The results of this study show that the level of stress reduction, especially after psychological and physical strain, is significantly influenced by a regulated air-conditioned environment.

We demonstrated the application of long-range wireless power transfer in a building management system using temperature sensors as an example. Our results

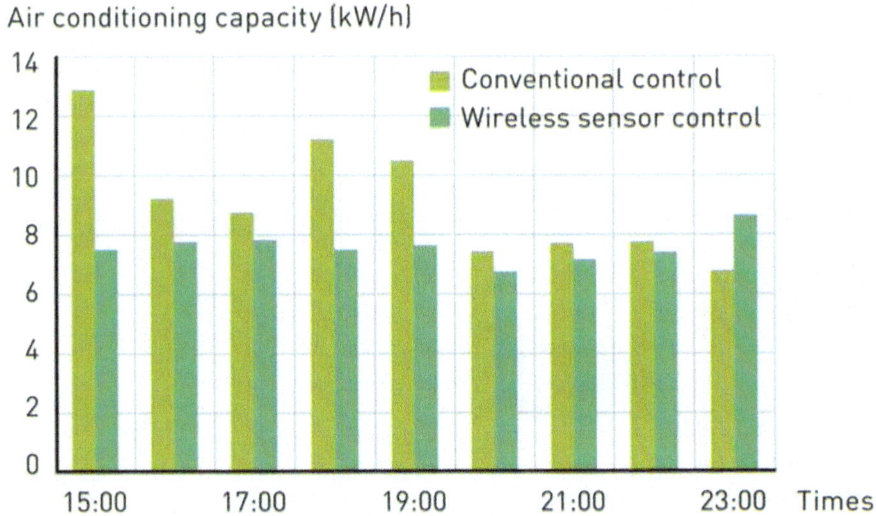

Fig. 10 Power Consumption Comparison. (*This measurement was conducted by Mitsubishi Electric Corp.)

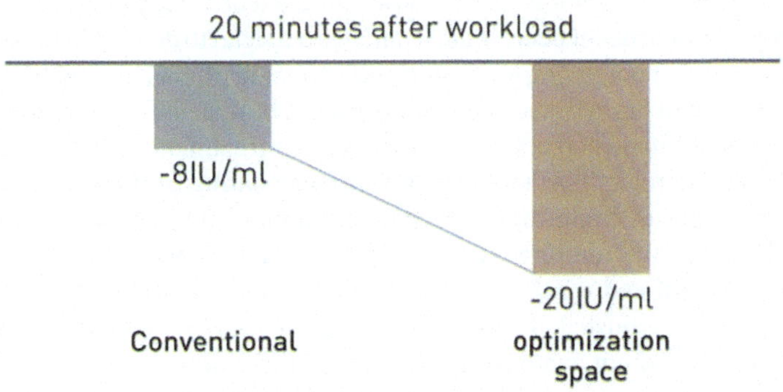

Fig. 11 Salivary amylase activity effect test

indicate that our system was able to improve electricity costs by 5% and reduce stress levels of people 2.5 times faster, while also providing a comfortable environment.

3 Wireless Photothermal Treatment of Brain Tumors

In this chapter, we present the development of integrated nanomedicine-bioelectronics for the remote-controlled, continuous, and long-term photothermal therapy of brain tumors in freely behaving mice without repeated craniotomies and

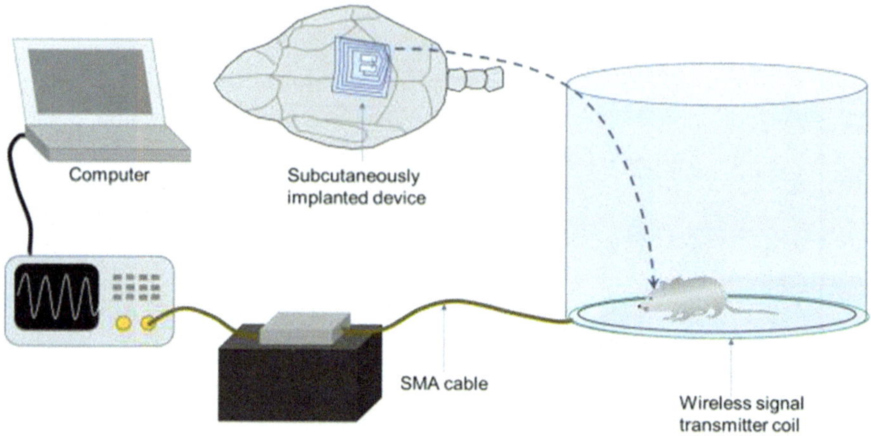

Fig. 12 Wirelessly powered implantable system for photothermal therapy of brain tumors in freely behaving mice

without administering anesthetics or analgesics during treatment (Fig. 12). Current treatment of brain tumors is based on resection of the tumor and postoperative chemotherapy or X-irradiation. Resection requires a technically difficult craniotomy, which can cause significant neurological damage and lead to death. On the other hand, x-ray therapy and chemotherapy cause significant side effects, including damage to surrounding normal brain tissue and other organs. The near-infrared surface plasmon properties of gold nanostars [5, 6] enabled precise treatment of deep brain tumors in freely behaving mice. Furthermore, the surface coating of the nanostars enabled selective treatment of brain tumors. After intratumor administration, the nanostar can diffuse within the tumor tissue and exclusively heat the tumor for treatment. This remote control and wireless versatility allow the photothermal intensity of the nanoparticles to be adjusted. The power and wavelength of the therapeutic light can target tumors in different anatomical locations in the brain.

Photothermal therapy using nanoparticles is focused on the treatment of tumors because it can accurately and selectively treat tumors. This is because irradiation and heating can be performed with minimal side effects on surrounding tissues. However, this technology is limited to a single surgical procedure for intraoperative treatment of tumors on the surface, and the target tumor must be exposed. Thus, a major technical challenge remains in open surgical treatment to activate light-triggered therapy in affected organs. Brain tumors are often located deep in the brain. Because of the large invasion of the brain and the complexity of craniotomy, conventional optical therapy has been difficult to treat.

In this chapter, we revisit the concept of photothermal therapy and show that we have developed a novel remote-controlled approach for continuous and long-term photothermal therapy of brain tumors in freely behaving mice without repeated craniotomies and without administering anesthetics or analgesics during treatment. This approach required the integration of a wirelessly powered light-emitting device.

that could be implanted subcutaneously in the skull to irradiate nanoparticles in the tumor microenvironment. We tuned the device to produce nanoparticles that penetrate tissue. Near-infrared (NIR) light of different wavelengths and intensities was irradiated to efficiently photothermally activate the nanoparticles at any site in the brain. On the other hand, gold nanostars with high absorption in the NIR region were developed, and their light absorption and photothermal effects were tuned to match the wireless signal transmission rate (frequency, duty cycle, etc.) and light power of the light-emitting device. Gold nanostars are widely used in cancer imaging (e.g., intraoperative Raman-guided tumor ablation) and photothermal tumor therapy because of their easy synthesis and tunable optical properties. We treated three stereotactic transplanted brain tumors with Nanostar administered intratumorally and repeated wireless therapy for up to 15 days, with little disruption to the normal activity of the animals during the treatment sessions. The treatments were effective, as evidenced by prolonged survival, histological analysis of tumor tissue, and electron microscopy. This was the first report of on-demand continuous photothermal therapy in awake animals, and it is expected to be a feasible modification for different tumors with low hurdles for clinical application.

Star-shaped gold nanoparticles were chosen as the photothermal material due to their effective surface plasmon properties in the near-infrared region, biocompatibility, and ease of synthesis and surface functionalization. The photothermal material was comprised of an active monolayer of Raman reporter molecules (trans-1, 2-bis(4-pyridyl)-ethylene (BPE)), coated with polyethylene glycol (PEG) molecules (molecular weight ~ 5 kDa) labeled with Cy5. The nanoparticles were also tagged with Cy5 and BPE molecules so that they could be tracked in the tumor and surrounding normal brain tissue using near-infrared fluorescence (NIRF) and surface-enhanced Raman spectroscopy (SERS).

3.1 Wirelessly Powered LED Luminescent Implant Device

Here, we present a novel wireless power transmission method (13.56 MHz) to operate a wirelessly powered light-emitting device, which can be implanted subcutaneously in the skull. We selected the 13.56 MHz (NFC band) for wireless power transmission because of its lower tissue absorption than the RF band and its ability to safely transmit high power to nanoparticles within radiation safety guidelines 25. A schematic of the entire measurement system is depicted in Fig. 13. While luminescent implants reported thus far have been designed for other applications and must be attached to the target tissue or inserted directly into the organ (e.g., brain), our device is designed to deliver tissue-transmitting NIR (longer wavelengths) to nanoparticles deep in the brain. Visible light has limited penetration into tissues, making it inefficient for this purpose. The fabricated device was coated with a 2-μm-thick parylene C layer before implantation to prevent moisture diffusion and extend its lifetime after implantation.

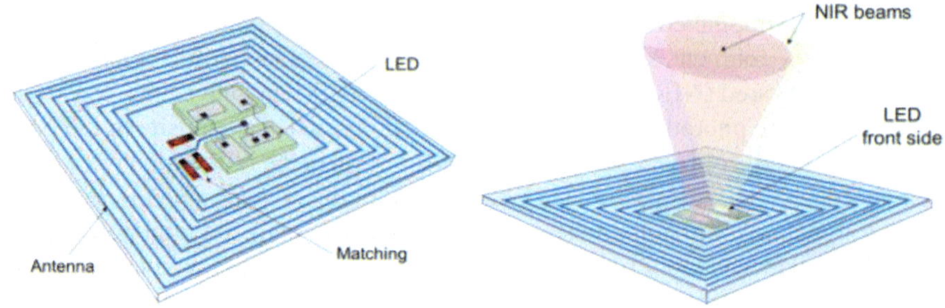

(a) Back-side of implant device. (b) Front-side of implant device.

Fig. 13 Schematic of the device, illustrating both back and front views, as well as the circuit design and components employed during fabrication

We propose a new approach for wireless powering of LEDs, in which a matching network, rectifier, and options are integrated with an energy-storing capacitor and a circuit to program the duty cycle of the LED (f). The duty cycle can be adjusted by continuous input or by modulating the incoming RF signal (red and blue waveforms in e) to generate a programmable duty cycle (Fig. 14). This approach is used to wirelessly activate the photothermal effect of nanoparticles. In the case of a single LED, the LED is turned on for half a cycle, resulting in a 1.5-fold increase in the photothermal effect, with a maximum duty cycle of 31.8%. On the other hand, in the back-to-back LED topology, each LED is turned on for half a cycle, allowing at least one LED to be turned on for half a cycle (Fig. 15). The figure illustrates the use of the device to evaluate the heat generation effect of nanoparticles when irradiated at 810 nm and 940 nm (Fig. 16). The figure shows the temperature variation of the nanoparticles with respect to the voltage amplitude of the sinusoidal input (peak voltage Vp). At a peak voltage of about 4 V (80 mW input power), a temperature difference of about 3 °C was observed in the nanoparticle droplets deposited on the coverslip, confirming a photothermal response. The wireless power transmission efficiency of photothermal therapy using free-running mice was adjusted and evaluated for safety. The figure shows the simulated and measured S11 of a back-to-back device equipped with an LED emitting light at a wavelength of 810 nm. The measured S11 of the antenna is also shown in Fig. 17, along with the ideal matching path (dashed line). The L-matching structure of the back-to-back device and the 52 pF capacitance that serves as the matching component were constructed by placing 22 pF and 30 pF capacitors in parallel. Figure 18a shows the variation of the matching network efficiency (ηMN) with input power (Pin) (assuming a capacitor quality factor Qc = 100 for the matching components). Figure 18b demonstrates the effect of the tissue on the power transmission efficiency (ratio of power received to power transmitted at the input of the device) after implantation in the skull of a mouse. It suggests that the power delivered to the RX antenna is reduced by ~ 5 dB due to tissue interaction; PLED and PTX represent the power received at the input of the device and the power transmitted from the transmitting coil, respectively. The optical power vs the polar position of the NIR-emitting device from the center of the wireless power transmitter is shown in Fig. 19.

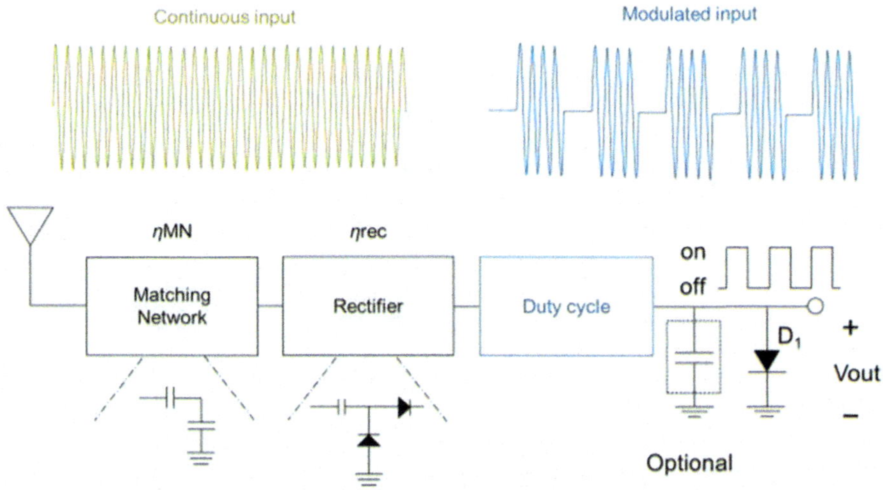

Fig. 14 Conventional approach for wireless LED powering integrating matching networks, rectifiers, and circuitry options for programming duty cycles and storage capacitors

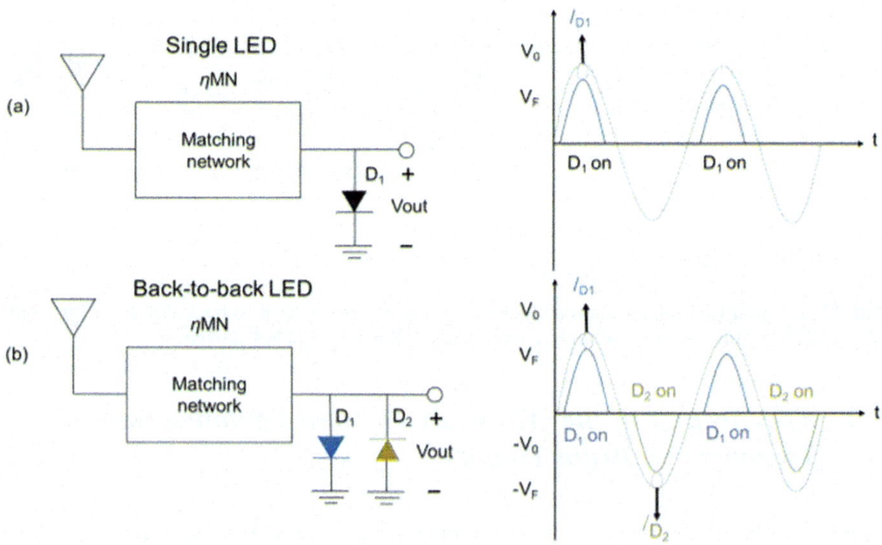

Fig. 15 Single (**a**) and back-to-back (**b**) LED circuit designs and use of nanoparticles for wireless activation of photothermal effects

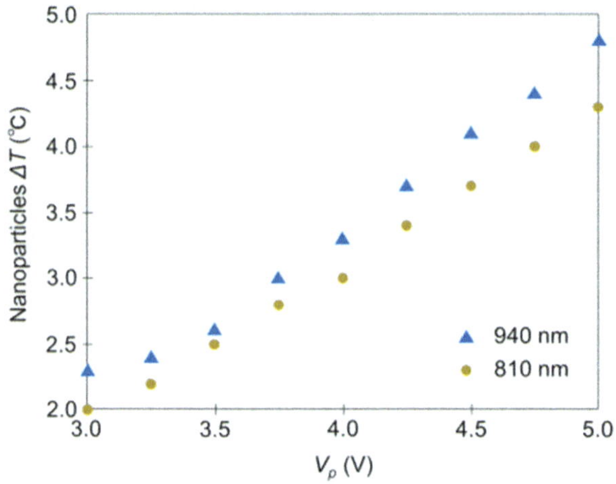

Fig. 16 Evaluation of the heating effect of nanoparticles during irradiation at 810 nm and 940 nm devices

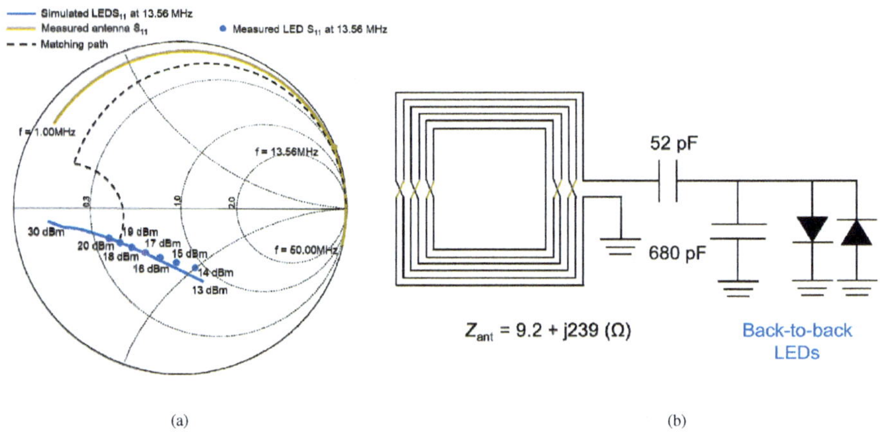

Fig. 17 **a** Smith Chart simulation and measurement of S11 of back-to-back devices with LEDs. **b** L-matching structure and matching components of the back-to-back device

3.2 Effectiveness of the Wirelessly Powered Photothermal Treatment on Brain Tumors

Figure 20 shows a multilevel power supply scheme used for day-long continuous treatment in a device using a 940 nm wavelength. In vivo open-skull thermal images of the brain tumor 24 h after intratumor injection of the nanoparticles are shown in Fig. 21a. When the temperature changes along the blue dotted line shown in Fig. 21b, i.e., the temperature of the surrounding normal brain tissue increases, only the tumor region containing nanoparticles is heated during NIR irradiation (~30 mW)

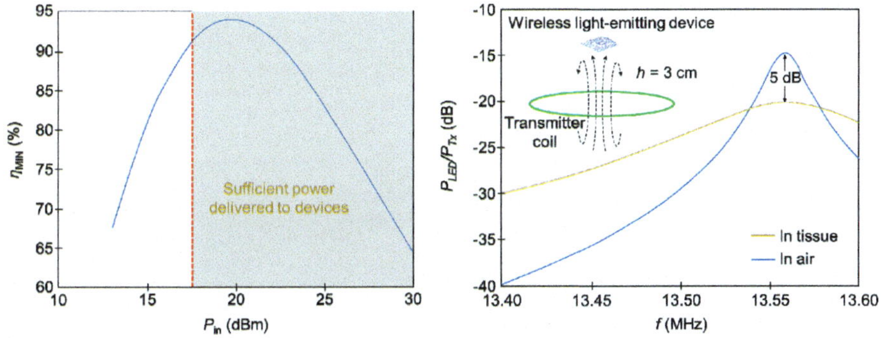

(a) (b)

Fig. 18 Performance of received power at implant device **a** Input power (Pin) vs matching network efficiency (η_{MIN}) **b** Wireless power transfer efficiency over the tissue

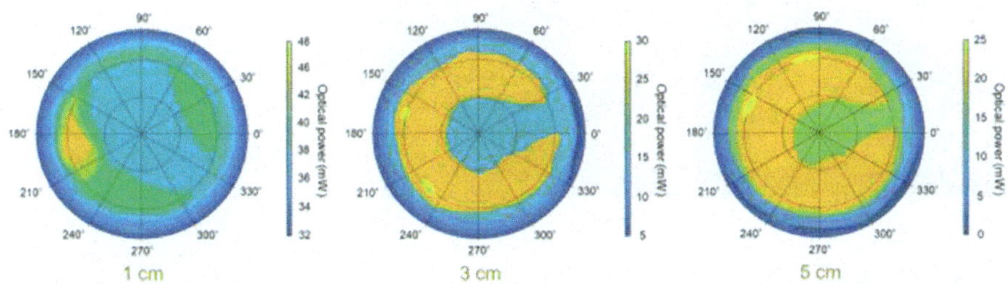

Fig. 19 The optical power vs the polar position of the NIR-emitting device from the center of the wireless power transmitter

due to the photothermal response of the nanoparticles. No temperature increase was observed in the surrounding normal brain tissue. The effectiveness of this method is demonstrated by the way the mouse eats during the wireless tumor treatment session. Finally, Fig. 22 shows the survival profiles of mice with human U87-eGFP-fLuc glioblastoma tumors treated with the 810 nm and 940 nm devices (treatment groups 1, 2, and 3) compared to control mice (n = 10 per group, 60 total). Nanoparticles (NPs) (1 μl, 0.5 nM) were administered into the tumor and 24 h later photothermal therapy was started (15 min/day for 15 days). Significant differences were found when compared under the following conditions: Control 1: NPs (–), implanted (–); Control 2: NPs (+), microfiber (+), irradiated (–); Control 3: NPs (+), device (+), irradiated (–); Treatment 1: NPs (+), microfiber (+), irradiated (810 nm) (+); Treatment 2: 810 nm device (+), irradiation (+); Treatment 3: NPs (+), 940 nm device (+), irradiation (+).

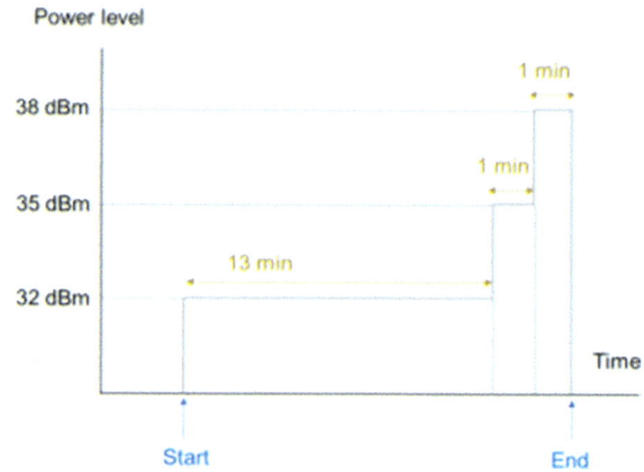

Fig. 20 Daily based multilevel powering scheme

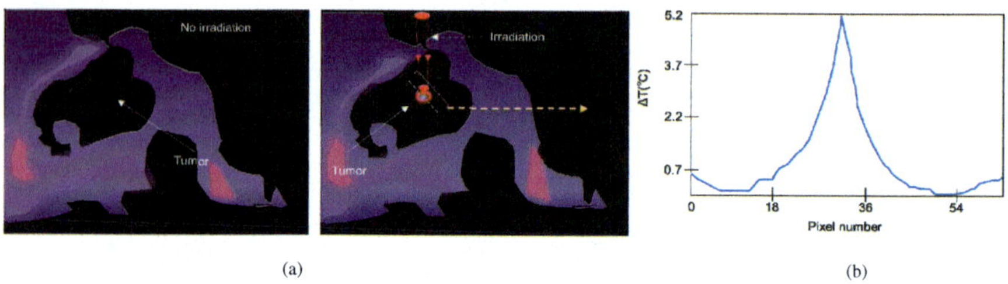

Fig. 21 Localized brain tumor treatment **a** In vivo image and NIR-emitting device location **b** Temperature changes at blue dotted line

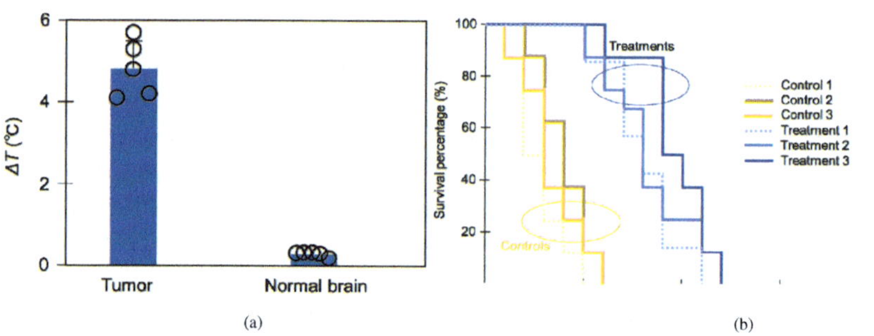

Fig. 22 Animal testing to confirm the effectiveness of wireless photothermal therapy of brain tumors **a** Normal brain versus tumor area **b** Survival of mice bearing tumors treated with wireless photothermal

4 Conclusion

In the first half of this paper, we presented experimental verification of the effectiveness of a wireless temperature sensor using AirPlug™, a long-distance wireless power transmission system developed by Aeterlink Corp. This technology makes it possible to install wireless environmental sensors closer to personal zones than was previously possible. This enables a more accurate estimation of the AC temperature setting, resulting in savings on electricity costs and providing a comfortable space for people. The experiment revealed that the stress level felt by people was about 2.5 times less in a space provided by wireless power transmission, as evidenced by the activity of amylase. This experiment serves as a guidepost for the practical application of AirPlug™.

In the second half of the paper, we show how brain tumors can be selectively treated with nanoparticles using an implantable device that enables photothermal therapy with wireless power delivery to selectively treat tumors. We designed an implantable and flexible bioelectronic platform that triggers a nanoparticle photothermal therapy response in brain tumors for on-demand and point-of-care treatment cycles. Our nanoparticles are designed to produce a uniform therapeutic action in the microenvironment of the brain tumor when activated by a wirelessly powered light-emitting device. This therapy can treat the tumor without heating the normal brain tissue surrounding the tumor. Daily treatments are performed wirelessly and do not require additional surgery or interfere with the normal activity of the mouse. These results suggest the effectiveness of our treatment method, especially when combined with standard therapies such as chemotherapy and radiation therapy, as well as investigational therapies such as immunotherapy, in the clinical treatment of GBM.

The extensive utilization of wireless power transfer in various applications strongly indicates its potential to become an integrated and indispensable aspect of our daily lives in society, and we anticipate its continued expansion in the future.

References

1. Dang, S., Amin, O., Shihada, B., Alouini, M-S.: What should 6G be? Nat. Elect., **3**, 20-29 (2020). (https://doi.org/10.48550/arXiv.1906.00741)
2. Shinohara, N., Zhou, J.: Far-field wireless power transfer and energy harvesting, Artech House, (2022). (ISBN: 9781630819125)
3. Agrawal, D.R., et al.: Conformal phased surfaces for wireless powering of bioelectronic microdevices, Nat. Biomed. Eng. **1**, 0043 (2017). (https://doi.org/10.1038/s41551-017-0043)
4. Cao, X., Dai, X., Liu, J.: Building energy consumption status worldwide and the state-of-the art technologies for zero-energy buildings during the past decade. Energy Buildings, **128**, 198–213 (2016). (https://doi.org/10.1016/j.enbuild.2016.06.089)
5. Vines, J. B., Yoon, J. H., Ryu, N. E., Lim, D. J., Park, H.: Gold nanoparticles for photothermal cancer therapy, Front. Chem. **7**, 167 (2019). (https://doi.org/10.3389/fchem.2019.00167)
6. Chatterjee, H., Rahman, D. S., Sengupta, M., Ghosh, S. K.: Gold nanostars in plasmonic photothermal therapy: the role of tip heads in the thermoplasmonic landscape, J. Phys. Chem. C **122**, 13082–13094 (2018). (https://doi.org/10.1021/acs.jpcc.8b00388)

Smart Sensor-Based Point-Of-Care Diagnostics in Ophthalmology: The Potential for Theranocloud as Combination of Theragnostic and Cloud Computing

Mouad Lamrani, Maryam Moghadas, Yogeshvar N. Kalia, and Verena Santer

1 Introduction

Theranostics, the combination of diagnostic and therapeutic tools, often also referred as Therangostic, is an emerging approach that leads to diagnostic tests that are directly linked to the application of specific therapies [1, 2]. The concept of the Theranocloud takes this a step further by integrating digitalization, smart communication, and artificial intelligence to create an intelligent medical device that allows for up-to-date consultation and automation of disease detection and diagnosis. Theranocloud is a term coined by combining Therapy, Diagnostics (Theranostics), and Cloud, and it has been registered in the International Trademark WIPO library by Menicon Co., Ltd. Although the Theranocloud does not yet exist, it has the potential to revolutionize healthcare by providing a centralized, intelligent medical device that can optimize treatment through its fine-tuned feedback loop (Fig. 1).

M. Lamrani (✉) · M. Moghadas · V. Santer
Menicon Co., Ltd., Geneva Branch, Nagoya, Japan
e-mail: m.lamrani@menicon.com

Fondation Pour Recherches Médicales, Avenue de La Roseraie 64, 1205 Geneva, Switzerland

M. Moghadas
e-mail: m.moghadas@meniconrd.com

V. Santer
e-mail: v.santer@meniconrd.com

Y. N. Kalia
School of Pharmaceutical Sciences, University of Geneva, CMU, 1 Rue Michel-Servet, 1205 Geneva, Switzerland
e-mail: yogi.kalia@unige.ch

K. Mitsubayashi (ed.), *Wearable Biosensing in Medicine and Healthcare*,
https://doi.org/10.1007/978-981-99-8122-9_19

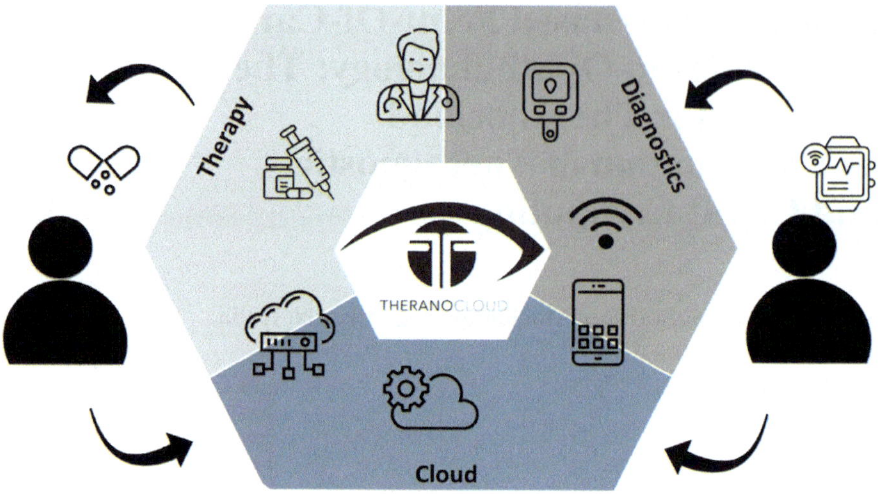

Fig. 1 Iconic representation of theranocloud concept with the three pillars therapy, diagnostics, and cloud—digitalization of healthcare. Icons made by Icon home from www.flaticon.com

The main applications of Theranostic tools can be summarized in three pillars: (i) the identification of patients or tissues that will respond to therapy (Efficacy), (ii) the avoidance of ineffective or harmful treatment for certain patients (Safety), and (iii) the monitoring the response to treatment (Care) [2]. Theranostic tools have been primarily developed for cancer treatment, and include radioactive iodine for thyroid cancer treatment and HER2 protein tests for breast cancer treatment [3–5]. More recently, Theranostic nanoparticles have been developed for targeted drug delivery [6–8] and its potential for further applications is largely reported in the same named Theranostics journal. In addition, non-invasive monitoring via wearable technologies has become increasingly important for long-term disease management [9–12].

While these technologies offer significant benefits, their potential can be further maximized through the development of a centralized, intelligent medical device that integrates all of these tools. This is the goal of the Theranocloud concept, and although it is not yet a reality, the healthcare industry stands to benefit greatly from its potential realization. By collecting and analyzing patient data in real time, the Theranocloud could optimize treatment plans, prevent adverse reactions in combination with the full record of patients' health story [13], and automate disease detection and diagnosis by the means of artificial intelligence (AI), ultimately improving patient outcomes and revolutionizing healthcare as we know it [14].

2 Artificial Intelligence in the Medical Field

Artificial intelligence (AI) is a general term for a set of computer-based data processing algorithms which aims to mimic human cognitive function [15]. In other words, it can be described as "intelligent behavior by analyzing the environment and taking actions—with some degree of autonomy- to achieve specific goals", as it has been defined by the European commission.

AI is an interdisciplinary science and present in most domains for example language recognition, automated manufacturing, finance, security, climate science, or traffic. It is considered as a key technological milestone for improvement in terms of effectiveness, reliability, and safety [16].

AI is also a mother term for the actions associated with intelligent beings performed by a machine; this term is englobing two subfields of data analysis namely Machine Learning (ML) and Deep Learning (DL), where the latter term is again a subclass of the former.

ML is a process often described by the ability of the algorithms to "learn" and improve their precision in tasks without being precisely prepared for it. Nevertheless, the training ("learning") of some of such data analysis processes is still based on categorized data labeled by domain experts, which makes the technique more reliant on human action [17, 18].

With the increase in computing power and the constantly growing volume of digital data, frequently referred to as "big data", we are evolving from ML techniques to DL. DL can imitate the human brain via the use of so-called artificial neural networks by stacking and wiring different layers of computations on top of each other like neurons. With this "brain like" behavior, DL can analyze unstructured datasets without the need of any expert annotation through multiple processing layers. It means that the input is transformed automatically from unstructured raw data (such as image pixels) through several iterations to abstract understandable information (such as eyes or nose in an image); obtained DL models can be applied to new data as well [19, 20]. The development of any DL algorithms consists of three essential steps: (1) training, on a given dataset, (2) validation on a different dataset with needed adjustments of the code, and finally (3) testing on an additional dataset to verify the performance of the software [21].

It is this last computational model that since its introduction in 2006 has brought major improvement such as driverless cars or fluent conversations with virtual assistants on the phone [22, 23].

When focusing on the domain of healthcare, the potential for DL is envisioned to generate more knowledge, such as answering retrospectively and then prospectively main clinical questions—creating fact based clinical evidences which will then be included in care protocols [24, 25]. In this sense, DL is envisioned to assist healthcare givers for clinical decision making and assist researchers to uncover unknown disease correlations and risk factors hidden in the vast medical data of extensive patient groups of different ethnicities, comorbidity, ages, etc. [15, 25]. Not only patient history and health monitoring data but also patients' biological data can be included in the data pool (e.g. genomics), which opens a new dimension for the personalized medicine [24].

As some practical cases, a breakthrough has been reported in the medical image analysis field for the diagnosis of skin cancer lesions by DL automated image analysis. For example, the algorithm developed by Esteva et al. was found as performant as the evaluation of 21 certified dermatologists in the binary classification of keratinocyte carcinoma vs. seborrheic keratosis and malignant melanoma vs benign nevus, by just requiring only a fraction of time [26]. In the case of cancer diagnosis or radiology, DL models can even detect clinically important features beyond what would be visible by the doctors' inspection [27]. However, in the case of skin cancer lesion analysis, DL software is not envisioned with the aim to substitute the doctors. Such a DL application could be universally used as a point-of-care (POC) tool for the individual as a form of app on the smartphone which can be used as a preventive tool by the individual at home. The aim of POC device is to approximate care to patient, for real-time, rapid, and accurate on-site detection. In case of finding suspicious lesions, the patient is reminded for a specialist appointment by the application [26].

In the field of ophthalmology, similar exemplary advances were made for the retinal fundus image analysis. The application of a DL algorithm was proposed by Tham et al. as a broad visual impairment screening tool, to identify disease related visual impairment such as cataract, glaucoma, diabetic retinopathy, and age-related macular degeneration (AMD). This is done by the simple screening of one single macular centered retinal photograph. Once again, the aim for this DL algorithm is complementary to the visual acuity test to identify patient groups in the community of the need of medical attention and to forward them to specialized eye care centers [28].

Since 2016, FDA has granted several approvals to AI based medical devices and algorithms for the fields of Radiology, Cardiology, and general practice [29]. Most prominent example is the company Arterys Inc, which offers 8 FDA cleared AI imaging algorithms for radiological techniques such as X-ray images, computed tomography, and magnetic resonance imaging scanner for thoracic and cardiovascular imaging [30]. All these tools, support the clinical decision and aim to shorten the time for proper analysis outcome.

Dramatic shift in DL assisted healthcare is envisioned with the increase of patient's compliance and collaboration, being active actor of the care received [27]. Specifically for chronic diseases the drug intake compliance is very important, as the behavioral adjustment needed for efficient care have only modest outcome because of patients' noncompliance.

Big data offers the chance to caregivers and patients to enrich medical records with personal data like exercise, diet but also income, education, or housing district, collected from wearable biosensors, smart watches, or smartphones [24]. These complementary data would enable AI to deliver targeted messaging alerts, content, or reminders to specifically selected patient groups which were found through DL analysis to be at risk of behavioral noncompliance depending on the personal lifestyle patterns [27, 31].

Some pilot studies in this field have already been conducted for drug regimen adherence. Drug intake DL algorithms in the form of reminder systems (SMS), DL patient support chat-bot [31], or mobile phone device with integrated AI platform showed significant increase in the patient compliance. When taking the example of the adherence to anticoagulation therapy in stroke patients, drug intake was improved from 50–67% by the use of a mobile app based on patient and medication recognition as well as drug ingestion confirmation [32].

Despite promising opportunities for AI to improve healthcare for society which are discussed in the preceding paragraphs, no technology comes without its limits and downsides.

One major hurdle for the AI revolution in healthcare is that most medical data and medication prescriptions are still processed in analog form, or in more advanced cases digitalized on hospital specific databases hindering the access of those data for any further research [33]. The data accessibility from large and various sources is crucial for the success of data driven AI revolution in healthcare. Therefore, any novel sensor device should be well equipped with connectivity feature to transfer data to/from a central database (e.g. hospital or patient medical record). In this way, this data can be used by DL models for early disease patterns, risk factors, disease prevention assessment, etc., as well as personalized treatment depending on the patient's body feedback to treatment [25]. The interconnection of sensors like smart watches, point-of-care sensors, or any other digital device capable of wireless data transmission and storage is called the Internet of Things (IoT) [34].

As we have experienced in the recent years the database interconnection of personal sensitive data has become most valuable for many companies. Due to the databases value in terms of profit it can be subject to fraud and data trafficking with the risk of patients' privacy violation [31, 35]. Issues of data sharing from public health institutions to private partners for DL development have so far yielded in poor patient privacy protection, involving different states and jurisdictions and undermined the importance for regulation on the data storage and ownership [35]. Furthermore, it was shown by Na et al. that even if participants' physical activity data from the National Health and Nutrition Examination Survey were anonymized, DL algorithm itself was capable of reidentifying 69–85% of participants, indicating that even current practices for patient data anonymization are obsolete and new protocols need to be implemented [36].

However, the issue of data protection is not the only ethical question regarding the novel frontier of AI in healthcare that must be addressed by the regulators.

The inherent "dark box" structure of AI algorithms (e.g. the inability to understand the analytics and "explainability" of the results behind DL, achieving high performance) leads to lack of transparency [27]. This is a major risk for the misuses of AI. If the calculations are not based on solid foundations, distorted, biased, or fully erroneous results might be yielded [33, 37]. This risk shows the need of a clear governance framework to protect the patients. Despite the implementation of AI in healthcare being still on its early stages, FDA approvals for algorithm tools are increasing and the policy making is lagging behind [29, 38]. A major unfamilar challenge for the lawmakers is the fact that DL algorithm are not locked systems. The inherent adaptive learning process (to improve the technique) is continuously modifying the product from the date of approval (compared to locked systems, which give same result as day 0 of approval) [29]. A starting point for a regulatory framework considering the adaptive nature of DL software was published by the FDA in 2019. The proposal is including constant safeguard checks on the product during its whole lifecycle [39]. However, ambiguity in the text is still conflicting its pursuit [29].

Digging further on possible ethical issues, one additional unresolved question is the accountability of DL systems for medical errors in patient diagnosis or treatment, which might occur for human doctors as well [27]. But finally, what is most important for the success of AI revolution in healthcare is the patients and care professionals trust and acceptance of the technology. As has been suggested in a consumers survey [40], the patients' trust in AI healthcare services can be built when a clear regulatory monitoring system is established which continuously monitors over safety and quality of the AI algorithms used, similar to the regular inspections and audits of healthcare institutions like hospitals. For the clinicians, factors such as easy integration into workflow, user-friendliness, and actual usefulness of the technology for an improved healthcare service were identified as main requirements to be addressed to AI software developers and also the policy making institutions [41].

Regulators will have to continuously adapt to the dynamically evolving AI research challenges. These adaptations will transform and improve care for the global population. And despite the hurdles in the implementation of AI that might be faced, no research entity or private company should ignore the technology. Fast and reliable data connectivity and collection technology as well as modern data security solutions should be included from now on in any laboratory device e.g. sensors or POC, to enable interconnectivity of the data for the possibility to enrich clinical data banks and develop innovative DL software with the final goal to generate new clinical discoveries.

3 Focus on Ophthalmology, Challenges in Diagnostics

Acknowledging that the 3 principal constituent parts of the Theranocloud concept including Therapy, Diagnostic, and Cloud have universal importance and effect on any part of the body, this chapter will mainly focus on the challenges of ocular diseases' diagnosis and sensing, drug delivery, as well as the advances that have already been made in cloud and AI technology for the improvement of ocular health tools.

Vision is a human being's most valued sense among other senses for daily activity such as social interaction, self-care, and mobility. Therefore, correct vision is one of the most prominent factors of quality of life for the individual and society as a whole [42-44].

The human eye is a complex isolated organ protected from the outside world and system by dynamic barriers such as a continuous lacrimal washout as the blood-retinal barrier [45]. The eye globe itself is commonly divided into two segments,

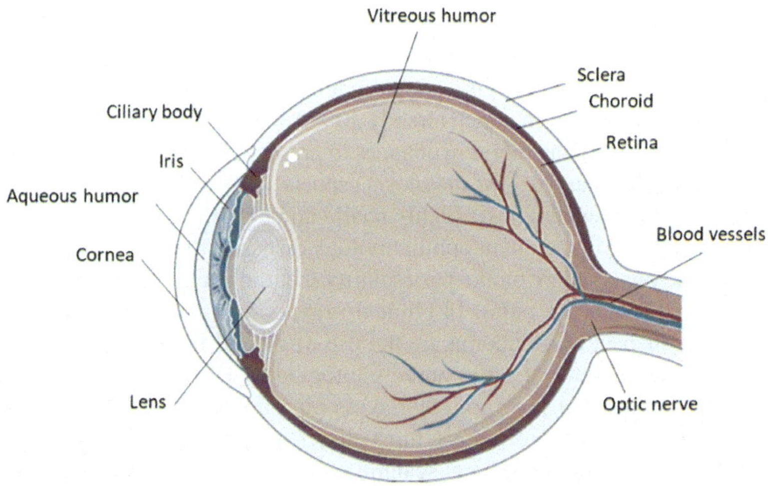

Fig. 2 Human eye globe cross-section evidencing the tissue components, adapted from Servier Medical Art, https://smart.servier.com/

namely anterior (cornea, aqueous humor, iris-ciliary body, and lens) and posterior (sclera, choroid, retina, vitreous humor, optic nerve) see Fig. 2 [46].

The vision process is enabled through the functionality and the interaction of all compartments of the eye; however, the posterior segment hosts the most crucial tissue for the photon detection, which is the retina with the photoreceptors [47]. Damage of the light sensitive neuronal retinal tissues is the heart of irreversible vision loss, summing up to cataract and under corrected refractive error (reversible causes of vision loss), represent global leading causes of blindness [48, 49]. In 2020, out of 33.6 million adults aged 50 and older who were blind, 6.3 million cases were due to glaucoma (11%), 5.6% were due to Age-related Macular Degeneration (AMD), and 2.5% due to Diabetic Retinopathy (DR) [50]. In addition, 13.2 millions of people over 50, suffer from moderate to severe vision impairment (MSVI) caused by those same diseases that year [50].

4 Glaucoma

Glaucoma is a neuropathy which leads to the degeneration of the optic nerve and is characterized by the gradual degeneration of Retinal Ganglion Cells (RGCs). RGCs cell body is situated in the retina and axons exit the eye in the optic nerve. The loss of such axons results in a "cupped" appearance of the optic nerve, which is used as clinical classificatory sign in ophthalmoscopic examination [51]. Though the pathogenesis of the disease is still poorly understood, commonly accepted risk factors for primary glaucoma are age, family history, race, and high intraocular pressure (IOP) [52].

The progression of glaucoma is asymptomatic until later stages of the disease with significant visual field loss in which already 20–50% of the RGCs are lost [21]. Early detection and treatment can increase chances to preserve the vision but unfortunately no perfect reference standard for diagnosis especially in early stages of the disease is known, which makes diagnosis highly challenging [52].

As for most posterior segment ophthalmology diseases, the diagnostics is majorly based on image analysis, given the particularity that the eye fundus can be inspected noninvasively through the pupil. In clinical praxis, most basic screening methods for the risk or suspect of glaucoma are the use of dilated funduscopic examination of the optical nerve head or optic nerve photography. However, the interpretation of the retinal fundus image or optic nerve cupping degree often leads to disagreement between specialists. Nowadays, most clinics are equipped with more sophisticated technologies such as confocal scanning laser ophthalmoscopy, scanning laser polarimetry, and optical coherence tomography (OCT), as complementary techniques to the retinal fundoscopy. The use of these techniques enables quantitative information to be obtained on structural changes in the optic nerve head [53] which by themselves or in combination of funduscopic examination support the clinician in the diagnosis.

AI Algorithms for Glaucoma Diagnosis

Given the advances of DL algorithms especially in the field of automated image analysis already described in previous sections it is natural that much research interest is spent in the development of DL software capable of studying the optic nerve head integrity and to classify the glaucomatous disc from ocular fundus photographs, or other retinal fundus imaging techniques such as optical coherence tomography (OCT) [21].

The challenge in the identification of glaucomatous optic disc damage becomes evident by simply considering the shear myriad of different optic disc anatomies present in the world population. In addition, the instruments and methods used for the measurement for the optic disc size vary and provide only estimations since correction factors are applied for the instrument and eye intrinsic magnification. In clinical practice, all those factors lead to only a modest overlap of expert interpretation/diagnosis [54]. However, DL algorithms, with access to big data of optical fundus images have the possibility to "see" and train on far more retinal fundus images than an expert ophthalmologist during his/her career. When observing the output of such DL algorithms for the binary classification of images to "glaucoma" versus "non glaucoma", Area Under the Curve (AUC) scores of above 0.95 have been reported [55, 56, 57], to be noted that AUC = 1 represents 100% success in classification [21]. Highest predictive scores were obtained when OCT imaging of the optic nerve was combined to additional patient data such as demographic data and risk factors [56]. As an alternative strategy for the detection of glaucomatous damage, Li et al. proposed a DL algorithm trained on glaucoma detection depending on the patients visual field impairment which was shown to outperform trained glaucoma experts [58].

However, despite the increasing accuracy of DL software for the glaucoma diagnosis, the image analysis algorithms have still poor scores when processing low quality images or early stages of glaucoma. Moreover, comorbidity of other ocular conditions such as myopia, age-related macular degeneration, or diabetic retinopathy are mostly excluded in DL model training datasets, as well as mostly homogenous ethnicity groups are usually tested. Such simplifications do not embody the real world scenarios, offering room for improvement in future [21].

POC IOP Sensors

Elevated IOP is commonly recognized as a potent risk factor for glaucoma. Despite this assumption in the vast population of glaucoma patients, 25–50% are found with "normal" IOP (below 21 mmHg) and substantial number of people with elevated IOP never develop glaucoma [52].

However, the lowering of IOP is so far the only proven treatment method shown to slow the disease progression [59] even in patients without elevated IOP. The current clinical practice is to lower the IOP of the patients with glaucoma to 20–50%, compared to their initial IOP, by drug treatment [60]. First line of drugs are usually eyedrops of prostaglandin analogues, which increase aqueous humor outflow and have few systemic side effects [61]. In parallel to the medication intake, continuous patient follow-ups are needed to evaluate IOP and disease evolution. This control visits are needed to finetune the medical treatment or make the decision for surgical intervention if required in severe cases.

IOP measurements are performed by the eye specialist during control visit using either the gold standard technique Goldmann applanation tonometry or other tonometer techniques based on for example air puff or rebound. All those tonometers are measuring the force needed to temporarily flatten part of the cornea, from which then the IOP is deduced [62]. Wearable sensors would give the opportunity for the continuous measurement of the IOP over 24 h to detect eventual IOP spikes which were found to occur often in the night [63]. These IOP spikes overnight were suggested to have major influence in disease progression and visual field loss [64]. However, such continuous measurement data are so far mostly out of reach for the caregiver. Therefore, IOP might often be underestimated for many patients due to only punctual measurements during daytime.

To counter this problem, the ICare Home (Icare Finland OY.), a hand-held POC tonometer based on rebound for at-home self-testing was developed [62]. The device was approved by FDA in 2017 giving reasonable comparable tonometry readouts to the Goldmann applanation tonometer for 73% of the measurements of trained patients [65]. Such device if used with high frequency by the patients over the day can be capable of spotting IOP peaks or fluctuations that are not detected during the one session of the specialist. Therefore, for the patients with unclear/rapid disease progression, such systems could be provided and frequent IOP data gathered but it needs the active participation and compliance by the patient. With the addition of an activities diary these data provide the caretaker more tools for the understanding of the disease state and treatment [65]. Ergonomy and usability have been improved in the ICare Home2 device for the generation of even more accurate data.

Another approach to fill the gap of the missing IOP data of the patients during the day or over 24 h are constant POC IOP measurement devices, many of them still at experimental stage. Most of such constant IOP measuring devices need surgical insertion, either anterior chamber or during cataract extraction [66]. Alternatively, intensive research was performed on contact lenses (CL) embedded IOP sensor which will be discussed in detail in a further book chapter. Therefore, it is only briefly mentioned in the following paragraph.

Wearable POC IOP sensors are based on the principle of corneal lens shape distortion due to IOP. They are still research prototypes. For example graphene embedded CL measures piezoelectric resistance to strain [67]. In passive microfluidic sensors included into CL the volume change can be detected by mobile phone camera [68, 69]. Color change of CL can be detected as well using photonic crystals responding to morphology change of the eyeball due to IOP [70].

So far, the most promising wearable POC IOP sensor was developed by Sensimed AG in Lausanne, Switzerland. The Triggerfish CL, a CE and FDA approved medical device, is based on disposable silicon CL with an embedded electromechanical system (strain indicator). It measures and transmits signal of corneal deformation e.g. circumferential changes of the corneoscleral junction to a wireless antenna placed around the eye [71–73]. 24 h monitoring of IOP through the Triggerfish device, measuring each 5 min for 30 s, has shown to be repeatable, safe, and with good tolerability [74–76]. Side effects of the wear such as corneal staining or conjunctival hyperemia were reported to disappear within 48 h. With those results Triggerfish can be considered useful for the assessment of the risk of glaucoma especially in normal pressure glaucoma [64, 77, 78]. The main drawback of the technology however is that the values of Triggerfish (millivolt equivalent) cannot be directly correlated to Goldmann tonometer IOP in mmHg [79]. Also, the patient tolerability of material needs to be addressed. Research is continuing in this field and soft CL prototypes were recently published in 2022 to overcome such problems [80].

The Theranocloud concept in action can be best described by the smart CL developed by Kim et al. [81] designed for the monitoring and control of IOP in glaucoma. The system is based on IOP resistance sensor for corneal radial deformation coupled with a flexible drug delivery system. The drug delivery system with timolol was designed as multicompartment storage which release was triggered by electrical signals as feedback to elevated IOP. Tests in vivo on rabbits showed a promising reduction of 22% IOP with treatment and the reduced IOP was shown lasting up to 18 h. In future IOP data and drug release patterns could be analyzed by AI and further personalize the treatment for the single person according to the very individual disease progression.

Smart Glaucoma Visual Field (VF) POC Monitoring

Glaucoma is a lifelong chronic disease; recurrent visits to the ocular specialist are needed for the follow-up of the disease progression. Patient care over time and treatment results can be improved when patients themselves are involved in the vigilance of disease propagation and compliance of drug intake. Moreover, in addition to patient empowerment, continuous monitoring at home can generate precious patient specific information, additional to the clinical evaluations performed by the specialist and guide the caregiver for the treatment [27].

Home based POC devices for the disease monitoring can be envisioned as decentralized healthcare devices which when linked to AI algorithms can identify high risk patients e.g. rapid progression of disease without straining the limited capacity of eye services [82].

One approach for at-home POC for glaucoma disease progression includes monitoring devices such as portable devices or computer screens as well as head-worn devices for the repeated visual field self-testing and detection of any loss.

Melbourne Rapid Fields iPad application is one of those POC screening tools which was tested on glaucoma patients for weekly visual filed telemedicine checkup over 1 year by Prea et al. [83]. The method claimed the capability of detecting cases of visual field loss progression. However major drawbacks in compliance and drop-out of patients were found. Another tablet based visual field monitoring device, called Eyecatcher, was tested over 6 months with only one monthly examination. This approach reduced the testing schedule. High patient test adherence of 98% was found and comparable VF (Visual Field) measures to standard in optical clinic were reported [84]. Nevertheless, challenges of patient compliance and accuracy of unsupervised VF investigation in terms of correct positioning of the screen in front of the eyes were acknowledged as main issues. As a possible solution to such problems, AI based image analysis through the camera of the tablet device was proposed. Video recordings of the self-test of the patients contain multiple indicators of low-quality testing such as gaze point, movement of camera, etc., to confirm the correctness of use [85]. Warning signs of incorrect posture or use as well as intelligent user instructions can be implemented in the software with reminder functions for enhancing patients' compliance.

Head held VF contrary to tablet applications have the advantage to assist the patients in the use and ensure correct positioning and have been shown to be in good correlation to standard laboratory equipment such as Humphrey Field Analyzes [86, 87]. Future miniaturization of such wearable VF monitors could be envisioned to be in the form of smart goggles to increase its practicality.

Glaucoma Molecular Biomarkers

Biomarkers are defined as "biological molecules found in blood, other body fluids, or tissues that can be used to follow body processes and diseases in humans and animals" by the European Medicines Agency (EMA).

As for any disease, numerous molecular biomarkers were found altered in glaucoma patients which could be used for the much-needed early detection, to report on disease progression or the response to treatment. Given the known neuronal injury of RGCs in glaucoma, most molecular biomarker research focuses on the cellular mechanisms involved in such injury [88]. Source of such molecular biomarker studies are body fluids as tear film, aqueous humor, vitreous body, and blood rather than RGC itself. Most accessible biofluids are tear film and blood, where the tear fluid can be sourced noninvasively from the patient via Schirmer strips or microcapillary tubes and the blood can be sampled as a part of standard medical procedures which can be performed by various caregivers not necessarily specialized in ophthalmology. Further to the molecular biomarker research, proteins and nucleic acids are quantified by highest accuracy and sensitive laboratory equipment [89, 90]. Fernandez-Vega Cueto et al. reported as many as 450 potential biomarkers for glaucoma in the literature [91].

In depth analysis of promising biomarkers is out of the scope of this chapter. However, the analysis of the tear film of glaucoma patients showed upregulated levels of inflammatory response proteins. Free radical scavenging and cell signaling proteins were also detected, when compared to healthy subjects [90]. Nevertheless upregulation of inflammatory proteins is often also found associated to other ocular inflammatory diseases such as dry eye [92], and tear biomarkers were found to be altered by the anti-hypertensive therapy for glaucoma patients, revealing the major pitfall of molecular biomarkers for glaucoma: lack of specificity.

Clinical validation of candidate biomarkers has to be performed over a large population, with race, age, and sex heterogeneity, to exclude all false positive outcomes [93]. This is the main reason why the transfer of biomarkers from discovery to clinical practice is still not breached. One strategy to overcome the lack of specificity is the use of combination of biomarkers to establish the disease fingerprint. For the detection of such fingerprint in the big data pool, AI algorithms would have enormous potential for the detection and we predict will soon show their capacity.

5 Age-Related Macular Degeneration (AMD)

AMD is a complex degenerative disease of the retina causing central vision impairment. Multiple factors such as genetics, lipid metabolism, oxidative stress, and mostly aging have been identified to contribute to its progression.

In the retina, as part of the aging process, so-called small drusen (white or yellow protein and lipid deposits) are formed; however when those drusen reach a diameter of 63 μm or more, early AMD is diagnosed [94]. One speaks of late AMD when the drusen increase substantially in size and retinal pigmented epithelium (RPE) cells abnormalities are found due to detachments. If presence of neovascularization in the retina is added to the clinical picture wet AMD is diagnosed. Such condition results into the most rapid decrease in visual acuity [95]. With the increase of life expectancy, AMD has emerged as severe health burden and leading cause of blindness for elderly population [96].

AI Algorithms for Age-Related Macular Degeneration (AMD) Diagnosis

Similar to the diagnosis of glaucoma, AMD is identified by the ophthalmologist through the examination of the retina, either by classical fundus photography or more advanced imaging techniques such as OCT and angiography. Therefore, DL computed image analysis suggests itself for the deconvolution of big data ocular fundus databases to classify AMD presence and severity based on the appearance of drusen, neovascularization, or even fluid leakage and retinal detachment [97].

In studies developing and testing AI algorithms for retinal fundus image analysis, accuracy ranging from 0.63 to 0.99 in the recognition and automatized grading of AMD was reported [98, 99, 100]. Again, other studies focus on OCT images of the ocular fundus for the detection of AMD. As examples, accuracy of more than 90%, with differentiation rate comparable to trained specialists, was found by one model [101]. Whereas in another study accuracy up to 97% was reported, though narrow training and test datasets were employed in this example [102].

Furthermore, not only the assistance for diagnosis and classification of AMD to support the caregivers can be envisioned. The automatized prediction of disease progression by AI for the identification of most vulnerable patient groups or to establish a routine checkup schedule can be supported. In a pioneer work, DL was investigated as a predictive tool for the progression of a late AMD patient's eye to wet AMD, the most severe form of the disease, based on 3D OCT images [103]. In another example, the combination of retinal fundus images and genetic factors reached an AUC of 0.85 to predict disease progression to late AMD [104]. Those studies give a first glimpse of the vast potential of AI to improve the chances for patients to receive personalized care potentially enhancing the prospect to maintain functional vision.

Therapy for Age-Related Macular Degeneration (AMD)

Wet AMD is the most rapidly deteriorating form of AMD; however, it is also the subclass of AMD disease with best therapeutic options currently available. Gold standard is the injection of anti-vascular endothelial growth factor (VEGF) biologics which suppress the neovascularization of the retina and herewith the progression of the disease [105]. Nonetheless, frequent intravitreal dosing (every 1–3 months) is considerably reducing the patient's compliance. A possible solution to prolong the intervals of administration is proposed by Genentech with the Ranibizumab Port Delivery System (PDS) [106, 107] which showed comparable visual and anatomical results to intravitreal injection. However, the delivery system is far from being perfect and suffering from the appearance of vitreous hemorrhages. Following the principle of the Theranocloud concept, there is a need for smart implantable Theranostic dosing system capable to administer drug according to the disease progression and with adjustable dosing can be envisioned for the future.

No game changer drug was found so far for dry AMD. Most investigated options are antioxidants, anti-inflammatory and neuroprotective drugs which similarly to the anti VEGF for wet AMD, might face the challenge of prolonged intravitreal administration to ease patients burden and compliance [108].

6 Diabetic Retinopathy (DR)

DR is one of the most frequently appearing complications of the diabetes mellitus. Taking into account the global epidemic numbers of diabetes affected people, which is estimated to be 592 million within the next 10 years; the large impact on vision among other complications in this population can be foreseen [109].

Type 1 or 2 Diabetes are metabolic diseases rooted in defect in insulin secretion and/or action which leads to characteristic hyperglycemia. Even when not acutely life-threatening, such chronic hyperglycemia is responsible for functional and structural alterations of the vessels, leading to dysfunction and failure of different organs by its associated macro and microvascular complications [110, 111].

In DR, visual loss is caused by microvascular lesions in the retina. DR is clinically classified into 2 stages: proliferative diabetic retinopathy (PDR) and non-proliferative diabetic retinopathy (NPDR), where the first presents vascular renewal in the retina and the second not [112]. NPDR is the earlier stage of the disease and is diagnosed by the presence of microvascular abnormalities and hemorrhages in the retina, whereas in the late stage as the name suggests pathological retinal neovascularization is found [113]. Such loose newly formed anomalous vessels lead to retinal detachment and vitreous hemorrhages which have direct impact on vision. Another severe complication of DR causing majority of the vision loss is the diabetic macular edema (DME). Such complication can occur in any stage of the disease and is caused by vascular leakage of the blood-retinal barrier of fluids into the neural retina causing retinal thickening and cysts in the region of the macula, responsible for central vision [112].

AI Algorithms for Diabetic Retinopathy (DR) Diagnosis

The gold standard for DR detection in a clinical examination is the pupil dilated fundoscopy, which when done in early stages was shown to prevent vision loss by prompt DR management [114]. In accordance to the International Council of Ophthalmology established recommendations for DR screening and management [115], annual screening programs are already implemented in national health services such as the United Kingdom as part of the diabetic care, showing the effectiveness in the reduced risk of vision loss among diabetic patients [116, 117].

Given the constantly increasing proportion of the world population in need of such screenings soon, the advantage of an automated AI based DR grading solution compared to the visit to a retinal specialist is indisputable. Therefore, much research is invested in the development of automated AI DR screening algorithms based on image analysis.

As one of the first AI/ML based diagnostic tools to gain approval from the FDA in 2018 is a AI/ML algorithm for the detection of more than mild DR [29]. The so-called IDx-DR device (Digital Diagnostics, Coralville, IA, USA) is an AI diagnostic system in which fundus camera images of the patients can be scanned for signals of DR and DME [118]. The device was shown to have a sensitivity of 87.2% and specificity of 90.7% in the identification of patients having mild to moderate DR and needing further evaluation by a specialized ophthalmologist [119]. Further, the

AI system showed 97.6% sensitivity for DR with DME. The automated tool was conceptualized to be used as a tool for primary care providers for the systematic screening of diabetic patients without medical record of DR [119].

The challenge to increase accessibility of cost-effective POC screening to the broad diabetic population, especially in low-income countries which suffer notoriously of shortage of costly high precision medical equipment and medical specialists, was tackled by Rajalakshimi et al. [120]. The classification of POC smartphone based fundus imaging device Remedio "Fundus on phone" (Remidio Innovative Solutions Pvt. Ltd, Bangalore, India) was validated by AI DR screening software (EyeArt™, EyeNuk Inc., Los Angeles, CA) designed to identify DR, moderate to severe DR and/or DME. Compared to the DR grading of ophthalmology professionals, AI software showed 95.8% sensitivity and 80.2% specificity to detect any DR in the patient retinal images despite the algorithm not being trained with any of the pictures obtained by the smartphone based fundus imaging device [121]. This combination of smartphone fundus imaging and DL shows enormous potential to bring the much-needed DR screening to a larger diabetic population without the need of trained retinal specialists or expensive machinery. In addition, due to the tremendous advances of AI technology using the DL approach, in 2020 the cloud-based EyeArt™ system was FDA cleared with 96% sensitivity and 88% specificity and in further studies could even report higher sensitivity for detection of DR compared to retinal specialists [122, 123].

Other DL algorithms for the automatic DR screening are being investigated, for example, the Automated Retinal Disease Assessment (ARDA) developed by Google, which is currently under clinical studies in USA and Thailand, showing first very promising outcomes in a multisite screening in Thailand. DL algorithm was reported to achieve comparable DR detection to retina specialists [124, 125].

Acknowledging such findings, it is believed to be key to couple a low-cost instrument POC with AI that can be used by non-physician healthcare personnel to establish a success screening program for the whole diabetic population. To complete the 3 pillars of the Theranocloud principle, in future the link with therapeutic drug delivery systems will be needed, for continued and automated patient support in order to slow disease progression.

Therapy for Diabetic Retinopathy (DR)

DR is a progressive disease; its risk of development or progression was found lowered when hyperglycaemia was mitigated by insulin. In addition, the increase of blood pressure was found as independent risk factor for DR, were 5 mmHg increase yielded to 11% increased probability to develop DR. Therefore, the tight blood pressure regulation is recommended [126, 127].

In the case of diagnosed proliferative DR, gold standard intervention is the laser photocoagulation, which stops the development of new blood vessels in the retina [128]. Laser photocoagulation was shown to be very useful clinical tool for dropping the progression of retinopathy and visual loss [126]. It is also effective for the most severe form of DR with DME. In the case of DME intravitreal injection of anti VEGF and steroids are the first line of therapy. However, this approach is suffering from low patient compliance given the recurrent injection is needed every 4–6 weeks bearing their own potentially devastating risks, as discussed in a previous section [129].

Combination of laser photocoagulation and intravitreal injection of anti VEGF and/or steroids were highlighted as options reporting even higher clinical efficacy [126].

7 Cataract

Cataract is provoked by the opacification of the lens in the eye, which progressively blocks the light transmission to the retina leaving the patient blind, if untreated. However, in comparison to the above discussed retinal diseases, this form of visual loss or even blindness is mostly reversible through surgery. Severe visual impairment is reached due to lack of awareness and treatment via socioeconomic conditions such as lack of resources in developing countries and/or old age in remote rural areas with insufficient infrastructures and care for vulnerable population [130]. Once the cataract has progressed to an advanced stage, surgical intervention becomes more difficult and less successful. Moreover, patients often develop a concomitant glaucoma which can lead to painful vision loss.

To address the challenge of elimination of avoidable blindness, in 1999 the joint initiative VISION 2020—the right to sight, was launched by World Health Organization (WHO) and the International Agency for the Prevention of Blindness, with focus on cataract [131]. The goal was to bring sustainable and sufficient cataract services to all communities by focusing on human resources and infrastructural development.

During the 20 years of implementation, major improvements in terms of increasing awareness, collection of funding from governmental and non-governmental sources, and creation of national plans for the reduction of the burden of avoidable blindness were achieved. Also, the number of cataract surgeries increased in most countries, championing Iran and Argentina. However, the highly ambitious initiative to achieve full elimination of avoidable blindness within 2020 could not be reached in the short timeframe. Main complication was the demographic shift in the last decade with the increase of population especially the elderly which strains further the system with the need of expansion. Issues of accumulation of overdue cataract surgeries with long waiting lists due to lack of trained personnel and means and the low quality of outcome of such mass surgery operations were seen as major setback findings needing for further improvement [132].

AI Algorithms for Cataract Diagnosis

Once more, computerized diagnosis via AI algorithm presents itself as innovative resource for a cost-effective screening done at home or by a non-physician health practitioner. Itinerating POC test stations can be imagined reaching remote and poor populations for rapid assessment of the presence and severity of cataract and to detect patients who most urgently need surgical intervention.

In classical clinical practice, cataract is diagnosed via a slit lamp examination of the lens opacity using the Lens Opacities Classification system III [133], which includes grading of the position of cataract and severity.

The slit lamp images were therefore an obvious choice to be fed into a DL algorithm for the automatic grading of cataract. In an early study of 2015, Gao et al. [134] analyzed 5378 slit lamp photographs for cataract grading with an accuracy of 70.7%. In a more extensive study on 37,638 slit lamp images, DL algorithm was able to distinguish cataract from normal lens with AUC 0.99 and detection of cataracts needing of medical attention with 0.91 AUC [135]. One additional milestone reached in that study was that the algorithm was also able to identify the slit lamp capture mode and take it to account when classifying the picture, rendering herewith the AI algorithm a universal tool. Given this advantage, a web telemedicine platform was established in which patients could upload their images and obtain remote real-time disease monitoring (AI system was supervised by clinicians, to avoid misdiagnosis). Doctors were alerted when advanced cataract was detected needing medical attention. This system was estimated to increase the ophthalmology service to population by 10.2 fold [135].

Other research was focused on the use of retinal fundus images, which already are vastly used for AI based detection of many retinal diseases mentioned in previous paragraphs. However, in the case of cataract, the invisibility of the retina structure was shown to be useful as indicator for the opacity of the lens. Hence the presence and severity of cataract was shown among others by Zhang et al. [136]. However, downside of this method is that it is not able to distinguish cataract from other factors which could influence the image quality such as poor-quality images during sample acquisition. In order to counter this problem, a so-called anti-interference AI model or algorithm capable to remove retinal fundus image noise has been developed which showed an improvement of cataract diagnosis in AI by 10% [137].

As culmination of the AI technology in ophthalmology, a DL algorithm capable of identifying several visual impairment diseases in the same retinal photograph can be envisioned instead of multiple single purpose designed algorithms specific for different eye diseases. A first proof of concept was performed by Tham et al. [28], who developed a DL algorithm capable of identifying disease related visual impaired patients who should be referred to receive expert care by analyzing a single retinal photograph and corresponding visual acuity data. Further capabilities by AI to distinguish disease subgroups can be expected from future research.

Other applications of AI for the cataract management, that are worth mention are, the support of cataract surgery in the calculation of intraocular lens power needed for each patient and AI based tools for surgical video analysis and support for postoperative management. However, a deep analysis of the technology is out of the scope of this chapter. For more information we refer to the very exhaustive review of Tognetto et al. [138].

8 POC Diagnostics for Posterior Segment Diseases and Cataract

About 3% of emergencies are ocular symptoms, which may rapidly lead to vision loss if not treated on the spot properly, especially for the most severe conditions such as retinal detachment. Classical ophthalmological exam consists of slit lamp examination, visual acuity and retinal fundus ophthalmoscopy which are performed by specialized practitioner on a conscious patient [139]. However, in the case of emergency medicine, specific ophthalmological equipment and experts are not always available. Therefore, a rapid method is needed to categorize the patients who require immediate intervention. Ocular ultrasound is a known tool for the observation of eye anatomy, diagnosis of ocular diseases or trauma. Nowadays, due to the advances in technology, Ocular Point of Care Ultrasound (POCUS) device is introduced for direct use on the patient bedside and most valued for the emergency care [139, 140]. In multiple center studies, it was shown that POCUS used by trained emergency practitioners lead to 96.9% sensitivity in detection of retinal detachment [141]. Additionally, promising results in the use of POCUS for the detection of other ophthalmological conditions such as vitreous haemorrhage, lens dislocation, intraocular foreign body, and globe rupture [142] as well as cataract [143] have been reported.

The further inclusion of DL assisted image processing into POCUS analysis is seen as a key element for the expansion of the technology for clinical use. However AI based algorithms suffer from the variety of devices in the market, not standardized analysis procedures, background noise and organ specificity, which leaves still enormous room for improvement [144].

Another advance in the diagnosis of posterior segment disease is the development of affordable POC smartphone based fundoscopy devices. Such devices have the advantage of being less pricy than standard medical equipments, easy to transport and usable by caregivers not necessarily experts in ophthalmology due to the wide knowledge in handling of smartphones nowadays. It is believed that such devices carry the potential to bring ophthalmic examination also to poor undertreated population. The valid performance of such smartphone based fundoscopes was shown in a large literature review by Vilela et al. [145] which evidenced the agreement of smartphone derived fundus pictures with gold standard retinal cameras independent of smartphone brand used or optics adapter.

9 Ocular Surface Diseases

The ocular surface is composed of the cornea and conjunctiva covered by the tear film, the lacrimal and meibomian glands, and eyelids. The correct functioning of the complex system is to enable sight, tear film and the cornea being the first transparent layers encountered by the entering light into the eye [146]. Only in severe cases can ocular surface diseases threaten the patient's vision.

One most common ocular surface disease is dry eye disease (DED), a chronic condition which was defined as "multifactorial disease of the ocular surface, characterized by a loss of homeostasis of the tear film. It is accompanied by ocular symptoms, such as tear film instability and hyperosmolarity, ocular surface inflammation and damage". Symptoms experienced by patients with DED are foreign body sensation, pruritus and burning sensation as well as photophobia and conjunctival redness. In severe cases corneal epithelial keratitis to ulceration can be found [147].

DED patients often suffer from comorbidity with other ocular surface conditions such as blepharitis (inflammation of the eyelids, caused by skin or bacterial conditions), or meibomian gland dysfunction (reduced functionality and irregular meibum secretion), in addition to allergic conjunctivitis [148].

Most common treatment for DED is the substitution of the defect in tear liquid secretion by artificial tears with ocular lubricant formulations. In severe cases, topical corticosteroid or Cyclosporine A treatment might be required in addition to autologous serum tears. If lid abnormalities such as blepharitis are present, lid scrubs or topical antibiotics might be of use. Whereas for the meibomian gland dysfunctions warm compresses are recommended in combination to ocular lubricant formulations [149].

A more severe ocular surface disease is corneal keratitis. The inflammation of the cornea can be caused by external insult such as infection as well as endogenous autoimmune factors. With the impairment of the corneal functionality by inflammation, opacity, corneal ulceration, and ocular pain occur which can impair vision up to vision loss [150]. Infectious keratitis has been shown to be most common cause of corneal blindness accounting for 3.2% of global blindness [151].

As for most of ocular disease, early detection and medical intervention provides the best visual outcomes. In case of infectious keratitis, the identification of the infectious agent is crucial (fungus, bacteria, parasite, or virus) since specific treatment needs to be applied. For the diagnosis, slit lamp examination is the most common medical analysis tool often followed by microbiological cultures or staining, in vivo confocal microscopy, or OCT imaging [152].

AI for the Diagnosis of Ocular Surface Diseases

DED like other ocular surface diseases is diagnosed in clinical practice by image analysis in combination with patient symptoms questionnaire. The most classical method is slit lamp examination of the cornea with dyes such as fluorescein or lissamine green to see epithelial or conjunctival staining. In addition, tear breakup time and conjunctival lid parallel folds, which are frequently seen in DED, can be determined as well. However, such slit lamp evaluation depends on the practitioners training and can be only verified if photographs or videos are taken, which is not a common practice [153]. More innovatively and without the use of any stains, OCT has been found to be capable of visualizing corneal defects and precorneal tear film, from which tear film breakup time and thickness can be derived as well [154].

Tear breakup time can also be analyzed noninvasively, without the use of any stain, via the Oculus Keratograph. The high-speed video keratographic method analyzes the changes of reflected patterns on the ocular surface after the blink. Edge deformation of those patterns are giving the indication for tear film to be disrupted and can be assessed by the operator as tear film breakup time [155].

Another frequently used imaging technique for ocular surface diseases diagnosis is meibography, which is the visualization of the meibomian glands via trans-illumination or infrared light. It is performed to investigate glands morphology and number by the examiner deducing their health status [156].

Other ocular surface imaging methods for the diagnosis of ocular surface diseases have been studied such as in vivo confocal microscopy, as well as aberrometery; interferometry to asses tear lipid layer thickness. Thermography considers the principle of elevated body temperature to be linked to inflammation [157].

All those methods briefly described above have the generation of images in common, which, as often discussed in this chapter, are raw materials which can be fed into automatic image analysis algorithms for the computerized assistance of diagnosis and patient case grading.

The potential use of computerized image analysis in ocular surface disease diagnosis was shown by Chase et al. [158] who developed a fully automated DED diagnosis DL algorithm using 27,180 OCT images. The outcome model resulted in 86% sensitivity for the diagnosis of DED compared to ophthalmologist diagnosis using gold standard methods such as tear breakup time, corneal and conjunctival staining, and patient DED questionnaires. It has to be noted that the algorithm is a black box model where no features on the OCT images were taught to the algorithm. Also, the output criteria on which AI was performing the classification are not known.

In another study, interferometry images, which are usually used for the visualization of the tear film thickness and lipid layer as well as tear film wetting, were introduced successfully to ML based algorithm to screen patients for DED [159].

Another method, which is largely investigated application of AI for ocular surface disease diagnosis, is the automated morphological analysis of the meibomian gland images. AI algorithms were found with performance of 95.4% accuracy in the demarcation of meibomian gland atrophy zones [160] and 84.4% sensitivity 71.7% specificity for the identification of meibomian gland morphological abnormalities such as ghost glands [161], both key indicators for severity of meibomian gland dysfunction. In addition, not only the accuracy of meibomian gland assessment was found to be increased by DL algorithms, but also the speed of analysis increased by 21-fold compared to an evaluation by an ophthalmology practitioner [162].

In latest research, DL automated quantitative evaluation of meibomian gland morphology was found to outperform the accuracy of meibomian gland specialists and supported ophthalmological examination by experts. This is due to automatic suppression of reflections which appear due to image acquisition on the meibographic photos and can lead to misinterpretation [163].

As an alternative, automated AI video analysis of ocular surface by OCULUS Keratograph 5 M equipment lead to the research project of automatic DED diagnosis by Zheng et al. [164]. The correlation of incomplete blinking frequency with DED was shown. This finding verifies to prior literature that found incomplete blinking or reduced blink rate promotes tear film instability and DED. However, high-speed cameras are needed to capture the blinking movement.

Like diagnosis of DED, the automated detection and classification of infectious corneal keratitis by AI on slit lamp images was proposed in recent research. An AI model published in 2021 by Li et al. [165] reported an accuracy of 98% in distinction of keratitis from healthy cornea or other corneal abnormalities. This algorithm was also tested on corneal images captured by smartphone in macro mode and once more high accuracy of 94% was obtained, showing the ability of the system to be employed as cost-effective screening tool. In another study, Hung et al. [166] reported accuracy of 0.8 and AUC of 0.85, respectively, for the classification of different keratitis types (bacterial or fungal keratitis) by automized DL algorithms, with the advantage to not only detect keratitis, but to further classify according to infectious agent subgroup. Such automatic diagnosis and distinction of keratitis type is poised to become a powerful clinical tool. Further research on the automatization of the distinction of bacterial and fungal keratitis reached similar accuracy of 0.83 but with increased AUC of 0.904 [167].

POC for the Diagnosis of Ocular Surface Diseases

With the attention to Near Patient Testing (NPT), POC devices are targeted to be used at home or by trained but not necessarily medical professionals to support diagnosis of ocular surface diseases while classical ocular surface exams are still performed by the ophthalmologist.

For the diagnosis of DED, tear film stability, quantity, and composition are indicative of the severity of the disease. Proteomic studies on DED have shown several biologically relevant biomarkers to be altered in the tear fluid by the disease state when compared to healthy eyes [92]. Numerous proteins were found to be significantly altered as to be named tear biomarker for DED disease [168]. However, in current clinical practice only few FDA approved POC devices exist so far.

One of these devices is InflammaDry (Quidel, San Diego, CA, USA), developed for the analysis of Matrix metalloproteinase 9 (MMP-9), a protein shown to be increased by inflammation, as it is the case for DED.

The lateral flow POC test qualitatively measures levels of MMP-9 in tears above 40 ng/ml which was found to correlate to patients' symptoms perception, conjunctival and corneal staining as well as very short tear breakup time (5 s or less). The POC was found to successfully identify DED induced ocular surface inflammation often correlated to autoimmune diseases [169, 170] and help the caregiver for the choice of treatment [171].

Despite the usefulness of MMP-9 qualitative POC test as additional tool for the diagnostic decision making by the specialist, the test alone cannot be used for DED diagnosis. This is a common problem encountered by tear protein biomarkers. The analysis of biomarker might indicate the presence of an inflammatory response (e.g. cytokines/chemokines) and often correlates well with clinical symptoms. However, it cannot distinguish for the origin of this inflammation [172] when bigger clinical picture of the patient is not taken into account with additional clinical measurements.

Future for the tear protein biomarker POC devices is believed to lie in the detection and/or quantification of several biomarkers whose patterns can be related to a specific disease [173]. As an instance of those next generation POC diagnostic tools, a dual biomarker detection lateral flow assay POC for IL6 and lactoferrin in the tears was developed. Such a test due to its simplicity in use can be imagined to be deployed at home. Its first test on DED tears and normal tears successfully distinguishment between the two groups of DED and control; however larger validation is needed [174].

Another approach for diagnosis of DED is measuring tear osmolarity. This measurement has been made easily accessible in recent years by FDA approved POC commercial device TearLab [175, 176]. The increase in tear film osmolarity values has been reported to correlate with increased DED severity grad. However, cut-off values in between normal and increased osmolarities were found to differ in several studies [176, 177]. Therefore, the patient specific variation of tear osmolarity was identified as an important parameter for DED diagnosis [178].

The breakthrough of the TearLab device is the easy-to-use tear sampling of few nanolitres with minimal risk of provoking reflex tears or discomfort. The collection of lacrimal tear fluid, despite not being invasive, has its own challenges because of the low volume of a few microliters, and its proximity to highly sensitive ocular tissue.

The alteration of the tear film composition due to DED was not only found to influence the osmolarity of the solution but also the tear film stability and evaporation. This tear film evaporation can be spotted by thermographic imaging of the ocular surface. Research has shown the ability of this method to differentiate between DED and healthy cases even in subclasses of the disease [179] and the correlation with tear film breakup time [180]. Main constraint to this technology to become a viable POC instrument is the high cost of the thermographic cameras. In future a feedback loop of DED symptoms detection by smart sensors via physical phenomena (osmolarity, thermal, or others) can be envisioned to be coupled to an automized eye drop dispensing system as example. Such combination would incorporate all principles of the Theranocloud concept and represent a milestone in modern personalized healthcare.

10 Eye Biomarkers for Systemic Diseases: Oculomics

The eye is not only the body's window to the external world but can be also used as window inside the body to observe the central nervous system and micro vascularization noninvasively. The combination of high-resolution retinal fundus images, OCT images, or OCT angiography with DL processing have shown capable to identify biomarkers of systemic disease. This new field was defined in 2020 as Oculomics [181].

The association between systemic vascular diseases and retinal vascular changes is no novel concept; however, now with the DL revolution in computing power, it is possible to extract so far unknown biomarkers for systemic diseases from the data rich retinal images on vast datasets [182]. Retinal fundus image processing by DL retrieved biomarkers for the risk assessment of cardiovascular diseases such as stroke, high blood pressure, or carotid atherosclerosis, investigating on large scale multiple key features such as vessel morphology, calibers, tortuosity on the retinal scans or retinal neuronal structure and choroidal layer thinning [183].

Intriguingly, DL algorithms were found able to predict the age and gender with AUC above 0.8, a capacity ophthalmologist do not have by purely looking at the retinal fundus image [182]

DL technology is poised to find correlations to many more of systemic diseases by extracting so far unknown features from the big data pool of retinal images which will define disease pattern for the conditions also called disease retinal fingerprint which is expected to revolutionize POC screening [184]. It can be foreseen the retinal scan to become a quick routine scan for the general health status of the patient thanks to the computing power of DL.

Even further the pool of disease biomarkers for systemic diseases does not end with the retinal scans, other sources such as tear fluid biomarkers have been shown to correlate to systemic diseases [172]. The combination of such biomarkers is what lies on the core of the Theranocloud principle for increased diagnostic and therapeutic options thanks to the AI revolution which will enable a paradigm shift in healthcare and in increased wellbeing for the society.

11 Conclusion

The combination of therapy, diagnostics, and cloud computing in modern medicine and healthcare, as exemplified by the innovative future concept of Theranocloud, is poised to bring significant benefits to patients and healthcare providers alike. By interconnecting clinical diagnosis point-of-care devices with each other and with cloud-based autonomous AI computing, we have the potential to generate and analyze big data that can lead to more accurate and personalized diagnoses. The resulting autonomous diagnostic tools have enormous potential to bring affordable healthcare services to underserved communities and facilitate at-home testing and real-time monitoring of patient health status.

The field of ophthalmology is an excellent example of how AI-supported image analysis of retinal fundus images can enable automated diagnosis algorithms to detect major blinding diseases such as glaucoma, age-related macular degeneration, diabetic retinopathy, and cataract. Combined with affordable, smart point-of-care imaging devices, these algorithms can expand ocular healthcare to underserved communities and help reduce the global burden of blindness.

Moreover, the use of rapid, cost-effective screening methods for potentially blinding diseases can help prescreen patients and reduce pressure on ophthalmological care centers. However, the successful implementation of these new models of healthcare will require collaboration among experts in various fields, including engineering, data science, pharmaceutical technology, ophthalmology, and optometry. Ultimately, the goal of the smart medical device system represented by the

Theranocloud is to deliver high-quality healthcare services for screening, preventive actions, diagnosis, therapy, disease management, and follow-up via digitalization.

References

1. Frangos, S., Buscombe, J.R.: Why should we be concerned about a "g"? Eur. J. Nucl. Med. Mol. Imaging **46**, 519–519 (2019). https://doi.org/10.1007/s00259-018-4204-z
2. Pene, F., Courtine, E., Cariou, A., Mira, J.-P.: Toward theragnostics. Crit. Care Med. **37**, S50-58 (2009). https://doi.org/10.1097/CCM.0b013e3181921349
3. Ehrhardt, J.D., Güleç, S.: A review of the history of radioactive iodine theranostics: the origin of nuclear ontology. Mol. Imaging Radionucl. Ther. **29**, 88–97 (2020). https://doi.org/10.4274/mirt.galenos.2020.83703
4. Jørgensen, J.T., Winther, H., Askaa, J., Andresen, L., Olsen, D., Mollerup, J.: A companion diagnostic with significant clinical impact in treatment of breast and gastric cancer. Front. Oncol. **11**, 676939 (2021). https://doi.org/10.3389/fonc.2021.676939
5. Velikyan, I., Wennborg, A., Frldwisch, J., Olofsson, H., Sandberg, D., Lubberink, M., Sandstrom, M., Lindman, H., Carlsson, J., Sorensen, J.: Theranostics at its best: clinical breast cancer imaging and quantification targeting HER2 receptors. Endocr. Abstr. **47**, (2016). https://doi.org/10.1530/endoabs.47.OC48
6. Chen, F., Ehlerding, E.B., Cai, W.: Theranostic nanoparticles. J. Nucl. Med. Off. Publ. Soc. Nucl. Med. **55**, 1919–1922 (2014). https://doi.org/10.2967/jnumed.114.146019
7. Gomes Marin, J.F., Nunes, R.F., Coutinho, A.M., Zaniboni, E.C., Costa, L.B., Barbosa, F.G., Queiroz, M.A., Cerri, G.G., Buchpiguel, C.A.: Theranostics in nuclear medicine: emerging and re-emerging integrated imaging and therapies in the era of precision oncology. Radiographics **40**, 1715–1740 (2020). https://doi.org/10.1148/rg.2020200021
8. DeNardo, G.L., DeNardo, S.J.: Concepts, consequences, and implications of theranosis. Semin. Nucl. Med. **42**, 147–150 (2012). https://doi.org/10.1053/j.semnuclmed.2011.12.003
9. Ahn, S., Woo, J.W., Lee, K., Park, S.Y.: HER2 status in breast cancer: changes in guidelines and complicating factors for interpretation. J. Pathol. Transl. Med. **54**, 34–44 (2020). https://doi.org/10.4132/jptm.2019.11.03
10. Shi, Z., Lu, Y., Shen, S., Xu, Y., Shu, C., Wu, Y., Lv, J., Li, X., Yan, Z., An, Z., Dai, C., Su, L., Zhang, F., Liu, Q.: Wearable battery-free theranostic dental patch for wireless intraoral sensing and drug delivery. Npj Flex. Electron. **6**, 1–11 (2022). https://doi.org/10.1038/s41528-022-00185-5
11. Manikkath, J., Subramony, J.A.: Toward closed-loop drug delivery: Integrating wearable technologies with transdermal drug delivery systems. Adv. Drug Deliv. Rev. **179**, 113997 (2021). https://doi.org/10.1016/j.addr.2021.113997
12. Zhang, Z., Zhang, R., Chang, C.-W., Guo, Y., Chi, Y.-W., Pan, T.: IWRAP: A theranostic wearable device with real-time vital monitoring and auto-adjustable compression level for venous thromboembolism. IEEE Trans. Biomed. Eng. **68**, 2776–2786 (2021). https://doi.org/10.1109/TBME.2021.3054335
13. Menachemi, N., Collum, T.H.: Benefits and drawbacks of electronic health record systems. Risk Manag. Healthc. Policy. **4**, 47–55 (2011). https://doi.org/10.2147/RMHP.S12985
14. Huang, J.-D., Wang, J., Ramsey, E., Leavey, G., Chico, T.J.A., Condell, J.: Applying artificial intelligence to wearable sensor data to diagnose and predict cardiovascular disease: a review. Sensors. **22**, 8002 (2022). https://doi.org/10.3390/s22208002
15. Jiang, F., Jiang, Y., Zhi, H., Dong, Y., Li, H., Ma, S., Wang, Y., Dong, Q., Shen, H., Wang, Y.: Artificial intelligence in healthcare: past, present and future. Stroke Vasc. Neurol. **2**, 230–243 (2017). https://doi.org/10.1136/svn-2017-000101
16. Sarker, I.H.: AI-based modeling: techniques, applications and research issues towards automation, intelligent and smart systems. Sn Comput. Sci. **3**, 158 (2022). https://doi.org/10.1007/s42979-022-01043-x

17. Yang, S., Zhu, F., Ling, X., Liu, Q., Zhao, P.: Intelligent health care: applications of deep learning in computational medicine. Front. Genet. **12**, (2021)
18. Nuzzi, R., Boscia, G., Marolo, P., Ricardi, F.: The impact of artificial intelligence and deep learning in eye diseases: a review. Front. Med. **8**, 710329 (2021). https://doi.org/10.3389/fmed.2021.710329
19. Du, A.X., Emam, S., Gniadecki, R.: Review of machine learning in predicting dermatological outcomes. Front. Med. **7**, (2020)
20. LeCun, Y., Bengio, Y., Hinton, G.: Deep learning. Nature **521**, 436–444 (2015). https://doi.org/10.1038/nature14539
21. Mayro, E.L., Wang, M., Elze, T., Pasquale, L.R.: The impact of artificial intelligence in the diagnosis and management of glaucoma. Eye **34**, 1–11 (2020). https://doi.org/10.1038/s41433-019-0577-x
22. Hinton, G.E., Osindero, S., Teh, Y.-W.: A Fast Learning Algorithm for Deep Belief Nets. Neural Comput. **18**, 1527–1554 (2006). https://doi.org/10.1162/neco.2006.18.7.1527
23. Sarker, I.H.: Deep learning: a comprehensive overview on techniques, taxonomy, applications and research directions. SN Comput. Sci. **2**, 420 (2021). https://doi.org/10.1007/s42979-021-00815-1
24. Murdoch, T.B., Detsky, A.S.: The inevitable application of big data to health care. JAMA **309**, 1351–1352 (2013). https://doi.org/10.1001/jama.2013.393
25. Jensen, P.B., Jensen, L.J., Brunak, S.: Mining electronic health records: towards better research applications and clinical care. Nat. Rev. Genet. **13**, 395–405 (2012). https://doi.org/10.1038/nrg3208
26. Esteva, A., Kuprel, B., Novoa, R.A., Ko, J., Swetter, S.M., Blau, H.M., Thrun, S.: Dermatologist-level classification of skin cancer with deep neural networks. Nature **542**, 115–118 (2017). https://doi.org/10.1038/nature21056
27. Davenport, T., Kalakota, R.: The potential for artificial intelligence in healthcare. Future Healthc. J. **6**, 94–98 (2019). https://doi.org/10.7861/futurehosp.6-2-94
28. Tham, Y.-C., Anees, A., Zhang, L., Goh, J.H.L., Rim, T.H., Nusinovici, S., Hamzah, H., Chee, M.-L., Tjio, G., Li, S., Xu, X., Goh, R., Tang, F., Cheung, C.Y.-L., Wang, Y.X., Nangia, V., Jonas, J.B., Gopinath, B., Mitchell, P., Husain, R., Lamoureux, E., Sabanayagam, C., Wang, J.J., Aung, T., Liu, Y., Wong, T.Y., Cheng, C.-Y.: Referral for disease-related visual impairment using retinal photograph-based deep learning: a proof-of-concept, model development study. Lancet Digit. Health. **3**, e29–e40 (2021). https://doi.org/10.1016/S2589-7500(20)30271-5
29. Benjamens, S., Dhunnoo, P., Meskó, B.: The state of artificial intelligence-based FDA-approved medical devices and algorithms: an online database. NPJ Digit. Med. **3**, 118 (2020). https://doi.org/10.1038/s41746-020-00324-0
30. Retson, T.A., Besser, A.H., Sall, S., Golden, D., Hsiao, A.: Machine learning and deep neural networks in thoracic and cardiovascular imaging. J. Thorac. Imaging **34**, 192–201 (2019). https://doi.org/10.1097/RTI.0000000000000385
31. Babel, A., Taneja, R., Mondello Malvestiti, F., Monaco, A., Donde, S.: Artificial Intelligence Solutions to Increase Medication Adherence in Patients With Non-communicable Diseases. Front. Digit. Health. **3**, 669869 (2021). https://doi.org/10.3389/fdgth.2021.669869
32. Labovitz, D.L., Shafner, L., Reyes Gil, M., Virmani, D., Hanina, A.: Using artificial intelligence to reduce the risk of nonadherence in patients on anticoagulation therapy. Stroke **48**, 1416–1419 (2017). https://doi.org/10.1161/STROKEAHA.116.016281
33. Lehne, M., Sass, J., Essenwanger, A., Schepers, J., Thun, S.: Why digital medicine depends on interoperability. Npj Digit. Med. **2**, 1–5 (2019). https://doi.org/10.1038/s41746-019-0158-1
34. Kelly, J.T., Campbell, K.L., Gong, E., Scuffham, P.: The internet of things: impact and implications for health care delivery. J. Med. Internet Res. **22**, e20135 (2020). https://doi.org/10.2196/20135
35. Murdoch, B.: Privacy and artificial intelligence: challenges for protecting health information in a new era. BMC Med. Ethics **22**, 122 (2021). https://doi.org/10.1186/s12910-021-00687-3
36. Na, L., Yang, C., Lo, C.-C., Zhao, F., Fukuoka, Y., Aswani, A.: Feasibility of re identifying individuals in large national physical activity data sets from which protected health information

has been removed with use of machine learning. JAMA Netw. Open **1**, e186040 (2018). https://doi.org/10.1001/jamanetworkopen.2018.6040
37. Papadakis, G.Z., Karantanas, A.H., Tsiknakis, M., Tsatsakis, A., Spandidos, D.A., Marias, K.: Deep learning opens new horizons in personalized medicine. Biomed. Rep. **10**, 215–217 (2019). https://doi.org/10.3892/br.2019.1199
38. Morley, J., Machado, C.C.V., Burr, C., Cowls, J., Joshi, I., Taddeo, M., Floridi, L.: The ethics of AI in health care: A mapping review. Soc. Sci. Med. 1982. **260**, 113172 (2020). https://doi.org/10.1016/j.socscimed.2020.113172
39. Health, C. for D. and R.: Artificial intelligence and machine learning in software as a medical device. FDA. (2022)
40. Esmaeilzadeh, P.: Use of AI-based tools for healthcare purposes: a survey study from consumers' perspectives. BMC Med. Inform. Decis. Mak. **20**, 170 (2020). https://doi.org/10.1186/s12911-020-01191-1
41. Gagnon, M.-P., Ngangue, P., Payne-Gagnon, J., Desmartis, M.: M-Health adoption by healthcare professionals: a systematic review. J. Am. Med. Inform. Assoc. **23**, 212–220 (2016). https://doi.org/10.1093/jamia/ocv052
42. Enoch, J., McDonald, L., Jones, L., Jones, P.R., Crabb, D.P.: Evaluating whether sight is the most valued sense. JAMA Ophthalmol. **137**, 1317–1320 (2019). https://doi.org/10.1001/jamaophthalmol.2019.3537
43. Langelaan, M., de Boer, M.R., van Nispen, R.M.A., Wouters, B., Moll, A.C., van Rens, G.H.M.B.: Impact of visual impairment on quality of life: a comparison with quality of life in the general population and with other chronic conditions. Ophthalmic Epidemiol. **14**, 119–126 (2007). https://doi.org/10.1080/09286580601139212
44. Chakravarthy, U., Biundo, E., Saka, R.O., Fasser, C., Bourne, R., Little, J.-A.: The economic impact of blindness in Europe. Ophthalmic Epidemiol. **24**, 239–247 (2017). https://doi.org/10.1080/09286586.2017.1281426
45. Huang, D., Chen, Y.-S., Rupenthal, I.D.: Overcoming ocular drug delivery barriers through the use of physical forces. Adv. Drug Deliv. Rev. **126**, 96–112 (2018). https://doi.org/10.1016/j.addr.2017.09.008
46. A brief overview of ocular anatomy and physiology. http://benthamscience.com/chapter/5273
47. Kolb, H.: How the Retina Works: Much of the construction of an image takes place in the retina itself through the use of specialized neural circuits. Am. Sci. **91**, 28–35 (2003)
48. Common Eye Disorders and Diseases|CDC. https://www.cdc.gov/visionhealth/basics/ced/index.html
49. Bourne, R.R.A., Flaxman, S.R., Braithwaite, T., Cicinelli, M.V., Das, A., Jonas, J.B., Keeffe, J., Kempen, J.H., Leasher, J., Limburg, H., Naidoo, K., Pesudovs, K., Resnikoff, S., Silvester, A., Stevens, G.A., Tahhan, N., Wong, T.Y., Taylor, H.R.: Vision loss expert group: magnitude, temporal trends, and projections of the global prevalence of blindness and distance and near vision impairment: a systematic review and meta-analysis. Lancet Glob. Health **5**, e888–e897 (2017). https://doi.org/10.1016/S2214-109X(17)30293-0
50. GBD 2019 blindness and vision impairment collaborators, vision loss expert group of the global burden of disease study: causes of blindness and vision impairment in 2020 and trends over 30 years, and prevalence of avoidable blindness in relation to VISION 2020: the Right to Sight: an analysis for the Global Burden of Disease Study. Lancet Glob. Health. **9**, e144–e160 (2021). https://doi.org/10.1016/S2214-109X(20)30489-7
51. Nickells, R.W., Howell, G.R., Soto, I., John, S.W.M.: Under pressure: cellular and molecular responses during glaucoma, a common neurodegeneration with axonopathy. Annu. Rev. Neurosci. **35**, 153–179 (2012). https://doi.org/10.1146/annurev.neuro.051508.135728
52. Weinreb, R.N., Aung, T., Medeiros, F.A.: The pathophysiology and treatment of glaucoma. JAMA **311**, 1901–1911 (2014). https://doi.org/10.1001/jama.2014.3192
53. Medeiros, F.A., Zangwill, L.M., Bowd, C., Weinreb, R.N.: Comparison of the GDx VCC scanning laser polarimeter, HRT II confocal scanning laser ophthalmoscope, and stratus OCT optical coherence tomograph for the detection of glaucoma. Arch. Ophthalmol. Chic. Ill **1960**(122), 827–837 (2004). https://doi.org/10.1001/archopht.122.6.827

54. Hoffmann, E.M., Zangwill, L.M., Crowston, J.G., Weinreb, R.N.: Optic disk size and glaucoma. Surv. Ophthalmol. **52**, 32–49 (2007). https://doi.org/10.1016/j.survophthal.2006.10.002
55. Shin, Y., Cho, H., Jeong, H.C., Seong, M., Choi, J.-W., Lee, W.J.: Deep learning-based diagnosis of glaucoma using wide-field optical coherence tomography images. J. Glaucoma **30**, 803–812 (2021). https://doi.org/10.1097/IJG.0000000000001885
56. Akter, N., Fletcher, J., Perry, S., Simunovic, M.P., Briggs, N., Roy, M.: Glaucoma diagnosis using multi-feature analysis and a deep learning technique. Sci. Rep. **12**, 8064 (2022). https://doi.org/10.1038/s41598-022-12147-y
57. An, G., Omodaka, K., Hashimoto, K., Tsuda, S., Shiga, Y., Takada, N., Kikawa, T., Yokota, H., Akiba, M., Nakazawa, T.: Glaucoma diagnosis with machine learning based on optical coherence tomography and color fundus images. J. Healthc. Eng. **2019**, 4061313 (2019). https://doi.org/10.1155/2019/4061313
58. Li, F., Wang, Z., Qu, G., Song, D., Yuan, Y., Xu, Y., Gao, K., Luo, G., Xiao, Z., Lam, D.S.C., Zhong, H., Qiao, Y., Zhang, X.: Automatic differentiation of Glaucoma visual field from non-glaucoma visual filed using deep convolutional neural network. BMC Med. Imaging **18**, 35 (2018). https://doi.org/10.1186/s12880-018-0273-5
59. Heijl, A., Leske, M.C., Bengtsson, B., Hyman, L., Bengtsson, B., Hussein, M.: Early manifest glaucoma trial group: reduction of intraocular pressure and glaucoma progression: results from the early manifest glaucoma trial. Arch. Ophthalmol. Chic. Ill **1960**(120), 1268–1279 (2002). https://doi.org/10.1001/archopht.120.10.1268
60. Sj, G., K, V., Mm, W., Kw, M., Jt, L., Pp, C., T, L., Sl, M.: Primary open-angle glaucoma preferred practice pattern®. Ophthalmol. **128**, (2021). https://doi.org/10.1016/j.ophtha.2020.10.022
61. Doucette, L.P., Walter, M.A.: Prostaglandins in the eye: Function, expression, and roles in glaucoma. Ophthalmic Genet. **38**, 108–116 (2017). https://doi.org/10.3109/13816810.2016.1164193
62. Brusini, P., Salvetat, M.L., Zeppieri, M.: How to measure intraocular pressure: an updated review of various tonometers. J. Clin. Med. **10**, 3860 (2021). https://doi.org/10.3390/jcm10173860
63. Barkana, Y., Anis, S., Liebmann, J., Tello, C., Ritch, R.: Clinical utility of intraocular pressure monitoring outside of normal office hours in patients with glaucoma. Arch. Ophthalmol. Chic. Ill **1960**(124), 793–797 (2006). https://doi.org/10.1001/archopht.124.6.793
64. Tojo, N., Abe, S., Ishida, M., Yagou, T., Hayashi, A.: The fluctuation of intraocular pressure measured by a contact lens sensor in normal-tension glaucoma patients and nonglaucoma subjects. J. Glaucoma **26**, 195–200 (2017). https://doi.org/10.1097/IJG.0000000000000517
65. Liu, J., De Francesco, T., Schlenker, M., Ahmed, I.I.: Icare home tonometer: a review of characteristics and clinical utility. Clin. Ophthalmol. Auckl. NZ. **14**, 4031–4045 (2020). https://doi.org/10.2147/OPTH.S284844
66. Yang, C., Huang, X., Li, X., Yang, C., Zhang, T., Wu, Q., Liu, D., Lin, H., Chen, W., Hu, N., Xie, X.: Wearable and implantable intraocular pressure biosensors: recent progress and future prospects. Adv. Sci. **8**, 2002971 (2021). https://doi.org/10.1002/advs.202002971
67. Zhang, Y., Chen, Y., Man, T., Huang, D., Li, X., Zhu, H., Li, Z.: High resolution non-invasive intraocular pressure monitoring by use of graphene woven fabrics on contact lens. Microsyst. Nanoeng. **5**, 39 (2019). https://doi.org/10.1038/s41378-019-0078-x
68. Agaoglu, S., Diep, P., Martini, M., Kt, S., Baday, M., Araci, I.E.: Ultra-sensitive microfluidic wearable strain sensor for intraocular pressure monitoring. Lab Chip **18**, 3471–3483 (2018). https://doi.org/10.1039/c8lc00758f
69. Araci, I.E., Agaoglu, S., Lee, J.Y., Rivas Yepes, L., Diep, P., Martini, M., Schmidt, A.: Flow stabilization in wearable microfluidic sensors enables noise suppression. Lab Chip **19**, 3899–3908 (2019). https://doi.org/10.1039/c9lc00842j
70. Maeng, B., Chang, H.-K., Park, J.: Photonic crystal-based smart contact lens for continuous intraocular pressure monitoring. Lab Chip **20**, 1740–1750 (2020). https://doi.org/10.1039/c9lc01268k

71. Leonardi, M., Leuenberger, P., Bertrand, D., Bertsch, A., Renaud, P.: First steps toward noninvasive intraocular pressure monitoring with a sensing contact lens. Invest. Ophthalmol. Vis. Sci. **45**, 3113–3117 (2004). https://doi.org/10.1167/iovs.04-0015
72. Mansouri, K., Weinreb, R.N.: Meeting an unmet need in glaucoma: continuous 24-h monitoring of intraocular pressure. Expert Rev. Med. Devices **9**, 225–231 (2012). https://doi.org/10.1586/erd.12.14
73. Jones, L., Hui, A., Phan, C.-M., Read, M.L., Azar, D., Buch, J., Ciolino, J.B., Naroo, S.A., Pall, B., Romond, K., Sankaridurg, P., Schnider, C.M., Terry, L., Willcox, M.: BCLA CLEAR—Contact lens technologies of the future. Contact Lens Anterior Eye. **44**, 398–430 (2021). https://doi.org/10.1016/j.clae.2021.02.007
74. Dunbar, G.E., Shen, B.Y., Aref, A.A.: The Sensimed triggerfish contact lens sensor: efficacy, safety, and patient perspectives. Clin. Ophthalmol. Auckl. NZ. **11**, 875–882 (2017). https://doi.org/10.2147/OPTH.S109708
75. Mansouri, K., Weinreb, R.N., Liu, J.H.K.: Efficacy of a contact lens sensor for monitoring 24-h intraocular pressure related patterns. PLoS ONE **10**, e0125530 (2015). https://doi.org/10.1371/journal.pone.0125530
76. Mansouri, K., Medeiros, F.A., Tafreshi, A., Weinreb, R.N.: Continuous 24-hour monitoring of intraocular pressure patterns with a contact lens sensor: safety, tolerability, and reproducibility in patients with glaucoma. Arch. Ophthalmol. Chic. Ill **1960**(130), 1534–1539 (2012). https://doi.org/10.1001/jamaophthalmol.2013.1350
77. Agnifili, L., Mastropasqua, R., Frezzotti, P., Fasanella, V., Motolese, I., Pedrotti, E., Di Iorio, A., Mattei, P.A., Motolese, E., Mastropasqua, L.: Circadian intraocular pressure patterns in healthy subjects, primary open angle and normal tension glaucoma patients with a contact lens sensor. Acta Ophthalmol. (Copenh.) **93**, e14-21 (2015). https://doi.org/10.1111/aos.12408
78. Yang, Z., Mansouri, K., Moghimi, S., Weinreb, R.N.: Nocturnal variability of intraocular pressure monitored with contact lens sensor is associated with visual field loss in glaucoma. J. Glaucoma **30**, e56 (2021). https://doi.org/10.1097/IJG.0000000000001727
79. Mottet, B., Aptel, F., Romanet, J.-P., Hubanova, R., Pépin, J.-L., Chiquet, C.: 24-hour intraocular pressure rhythm in young healthy subjects evaluated with continuous monitoring using a contact lens sensor. JAMA Ophthalmol. **131**, 1507–1516 (2013). https://doi.org/10.1001/jamaophthalmol.2013.5297
80. Zhang, J., Kim, K., Kim, H.J., Meyer, D., Park, W., Lee, S.A., Dai, Y., Kim, B., Moon, H., Shah, J.V., Harris, K.E., Collar, B., Liu, K., Irazoqui, P., Lee, H., Park, S.A., Kollbaum, P.S., Boudouris, B.W., Lee, C.H.: Smart soft contact lenses for continuous 24-hour monitoring of intraocular pressure in glaucoma care. Nat. Commun. **13**, 5518 (2022). https://doi.org/10.1038/s41467-022-33254-4
81. Kim, T.Y., Mok, J.W., Hong, S.H., Jeong, S.H., Choi, H., Shin, S., Joo, C.-K., Hahn, S.K.: Wireless theranostic smart contact lens for monitoring and control of intraocular pressure in glaucoma. Nat. Commun. **13**, 6801 (2022). https://doi.org/10.1038/s41467-022-34597-8
82. Foot, B., MacEwen, C.: Surveillance of sight loss due to delay in ophthalmic treatment or review: frequency, cause and outcome. Eye Lond. Engl. **31**, 771–775 (2017). https://doi.org/10.1038/eye.2017.1
83. Prea, S.M., Vingrys, A.J., Kong, G.Y.X.: Test Reliability and Compliance to a Twelve-Month Visual Field Telemedicine Study in Glaucoma Patients. J. Clin. Med. **11**, 4317 (2022). https://doi.org/10.3390/jcm11154317
84. Jones, P.R., Campbell, P., Callaghan, T., Jones, L., Asfaw, D.S., Edgar, D.F., Crabb, D.P.: Glaucoma home monitoring using a tablet-based visual field test (Eyecatcher): an assessment of accuracy and adherence over 6 months. Am. J. Ophthalmol. **223**, 42–52 (2021). https://doi.org/10.1016/j.ajo.2020.08.039
85. Jones, P.R., Demaria, G., Tigchelaar, I., Asfaw, D.S., Edgar, D.F., Campbell, P., Callaghan, T., Crabb, D.P.: The human touch: using a webcam to autonomously monitor compliance during visual field assessments. Transl. Vis. Sci. Technol. **9**, 31 (2020). https://doi.org/10.1167/tvst.9.8.31

86. Matsumoto, C., Yamao, S., Nomoto, H., Takada, S., Okuyama, S., Kimura, S., Yamanaka, K., Aihara, M., Shimomura, Y.: Visual field testing with head-mounted perimeter "imo." PLoS ONE **11**, e0161974 (2016). https://doi.org/10.1371/journal.pone.0161974
87. Alawa, K.A., Nolan, R.P., Han, E., Arboleda, A., Durkee, H., Sayed, M.S., Aguilar, M.C., Lee, R.K.: Low-cost, smartphone-based frequency doubling technology visual field testing using a head-mounted display. Br. J. Ophthalmol. **105**, 440–444 (2021). https://doi.org/10.1136/bjophthalmol-2019-314031
88. Beykin, G., Norcia, A.M., Srinivasan, V.J., Dubra, A., Goldberg, J.L.: Discovery and clinical translation of novel glaucoma biomarkers. Prog. Retin. Eye Res. **80**, 100875 (2021). https://doi.org/10.1016/j.preteyeres.2020.100875
89. Beykin, G., Goldberg, J.L.: Molecular biomarkers for glaucoma. Curr. Ophthalmol. Rep. **7**, 171–176 (2019). https://doi.org/10.1007/s40135-019-00213-0
90. Nättinen, J., Aapola, U., Nukareddy, P., Uusitalo, H.: Clinical tear fluid proteomics-a novel tool in glaucoma research. Int. J. Mol. Sci. **23**, 8136 (2022). https://doi.org/10.3390/ijms23158136
91. Fernández-Vega Cueto, A., Álvarez, L., García, M., Álvarez-Barrios, A., Artime, E., Fernández-Vega Cueto, L., Coca-Prados, M., González-Iglesias, H.: Candidate glaucoma biomarkers: from proteins to metabolites, and the pitfalls to clinical applications. Biology. **10**, 763 (2021). https://doi.org/10.3390/biology10080763
92. Willcox, M.D.P., Argüeso, P., Georgiev, G.A., Holopainen, J.M., Laurie, G.W., Millar, T.J., Papas, E.B., Rolland, J.P., Schmidt, T.A., Stahl, U., Suarez, T., Subbaraman, L.N., Uçakhan, O.Ö., Jones, L.: TFOS DEWS II Tear Film Report. Ocul. Surf. **15**, 366–403 (2017). https://doi.org/10.1016/j.jtos.2017.03.006
93. Drucker, E., Krapfenbauer, K.: Pitfalls and limitations in translation from biomarker discovery to clinical utility in predictive and personalised medicine. EPMA J. **4**, 7 (2013). https://doi.org/10.1186/1878-5085-4-7
94. Ferris, F.L., Wilkinson, C.P., Bird, A., Chakravarthy, U., Chew, E., Csaky, K., Sadda, S.R.: Clinical classification of age-related macular degeneration. Ophthalmology **120**, 844–851 (2013). https://doi.org/10.1016/j.ophtha.2012.10.036
95. Deng, Y., Qiao, L., Du, M., Qu, C., Wan, L., Li, J., Huang, L.: Age-related macular degeneration: Epidemiology, genetics, pathophysiology, diagnosis, and targeted therapy. Genes Dis. **9**, 62–79 (2022). https://doi.org/10.1016/j.gendis.2021.02.009
96. Blasiak, J., Sobczuk, P., Pawlowska, E., Kaarniranta, K.: Interplay between aging and other factors of the pathogenesis of age-related macular degeneration. Ageing Res. Rev. **81**, 101735 (2022). https://doi.org/10.1016/j.arr.2022.101735
97. Heo, T.-Y., Kim, K.M., Min, H.K., Gu, S.M., Kim, J.H., Yun, J., Min, J.K.: Development of a deep-learning-based artificial intelligence tool for differential diagnosis between dry and neovascular age-related macular degeneration. Diagnostics. **10**, 261 (2020). https://doi.org/10.3390/diagnostics10050261
98. Burlina, P.M., Joshi, N., Pekala, M., Pacheco, K.D., Freund, D.E., Bressler, N.M.: Automated grading of age-related macular degeneration from color fundus images using deep convolutional neural networks. JAMA Ophthalmol. **135**, 1170–1176 (2017). https://doi.org/10.1001/jamaophthalmol.2017.3782
99. Ting, D.S.W., Cheung, C.Y.-L., Lim, G., Tan, G.S.W., Quang, N.D., Gan, A., Hamzah, H., Garcia-Franco, R., San Yeo, I.Y., Lee, S.Y., Wong, E.Y.M., Sabanayagam, C., Baskaran, M., Ibrahim, F., Tan, N.C., Finkelstein, E.A., Lamoureux, E.L., Wong, I.Y., Bressler, N.M., Sivaprasad, S., Varma, R., Jonas, J.B., He, M.G., Cheng, C.-Y., Cheung, G.C.M., Aung, T., Hsu, W., Lee, M.L., Wong, T.Y.: Development and validation of a deep learning system for diabetic retinopathy and related eye diseases using retinal images from multiethnic populations with diabetes. JAMA. **318**, 2211–2223 (2017). https://doi.org/10.1001/jama.2017.18152
100. Dong, L., Yang, Q., Zhang, R.H., Wei, W.B.: Artificial intelligence for the detection of age-related macular degeneration in color fundus photographs: A systematic review and meta-analysis. eClinical Medicine. **35**, (2021). https://doi.org/10.1016/j.eclinm.2021.100875

101. Hwang, D.-K., Hsu, C.-C., Chang, K.-J., Chao, D., Sun, C.-H., Jheng, Y.-C., Yarmishyn, A.A., Wu, J.-C., Tsai, C.-Y., Wang, M.-L., Peng, C.-H., Chien, K.-H., Kao, C.-L., Lin, T.-C., Woung, L.-C., Chen, S.-J., Chiou, S.-H.: Artificial intelligence-based decision-making for age-related macular degeneration. Theranostics. **9**, 232–245 (2019). https://doi.org/10.7150/thno.28447
102. He, T., Zhou, Q., Zou, Y.: Automatic detection of age-related macular degeneration based on deep learning and local outlier factor algorithm. Diagn. Basel Switz. **12**, 532 (2022). https://doi.org/10.3390/diagnostics12020532
103. Yim, J., Chopra, R., Spitz, T., Winkens, J., Obika, A., Kelly, C., Askham, H., Lukic, M., Huemer, J., Fasler, K., Moraes, G., Meyer, C., Wilson, M., Dixon, J., Hughes, C., Rees, G., Khaw, P.T., Karthikesalingam, A., King, D., Hassabis, D., Suleyman, M., Back, T., Ledsam, J.R., Keane, P.A., De Fauw, J.: Predicting conversion to wet age-related macular degeneration using deep learning. Nat. Med. **26**, 892–899 (2020). https://doi.org/10.1038/s41591-020-0867-7
104. Yan, Q., Weeks, D.E., Xin, H., Swaroop, A., Chew, E.Y., Huang, H., Ding, Y., Chen, W.: Deep-learning-based prediction of late age-related macular degeneration progression. Nat. Mach. Intell. **2**, 141–150 (2020). https://doi.org/10.1038/s42256-020-0154-9
105. Kaiser, S.M., Arepalli, S., Ehlers, J.P.: Current and future anti-vegf agents for neovascular age-related macular degeneration. J. Exp. Pharmacol. **13**, 905–912 (2021). https://doi.org/10.2147/JEP.S259298
106. Loewenstein, A., Laganovska, G., Bressler, N.M., Vanags, J., Alster, Y., De Juan, E., Stewart, J.M., Kardatzke, D., Singh, N., Erickson, S.: Phase 1 clinical study of the port delivery system with ranibizumab for continuous treatment of neovascular age-related macular degeneration. Invest. Ophthalmol. Vis. Sci. **61**, 4201 (2020)
107. Chen, E.R., Kaiser, P.K.: Therapeutic potential of the ranibizumab port delivery system in the treatment of AMD: Evidence to date. Clin. Ophthalmol. Auckl. NZ. **14**, 1349–1355 (2020). https://doi.org/10.2147/OPTH.S194234
108. Cabral de Guimaraes, T.A., Daich Varela, M., Georgiou, M., Michaelides, M.: Treatments for dry age-related macular degeneration: therapeutic avenues, clinical trials and future directions. Br. J. Ophthalmol. **106**, 297–304 (2022). https://doi.org/10.1136/bjophthalmol-2020-318452
109. Guariguata, L., Whiting, D.R., Hambleton, I., Beagley, J., Linnenkamp, U., Shaw, J.E.: Global estimates of diabetes prevalence for 2013 and projections for 2035. Diabetes Res. Clin. Pract. **103**, 137–149 (2014). https://doi.org/10.1016/j.diabres.2013.11.002
110. American Diabetes Association: Diagnosis and classification of diabetes mellitus. Diabetes Care **33**, S62–S69 (2010). https://doi.org/10.2337/dc10-S062
111. Liu, R., Li, L., Shao, C., Cai, H., Wang, Z.: The impact of diabetes on vascular disease: progress from the perspective of epidemics and treatments. J. Diabetes Res. **2022**, 1531289 (2022). https://doi.org/10.1155/2022/1531289
112. Duh, E.J., Sun, J.K., Stitt, A.W.: Diabetic retinopathy: current understanding, mechanisms, and treatment strategies. JCI Insight. 2, e93751. https://doi.org/10.1172/jci.insight.93751
113. Wang, W., Lo, A.C.Y.: Diabetic retinopathy: pathophysiology and treatments. Int. J. Mol. Sci. **19**, 1816 (2018). https://doi.org/10.3390/ijms19061816
114. Schoenfeld, E.R., Greene, J.M., Wu, S.Y., Leske, M.C.: Patterns of adherence to diabetes vision care guidelines: Baseline findings from the Diabetic Retinopathy Awareness Program. Ophthalmology **108**, 563–571 (2001). https://doi.org/10.1016/S0161-6420(00)00600-X
115. Wong, T.Y., Sun, J., Kawasaki, R., Ruamviboonsuk, P., Gupta, N., Lansingh, V.C., Maia, M., Mathenge, W., Moreker, S., Muqit, M.M.K., Resnikoff, S., Verdaguer, J., Zhao, P., Ferris, F., Aiello, L.P., Taylor, H.R.: Guidelines on diabetic eye care: the international council of ophthalmology recommendations for screening, follow-up, referral, and treatment based on resource settings. Ophthalmology **125**, 1608–1622 (2018). https://doi.org/10.1016/j.ophtha.2018.04.007
116. Lanzetta, P., Sarao, V., Scanlon, P.H., Barratt, J., Porta, M., Bandello, F., Loewenstein, A., Eldem, B., Hunyor, A., Joussen, A., Koh, A., Korobelnik, J.-F., Lanzetta, P., Loewenstein, A., Lövestam-Adrian, M., Navarro, R., Okada, A.A., Pearce, I., Rodríguez, F.J., Staurenghi, G.,

Wolf, S., Wong, D.T.: The Vision Academy: Fundamental principles of an effective diabetic retinopathy screening program. Acta Diabetol. **57**, 785–798 (2020). https://doi.org/10.1007/s00592-020-01506-8

117. Scanlon, P.H.: The English national screening programme for diabetic retinopathy 2003–2016. Acta Diabetol. **54**, 515–525 (2017). https://doi.org/10.1007/s00592-017-0974-1
118. van der Heijden, A.A., Abramoff, M.D., Verbraak, F., van Hecke, M.V., Liem, A., Nijpels, G.: Validation of automated screening for referable diabetic retinopathy with the IDx-DR device in the Hoorn Diabetes Care System. Acta Ophthalmol. (Copenh.) **96**, 63–68 (2018). https://doi.org/10.1111/aos.13613
119. Abràmoff, M.D., Lavin, P.T., Birch, M., Shah, N., Folk, J.C.: Pivotal trial of an autonomous AI-based diagnostic system for detection of diabetic retinopathy in primary care offices. Npj Digit. Med. **1**, 1–8 (2018). https://doi.org/10.1038/s41746-018-0040-6
120. Rajalakshmi, R., Arulmalar, S., Usha, M., Prathiba, V., Kareemuddin, K.S., Anjana, R.M., Mohan, V.: Validation of smartphone based retinal photography for diabetic retinopathy screening. PLoS ONE **10**, e0138285 (2015). https://doi.org/10.1371/journal.pone.0138285
121. Rajalakshmi, R., Subashini, R., Anjana, R.M., Mohan, V.: Automated diabetic retinopathy detection in smartphone-based fundus photography using artificial intelligence. Eye **32**, 1138–1144 (2018). https://doi.org/10.1038/s41433-018-0064-9
122. Ipp, E., Liljenquist, D., Bode, B., Shah, V.N., Silverstein, S., Regillo, C.D., Lim, J.I., Sadda, S., Domalpally, A., Gray, G., Bhaskaranand, M., Ramachandra, C., Solanki, K.: EyeArt study group: pivotal evaluation of an artificial intelligence system for autonomous detection of referrable and vision-threatening diabetic retinopathy. JAMA Netw. Open **4**, e2134254 (2021). https://doi.org/10.1001/jamanetworkopen.2021.34254
123. Lim, J.I., Regillo, C.D., Sadda, S.R., Ipp, E., Bhaskaranand, M., Ramachandra, C., Solanki, K.: Artificial intelligence detection of diabetic retinopathy: subgroup comparison of the eyeart system with ophthalmologists' dilated examinations. Ophthalmol. Sci. **3**, (2023). https://doi.org/10.1016/j.xops.2022.100228
124. Gulshan, V., Peng, L., Coram, M., Stumpe, M.C., Wu, D., Narayanaswamy, A., Venugopalan, S., Widner, K., Madams, T., Cuadros, J., Kim, R., Raman, R., Nelson, P.C., Mega, J.L., Webster, D.R.: Development and validation of a deep learning algorithm for detection of diabetic retinopathy in retinal fundus photographs. JAMA **316**, 2402–2410 (2016). https://doi.org/10.1001/jama.2016.17216
125. Ruamviboonsuk, P., Tiwari, R., Sayres, R., Nganthavee, V., Hemarat, K., Kongprayoon, A., Raman, R., Levinstein, B., Liu, Y., Schaekermann, M., Lee, R., Virmani, S., Widner, K., Chambers, J., Hersch, F., Peng, L., Webster, D.R.: Real-time diabetic retinopathy screening by deep learning in a multisite national screening programme: a prospective interventional cohort study. Lancet Digit. Health. **4**, e235–e244 (2022). https://doi.org/10.1016/S2589-7500(22)00017-6
126. Fong, D.S., Aiello, L., Gardner, T.W., King, G.L., Blankenship, G., Cavallerano, J.D., Ferris, F.L., III., Klein, R.: For the American diabetes association: retinopathy in diabetes. Diabetes Care **27**, s84–s87 (2004). https://doi.org/10.2337/diacare.27.2007.S84
127. Ansari, P., Tabasumma, N., Snigdha, N.N., Siam, N.H., Panduru, R.V.N.R.S., Azam, S., Hannan, J.M.A., Abdel-Wahab, Y.H.A.: Diabetic retinopathy: An overview on mechanisms, pathophysiology and pharmacotherapy. Diabetology. **3**, 159–175 (2022). https://doi.org/10.3390/diabetology3010011
128. Everett, L.A., Paulus, Y.M.: Laser therapy in the treatment of diabetic retinopathy and diabetic macular edema. Curr. Diab. Rep. **21**, 35 (2021). https://doi.org/10.1007/s11892-021-01403-6
129. Udaondo, P., Parravano, M., Vujosevic, S., Zur, D., Chakravarthy, U.: Update on current and future management for diabetic maculopathy. Ophthalmol. Ther. **11**, 489–502 (2022). https://doi.org/10.1007/s40123-022-00460-8
130. Allen, D., Vasavada, A.: Cataract and surgery for cataract. BMJ **333**, 128–132 (2006)
131. Foster, A.: Cataract and "Vision 2020-the right to sight" initiative. Br. J. Ophthalmol. **85**, 635–637 (2001). https://doi.org/10.1136/bjo.85.6.635

132. Abdulhussein, D., Abdul Hussein, M.: WHO Vision 2020: Have We Done It? Ophthalmic Epidemiol. 0, 1–9 (2022). https://doi.org/10.1080/09286586.2022.2127784
133. Chylack, L.T., Wolfe, J.K., Singer, D.M., Leske, M.C., Bullimore, M.A., Bailey, I.L., Friend, J., McCarthy, D., Wu, S.Y.: The lens opacities classification system III. The longitudinal study of cataract study group. Arch. Ophthalmol. Chic. Ill 1960. **111**, 831–836 (1993). https://doi.org/10.1001/archopht.1993.01090060119035
134. Gao, X., Lin, S., Wong, T.Y.: Automatic feature learning to grade nuclear cataracts based on deep learning. IEEE Trans. Biomed. Eng. **62**, 2693–2701 (2015). https://doi.org/10.1109/TBME.2015.2444389
135. Wu, X., Huang, Y., Liu, Z., Lai, W., Long, E., Zhang, K., Jiang, J., Lin, D., Chen, K., Yu, T., Wu, D., Li, C., Chen, Y., Zou, M., Chen, C., Zhu, Y., Guo, C., Zhang, X., Wang, R., Yang, Y., Xiang, Y., Chen, L., Liu, C., Xiong, J., Ge, Z., Wang, D., Xu, G., Du, S., Xiao, C., Wu, J., Zhu, K., Nie, D., Xu, F., Lv, J., Chen, W., Liu, Y., Lin, H.: Universal artificial intelligence platform for collaborative management of cataracts. Br. J. Ophthalmol. **103**, 1553–1560 (2019). https://doi.org/10.1136/bjophthalmol-2019-314729
136. Zhang, H., Niu, K., Xiong, Y., Yang, W., He, Z., Song, H.: Automatic cataract grading methods based on deep learning. Comput. Methods Programs Biomed. **182**, 104978 (2019). https://doi.org/10.1016/j.cmpb.2019.07.006
137. Wu, X., Xu, D., Ma, T., Li, Z.H., Ye, Z., Wang, F., Gao, X.Y., Wang, B., Chen, Y.Z., Wang, Z.H., Chen, J.L., Hu, Y.T., Ge, Z.Y., Wang, D.J., Zeng, Q.: Artificial intelligence model for anti interference cataract automatic diagnosis: a diagnostic accuracy study. Front. Cell Dev. Biol. **10**, 906042 (2022). https://doi.org/10.3389/fcell.2022.906042
138. Tognetto, D., Giglio, R., Vinciguerra, A.L., Milan, S., Rejdak, R., Rejdak, M., Zaluska-Ogryzek, K., Zweifel, S., Toro, M.D.: Artificial intelligence applications and cataract management: A systematic review. Surv. Ophthalmol. **67**, 817–829 (2022). https://doi.org/10.1016/j.survophthal.2021.09.004
139. Lahham, S., Shniter, I., Thompson, M., Le, D., Chadha, T., Mailhot, T., Kang, T.L., Chiem, A., Tseeng, S., Fox, J.C.: Point-of-care ultrasonography in the diagnosis of retinal detachment, vitreous hemorrhage, and vitreous detachment in the emergency department. JAMA Netw. Open **2**, e192162 (2019). https://doi.org/10.1001/jamanetworkopen.2019.2162
140. Roque, P.J., Hatch, N., Barr, L., Wu, T.S.: Bedside ocular ultrasound. Crit. Care Clin. **30**(227–241), v (2014). https://doi.org/10.1016/j.ccc.2013.10.007
141. Lahham, S., Ali, Q., Palileo, B.M., Lee, C., Fox, J.C.: Role of point of care ultrasound in the diagnosis of retinal detachment in the emergency department. Open Access Emerg. Med. OAEM. **11**, 265–270 (2019). https://doi.org/10.2147/OAEM.S219333
142. Propst, S.L., Kirschner, J.M., Strachan, C.C., Roumpf, S.K., Menard, L.M., Sarmiento, E.J., Hunter, B.R.: Ocular point-of-care ultrasonography to diagnose posterior chamber abnormalities: a systematic review and meta-analysis. JAMA Netw. Open **3**, e1921460 (2020). https://doi.org/10.1001/jamanetworkopen.2019.21460
143. Dornhofer, K., Alkhattabi, M., Lahham, S.: Point-of-care ultrasound detection of cataract in a patient with vision loss: a case report. Clin. Pract. Cases Emerg. Med. **4**, 355–357 (2020). https://doi.org/10.5811/cpcem.2020.4.46597
144. Shokoohi, H., LeSaux, M.A., Roohani, Y.H., Liteplo, A., Huang, C., Blaivas, M.: Enhanced point-of-care ultrasound applications by integrating automated feature-learning systems using deep learning. J. Ultrasound Med. Off. J. Am. Inst. Ultrasound Med. **38**, 1887–1897 (2019). https://doi.org/10.1002/jum.14860
145. Vilela, M.A., Valença, F.M., Barreto, P.K., Amaral, C.E., Pellanda, L.C.: Agreement between retinal images obtained via smartphones and images obtained with retinal cameras or fundoscopic exams—systematic review and meta-analysis. Clin. Ophthalmol. Auckl. NZ. **12**, 2581–2589 (2018). https://doi.org/10.2147/OPTH.S182022
146. Bron, A.J., de Paiva, C.S., Chauhan, S.K., Bonini, S., Gabison, E.E., Jain, S., Knop, E., Markoulli, M., Ogawa, Y., Perez, V., Uchino, Y., Yokoi, N., Zoukhri, D., Sullivan, D.A.: TFOS DEWS II pathophysiology report. Ocul. Surf. **15**, 438–510 (2017). https://doi.org/10.1016/j.jtos.2017.05.011

147. Craig, J.P., Nichols, K.K., Akpek, E.K., Caffery, B., Dua, H.S., Joo, C.-K., Liu, Z., Nelson, J.D., Nichols, J.J., Tsubota, K., Stapleton, F.: TFOS DEWS II definition and classification report. Ocul. Surf. **15**, 276–283 (2017). https://doi.org/10.1016/j.jtos.2017.05.008
148. Schechter, B., Mah, F.: Optimization of the ocular surface through treatment of ocular surface disease before ophthalmic surgery: a narrative review. Ophthalmol. Ther. **11**, 1001–1015 (2022). https://doi.org/10.1007/s40123-022-00505-y
149. Jones, L., Downie, L.E., Korb, D., Benitez-del-Castillo, J.M., Dana, R., Deng, S.X., Dong, P.N., Geerling, G., Hida, R.Y., Liu, Y., Seo, K.Y., Tauber, J., Wakamatsu, T.H., Xu, J., Wolffsohn, J.S., Craig, J.P.: TFOS DEWS II management and therapy report. Ocul. Surf. **15**, 575–628 (2017). https://doi.org/10.1016/j.jtos.2017.05.006
150. Ting, D.S.J., Ho, C.S., Deshmukh, R., Said, D.G., Dua, H.S.: Infectious keratitis: an update on epidemiology, causative microorganisms, risk factors, and antimicrobial resistance. Eye **35**, 1084–1101 (2021). https://doi.org/10.1038/s41433-020-01339-3
151. Flaxman, S.R., Bourne, R.R.A., Resnikoff, S., Ackland, P., Braithwaite, T., Cicinelli, M.V., Das, A., Jonas, J.B., Keeffe, J., Kempen, J.H., Leasher, J., Limburg, H., Naidoo, K., Pesudovs, K., Silvester, A., Stevens, G.A., Tahhan, N., Wong, T.Y., Taylor, H.R.: Vision loss expert group of the global burden of disease study: global causes of blindness and distance vision impairment 1990–2020: a systematic review and meta-analysis. Lancet Glob. Health **5**, e1221–e1234 (2017). https://doi.org/10.1016/S2214-109X(17)30393-5
152. Austin, A., Lietman, T., Rose-Nussbaumer, J.: Update on the management of infectious keratitis. Ophthalmology **124**, 1678–1689 (2017). https://doi.org/10.1016/j.ophtha.2017.05.012
153. Schmidl, D., Schlatter, A., Chua, J., Tan, B., Garhöfer, G., Schmetterer, L.: Novel approaches for imaging-based diagnosis of ocular surface disease. Diagnostics. **10**, 589 (2020). https://doi.org/10.3390/diagnostics10080589
154. Han, S.B., Liu, Y.-C., Noriega, K.M., Mehta, J.S.: Applications of anterior segment optical coherence tomography in cornea and ocular surface diseases. J. Ophthalmol. **2016**, 4971572 (2016). https://doi.org/10.1155/2016/4971572
155. Di Cello, L., Pellegrini, M., Vagge, A., Borselli, M., Ferro Desideri, L., Scorcia, V., Traverso, C.E., Giannaccare, G.: Advances in the noninvasive diagnosis of dry eye disease. Appl. Sci. **11**, 10384 (2021). https://doi.org/10.3390/app112110384
156. Fineide, F., Arita, R., Utheim, T.P.: The role of meibography in ocular surface diagnostics: A review. Ocul. Surf. **19**, 133–144 (2021). https://doi.org/10.1016/j.jtos.2020.05.004
157. Craig, J.P., Singh, I., Tomlinson, A., Morgan, P.B., Efron, N.: The role of tear physiology in ocular surface temperature. Eye Lond. Engl. **14**(Pt 4), 635–641 (2000). https://doi.org/10.1038/eye.2000.156
158. Chase, C., Elsawy, A., Eleiwa, T., Ozcan, E., Tolba, M., Abou Shousha, M.: Comparison of autonomous AS-OCT deep learning algorithm and clinical dry eye tests in diagnosis of dry eye disease. Clin. Ophthalmol. Auckl. NZ. **15**, 4281–4289 (2021). https://doi.org/10.2147/OPTH.S321764
159. da Cruz, L.B., Souza, J.C., de Sousa, J.A., Santos, A.M., de Paiva, A.C., de Almeida, J.D.S., Silva, A.C., Junior, G.B., Gattass, M.: Interferometer eye image classification for dry eye categorization using phylogenetic diversity indexes for texture analysis. Comput. Methods Programs Biomed. **188**, 105269 (2020). https://doi.org/10.1016/j.cmpb.2019.105269
160. Wang, J., Yeh, T.N., Chakraborty, R., Yu, S.X., Lin, M.C.: A Deep Learning Approach for Meibomian Gland Atrophy Evaluation in Meibography Images. Transl. Vis. Sci. Technol. **8**, 37 (2019). https://doi.org/10.1167/tvst.8.6.37
161. Wang, J., Li, S., Yeh, T.N., Chakraborty, R., Graham, A.D., Yu, S.X., Lin, M.C.: Quantifying meibomian gland morphology using artificial intelligence. Optom. Vis. Sci. Off. Publ. Am. Acad. Optom. **98**, 1094–1103 (2021). https://doi.org/10.1097/OPX.0000000000001767
162. Yu, Y., Zhou, Y., Tian, M., Zhou, Y., Tan, Y., Wu, L., Zheng, H., Yang, Y.: Automatic identification of meibomian gland dysfunction with meibography images using deep learning. Int. Ophthalmol. **42**, 3275–3284 (2022). https://doi.org/10.1007/s10792-022-02262-0

163. Saha, R.K., Chowdhury, A.M.M., Na, K.-S., Hwang, G.D., Eom, Y., Kim, J., Jeon, H.-G., Hwang, H.S., Chung, E.: Automated quantification of meibomian gland dropout in infrared meibography using deep learning. Ocul. Surf. **26**, 283–294 (2022). https://doi.org/10.1016/j.jtos.2022.06.006
164. Zheng, Q., Wang, L., Wen, H., Ren, Y., Huang, S., Bai, F., Li, N., Craig, J.P., Tong, L., Chen, W.: Impact of incomplete blinking analyzed using a deep learning model with the keratograph 5M in dry eye disease. Transl. Vis. Sci. Technol. **11**, 38 (2022). https://doi.org/10.1167/tvst.11.3.38
165. Li, Z., Jiang, J., Chen, K., Chen, Q., Zheng, Q., Liu, X., Weng, H., Wu, S., Chen, W.: Preventing corneal blindness caused by keratitis using artificial intelligence. Nat. Commun. **12**, 3738 (2021). https://doi.org/10.1038/s41467-021-24116-6
166. Hung, N., Shih, A.K.-Y., Lin, C., Kuo, M.-T., Hwang, Y.-S., Wu, W.-C., Kuo, C.-F., Kang, E.Y.-C., Hsiao, C.-H.: Using slit-lamp images for deep learning-based identification of bacterial and fungal keratitis: model development and validation with different convolutional neural networks. Diagnostics. **11**, 1246 (2021). https://doi.org/10.3390/diagnostics11071246
167. Ghosh, A.K., Thammasudjarit, R., Jongkhajornpong, P., Attia, J., Thakkinstian, A.: Deep learning for discrimination between fungal keratitis and bacterial keratitis: deepkeratitis. Cornea **41**, 616–622 (2022). https://doi.org/10.1097/ICO.0000000000002830
168. Tamhane, M., Cabrera-Ghayouri, S., Abelian, G., Viswanath, V.: Review of biomarkers in ocular matrices: challenges and opportunities. Pharm. Res. **36**, 40 (2019). https://doi.org/10.1007/s11095-019-2569-8
169. Messmer, E.M., von Lindenfels, V., Garbe, A., Kampik, A.: Matrix metalloproteinase 9 testing in dry eye disease using a commercially available point-of-care immunoassay. Ophthalmology **123**, 2300–2308 (2016). https://doi.org/10.1016/j.ophtha.2016.07.028
170. Chotikavanich, S., de Paiva, C.S., Li, D.Q., Chen, J.J., Bian, F., Farley, W.J., Pflugfelder, S.C.: Production and activity of matrix metalloproteinase-9 on the ocular surface increase in dysfunctional tear syndrome. Invest. Ophthalmol. Vis. Sci. **50**, 3203–3209 (2009). https://doi.org/10.1167/iovs.08-2476
171. Lanza, N.L., Valenzuela, F., Perez, V.L., Galor, A.: The matrix metalloproteinase 9 point-of-care test in dry eye. Ocul. Surf. **14**, 189–195 (2016). https://doi.org/10.1016/j.jtos.2015.10.004
172. Hagan, S., Martin, E., Enríquez-de-Salamanca, A.: Tear fluid biomarkers in ocular and systemic disease: potential use for predictive, preventive and personalised medicine. EPMA J. **7**, 15 (2016). https://doi.org/10.1186/s13167-016-0065-3
173. Enríquez-de-Salamanca, A., Calonge, M.: Cytokines and chemokines in immune-based ocular surface inflammation. Expert Rev. Clin. Immunol. **4**, 457–467 (2008). https://doi.org/10.1586/1744666X.4.4.457
174. Zhang, S., Echegoyen, J.: Point of care diagnosis of dry eye disease with a sensitive immunoassay for dual biomarker detection. Biochem. Biophys. Rep. **32**, 101396 (2022). https://doi.org/10.1016/j.bbrep.2022.101396
175. Versura, P., Campos, E.C.: TearLab® Osmolarity System for diagnosing dry eye. Expert Rev. Mol. Diagn. **13**, 119–129 (2013). https://doi.org/10.1586/erm.12.142
176. Suzuki, M., Massingale, M.L., Ye, F., Godbold, J., Elfassy, T., Vallabhajosyula, M., Asbell, P.A.: Tear osmolarity as a biomarker for dry eye disease severity. Invest. Ophthalmol. Vis. Sci. **51**, 4557–4561 (2010). https://doi.org/10.1167/iovs.09-4596
177. Sullivan, B.D., Whitmer, D., Nichols, K.K., Tomlinson, A., Foulks, G.N., Geerling, G., Pepose, J.S., Kosheleff, V., Porreco, A., Lemp, M.A.: An objective approach to dry eye disease severity. Invest. Ophthalmol. Vis. Sci. **51**, 6125–6130 (2010). https://doi.org/10.1167/iovs.10-5390
178. Wolffsohn, J.S., Arita, R., Chalmers, R., Djalilian, A., Dogru, M., Dumbleton, K., Gupta, P.K., Karpecki, P., Lazreg, S., Pult, H., Sullivan, B.D., Tomlinson, A., Tong, L., Villani, E., Yoon, K.C., Jones, L., Craig, J.P.: TFOS DEWS II Diagnostic methodology report. Ocul. Surf. **15**, 539–574 (2017). https://doi.org/10.1016/j.jtos.2017.05.001

179. Abreau, K., Callan, C., Kottaiyan, R., Zhang, A., Yoon, G., Aquavella, J.V., Zavislan, J., Hindman, H.B.: Temperatures of the ocular surface, lid, and periorbital regions of Sjögren's, evaporative, and aqueous-deficient dry eyes relative to normals. Ocul. Surf. **14**, 64–73 (2016). https://doi.org/10.1016/j.jtos.2015.09.001
180. Li, W., Graham, A.D., Selvin, S., Lin, M.C.: Ocular Surface Cooling Corresponds to Tear Film Thinning and Breakup. Optom. Vis. Sci. **92**, e248 (2015). https://doi.org/10.1097/OPX.0000000000000672
181. Wagner, S.K., Fu, D.J., Faes, L., Liu, X., Huemer, J., Khalid, H., Ferraz, D., Korot, E., Kelly, C., Balaskas, K., Denniston, A.K., Keane, P.A.: Insights into systemic disease through retinal imaging-based oculomics. Transl. Vis. Sci. Technol. **9**, 6 (2020). https://doi.org/10.1167/tvst.9.2.6
182. Poplin, R., Varadarajan, A.V., Blumer, K., Liu, Y., McConnell, M.V., Corrado, G.S., Peng, L., Webster, D.R.: Prediction of cardiovascular risk factors from retinal fundus photographs via deep learning. Nat. Biomed. Eng. **2**, 158–164 (2018). https://doi.org/10.1038/s41551-018-0195-0
183. Wu, J.-H., Liu, T.Y.A.: Application of deep learning to retinal-image-based oculomics for evaluation of systemic health: a review. J. Clin. Med. **12**, 152 (2023). https://doi.org/10.3390/jcm12010152
184. Cheung, C.Y., Mok, V., Foster, P.J., Trucco, E., Chen, C., Wong, T.Y.: Retinal imaging in Alzheimer's disease. J. Neurol. Neurosurg. Psychiatry **92**, 983–994 (2021). https://doi.org/10.1136/jnnp-2020-325347

Biofluorometric Gas Imaging for Wearable Human-Borne VOCs Monitoring

Kenta Iitani, Koji Toma, Takahiro Arakawa, and Kohji Mitsubayashi

1 Introduction

Volatile organic compounds (VOCs) in blood, released as exhaled and transdermal gases, have been reported to reflect disease and metabolic status [1, 2]. In contrast to commonly used invasive blood tests, non-invasive human-borne VOC measurement has advantages such as the possibility of frequent testing and the prevention of puncture risk. Transdermal gases are considered to be the best samples for biochemical monitoring of our body because of their steady emission [3]. However, transdermal gases are more difficult to sample than exhaled breath VOCs, and their extremely low concentrations pose a measurement challenge.

In transdermal gas measurement, the method of analyzing condensed gas samples obtained by sealing the target body part for a certain period of time with a gas chromatography-mass spectrometer is common [4–6]. This method is limited to the measurement of temporal and spatial average values of VOCs emitted from a unit area of the body surface. On the other hand, transdermal gases are emitted from the entire body surface, which has an area of about 1.5–2.0 m^2, and various factors such as epidermal thickness [7], type [8], distribution [8], and activity of sweat glands [9], and distribution of arteries, veins, and capillaries [10] affect the skin

K. Iitani · K. Mitsubayashi (✉)
Department of Biomedical Devices and Instrumentation, Institute of Biomaterials and Bioengineering, Tokyo Medical and Dental University (TMDU), 2-3-10 Kanda-Surugadai, Chiyoda-Ku, Tokyo 101-0062, Japan
e-mail: m.bdi@tmd.ac.jp

K. Toma
Department of Electronic Engineering, Shibaura Institute of Technology, 3-7-5 Toyosu, Koto-Ku, Tokyo 135-8547, Japan

T. Arakawa
Department of Electric and Electronic Engineering, Tokyo University of Technology, 1404-1 Katakura, Hachioji City, Tokyo 192-0982, Japan

K. Mitsubayashi (ed.), *Wearable Biosensing in Medicine and Healthcare*,
https://doi.org/10.1007/978-981-99-8122-9_20

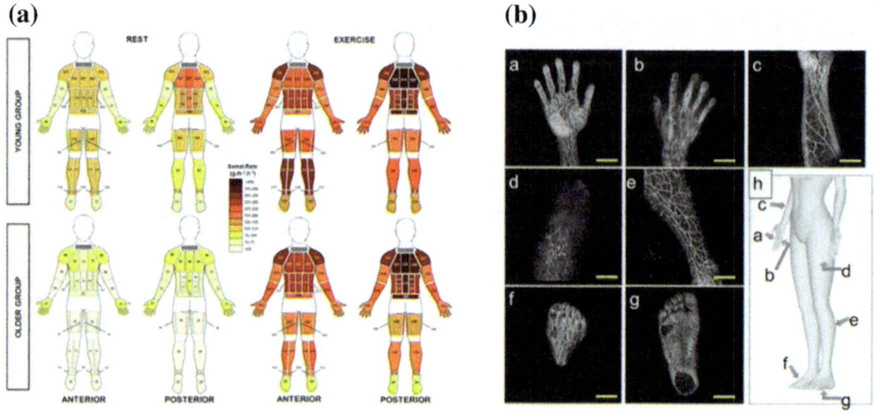

Fig. 1 **a** Activity maps of perspiration on young and older persons [9]. **b** blood capillary images of various body parts [10]. Both images are referenced from Open Access papers

characteristics. Also, those skin characteristics have large regional and individual differences (see Fig. 1). The diversity of skin properties is thought to influence the emission of VOCs in skin gases: indeed, the concentration of several VOCs fluctuates at sites where the stratum corneum (skin barrier) is disrupted by the tape-stripping [11]. Therefore, measurement of the distribution of transdermal VOCs may enable the evaluation of skin diseases and skin conditions that are difficult to diagnose visually. In addition, determining the optimal measurement site is an important task for metabolic monitoring by wearable transdermal VOC measurement devices. It is important to emphasize here that spatiotemporal information would be essential to understand the continuous release of VOCs from the skin.

We have developed an enzyme-based gas-phase biosensor "bio-sniffer" with high selectivity and sensitivity and a gas imaging system based on the technologies of bio-sniffer [12]. This chapter first describes the measurement principle of this gas imaging system, the system configuration, examples of optimization, quantitative characteristics, and gas selectivity. Then, we introduce the experimental results that led to the conclusion that the ear is the optimal measurement site.

2 Gas Imaging Principle

Figure 2a shows the common gas-detection principle of biofluorometry based on NAD-dependent enzymes. Substrates that react with oxidized NAD (NAD^+) are oxidized to produce reduced NAD (NADH). In this reaction scheme, essentially only NADH emits the fluorescence around the wavelength of 460 nm upon irradiating with 340 nm of ultraviolet (UV), as shown in Fig. 2b. Therefore, the progress of the

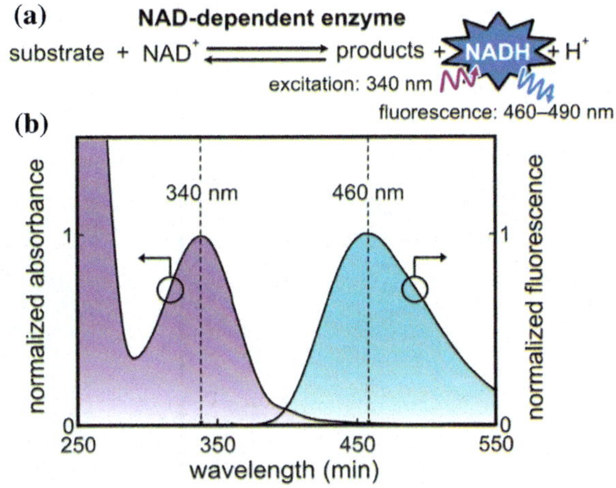

Fig. 2 **a** Common gas-detection principle using biofluorometry based on NAD-dependent enzymes. **b** Absorbance and fluorescence spectra of NADH

enzymatic reaction can be monitored, and its concentration can be quantified by measuring the fluorescence intensity of NADH.

More than 60,000 enzyme reaction schemes involving NADH have been reported so far (BRENDA search, Ligand NAD^+, Substrate in Enzyme-catalyzed reactions, 63,283 results, March/20/2023). Since the same detection system and principle can be applied to different NAD-dependent enzymes in this method, it has the versatility to simply swap the enzymes according to the VOC species to be measured. For some NAD-dependent enzymes, the direction of redox progression can be controlled by the pH of the reaction field and the coenzyme added. Figure 3 shows the pH dependence of the reaction direction of alcohol dehydrogenase (ADH) as an example. Ethanol oxidation in the presence of NAD^+ and acetaldehyde reduction in the presence of NADH are dominant in alkaline and acidic environments, respectively. The two VOCs, ethanol and acetaldehyde, can be selectively measured with a single enzyme, ADH, by measuring the increase or decrease of fluorescence intensity of NADH. Table 1 shows examples of gas sensors based on the biofluorometry developed by our group. As can be seen, there is a wide variety of VOC species that can be measured by this method.

3 Setup of the Gas Imaging System

Imaging the VOC distribution using the biofluorometric principle requires a two-dimensional (2D) sensor surface with a uniform distribution of NAD-dependent enzymes and their coenzymes. When target VOC is applied onto this 2D sensor surface, the fluorescence intensity changes in accordance with the VOC concentration at the application point. Therefore, the fluorescence distribution on the 2D sensor surface reflects the target VOC concentration distribution. Hereafter, the system

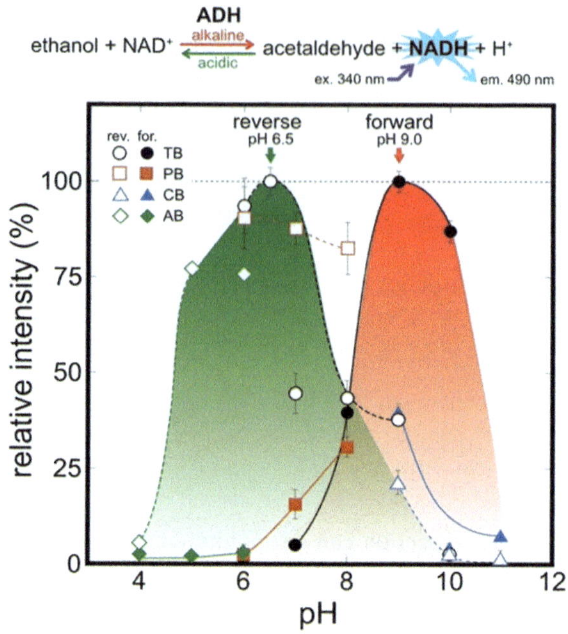

Fig. 3 pH dependency of redox reactions of ADH. Tris–HCl buffer; TB (○ and ●), phosphate buffer; PB (□ and ■), carbonate-bicarbonate buffer; CB (△ and ▲), acetate buffer, AB (◇ and ◆). Image adapted from Elsevier [13]

Table 1 Examples of VOCs that can be measured by the biofluorometry, their measurement significance, and the enzyme used

VOCs		Significance	Enzyme	Refs
Ethanol	Alcohol	Alcohol metabolisms	ADH	[14]
Methanol		Intestinal microflora	Alcohol oxidase and formaldehyde dehydrogenase (FALDH)	[15]
2-propanol		Diabetes, lipid metabolism	Secondary alcohol dehydrogenase (S-ADH)	[16]
Acetaldehyde	Aldehyde	Cancer risk factor	ADH	[17]
Formaldehyde		Cancer risk factor	FALDH	[18]
Acetone	Ketone	Diabetes, lipid metabolism	S-ADH	[19]
Dimethyl sulfide	Sulfides	Bad breath, oral environment	Flavin containing monooxygenase	[20]
Methyl mercaptan			Monoamine oxidase type-A	[21]

configuration will be explained using the example of an imaging system for ethanol gas.

Figure 4 shows the overview of the system. The system consists of a 2D mako which has been inspired by the mako (

真弧

), an instrument used in the archaeological field to copy the outline of excavated objects without damaging them [22], an ADH-immobilized mesh, bandpass filters (BPFs, 340 nm and 490 nm), a UV-light emitting diodes (LEDs) ring light, and a consumer mirrorless digital camera. In this system, the ADH-immobilized mesh pasted on the 2D mako was placed in front of the optics for fluorescence observation. Transdermal gas was introduced into the ADH-immobilized mesh by pushing on the targeted body part from the side where the 2D mako was attached, as shown in Fig. 4. Excitation light irradiation to the ADH-immobilized mesh was provided from the UV-LED ring light attached to the camera lens, and the fluorescence distribution generated on the ADH-immobilized mesh was captured by the camera. Note that the areal uniformity of the excitation light intensity was 95.8% at a distance of 60 mm from the excitation light, and thus it did not affect the fluorescence intensity distribution. In the lower right of Fig. 4, the fluorescence observation system is shown from the front. The perforated excitation BPF and the circular BPF were placed in the same plane. This system allows imaging of the spatiotemporal distribution of ethanol concentration in real time.

The 2D mako used in the system deformed the surface shape of the ADH-immobilized mesh to match the complex body surface shape for equalizing the

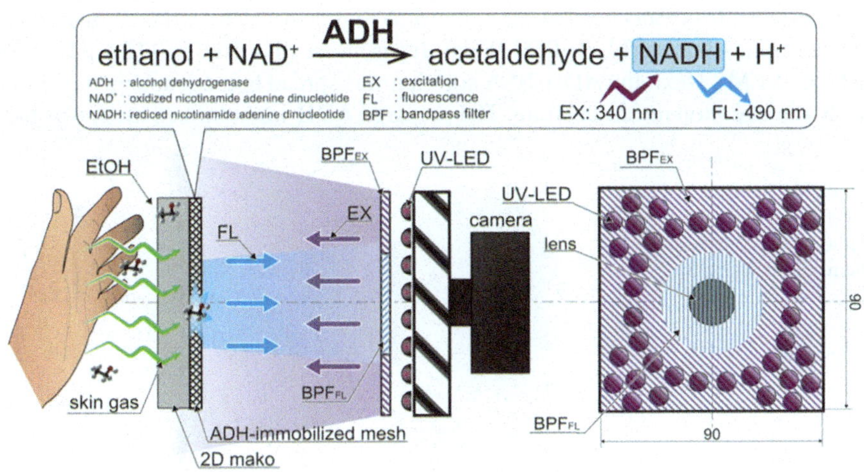

Fig. 4 Overview of the gas imaging system. Image adapted from American Chemical Society [23]

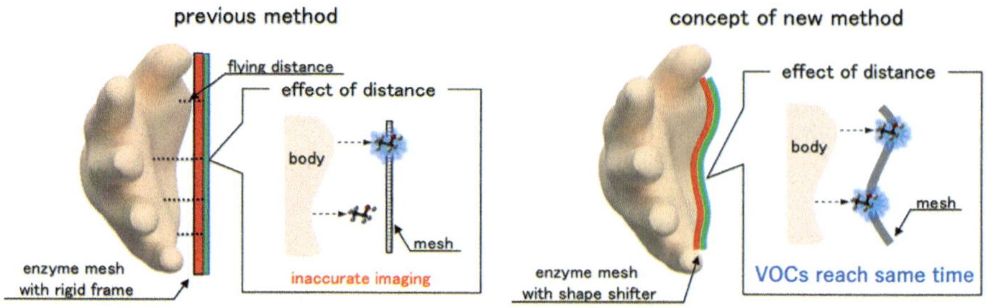

Fig. 5 Conceptual scheme of the effect of distance between the body surface and the ADH-immobilized mesh. Image adapted from American Chemical Society [23]

distance between the body surface and the ADH-immobilized mesh. Figure 5 shows an illustration of the effect on spatiotemporal imaging with and without mesh fitting by using the 2D mako. The left and right sides show the case with rigid planar sheet-like ADH-immobilized mesh and mesh fitting, respectively. Without surface shape fitting, the unevenness of the body surface causes a difference in the reach time of VOCs simultaneously emitted from the body surface to the mesh. Therefore, correct spatiotemporal imaging will be impossible. In contrast, the use of mesh fitting allows the travel distance of VOCs to be unified.

Figure 6 shows the working mechanisms of the 2D mako. The 2D mako had a structure in which holes were made in an acrylic plate and metal tubes (I.D. 1.8 mm, O.D. 2.0 mm, length 10 mm) were densely arranged by passing through the holes. Since the metal tubes can slide vertically with respect to the acrylic plate using the through hole as a guide, the curved surface shape of the object can be fitted on the ADH-immobilized mesh by pushing on the object from the opposite side of the ADH-immobilized mesh fixed.

The ADH-immobilized mesh was prepared with the following process. First, ADH (60 units/cm^2, from *Saccharomyces cerevisiae*, Sigma-Aldrich) and 0.25 mg/cm^2 of bovine serum albumin (FUJIFILM Wako, Japan) were dissolved in 112.5 μL of PB (pH 8.0, 0.1 M). Next, the ADH-BSA solution was spread homogenously on 15 mm by 15 mm of cotton mesh substrate. The prepared ADH-immobilized mesh was then

Fig. 6 Mechanisms of 2D Mako in action. Image adapted from American Chemical Society [23]

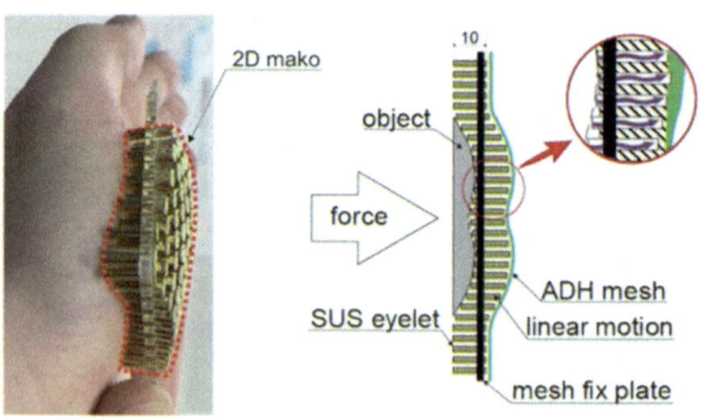

soaked with NAD^+ solution immediately before the gas imaging experiment (Ohki Healthcare Holdings, Japan), and then stored at 4 °C for 60 min. After that, 2.5% glutaraldehyde solution was dropped onto the ADH-absorbed mesh. Crosslinking was performed in the 4 °C.

4 Imaging of Standard Ethanol Gas

Imaging of standard ethanol gas was performed with the ADH-immobilized mesh that was wetted by NAD^+ solution immediately before the gas imaging experiment. Standard ethanol gas was applied from the 2D mako side of the ADH-immobilized mesh. The gas application system was constructed as shown in Fig. 7. The standard gas generator was purchased from Gastec, Japan, which allows the generation of known concentrations of ethanol gas based on the diffusion tube or permeation tube method. The dry and clean dilution air for the standard gas generator was prepared by filtering compressed room air. The standard gas generator controls the concentration of ethanol gas by a flow rate of dilution air and the temperature of the water bath. Basically, the temperature of the thermostatic bath sets a concentration range of standard gas, and the gas concentration is determined by the dilution gas flow rate. Therefore, to maintain a constant gas flow rate introduced into the gas imaging system, it is necessary to control the flow rate of the gas taken from the standard gas generator. In the experiment, the flow rate of gas application was fixed at 100 mL/min. Four finger valves were used to select the gas to be applied onto the ADH-immobilized mesh from ethanol gas or dry clean air.

The timeline of the gas application was as follows. Dry clean air was applied from –100 s to 20 s after the camera recording. Then, the application gas was changed to

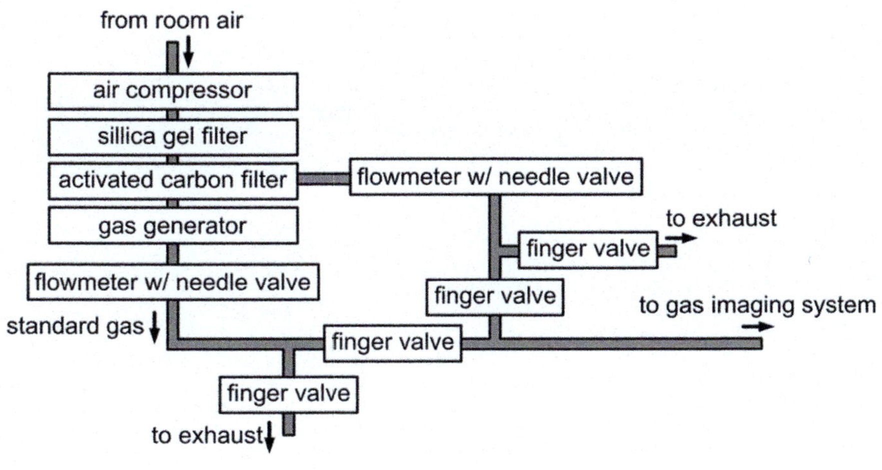

Fig. 7 Gas application system's setup

standard ethanol gas for 20 s. After that, the gas was switched to dry clean air again. The camera recording was stopped 170 s after this change. The captured images were processed by ImageJ with a custom Python script to calculate the mean intensity and its change rate. The fluorescence changing rate can be calculated by time-domain image differential analysis [24].

Figure 8 shows an example of calculated fluorescence intensity and fluorescence changing rate. The results were obtained by applying 100 ppm of gaseous EtOH. Changes in fluorescence intensity and fluorescence changing rate were significantly increased at the center, the gas application point. The fluorescence change rate, calculated by differential analysis, showed that the output occurs only during gas application and recovered to the initial value after the gas flow was stopped. From these results, it can be concluded that the system is capable of measuring spatiotemporal changes in ethanol concentration.

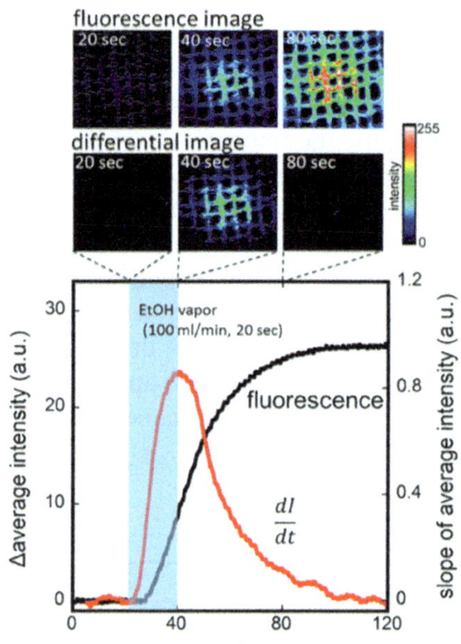

Fig. 8 An example of signals of fluorescence intensity and differential values. Image adapted from American Chemical Society [24]

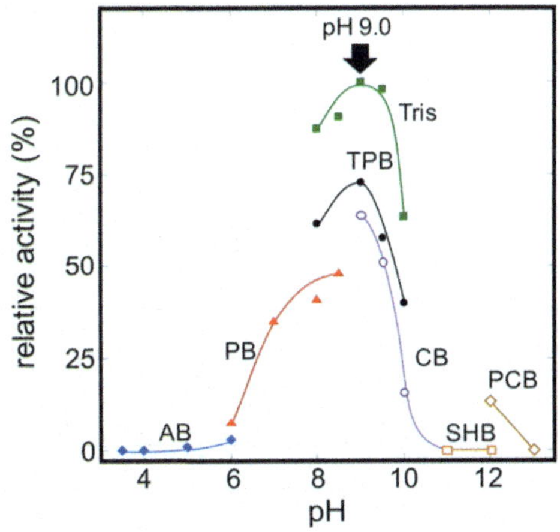

Fig. 9 pH dependence of systems output. Image adapted from American Chemical Society [25]

5 Effect of Optimization of an Enzyme Reaction Environment

5.1 Effect of pH Value

The enzymatic reaction is known to vary with ambient conditions. In this system, for example, the NAD^+ concentration, pH value, and buffer species of the NAD^+ solution that was soaked into the ADH-immobilized mesh were varied. Figure 9 shows the dependence of ADH activities in response to gaseous ethanol of the buffer pH value and buffer species. The applied ethanol concentration was 25 ppm. Tris–HCl buffer at a pH of 9.0 was optimum. When trisodium phosphate buffer or carbonate-bicarbonate buffer is used, the output is smaller than that of Tris–HCl, even though the pH value is adjusted to 9.0. This may be due to the interaction between the salt in the buffer and the enzyme.

5.2 Effect of NAD^+ Concentration

NAD^+ concentration on ADH-immobilized mesh also affected ADH activity. Figure 10a shows the time courses of mean intensity obtained by applying various concentrations of gaseous ethanol to the ADH-immobilized mesh with 0.1 and 10 mM of NAD^+ solution. The time required for the output value to reach 90% of its maximum value (T90) changed little regardless of the NAD^+ concentration. On the other hand, the change in fluorescence intensity depended on the NAD^+ concentration and was maximized when the NAD^+ concentration was 10 mM, as shown in Fig. 10b. We currently believe that concentrations lower than 10 mM produce insufficient NAD^+ for the amount of ethanol loaded, while concentrations higher than

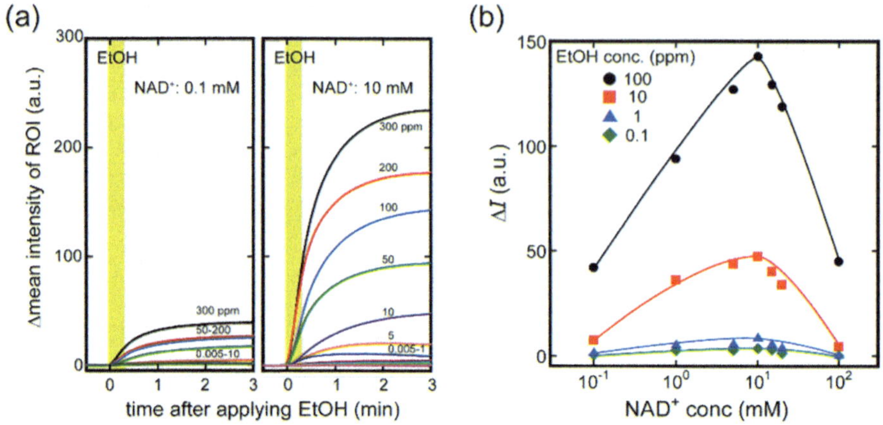

Fig. 10 **a** Typical signals by applying gaseous ethanol on different concentrations of NAD^+ solution (left: 0.1 mM, right: 10 mM). **b** Effect of preload NAD^+ solution's concentration on the output signal. Image adapted from American Chemical Society [25]

10 mM produce such results because the relatively large amount of NAD^+ remaining after the reaction inhibits the photoexcitation of NADH.

6 Characteristics of the System

6.1 Gas Selectivity

The selectivity of the system depends on the substrate specificity of the enzyme. Even enzymes belonging to the same category and having the same name may have different substrate specificities depending on their origin. We compared two different origins of ADH to determine which enzyme to use in the system. Figure 11 left and right show the results of applying VOCs in exhaled breath and transdermal gas on ADH from *yeast* (Oriental yeast, Japan) and ADH from *Saccharomyces cerevisia*e (Sigma-Aldrich, USA), respectively. From the results, neither ADH react to human-borne VOC components other than ethanol, and high selectivity based on the substrate specificity of ADH was obtained. It is important to note that the exhaled breath and transdermal gas after drinking contain higher concentrations of acetaldehyde gas than under normal conditions. Therefore, we examined the output to a mixture of 50 ppm ethanol and 1 ppm acetaldehyde gas and found that the output to ethanol gas was much lower for ADH from *yeast* than when ethanol gas alone was loaded. Based on these results, our system uses ADH from *Saccharomyces cerevisiae.*

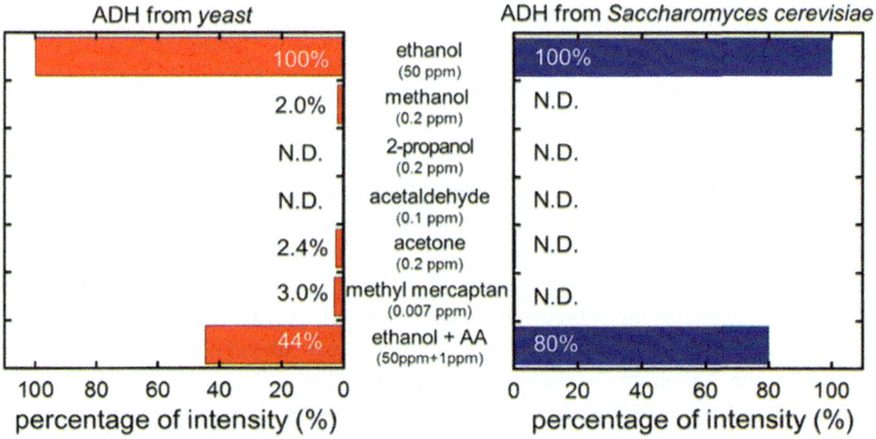

Fig. 11 Comparison of the output signal obtained by typical breath components for ADH enzymes derived from yeast and Saccharomyces cerevisiae. Image adapted from American Chemical Society [24]

6.2 Dynamic Range

Figure 12 shows the dynamic ranges of the system. Whether the intensity of differential analysis was used, the dynamic range was 0.01–100 ppm. This wide dynamic range allows measuring ethanol concentration of both exhaled breath and transdermal gas after drinking.

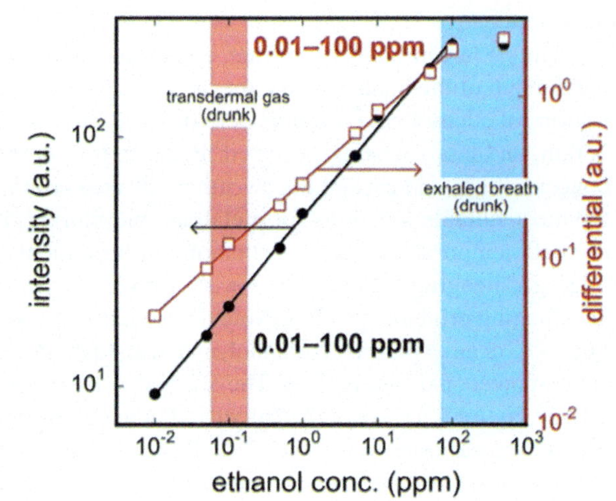

Fig. 12 The dynamic range of the developed system against gaseous ethanol. Image adapted from American Chemical Society [23]

7 Results of Gas Imaging on Different Body Surfaces

The human subject experiment was performed after permission from the ethics review committee of Tokyo Medical and Dental University (approved number M2018-160). Before the experiment, the subjects got an explanation about the method and purpose of the experiment. Then, they signed written informed consent if they agreed to join the study. We then assessed the activity of aldehyde dehydrogenase 2 (ALDH2) in each subject using an alcohol patch test method. Subjects who were determined to have no ALDH2 activity were excluded from subsequent experiments. Prior to the drinking experiment, all subjects had their body temperature, blood pressure, and heart rate measured and filled out a questionnaire. All subjects were also prohibited from drinking, smoking, or taking medication within 72 h prior to the drinking experiment. The amount of alcohol taken orally was normalized to 0.4 g of ethanol per kg of body weight to be constant among subjects of different body weights. Subjects were asked to drink a set amount of alcohol over a 15-min period. After drinking, the body part of the subject to be observed was pushed over a 90 mm by 90 mm of ADH-immobilized mesh moistened with NAD^+ solution over a 2D mako. The fluorescence distribution generated because of the transdermal gas application was captured as a video by using a camera, and then processed after experiments.

The readers can see the spatiotemporal change of transdermal ethanol emitted around the palm after drinking alcohol by visiting this URL.[1] The emission of gaseous ethanol from the palm was started at 2 min after drinking. Later, around 30 min after drinking, high concentrations of ethanol were observed to be released intermittently over a short period of time. This may be due to the release of ethanol in the sweat under the influence of psychogenic sweating occurring in the palms. Thus, body parts with less activity of sweating are considered more suitable for transdermal VOCs measurements.

Figure 13a compares sweat rate, epidermis thickness, and ethanol distribution on the dorsum of the hand, fingers, toe, and ear. Among these body parts, high concentrations of ethanol gas emission were observed around the ears, where the epidermis is thin, and less sweating occurred. Because transdermal gases are emitted at high concentrations and cause little sweating, the area around the ear is considered one of the most suitable locations for metabolic monitoring. Figure 13b shows the results of spatiotemporal imaging of ethanol and acetaldehyde around the ear. Acetaldehyde gas imaging was executed with a reverse reaction of ADH described in the gas imaging principle section. Note that because this system allows imaging of only one VOC at an experiment, ethanol, and acetaldehyde imaging was performed on the same subject on different days. The original video of these images can be seen by this URL.[2] Figure 12c shows the calculated concentration of ethanol and acetaldehyde around the ear hole. As in the alcohol metabolism kinetics, the ethanol concentration increased first, followed by an increase in acetaldehyde concentration. The time difference between the ethanol and acetaldehyde concentrations reaching their peaks

[1] https://pubs.acs.org/doi/suppl/10.1021/acssensors.9b01658/suppl_file/se9b01658_si_002.mp4.

[2] https://pubs.acs.org/doi/suppl/10.1021/acssensors.9b01658/suppl_file/se9b01658_si_003.mp4.

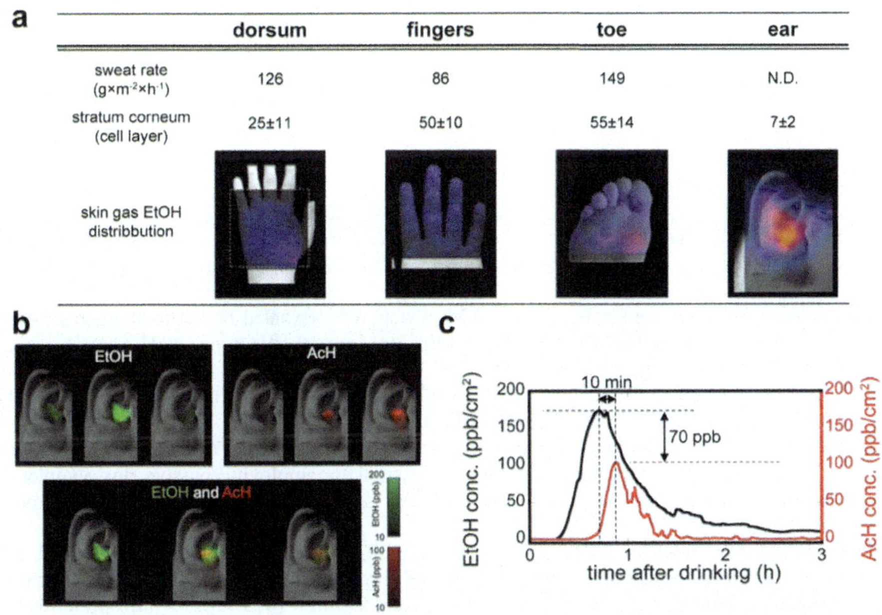

Fig. 13 **a** Comparison of sweat rate, the thickness of the epidermis (number of cell layers), and ethanol distribution on the dorsum of the hand, hand fingers, toe, and ear. **b** Images of ethanol (upper left), acetaldehyde (upper right), and overlayed image of concentration distribution. **c** The calculated concentration of ethanol and acetaldehyde around the ear hole. Images adapted from American Chemical Society [23]

was approximately 10 min. Based on these results, we believe that monitoring alcohol metabolism by transdermal VOCs is feasible around the ear.

8 Concluding Remarks

In this chapter, we introduced a gas imaging system that can evaluate how transdermal VOCs are released from our body based on spatiotemporal VOCs imaging, toward evaluation of metabolism using wearable transdermal VOCs sensors. Experimental results had shown that the ear, with its low activity of sweating and active emission of VOCs, was a suitable measurement target for transdermal VOC sensors. However, the body areas studied so far are limited compared to the area of skin covering the entire body. Further evaluation of the relationship between various skin characteristics and VOCs release dynamics will bring us closer to the realization of metabolism monitoring based on transdermal VOCs without making the subject aware of the measurement. In the future, the findings from this research will be applied to wearable transdermal VOC sensors. At the same time, we hope to apply imaging measurements to other VOCs such as acetone, which is related to lipid metabolisms, in addition to ethanol and acetaldehyde.

References

1. Shirasu, M., Touhara, K.: The scent of disease: volatile organic compounds of the human body related to disease and disorder. J. Biochem. **150**, 257–266 (2011). https://doi.org/10.1093/jb/mvr090
2. Vasilescu, A., Hrinczenko, B., Swain, G.M., Peteu, S.F.: Exhaled breath biomarker sensing. Biosens. Bioelectron. **182**, 113193 (2021). https://doi.org/10.1016/j.bios.2021.113193
3. Iitani, K., Ramamurthy, S.S., Ge, X., Rao, G.: Transdermal sensing: in-situ non-invasive techniques for monitoring of human biochemical status. Curr. Opin. Biotechnol. **71**, 198–205 (2021). https://doi.org/10.1016/j.copbio.2021.08.005
4. Sekine, Y., Sato, S., Kimura, K., Sato, H., Nakai, S., Yanagisawa, Y.: Detection of tobacco smoke emanating from human skin surface of smokers employing passive flux sampler—GCMS system. J. Chromatogr. B **1092**, 394–401 (2018). https://doi.org/10.1016/j.jchromb.2018.06.038
5. Grabowska-Polanowska, B., Miarka, P., Skowron, M., Sułowicz, J., Wojtyna, K., Moskal, K., Śliwka, I.: Development of sampling method and chromatographic analysis of volatile organic compounds emitted from human skin. Bioanalysis **9**, 1465–1475 (2017). https://doi.org/10.4155/bio-2017-0128
6. Roodt, A.P., Naudé, Y., Stoltz, A., Rohwer, E.: Human skin volatiles: Passive sampling and GC × GC-ToFMS analysis as a tool to investigate the skin microbiome and interactions with anthropophilic mosquito disease vectors. J. Chromatogr. B **1097–1098**, 83–93 (2018). https://doi.org/10.1016/j.jchromb.2018.09.002
7. Ya-Xian, Z., Suetake, T., Tagami, H.: Number of cell layers of the stratum corneum in normal skin - relationship to the anatomical location on the body, age, sex and physical parameters. Arch. Dermatol. Res. **291**, 555–559 (1999). https://doi.org/10.1007/s004030050453
8. Wilke, K., Martin, A., Terstegen, L., Biel, S.S.: A short history of sweat gland biology. Int. J. Cosmet. Sci. **29**, 169–179 (2007). https://doi.org/10.1111/j.1467-2494.2007.00387.x
9. Coull, N.A., West, A.M., Hodder, S.G., Wheeler, P., Havenith, G.: Body mapping of regional sweat distribution in young and older males. Eur. J. Appl. Physiol. **121**, 109–125 (2021). https://doi.org/10.1007/s00421-020-04503-5
10. Nagae, K., Asao, Y., Sudo, Y., Murayama, N., Tanaka, Y., Ohira, K., Ishida, Y., Otsuka, A., Matsumoto, Y., Saito, S., Furu, M., Murata, K., Sekiguchi, H., Kataoka, M., Yoshikawa, A., Ishii, T., Togashi, K., Shiina, T., Kabashima, K., Toi, M., Yagi, T.: Real-time 3D photoacoustic visualization system with a wide field of view for imaging human limbs. F1000Research **7**, 1813 (2018). https://doi.org/10.12688/f1000research.16743.1
11. Duffy, E., Jacobs, M.R., Kirby, B., Morrin, A.: Probing skin physiology through the volatile footprint: Discriminating volatile emissions before and after acute barrier disruption. Exp. Dermatol. **26**, 919–925 (2017). https://doi.org/10.1111/exd.13344
12. Mitsubayashi, K., Toma, K., Iitani, K., Arakawa, T.: Gas-phase biosensors: A review. Sens. Actuators B Chem. **367**, 132053 (2022). https://doi.org/10.1016/j.snb.2022.132053
13. Iitani, K., Hayakawa, Y., Toma, K., Arakawa, T., Mitsubayashi, K.: Switchable sniff-cam (gas-imaging system) based on redox reactions of alcohol dehydrogenase for ethanol and acetaldehyde in exhaled breath. Talanta **197**, 249–256 (2019). https://doi.org/10.1016/j.talanta.2018.12.070
14. Kudo, H., Sawai, M., Suzuki, Y., Wang, X., Gessei, T., Takahashi, D., Arakawa, T., Mitsubayashi, K.: Fiber-optic bio-sniffer (biochemical gas sensor) for high-selective monitoring of ethanol vapor using 335nm UV-LED. Sens. Actuators B Chem. **147**, 676–680 (2010). https://doi.org/10.1016/j.snb.2010.03.066

15. Toma, K., Iwasaki, K., Zhang, G., Iitani, K., Arakawa, T., Iwasaki, Y., Mitsubayashi, K.: Biochemical methanol gas sensor (MeOH Bio-Sniffer) for non-invasive assessment of intestinal flora from breath methanol. Sensors **21**, 4897 (2021). https://doi.org/10.3390/s21144897
16. Chien, P.-J., Suzuki, T., Tsujii, M., Ye, M., Toma, K., Arakawa, T., Iwasaki, Y., Mitsubayashi, K.: Bio-sniffer (gas-phase biosensor) with secondary alcohol dehydrogenase (S-ADH) for determination of isopropanol in exhaled air as a potential volatile biomarker. Biosens. Bioelectron. **91**, 341–346 (2017). https://doi.org/10.1016/j.bios.2016.12.050
17. Iitani, K., Chien, P.-J., Suzuki, T., Toma, K., Arakawa, T., Iwasaki, Y., Mitsubayashi, K.: Fiber-optic bio-sniffer (biochemical gas sensor) using reverse reaction of alcohol dehydrogenase for exhaled acetaldehyde. ACS Sens. **3**, 425–431 (2018). https://doi.org/10.1021/acssensors.7b00865
18. Kudo, H., Suzuki, Y., Gessei, T., Takahashi, D., Arakawa, T., Mitsubayashi, K.: Biochemical gas sensor (bio-sniffer) for ultrahigh-sensitive gaseous formaldehyde monitoring. Biosens. Bioelectron. **26**, 854–858 (2010). https://doi.org/10.1016/j.bios.2010.07.099
19. Ye, M., Chien, P.-J., Toma, K., Arakawa, T., Mitsubayashi, K.: An acetone bio-sniffer (gas phase biosensor) enabling assessment of lipid metabolism from exhaled breath. Biosens. Bioelectron. **73**, 208–213 (2015). https://doi.org/10.1016/j.bios.2015.04.023
20. Saito, H., Hashimoto, Y., Minamide, T., Kon, T., Toma, K., Arakawa, T., Kohji, M.: Fiber optic biosniffer (biochemical gas sensor) for gaseous dimethyl sulfide. Sens. Mater. 1295 (2016). https://doi.org/10.18494/SAM.2016.1370
21. Mitsubayashi, K., Minamide, T., Otsuka, K., Kudo, H., Saito, H.: Optical bio-sniffer for methyl mercaptan in halitosis. Anal. Chim. Acta **573–574**, 75–80 (2006). https://doi.org/10.1016/j.aca.2006.01.062
22. Banning, E.B.: The archaeologist's laboratory: the analysis of archaeological evidence. Springer, Cham (2020)
23. Iitani, K., Toma, K., Arakawa, T., Mitsubayashi, K.: Transcutaneous blood VOC imaging system (skin-gas cam) with real-time bio-fluorometric device on rounded skin surface. ACS Sens. **5**, 338–345 (2020). https://doi.org/10.1021/acssensors.9b01658
24. Arakawa, T., Sato, T., Iitani, K., Toma, K., Mitsubayashi, K.: Fluorometric biosniffer camera "sniff-cam" for direct imaging of gaseous ethanol in breath and transdermal vapor. Anal. Chem. **89**, 4495–4501 (2017). https://doi.org/10.1021/acs.analchem.6b04676
25. Iitani, K., Toma, K., Arakawa, T., Mitsubayashi, K.: Ultrasensitive sniff-cam for biofluorometric-imaging of breath ethanol caused by metabolism of intestinal flora. Anal. Chem. **91**, 9458–9465 (2019). https://doi.org/10.1021/acs.analchem.8b05840

SPRINGER NATURE

GPSR Compliance

The European Union's (EU) General Product Safety Regulation (GPSR) is a set of rules that requires consumer products to be safe and our obligations to ensure this.

If you have any concerns about our products, you can contact us on ProductSafety@springernature.com

In case Publisher is established outside the EU, the EU authorized representative is:

Springer Nature Customer Service Center GmbH
Europaplatz 3
69115 Heidelberg, Germany

The manufacturer's authorised representative in the EU is Springer Nature Customer Service Centre GmbH, Europaplatz 3, 69115 Heidelberg, Germany. If you have any concerns regarding our products, please contact ProductSafety@springernature.com

Printed and bound by CPI Group (UK) Ltd, Croydon, CR0 4YY
01/10/2026
02225868-0016